# VIROLOGY MONOGRAPHS

# DIE VIRUSFORSCHUNG IN EINZELDARSTELLUNGEN

CONTINUING / FORTFÜHRUNG VON
HANDBOOK OF VIRUS RESEARCH
HANDBUCH DER VIRUSFORSCHUNG
FOUNDED BY / BEGRÜNDET VON
R. DOERR

EDITED BY / HERAUSGEGEBEN VON

S. GARD · C. HALLAUER · K. F. MEYER

4

1968

SPRINGER-VERLAG WIEN GMBH

# THE INFLUENZA VIRUSES

BY

L. HOYLE

1968

SPRINGER-VERLAG WIEN GMBH

ISBN 978-3-7091-2048-4     ISBN 978-3-7091-2046-0(eBook)
DOI 10.1007/978-3-7091-2046-0

# The Influenza Viruses

By

## L. Hoyle

Director, Public Health Laboratory, Northampton, England

With 58 Figures

## Table of Contents

Figures 16 and 53 from Virology 21, 278, 1963 and 14, 398, 1961 are reproduced
by permission of Academic Press Inc.

Figures 15, 51 and 52 from the Journal of Experimental Medicine 104, 171, 1956
and 114, 825 1961 are reproduced by permission of the Rockefeller University Press.

Figure 55 is reproduced by permission of Dr. B. S. BERLIN from the Journal of
Immunology 96, 424, 1966, Williams & Wilkins Co., Baltimore.

# Preface

Virology has developed with such rapidity in recent years that it is increasingly difficult to maintain an adequate knowledge of more than a small part of the science. There is therefore a need for reference works which assess the established knowledge and make some record of past achievement. It is hoped that this work will fulfil this purpose in the case of the influenza viruses. The work is primarily intended for the virologist, and the clinical, pathological and epidemiological aspects of influenza are dealt with only in so far as they relate to the properties of the viruses.

An attempt has been made to review work up to the end of 1966. The list of some 2500 references is not complete, but it is hoped that no important aspect of the field has been overlooked.

My thanks are due to the many virologists who have discussed their work with me, to those who have provided electron photomicrographs and other data, to the Library staff of the Central Public Health laboratory, and to my secretary Miss A. W. H. BAILEY.

LESLIE HOYLE

# I. Historical Survey

Influenza is a disease of considerable antiquity, numerous epidemics with the characteristics of influenza having been described in past centuries, but the scientific investigation of the disease commenced with the studies of PFEIFFER (P$_{58}$) in the pandemic of 1890 when the influenza bacillus, *Haemophilus influenzae*, was isolated and was believed to be the cause of the disease.

In the pandemic of 1918, however, although *H. Influenzae* played a considerable role as a secondary invader, the balance of evidence indicated that it was not the primary cause, and many workers claimed that the disease was due to a filterable virus. In the light of modern knowledge of the properties of the virus much of the work in 1918 on the transmission of the disease by filtrates must be regarded with scepticism. Even under ideal conditions attempts to transmit the disease to man have not always been successful, and under the conditions of a pandemic of the extent of the 1918 outbreak it must have been difficult to find volunteers who had no exposure to the virus and were completely susceptible. Some successful transmissions probably were achieved, as for example the production of febrile illness in monkeys by NICOLLE and LEBAILLY (N$_{18}$) using intranasal inoculation of sputum, but claims to produce disease in rabbits which are now known to be completely resistant, or to produce disease by subcutaneous inoculation must be rejected in the light of modern knowledge. In spite of the large amount of work done in 1918—1919 the virus aetiology of the disease was not established, and in 1921 OLITSKY and GATES (O$_8$) isolated a small anaerobic organism, *Dialister Pneumosintes* which they suggested might be the cause. This organism however, proved to be a normal inhabitant of the respiratory tract.

The first undoubted demonstration of an influenza virus was achieved in 1901 when CENTANNI and SAVUNOZZI (C$_{20}$) showed that fowl plague was a virus disease. Only four other filterable virus diseases were known at that date, and fowl plague was intensively studied, but the characters of the disease were so different from those of human influenza that no one suspected a relationship and it was not until 1955 that SCHÄFER ($S_{29}$) showed that the fowl plague virus was antigenically related to human influenza A.

The work of SHOPE in 1931 on swine influenza was in a different category. Not only did the swine disease closely resemble human influenza but many veterinarians believed that the disease had originated from man at the time of the 1918 pandemic. SHOPE ($S_{107}$) showed that swine influenza was primarily due to a filterable virus, but that the complete disease picture was only found when an organism, *Haemophilus influenzae suis*, was also present. This work immediately suggested that a similar state of affairs might exist in human influenza.

## 1. Isolation of the Human Virus

In 1933 SMITH, ANDREWES and LAIDLAW ($S_{167}$) successfully transmitted human influenza to ferrets by intranasal inoculation of human nasopharyngeal washings. The ferrets developed a febrile illness transmissible in series by the inoculation of extracts of the nasal mucosa and turbinates. The responsible agent was shown to be filterable and to be neutralized by the serum of human influenza convalescents. These results were soon confirmed. In 1934 FRANCIS ($F_{86}$) isolated strains of virus in epidemics in Puerto Rico and Philadelphia, BURNET ($B_{154}$) isolated a strain in Australia, and PETTIT, MUDD and PEPPER ($P_{57}$) isolated a strain in Alaska. In 1936 strains were isolated in Leningrad ($S_{202}$) and in New Haven ($B_{125}$). The original workers isolated further strains in 1937 and strains were also isolated in France ($D_{92}$) and in Manchester ($H_{202}$). The neutralization test showed all these strains to be antigenically similar, but to differ from the swine virus which was also shown to produce disease in ferrets.

## 2. Neutralization and Complement Fixation

The pathogenicity of the virus for mice was discovered by ANDREWES, LAIDLAW and SMITH ($A_{80}$) and this discovery greatly facilitated the study of the distribution of serum antibody by the neutralization test. A description of a complement fixation reaction in influenza was first published by SMITH ($S_{162}$) but he did not pursue the subject and our knowledge of the reaction is due to the work of FAIRBROTHER and HOYLE ($F_9$, $H_{186-187}$, $H_{202-205}$) who developed a complement fixation reaction using extracts of infected mouse lung and later of infected egg membranes as antigen. This reaction proved to be a reliable test for the serological diagnosis of influenza. It was shown that the effective antigen in the mouse lung extracts was not the virus particle but an agent of smaller size, which could not be deposited by the centrifuges available at the time and came to be known as soluble antigen. The soluble antigens of the human and swine viruses were shown to be identical ($H_{203}$).

In 1937 FRANCIS and MAGILL ($XF_9$) and HYDE and CHAPMAN ($H_{231}$) showed that influenza virus could be isolated from man by inoculation of the chorioallantoic membrane of fertile eggs, and in 1940 BURNET ($B_{158}$) introduced the technique of amniotic inoculation. The fertile egg soon replaced the ferret for primary isolation of virus and BEVERIDGE and BURNET ($B_{65}$) developed techniques in the use of fertile eggs which became of major importance in influenza research.

In 1940 FRANCIS ($F_{90}$) isolated a strain of virus, the LEE strain, which was serologically completely different from all previous strains both when tested by neutralization and by complement fixation methods. It was therefore called influenza B, the original SWINE, WS, and PR8 strains being referred to as virus A. The essential difference between the A and B viruses was their difference in the complement fixing soluble antigen ($F_{91}$). A third type of influenza virus, influenza C, was first isolated by TAYLOR in 1947 ($T_{39}$).

### 3. The Haemagglutination Reaction

In 1941 HIRST ($H_{125}$) and McCLELLAND and HARE ($M_{54}$) made a discovery of major importance when they found that influenza virus particles would agglutinate the red cells of fowls and other animal species. HIRST showed that the virus particles were first adsorbed by the red cells and subsequently eluted as a result of what was clearly an enzymic reaction. The virus content of preparations could readily be estimated by the haemagglutination test and the presence of serum antibody could be demonstrated by the inhibition of haemagglutination which occurred in its presence. In 1944 SALK ($S_4$) introduced a simple modification of the haemagglutination test which became extensively used in studies of serum antibody. In contrast to the complement fixation test which showed type specificity, distinguishing only between virus A and virus B, the haemagglutination-inhibition test showed strain specificity, revealing differences between individual strains of virus A which were demonstrated only with difficulty by complement fixation methods.

### 4. Protective Vaccination

Soon after the first isolation of the influenza virus many studies were made of the possibility of protecting animals and man against infection by the use of killed virus vaccines, mostly with very crude preparations of virus, but during the war years intensive work was done in America on the production of vaccine in bulk with the object of protecting the armed forces of the U.S.A. against a possible pandemic at the end of the war. STANLEY and SALK were particularly associated with this work. Virus was concentrated and partially purified by centrifugation methods or by adsorption-elution from red blood cells. It was found desirable to use recently isolated strains of virus to give satisfactory protection, but even with old strains of virus some protection was achieved. By 1946 it seemed probable that with further development of methods of purifying and concentrating virus and by the use of recently isolated strains protective vaccination might play an increasingly useful role in the control of influenza. These hopes were shattered in 1947 — a year which proved a turning point in influenza virus research. In 1947—1948 there occurred a pandemic of influenza due to a strain of virus which, while recognisable as an A strain by complement fixation methods, differed so completely from all previously isolated A strains that workers using the strain-specific haemagglutination test found difficulty in identifying it. The new strain was first isolated in Australia in late 1946 but its peculiar characters were not recognised until its appearance in the U.S.A. in 1947. It became known as the A prime or A1 virus. Vaccination with potent preparations of the earlier strains of virus A failed completely to give protection or to produce specific antibody against the new strain. The results of vaccination were so disappointing that one virologist is said to have remarked that it would have done the recipients more good to have eaten the eggs used to make the vaccine!

## 5. Inhibitors of Haemagglutination

Apart from the difference in antigenic structure the 1947 strain differed from previous strains in other ways. In 1946 FRANCIS had discovered in normal sera a substance — the Francis inhibitor — which interfered with haemagglutination of red cells by heated virus, while unheated virus was almost unaffected ($F_{94}$). The new strain of virus proved to be very sensitive to the action of inhibitors present in human and animal sera and this caused difficulty in the use of the haemagglutination-inhibition test as a measure of serum antibody.

Studies of the Francis inhibitor indicated it to be of mucoprotein nature and it was soon found that mucoproteins from many sources possessed inhibitory action. The enzymic nature of the interaction of virus with red cells and with mucoprotein inhibitors was intensively studied by Australian workers. BURNET ($B_{170}$) discovered that filtrates of V cholerae cultures contained an enzyme RDE which was able to destroy the virus receptors in red cells and the inhibitory powers of mucoproteins. The chemical basis of the reaction of virus enzyme and RDE with mucoproteins was studied by GOTTSCHALK ($G_{52-54}$) and the enzyme was finally identified by German workers ($K_{59}$, $F_7$) as a neuraminidase which split acetyl neuraminic acid from the polysaccharide of the red cell surface. MULDER and his colleagues were able to overcome many of the difficulties caused by inhibitors in the estimation of serum antibody by the haemagglutination test by treatment of the sera with filtrates of V cholerae which destroyed inhibitors without affecting antibody titres ($M_{140}$). But in 1951 CHU discovered a serum inhibitor which differed from the Francis inhibitor in being heat labile and resistant to RDE ($C_{53}$).

While the haemagglutination test continued to be much used in serological diagnosis difficulties with inhibitors led to a renewed interest in the complement fixation reaction and HENLE, LIEF and FABIYI ($H_{90}$) developed complement fixing antigens which were free from the group S antigen and could be used to detect differences between different strains of virus. In the pandemic of Asian influenza which occurred in 1956—1957 the strain-specific complement fixation test proved more satisfactory than the haemagglutination test. The Asian strain A2 differed from previous strains of virus A and was particularly sensitive to serum inhibitors.

## 6. Intracellular Reproduction of the Influenza Virus

The year 1947 had been marked by a defeat in the campaign to develop protective vaccination, but the same year saw the beginning of a considerable advance in our knowledge of the nature of the virus and the mode of its reproduction.

As a result of studies of the interference phenomenon HENLE and his associates ($H_{87}$) had developed a technique which enabled them to restrict the growth of the virus in eggs to a single cycle, subsequent cycles being prevented by the inoculation of ultraviolet irradiated virus, and under these conditions one step growth curves occurred in which virus was released from infected cells six hours after inoculation in an amount some 50—60 times that originally inoculated. In 1947 GARD and VON MAGNUS ($M_{19}$, $G_7$) discovered the so-called "VON MAGNUS phenomenon". It was found that if eggs were inoculated with small doses of virus the virus released from the cells into the allantoic fluid had a high infectivity in relation to haemagglutinin titre, but if serial passages were made with large inocula, after a few passages while the yield of virus as measured by haemagglutination was unchanged, the virus had an infectivity titre a thousand-fold lower than if small inocula were used. The progeny resulting from the introduction of large infecting doses appeared to be in some way incomplete.

So far nothing was known of the intracellular phase of virus reproduction, but in 1948 and 1950 HOYLE ($H_{188}$, $H_{191}$) described studies of the intracellular phase. By inoculating eggs to the allantoic sac with amounts of virus large enough to infect simultaneously all the cells lining the sac, and then making extracts of the chorio-allantoic membrane at intervals, it was shown that no increase in infective virus could be detected until six hours after inoculation when infective virus was released from the cells in large amount, but that complement fixing soluble antigen appeared in the cells 2 hours after inoculation and rapidly increased in amount, while non-infective haemagglutinin appeared 3—4 hours after inoculation. Not merely was there no increase in infective virus in the early hours but even the original inoculum could not be detected, the virus becoming "eclipsed". HOYLE suggested that there was an intracellular multiplying form of the virus which differed from the extracellular infective form, and that on entry into the cell the virus particle disintegrated with the release of sub-units which were then reproduced in the cell and finally assembled to produce complete virus at the moment of excretion. The soluble antigen was considered to be such a sub-unit. The new ideas aroused great controversy. Before 1948 it was almost universally believed by animal virus workers that viruses had evolved from bacteria by a process of increasing parasitism, adopting an intracellular habitat and gradually losing their enzymes until they became completely dependent on the host cell, only retaining the ability to multiply by some process of growth and fission. This was the LAIDLAW-GREEN hypothesis ($XG_1$, $XL_1$). It had never been accepted by plant virologists who had long regarded viruses as protein molecules, or by bacteriophage workers who had tended to regard phages as self reduplicating enzymes or genes. In spite of this the LAIDLAW-GREEN hypothesis was supported with a fervour more appropriate to a religious dogma than a scientific theory. The minds of many of the older virologists seemed unable to penetrate the cell wall and HOYLE's ideas were dismissed as wildly speculative and based on inadequate data, a criticism which could with far more justice have been applied to the LAIDLAW-GREEN hypothesis.

However, evidence in support of the new ideas rapidly accumulated. In 1949 HENLE showed that sub-units with both complement fixing and haemagglutinating properties appeared in the cells before the production of infective virus ($H_{66\ 68}$). SCHLESINGER ($S_{47}$) showed that in mouse brain influenza virus went through an incomplete growth cycle, complement fixing antigen and haemagglutinin being produced without any increase in infective virus. In 1951 BURNET and LIND ($B_{200}$) discovered the phenomenon of recombination, in which from cells inoculated simultaneously with two different strains of virus, virus particles could be recovered with properties partly derived from one of the inoculated strains and partly from the other. This work led to a considerable study of genetic interactions between influenza viruses by the Australian group and later by

HIRST, KILBOURNE and others in America. FAZEKAS DE ST GROTH made a study of the growth of virus in mouse lungs and showed that production of soluble antigen preceded increase in infective virus. SCHÄFER and his colleagues in Tübingen ($S_{28}$, $S_{29}$, $S_{32-37}$) made a study of the multiplication of fowl plague virus and obtained results essentially the same as those of HOYLE with influenza virus, and it was later shown that the group antigen of fowl plague virus was serologically similar to the S antigen of influenza virus A, so that fowl plague virus came to be regarded as a strain of virus A. In the early days of influenza virus research ELFORD, ANDREWES and TANG by the use of gradocol membranes ($E_{36}$) had determined the particle size of the virus and electron microscope studies had shown it to have a spherical morphology. In 1948 HEINMETZ ($H_{59}$) and in 1949 CHU, DAWSON and ELFORD ($C_{56}$) described filamentous forms in some strains of virus and by the technique of adsorption on red cell ghosts it was shown that the filaments were adsorbed by red cells.

The nature of these filaments was elucidated by HOYLE using the dark-field microscope to study the release of virus from infected cells. It was found that virus material was released from cells enclosed in an outer membrane derived from the wall of the host cell, and that in some cases this outer membrane became drawn out into a tube which ultimately became a virus filament. WYCKOFF ($W_{69}$), MURPHY and BANG ($M_{150}$) and MORGAN, ROSE and MOORE ($M_{117}$) came to similar conclusions using electron microscopy. HOYLE ($H_{193}$) showed that virus particles could be disintegrated by shaking with ether with the production of separate haemagglutinating and complement fixing sub-units. The complement fixing sub-unit was shown to be serologically identical with the soluble antigen which was thus shown to be an integral component of the infective particle.

Early studies of the chemical composition of influenza virus by TAYLOR ($T_{33}$) and KNIGHT ($K_{67}$) had suggested the presence of both types of nucleic acid, but with more improved methods of purification ADA and PERRY ($A_{27}$) and FRISCH-NIGGEMEYER and HOYLE ($F_{154}$) showed that the virus particle contained only ribonucleic acid. By fractionating $P^{32}$ labelled virus with ether it was shown ($H_{211}$) that the soluble antigen fraction contained all the virus RNA and ADA ($A_{17}$) and HOYLE and FRISCH-NIGGEMEYER estimated the RNA content of soluble antigen at 5.3%.

By inoculating cells with virus labelled with $P^{32}$ or $S^{35}$ ($H_{207}$, $H_{206}$) it was shown that on entry into the cell the virus particle disintegrated, the protein remaining in the cytoplasm while the nucleic acid entered the cell nucleus, and LIU ($L_{99}$) by the use of fluorescent antibody staining showed that the group S antigen was synthesised in the cell nucleus while the specific V antigen was produced in the cytoplasm. Similar results were obtained by SCHÄFER with fowl plague virus.

In 1947 LURIA ($XL_3$) working with bacteriophage had discovered the reactivation phenomenon and with DELBRÜCK, DOERMAN and others demonstrated the occurrence of an eclipse phase in phage reproduction. This was followed by an extensive study of the genetic aspects of bacteriophage reproduction. The methods developed by the phage and influenza virus workers were applied to other virus diseases and it soon became apparent that the existence of an eclipse phase and the appearance of non-infective serologically active sub-units in the early stages of the cycle were characteristic features of virus reproduction. The discovery that some of the animal viruses could be crystallised, and that in some cases infection could be transmitted by pure virus nucleic acid made it certain that no process of growth and fission was involved in virus reproduction. Finally the development of the negative staining electron microscope technique by BRENNER and HORNE ($B_{124}$) which made possible the direct visualisation of the sub-unit structure of viruses terminated all argument as to the essential accuracy of the new ideas. When the controversy was finally over the study of viruses was no longer regarded as a branch of bacteriology, the similarity of plant, animal and bacterial viruses was established, and virology had become a science on its own right.

## 7. Epidemiological Studies

The failure of the vaccination campaign in 1947 stimulated studies on the epidemiology of influenza and on variation in influenza viruses. As a result of the efforts of ANDREWES, FRANCIS and others a system of influenza surveillance was set up by the World Health Organisation. Diagnostic laboratories were established in many countries with the object of rapidly isolating and identifying influenza viruses and of detecting changes in antigenic structure so that new strains could be rapidly included in vaccines. The system met a major test in 1956—1957 when a new strain of virus A, the Asian or A2 strain, appeared and spread as a pandemic. No epidemic in history has been so extensively studied as the Asian pandemic of 1957. Strains of virus were isolated and their characters examined in laboratories in almost every civilised state, but attempts to protect by vaccination were without significant effect on the spread of the disease, mainly because of the technical difficulty of producing anything approaching the amounts of vaccine that would have been needed in the available time. However, some observations were made which suggested that protective vaccination might not be entirely impossible. From studies of the age incidence of antibodies to the swine virus in human sera LAIDLAW ($L_{14}$) and SHOPE ($S_{110}$) had suggested that the 1918 pandemic of human influenza might have been due to a strain resembling the swine virus which had since died out in man but persisted unchanged in swine. Similar studies of the age distribution of antibody to the Asian virus of 1956—1957 indicated the possibility that this virus might have been associated with the 1890 pandemic. MULDER, DAVENPORT, HENNESSY, FRANCIS and others were responsible for these concepts and HILLEMAN suggested that there might be a limit to the range of possible antigenic structures in influenza A viruses, and that it might ultimately be possible to include all the antigenic components in a polyvalent vaccine.

## 8. Vaccination with Live Virus

Russian workers considered that protective vaccination using killed virus would never be possible except for small groups and that the only hope of successful large scale vaccination lay in the use of live virus. Although live virus vaccines had been used by BURNET ($B_{163}$) our knowledge of these vaccines is almost entirely due to the work of Russian virologists. Live vaccines appear to have been first used by CHALKINA ($C_{22}$) in 1938 but their use on a large scale commenced in 1948. SMORODINTSEV, CHALKINA, ANSHELES and others in Leningrad, and ZHDANOV, SOLOV'YEV, SOKOLOV, FADEYEVA and others in Moscow were particularly associated with the work. Live virus was inoculated intranasally or by aerosol inhalation, the invasiveness for the human respiratory tract being maintained by passage of the virus in human tissue cultures, while subsequent passage for a few generations in eggs produced sufficient attenuation to render the vaccine safe for use in man. RITOVA and ZAKSTELSKAYA ($R_{37}$) introduced a method of producing polyvalent strains by recombination techniques, and SMORODINTSEV and SOKOLOV developed dried vaccines which were insufflated as a powder.

Large numbers of people were inoculated with live vaccines (over half a million in 1953—1954) but the results were very variable. In some cases considerable protection was achieved, but in other cases the vaccine was unsuccessful, either because of the use of unsuitable strains in the vaccine or failure of the strains to infect the respiratory mucosa. The vaccines in general proved to be safe for use in adults, but considerable difficulties occurred in children who often reacted strongly to the vaccine.

Russian workers also studied passive protection against influenza by the intranasal inoculation of immune serum. Horse serum was used and successful results were obtained, but the short duration of the immunity and the danger of sensitisation to horse serum rendered the procedure of small value.

The concentration by Russian workers on the practical aspects of influenza control led to studies on the rapid diagnosis of influenza. SMORODINTSEV ($S_{181}$) was able to detect complement fixing antigen in the blood in the early stages of the disease. SOLOV'YEV, PETERSEN and others studied the detection of haemagglutinin in nasopharyngeal washings ($P_{52}$, $S_{241}$). KOLDATITSKAYA ($K_{83}$, $K_{84}$) used cytological diagnosis by examination of nasal smears, and this procedure was later improved by the use of staining with fluorescent antibody.

## 9. Animal Influenza

A major problem in the epidemiology of influenza was the apparent disappearance of the virus in inter-epidemic periods. In 1956 HELLER, ESPMARK and VIRIDEN ($H_{60}$) found that the serum of horses convalescent from an influenza-like illness contained antibodies reacting with the soluble S antigen of influenza virus A, and in the following year Czechoslovak workers confirmed the existence of equine influenza by the isolation of an A type virus (equi-1) from affected horses. This work suggested the possible existence of an animal reservoir of influenza and led to a search for influenza viruses in animals. In 1963 equine influenza due to equi-1 virus occurred in England, while in America an extensive outbreak occurred due to a strain of virus (equi-2) distinct from the English and Czechoslovak strains.

In 1955 SCHÄFER had shown that the group antigen of fowl plague virus was similar to the S antigen of influenza virus A, and it was found that a number of avian viruses of A type existed. Some of these, virus N isolated in 1949, and a strain isolated from turkeys in 1963 were serologically related to fowl plague, but others such as the duck viruses 1 and 2 isolated in Czechoslovakia and England were distinct. A further A type virus was isolated from chickens in Scotland in 1959 and from South Africa terns in 1961.

The swine and human viruses were known to be serologically related, and there was much evidence that human strains would infect swine, but the avian and equine strains, apart from the common S antigen, appeared to be serologically distinct from the human-porcine group, and no evidence of ability to cross-infect has been found between the human and the avian or equine strains. There is, however, some serological relation between the avian and equine viruses. In 1965 Tumova and Pereira ($T_{70}$) showed that genetic interactions between the groups could be obtained, UV inactivated strains of avian viruses being reactivated by human, porcine or equine viruses. At the time of writing no virus of B type has been isolated from animals.

## 10. Chemotherapy

Many attempts have been made to develop a chemotherapeutic remedy for influenza, but the vast majority of substances tested have proved without value. Ackermann, Tamm and others showed that many amino acid or nucleotide analogues were able to retard or suppress the growth of influenza viruses under experimental conditions, and these agents have proved of great value in studies of the biochemical mechanisms of virus reproduction, but the effective doses of these agents usually proved to be so close to the toxic doses that they had no chemotherapeutic value.

A number of substances appear to be able to retard or prevent infection of cells. Haemagglutinin inhibitors probably play some part in natural resistance to influenza by temporarily blocking the virus and preventing union with the cells. Destruction of cell receptors by V. cholerae enzyme ($S_{271}$) also prevents infection under experimental conditions. Certain amines appear to be able to prevent penetration of cells and amantadine ($D_{32}$) has some therapeutic value. But none of these substances is equal to antibody in preventing infection.

One of the curious features of influenza virus infection is that virus production ceases before antibody production has commenced, even in the absence of cell destruction, and it is a general characteristic of animal virus growth that the total production of virus is very small in comparison with the massive yield of virus obtained in many plant virus infections. The work of Isaacs and his associates ($I_{27-29}$) threw much light on this matter. Isaacs showed that towards the end of the period of active virus growth a protein substance, interferon, appeared in the cells. Interferon proved to be able to protect against many virus infections and to be non-toxic. It is, however, difficult to produce in large amount and suffers from the disadvantage as a chemotherapeutic agent in being host-cell specific in its action. Interferon may, however, be produced in cells by the inoculation of foreign nucleic acid and it is possible that substances causing the production of interferon in cells might prove of therapeutic value. The antiviral agent Helenine discovered by Shope ($S_{120-122}$) probably acts in this way.

## 11. Quantitative Relationships

Many workers have devised methods of increasing the precision of the various serological tests used in influenza virus work, and have attempted to relate the serological titres to the absolute numbers of virus particles present in a suspension.

Accurate infectivity titrations were developed in the early days of influenza virus research but were laborious and expensive in material. In 1949 BERNKOPF ($B_{57}$) introduced the de-embryonated egg technique which proved very useful in quantitative studies and in 1951 FULTON and ARMITAGE ($F_{163}$) developed a simplified tissue culture technique.

In 1951 MAGILL ($M_8$) made a quantitative analysis of the haemagglutination reaction, and in 1952 FAZEKAS DE ST GROTH, CAIRNS and EDNEY ($F_{33}$, $F_{39}$) studied the quantitative relationships between virus and host cell in the course of the growth cycle. Very precise methods of haemagglutinin titration were introduced by LEVINE, PUCK, and SAGIK ($L_{67}$), HORSFALL ($H_{166}$) and DRESCHER ($D_{67}$, $D_{68}$) and the relations between infectivity titre, haemagglutinin titre and absolute particle counts were determined by these workers and by DONALD and ISAACS ($D_{61}$).

# II. Nomenclature

In the early years of influenza virus research strains of virus isolated were given arbitrary labels by the investigators responsible for their isolation. In some cases the initials or name of the patient were used — WS, LEE, DSP; in others names indicative of the place of isolation — PR8, PHILA, MEL; or even mere laboratory numbers — 1233. Later some special property might be indicated — Neuro WS. Most of the well-established laboratory strains will probably always be referred to in this way, but the unsatisfactory nature of these arbitrary labels was soon realised, and as a result of the activities of the influenza centres of the World Health Organization a more satisfactory nomenclature was evolved, and has achieved wide acceptance.

In this system the serological type, based on the complement fixing property of the ribonucleoprotein component (S antigen) is first indicated, followed by the natural host, the serological subtype (V antigen), the place of isolation, the serial number of the strain isolated, and the year of isolation. Thus the original WS strain would become A/Human O/London/1/33, and the original Asian strain A/Human 2/Singapore/1/57.

The system is not perfect. The place of isolation may be given either as the state or the city of isolation according to the judgement of the investigator. The serial number of the strain is unnecessary for almost all purposes and is commonly omitted. The natural host is usually omitted if the work described deals only with human influenza. The fowl plague virus may infect a wide range of birds and it may prove better to use the broad term "avian" for the natural host in the case of viruses isolated from birds ($P_{35}$).

# III. Classification

The classification of viruses has been the subject of much discussion in recent years, and most workers have felt that available information was insufficient for permanent decisions to be made, but ANDREWES, BANG and BURNET ($A_{73}$)

considered that the viruses of influenza, Newcastle disease, fowl plague and mumps possessed sufficient common properties to justify their inclusion in a defined group, the myxoviruses. The group has the following characteristics.

## 1. The Myxovirus Group

a) Morphology: Spheres ranging from 60—200 mμ diameter and filamentous forms of the same diameter but of length up to several microns.

b) Sensitivity to ether.

c) Adsorption and elution from fowl and certain other types of red cells as a result of the presence of an enzyme (neuraminidase) attacking mucoproteins.

d) Presence of two types of antigen, S antigen present in infected tissues and of smaller particle size than the infective virus and having a specificity distinguishing the individual members of the group from each other, and V antigen associated with the virus particle and showing strain specificity.

e) Production of disease of the respiratory tract of mammals and birds and transmitted by respiratory secretions.

f) Growth in the amniotic cavity of fertile hen's eggs.

g) Ability to produce transmissible pneumonia in mice and hamsters, and after adaptation to produce encephalitis in rodents.

As a result of the development of methods of purification and fractionation, and the application of chemical analysis to the virus particles, together with the use of improved techniques of electron microscopy, it was found that the myxoviruses possessed an outer envelope consisting of protein, lipid and carbohydrate and carrying the strain-specific V antigen, and an inner helical ribonucleoprotein component having the antigenic properties of the S antigen. Certain other viruses, measles, rinderpest and distemper, possess a similar morphology and chemical constitution so that the exact limits of the myxovirus group has become uncertain.

## 2. Sub-division of the Myxovirus Group

It was realised that the characteristics of the inner ribonucleoprotein helix must be of fundamental importance in classification and in 1962 WATERSON ($W_{91}$) divided the myxovirus group into two sub-groups on this basis, one having an internal helix of diameter 90 Å and the other of 180 Å. The first group included the viruses of influenza, fowl plague and virus N, while the second included Newcastle disease virus, mumps, the para-influenza viruses, measles, rinderpest and distemper. The two groups were also distinguished by certain features of their multiplication cycles, the influenza group showing the phenomena of multiplicity reactivation, genetic recombination and production of incomplete virus which did not occur with the second group. The influenza-fowl plague viruses were also distinguished by the synthesis of their ribonucleoprotein components in the cell nucleus.

## 3. Classification of the Influenza Subgroup of the Myxoviruses

While there is no doubt that from the viewpoint of academic virus classification the physico-chemical properties of the ribonucleoprotein component of the viruses must be regarded as of paramount importance, the practical virologist would certainly regard serology as the most satisfactory basis for classification. The identification of an influenza virus depends in the end on the demonstration that its S antigen corresponds to that of one of the classical types. The S antigen, which is the protein component of the ribonucleoprotein helix, appears in large amount in all tissues infected with virus and appears to be completely specific, having never been encountered in the absence of influenzal infection. The presence of S antibody in serum is also regarded as absolute evidence of previous infection by influenza virus. The specificity of the antigen appears to depend on the primary structure of the protein molecule, the antigenic determinants being the aromatic amino acids tyrosine, tryptophane and histidine ($H_{208}$). The antigen is very stable and readily detected by the complement fixation reaction. These characteristics make it an ideal basis for an initial division of the influenza group into three types A, B and C with S antigens corresponding to those of the classical type strains WS, LEE and 1233. There is no cross-reaction whatever between the three types.

The three major types may be further subdivided according to the serological characteristics of the V antigen complex ($H_{113}$). This antigen complex is associated with the outer component of the virus particle and appears to consist usually of one dominant antigen and a number of minor components. The viruses may be classified into sub-types according to the predominant V antigen, but there are many complicated inter-relations due to the presence of the minor components. The properties of the V antigen may be studied by the use of the haemagglutination-inhibition test or by the complement fixation reaction using the intact virus particle as antigen.

Only one type of C virus has so far been detected, but with the B virus differences in the V antigen complex have been found in strains isolated in different years. Most workers agree that there have been three prevalent types of B virus, the LEE type prevalent from 1940—1948, a type BON which became dominant in about 1949 and persisted for some years and a third type prevalent in 1959—1961 ($M_{13}$, $M_{136}$, $T_{24}$, $T_{67}$, $R_{41}$, $C_{12}$). But HENNESSY, MINUSE and DAVENPORT ($H_{104}$) while recognizing the existence of the three groups considered that they were not sufficiently different to warrant classifying as distinct families.

With the A group of viruses the position is much more complex. There have been three major subtypes in the human A viruses, the original A type prevalent from 1933—1945, the A1 type from 1946—1955, and the A2 type from 1956 to the present date. But there also exist many animal strains of virus A distinguished by their different V antigens, and one of these, the swine virus, is believed to have been responsible for the human pandemic of 1918. Complicated cross relationships occur between the V antigens of both human and animal strains. The A viruses appear to fall naturally into two major groups, the human-porcine group, and the avian-equine group. The human and porcine strains have related V antigens and may be able to cross-infect, but do not infect birds or horses. The avian and equine viruses do not cross-infect but may possess antigens in common. They do not infect man.

A suggested classification of the influenza group of viruses is given in Fig. 1. The viruses are first divided into three types according to the type of S antigen present. The A strains are then sub-divided into human-porcine and equine-

*A Viruses:* S Antigen type A

Natural Host and V Antigen subtypes

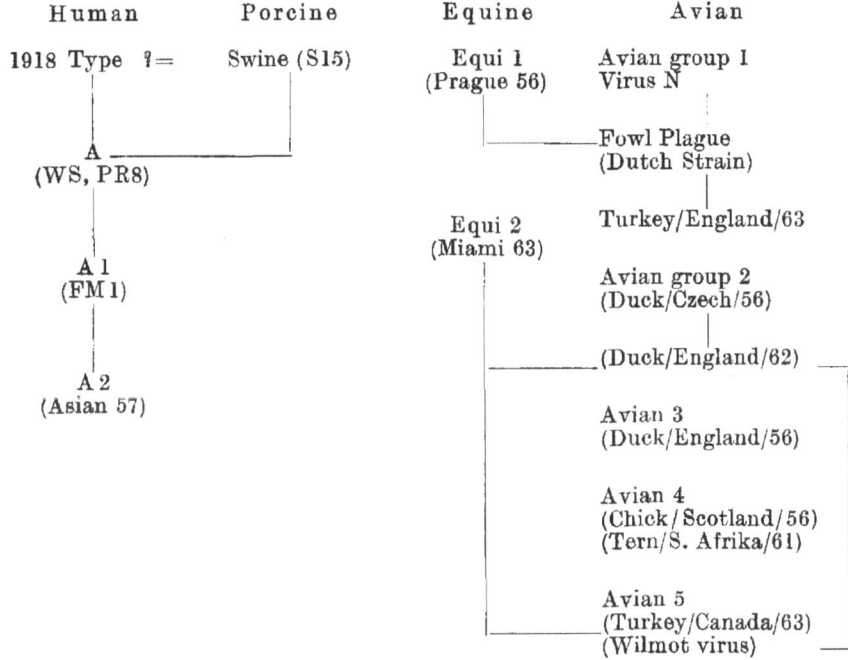

Fig. 1. Serological classification of influenza viruses.

avian groups. All groups are then further sub-divided into subtypes according to the dominant V antigen present. Inter-relations between the V antigen complexes are indicated by lines.

Data on some of the more important laboratory strains of influenza Viruses:

### a) Human A Viruses

1. *WS* (1933). The first human influenza virus, isolated in ferrets by SMITH, ANDREWES and LAIDLAW in England ($S_{167}$). Two neurotropic variants, *NWS* obtained by STUART-HARRIS ($S_{286}$), and *WSN* by FRANCIS and MOORE ($F_{110}$) have been much used in genetic studies.

2. *PR8* (1934). Isolated in Puerto Rico by FRANCIS ($F_{86}$). The most extensively studied strain, usually regarded as the prototype virus A and widely used in serological work and in vaccines because of its broad antigenic structure.

3. *MEL* (1935). Isolated by BURNET ($B_{154}$) in Australia, the first strain adapted to eggs. Used in studies of the recombination phenomenon.

4. *A36* (1936). The first Russian isolate by SMORODINTSEV and others ($S_{202}$). Similar to PR8.

5. *SHKLYAVER* (1938). Russian strain related to WS.

6. *WEISS* (1943). American strain much used in vaccines.

7. *DSP* (1943). English strain isolated by Hoyle and used in studies of growth cycle, chemical structure and morphology.

### b) Human A1 Viruses

1. *CAM* (1946). Isolated in Australia, the first isolation of A1 virus.

2. *FM1* (1947). The prototype A1 virus. Isolated in America and much used in vaccines.

3. *CUPPETT* (1950). Also known as FW/1/50. American A1 strain.

4. *KLIM* (1949), *PAN* (1952), and *ROSTOCK* (1953). Russian A1 strains used in serological and vaccine studies.

5. *A/Sweden/3/50*. The prototype Scandinavian A1 virus.

6. *A/England/1/51*. The prototype Liverpool A1 virus.

### c) Human A2 Viruses

1. *A/FE* (1957). A2 strain isolated in Formosa.

2. *A2/Singapore/1/57*. The prototype A2 strain.

3. *A2/Netherlands/65/63* and *A2/England/12/64*. Representative of later A2 strains.

4. *IKSHA*. Russian A2 strain used in live vaccines.

### d) Swine A Viruses

1. *S15* (1931). The prototype swine virus isolated by Shope ($S_{107}$).

2. *1976* (1931). Similar to S15.

3. *S/CAM* (1939). English isolate of swine virus, differing from S15.

4. *OTI* (1939). Strain isolated in Korea and more closely related to human viruses than to S15.

5. *S/Wisconsin/61*. Characteristic of recent swine viruses.

### e) Equine A Viruses

1. *A/equi/Prague/56*. The first isolation of an equine influenza virus by Sovinova and others in Czechoslovakia ($S_{247}$).

2. *A/equi-1/England/1/63*. Similar to A/equi-1/Prague/56.

3. *A/equi-2/Miami/63*. The prototype equi-2 virus. Isolated by Waddell, Teigland and Sigel ($W_1$).

4. *A/equi-2/Uruguay/63*. South American equi-2 virus.

### f) Avian A viruses

1. Fowl plague viruses. The *"Brescia-Ascoli"* strain was isolated by Centanni in 1902. The *"Dutch"* strain isolated in Indonesia in 1927 and the *"Rostock"* strain have been most widely used.

The KP (Klassische Geflügelpest) virus used by Schäfer and others in studies of virus reproduction is the *"Rostock"* strain.

2. *VIRUS N* (1949). Isolated by DINTER from chickens in Germany ($D_{56}$). Related to classical fowl plague virus but much less virulent.

3. *A/AVIAN 1/England/1/63*. Isolated from turkeys by WELLS ($W_{33}$). Related to fowl plague virus.

4. *ANATIS 1 (A/AVIAN/2/Czechoslovakia/56)*. Isolated by KOPPEL and others ($K_{87}$) from ducklings in Slovakia.

5. *A/AVIAN 2/England/1/62*. Similar to Anatis 1.

6. *ANATIS 2 (A/AVIAN 2/England/4/56)*. Duck virus distinct from Anatis 1.

7. *SMITH* virus *(A/CHICK/Scotland/59)*, *A/AVIAN 4/Scotland/1/59*. First isolation of the Chick-Tern virus ($A_{83}$).

8. *TERN* virus *(A/AVIAN 4/S. Africa/61)*. Isolated by BECKER ($B_{33}$) from South African terns.

9. *WILMOT* virus (1963). Isolated from turkeys in Canada ($L_{17}$). Not related to other avian viruses except slightly to A/AVIAN 2/England/62.

### g) Human B Viruses

1. *LEE* (1940). The most studied strain of virus B, isolated by FRANCIS in America ($F_{90}$).

2. *MONTGOMERY* (1940). American isolate serologically close to LEE.

3. *PADDINGTON* (1943). English strain isolated by HIMMELWEIT ($H_{121}$). Serologically remote from LEE, but related to BON.

4. *BON* (1943). Isolated in Australia and much used in vaccines. Many strains serologically related to BON were isolated between 1943 and 1952, and include *CHADDICK* and *ALLEN* (U.S.A., 1945), *MIL B* (Australia, 1945), *KRI* (Russia, 1945), *CRAWLEY* (England, 1946), *WARNER* (U.S.A., 1948), *LONGHWAY* (U.S.A., 1952) and many others. These strains are usually regarded as a group B1 distinct from the original B (LEE). The *KRI* strain was used in vaccine production in Russia.

5. *B/Great Lakes/54* and *B/Johannesburg/58*. Representative of strains isolated between 1954 and 1959 intermediate between BON and the 1961 strains.

6. *B/Arizona/61* and *B/Taiwan/62*. Serologically remote from LEE and BON, representative of a group B2.

### h) Human C Viruses

1. *1233* (1947). First isolation of influenza C by TAYLOR ($T_{39}$).

2. *J.J.* (1950). Isolated by FRANCIS, QUILLIGAN and MINUSE ($F_{115}$). Identical to 1233.

# IV. Isolation of the Influenza Viruses

## 1. Source of Material

The most suitable materials for isolation of influenza virus from man are nasopharyngeal washings or throat garglings made with equal parts of saline and nutrient broth. Sputum emulsified in broth-saline is also very satisfactory. In children nasopharyngeal mucus may be collected on a swab and emulsified.

Isolation is usually only successful in the first 48 hours of disease. From fatal cases of influenzal pneumonia isolation may be attempted using 10% lung tissue suspension in saline or suspension of tracheal mucosa, but the results are only occasionally successful.

If material cannot be inoculated immediately it should be frozen in solid $CO_2$ for storage or transport. The fluid may be filtered before inoculation but in ferrets better results are obtained with unfiltered material. Unfiltered material has also been used for inoculation of the allantoic sac of fertile eggs ($R_{26}$) but is unsuitable for amniotic inoculation. BURNET and STONE ($B_{218}$) introduced the use of antibiotics, adding 100 units of penicillin per ml to the washings and inoculating the amniotic sac, 0.1 ml of 5% sodium sulphadiazine being introduced into the sac before inoculation. ROSE, MOLLOY and O'NEILL ($R_{53}$) also used penicillin, and FLORMAN, WEISS and COUNCIL ($F_{79}$) introduced streptomycin. A combination of penicillin and streptomycin is very satisfactory, amounts of 100—1000 units per ml of each antibiotic being added to the inoculum.

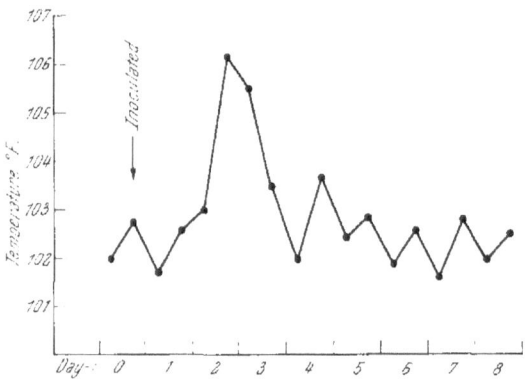

Fig. 2. Ferret influenza .
Rectal temperature of a ferret inoculated with material from a case of human influenza.

## 2. Isolation in Ferrets

Influenza virus A was first isolated from man by SMITH, ANDREWES and LAIDLAW in 1933. They inoculated ferrets intranasally under light ether anaesthesia with nasopharyngeal washings filtered through gradocol membranes of APD 0.6 μ. Other workers also obtained successful isolations in ferrets by inoculation of filtered or unfiltered material ($F_{86}$, $B_{154}$, $P_{57}$, $S_{202}$, $B_{125}$, $D_{92}$, $H_{202}$).

Ferrets are very susceptible to infection by influenza virus A. Intranasal inoculation under ether anaesthesia of 1 ml of unfiltered human nasopharyngeal washings results in a febrile response 48 hours after inoculation. The normal temperature of the ferret varies between 102°F and 103°F, and infected animals develop fever of short duration with temperatures of 104°F—107°F (Fig. 2). The animals are listless, sneeze frequently, have a mucoid nasal discharge, obstructed breathing, and loss of appetite. Symptoms persist for 2—3 days after which the animal recovers. The condition is very infectious and is readily transmitted to normal ferrets kept in the same cage. On serial transmission by inoculation of suspensions of the nasal mucosa of infected ferrets the disease becomes more severe and animals may die with lung consolidation.

Influenza virus B can also be isolated by ferret inoculation ($F_{90}$, $H_{121}$) but the B virus appears to be less pathogenic for the ferret than virus A, and in some outbreaks could not be isolated even when serial passage was employed ($D_{91}$, $H_{134}$). In some cases, however, the occurrence of inapparent infection could be shown by the demonstration of an antibody rise in the serum 3—4 weeks after inoculation.

### 3. Isolation in Mice

Intranasal inoculation of material from cases of human influenza produces no evidence of illness in mice, but if the mice are killed 2—4 days after inoculation and serial passages made with lung extracts, after a series of such "blind passages" the virus becomes adapted and lung lesions are produced ($F_{106}$, $Z_{42}$, $Z_{12}$, $S_{331}$). BURNET and FOLEY ($B_{197}$) found that infection of mice could be detected by the demonstration of immunity if the mice were challenged by intranasal inoculation of stock virus 14 days after inoculation.

There are several serious disadvantages in the use of the blind passage technique for isolation of virus in mice. Serial passage may result in the appearance of lung lesions due to bacteria, especially organisms of the Pasteurella group, or in the activation of latent viruses such as ectromelia. But the most serious difficulty is a result of the great sensitivity of mice to infection by mouse adapted laboratory strains of virus such as WS or PR8. ANDREWES, GLOVER, HIMMELWEIT and SMITH ($A_{75}$) describe 14 incidents in which the apparent isolation of influenza virus from man by mouse inoculation was apparently due to the accidental contamination with WS virus of the material used for intranasal inoculation. The strains isolated were antigenically identical to WS virus and were pathogenic to mice on first isolation. FINLAND and BARNES ($F_{66}$) in 1949 isolated seven strains of virus from human material which were similar to PR8. No antibody increase was found in the patients sera and it was considered that the strains were laboratory contaminants. The danger of accidental laboratory infection of mice has also been noted by KRAVCHENKO and SEKRETTA ($K_{100}$) and by PARNES ($P_{14}$). WARD and EDDY ($W_{18}$) found 5 of 35 strains isolated in Baltimore in 1946—1948 to be pathogenic for mice on first isolation and suggested that pathogenicity for mice may be a special characteristic of some strains, but the strains were serologically related to strains isolated in 1935 and may well have been laboratory contaminants.

### 4. Isolation in Rats

CHALKINA and GULAMOV ($C_{26}$) obtained 41% successful isolations of virus from man by inoculation of white rats. They state that rats can be infected by contact with an infected animal in the same cage.

### 5. Isolation in Hamsters

Influenza virus produces an inapparent infection in hamsters, but on serial passage the virus becomes adapted and lung lesions and death may occur ($F_{146}$). LOZOVAYA ($L_{122}$) states that the reactivity of the animal depends on the time of the year, in summer lung changes are produced, but during hibernation in winter the virus grows in the lungs but produces no visible pathological effects. TAYLOR and PARODI ($T_{42}$) were able to detect influenza virus in human throat washings by inoculation of hamsters followed by the demonstration of antibody in the serum of the animal. Virus could be adapted to mice after one hamster passage.

### 6. Isolation in Eggs

The use of the developing fertile hen's egg has proved the most useful technique for the isolation and cultivation of influenza viruses. The methods of inoculation of fertile eggs are described by BEVERIDGE and BURNET ($B_{65}$). Virus may be introduced into the allantoic, amniotic, or yolk sacs, on to the external surface of the chorioallantoic membrane, directly into the embryo, or intravenously.

Inoculation of the external surface of the chorioallantoic membrane ($XF_9$) has not been much used for primary isolation of virus from man. Inoculation of the yolk sac of 3—6 day old fertile eggs may be used ($B_{189}$, $B_{65}$), haemagglutinin being demonstrable 9 days after inoculation, but the yolk sac is very susceptible to bacterial and fungal infection and many deaths may occur. HIRST ($H_{128}$) made inoculations directly into the embryo, and made passages with 10% suspensions of embryo trachea. Successful isolation was obtained with 75% of washings known to be positive by ferret inoculations, but the method has no special advantages and has not been used by other workers.

Influenza viruses are readily adapted to growth in the allantoic sac, and by the inoculation of unfiltered throat washings RICKARD, THIGPEN and CROWLEY ($R_{26}$) isolated 9 strains from 20 cases. Most workers, however, have found the allantoic route of inoculation very unsatisfactory for primary isolation ($H_{128}$, $B_{190}$, $E_8$). BURNET and STONE ($B_{216}$) found that with virus in the O phase very large doses were required to infect the allantoic sac.

Inoculation of the amniotic sac, introduced by BURNET ($B_{158}$) is by far the most successful method of isolation of influenza viruses from man by the use of fertile eggs ($B_{197}$, $B_{190}$, $H_{132}$, $B_{218}$, $B_{221}$, $D_{91}$, $H_{134}$, $V_7$, $N_2$, $N_1$, $H_{216}$, $F_{137}$). Influenza A, B and C viruses can all be isolated by this method.

*Technique of virus isolation by inoculation of the amniotic sac:* Fertile eggs are incubated at $38°C—39°C$ for 10—13 days with the broad end uppermost. A circle of shell is then cut away over the air sac, and the shell membrane covering the chorioallantoic membrane moistened with sterile liquid paraffin to render it transparent. A small cut is then made through the chorioallantoic membrane avoiding any large vessel. The amnion is seized with a pair of fine forceps introduced through the incision, drawn up and the inoculum introduced with a fine hypodermic syringe. An alternative method is to use a finely drawn Pasteur pipette which when broken gives a very sharp point and can be introduced into the amnion without making any preliminary incision through the chorioallantoic membrane and without the use of forceps. A small air bubble should be introduced along with the inoculum. If the air bubble remains in the amniotic sac the inoculation has been successful. If the inoculum consists of unfiltered material penicillin and streptomycin should be added, 250 units of each antibiotic per ml. A suitable volume of inoculum is 0.2 ml. After inoculation eggs are capped or sealed with sellotape, incubated at $35°C$, and examined after 4 days. From 4 to 6 eggs should be inoculated with each fluid. After incubation the egg-shell at the narrow end is broken away to give a large opening and the contents of the egg can then be turned out into a sterile Petri dish without damage to amnion, embryo or yolk sac. Amniotic fluid is then collected and the embryo lungs and trachea dissected out. The presence of virus is tested by carrying out haemagglutination tests with amniotic fluid diluted 1:10, and by complement fixation tests with saline extracts of the embryo lung and tracheal tissue. If the results are negative a further amniotic passage may be made. Haemagglutination tests should be made with both chick and guinea pig red cells. Influenza B viruses agglutinate both types of cell equally, virus A strains agglutinate guinea pig cells more powerfully than chick cells, while influenza C strains only agglutinate chick cells. Complement fixation tests should be made with human convalescent sera containing antibody to the S antigens of the viruses and may give positive results when haemagglutination tests are negative. BEVERIDGE and BURNET (B$_{65}$) state that virus A strains are best isolated in 13—14 day old eggs, while virus B strains grow better in 8—10 day eggs.

After the initial amniotic passage A and B viruses usually grow readily in the allantoic sac, but virus C only grows in the amniotic cavity.

## 7. Isolation in Tissue Cultures

FADEYEVA (F$_2$) and RITOVA (R$_{32}$, R$_{37}$) used human embryo lung tissue cultures for primary isolation and for culture of strains to be used for preparation of live virus vaccine. Minced human embryo lung was washed with Tyrode's solution and exposed to the inoculum for an adsorption period of 20 minutes. The tissue was then transferred to Carrel flasks, heparinised chick or human plasma added and coagulated by addition of chick embryo extract. Nutrient fluid consisting of Tyrode solution 3 parts and heated chick or human serum 1 part was then added, and the cultures incubated 2—3 days. Penicillin and streptomycin were added to the nutrient fluid if unsterile inocula were used. Human plasma or serum was selected from individuals with no virus neutralising antibody. Minced chorioallantoic membrane cultures have been used (H$_{174}$) but have no advantages over the use of intact eggs. Human embryo kidney cells have been used (M$_{105}$), but the most extensive work on the primary isolation of influenza virus has been made with monkey kidney tissue cultures (M$_{106}$, T$_{11}$, S$_{99}$, R$_{68}$, M$_{109}$, S$_{25}$, S$_{263}$, C$_{12}$). Some workers have used nutrient fluids containing bovine amniotic fluid, embryo extract and heated horse serum, but it is best to avoid the use of these materials which may contain inhibitors, and satisfactory results can be obtained with cultures in medium 199. Incubation at 33°C is recommended by STERN and TIPPET (S$_{263}$). Some strains of virus produce cytopathogenic effects in monkey kidney cultures, but many do not and in these cases the presence of virus may be detected by the appearance of haemagglutinin in the culture fluids. The most satisfactory method of detecting the presence of virus is, however, the haemadsorption technique of VOGEL and SHELOKOV (V$_{25}$), which is more sensitive than the detection of haemagglutinin, and can be observed at an earlier stage of incubation.

### The Haemadsorption Technique

VOGEL and SHELOKOV (V$_{25}$, S$_{99}$) found that red blood cells are adsorbed by monkey kidney cells infected with influenza virus.

The nutrient fluid of monkey kidney tissue cultures is poured off and a dilute suspension (0.4—2%) of chick red blood cells is added to the tubes and allowed to sediment on to the cell sheet. After 10—30 minutes unadsorbed red cells are removed by gentle washing with saline. The presence of virus is indicated by the presence of clumps of adsorbed red cells adhering to the cell sheet. HOTCHIN and his associates (H$_{176-178}$) distinguished two types of haemadsorption, in one the red cells were attached to virus filaments protruding from the cell surface, the red cell and host cell being separated by a distance of up to 1 μ, while in the other the red cells were directly adherent to the cell, indicating the presence of influenza antigen on the cell surface. Haemadsorption can be observed in many other types of tissue culture (B$_{113}$, R$_{68}$, S$_{233}$).

All workers are agreed that the method is highly sensitive for the detection of virus infection, but the presence of simian haemagglutinating virus in monkey kidney cultures may give rise to errors. The haemadsorption reaction is inhibited by homologous immune sera, and PEREIRA ($P_{32}$) applied it to the development of a highly strain-specific neutralization test which could be used to type Q phase strains not typable by HI tests.

## 8. Use of Virus Concentration Methods in Isolation

Several workers have attempted to increase the efficiency of virus isolation by adsorption of garglings with red blood cells followed by inoculation of the sedimented red cells to eggs ($P_{45}$, $L_{94}$, $A_{101}$). By this method ARI ($A_{101}$) obtained 5 positive results from 17 washings, while only 2 positives were obtained from 20 cases with untreated washings. However, serial passage of the unconcentrated specimens yielded a further 6 positives, while no further positives were obtained with the concentrated specimens.

TAKEMOTO ($T_9$) adsorbed the virus with Nalcite HCR-X12 resin and eluted with 10% NaCl and obtained more positive results on amniotic inoculation than with untreated material.

Attempts have also been made to improve isolation efficiency by depressing the host resistance. MELNIKOVA ($M_{76}$) found that embryonated eggs irradiated with 250—500 r doses of X rays gave a greater percentage of virus isolations than normal eggs.

## 9. Relative Efficiency of Various Methods of Virus Isolation

The most generally successful method of virus isolation is the inoculation of fertile eggs by the amniotic route. Inoculation to the allantoic sac is much less effective ($B_{190}$, $H_{132}$).

Inoculation of ferrets is very satisfactory with virus A, and the detection of virus by the demonstration of antibody rises in the ferret sera may be more successful than the use of amniotic inoculation ($E_8$). The ferret is much less susceptible to virus B ($D_{91}$).

Several workers have compared the efficiency of amniotic inoculation of eggs with the use of monkey kidney tissue cultures ($M_{106}$, $T_{11}$, $M_{109}$, $R_{68}$, $S_{263}$). The egg proved superior with A1 and A2 viruses, but B virus was more easily isolated in monkey kidney cultures. Virus C can be isolated by both methods ($M_{104}$).

Whatever method is used the results of attempts at isolation vary considerably in different epidemics. In rapidly spreading pandemics of influenza virus is usually easily isolated, but in small localized outbreaks the results may be poor or even entirely negative. In the early stages of an epidemic virus may be difficult to isolate, while isolations become more frequent as the epidemic reaches its peak. Although serological results may indicate the occurrence of infection, virus is practically never isolated in inter-epidemic periods.

# V. Cultivation of Influenza Viruses in Fertile Eggs

## 1. Cultivation on the External Surface of the Chorioallantoic Membrane

This was the first method used for cultivation of influenza virus in eggs ($S_{161}$, $B_{155}$, $H_{204}$). In early passages no lesions are produced, and FULTON and ISAACS ($F_{165}$) showed that apparently only one cycle of growth occurred, but on serial passage many strains become adapted and produce small focal lesions up to 1 mm in diameter and a fatal haemorrhagic infection of the embryo may occur ($B_{65}$).

A very satisfactory technique for inoculation of the chorionic surface of the membrane is that described by HOYLE ($XII_4$). Twelve day old eggs are used, preferably incubated with the narrow end uppermost. A small square of egg shell is removed at the narrow end, and a small hole made into the air sac at the broad end of the egg. The egg shell membrane at the narrow end is then perforated with a needle and suction applied at the broad end when the chorioallantoic membrane falls away from the narrow end leaving an artificial air sac. The inoculum is then introduced on to the surface of the depressed membrane. The eggs are then capped or sealed with sellotape and incubated.

## 2. Cultivation in the Amniotic Sac

This method, described above, is the best procedure for primary isolation of influenza viruses from man, but is not used for routine culture of established strains, since on adaptation to the egg the virus produces a great reduction in the amount of amniotic fluid and larger yields of virus can be produced more easily by culture in the allantoic sac. Influenza C strains, however, will only grow in the amniotic sac.

## 3. Cultivation in the Embryo and Yolk Sac

These procedures have no special virtues and are not much used.

## 4. Cultivation in the Allantoic Sac

Virus isolated by amniotic inoculation is usually easily adapted to growth in the allantoic sac, and inoculation of the allantoic sac is the most valuable method of cultivating influenza A and B viruses and is almost exclusively used for the production of large amounts of virus for chemical studies or vaccine production. The technique is very simple.

Fertile eggs are incubated at $38°C—39°C$ for $10—12$ days. They are then candled, the outline of the air sac marked, and an inoculation point marked on the shell a short distance away from the air sac at a point free from large blood vessels. A hole is made through the shell into the air sac, and after sterilising the shell with alcohol a hole is made through the shell at the inoculation point. The inoculum is introduced with a syringe or drawn out Pasteur pipette, $0.1—0.5$ ml may be inoculated. The hole is then sealed with paraffin wax or collodion, and the eggs incubated at $35°C$. Infected allantoic fluid is harvested at $24—48$ hours. Optimal yields of virus are produced by the inoculation of 0.1 ml of infected allantoic fluid diluted $1:100$ or $1:1000$, the eggs being harvested at $36—40$ hours. MILLER ($M_{85}$) recommended inoculating 0.1 ml of a $10^{-5}$ dilution and harvesting at $36—48$ hours. Larger inocula may yield virus with a lower infectivity titre (incomplete virus).

Infected allantoic fluid prepared in this way contains some $10^9$ $TD_{50}$ of virus per ml, has a haemagglutinin titre of $1000—4000$ by the Salk method, and by the use of efficient methods of purification it is possible to obtain from each egg about 100 micrograms of purified virus.

## 5. Cultivation in De-embryonated Eggs

This technique, introduced by BERNKOPF ($B_{57}$, $B_{58}$) has proved very useful in studies of the growth cycle of influenza viruses, in studies of the effects of chemotherapeutic agents, and for the production of virus heavily labelled with radioactive isotopes.

In the original technique 14—15 day eggs were used. A large circle of shell was cut out over the air sac, the free-lying egg shell membrane and underlying chorio-allantoic were then cut away, the embryo and yolk sac were then emptied out through the hole, their connections with the chorioallantoic membrane being cut with scis-sors. The remainder of the chorioallantoic membrane remained attached to the egg shell and was thoroughly washed with saline to remove albumen and blood. After draining the egg was filled with 10—40 ml of Tyrode's solution containing 10 units of penicillin and 40 μg streptomycin per ml, and inoculation made into this fluid. The hole in the egg was then sealed with a rubber cap and the eggs incubated in a roller machine making 6 revolutions per hour.

An alternative technique which permits the use of 11—13 day eggs is to cut away the narrow end of the egg with scissors and evacuate the albumen, embryo and yolk sac through the narrow end. The rest of the procedure is as in the original method.

The yield of virus in de-embryonated eggs is less than in intact eggs. FINTER and others ($F_{71}$) developed a method in which infection is allowed to proceed to any desired stage in intact eggs which are then de-embryonated and incubated for a further 1—3 hours. By this means it was possible to study the yield of virus at different stages, the yields in the short period after de-embryonation being similar to those in intact eggs.

# VI. Tissue Culture Methods for Influenza Virus Cultivation

Early tissue culture studies using minced chick embryo in Tyrode solution and similar methods ($S_{161}$, $M_{11}$, $N_{22}$, $P_{26}$, $P_3$) are now only of historical interest. With the development of modern methods of tissue culture the growth of influenza viruses has been studied in a great variety of cell types. Some of the major studies are listed in Table 1.

It appears that influenza viruses, with the exception of influenza C, may be grown in a wide variety of tissue cultures by most of the established tech-niques. It is, however, desirable to avoid the use in the culture fluid of materials such as animal sera, amniotic fluid and embryo extract which may contain in-hibitors. Four types of result may be obtained in tissue cultures with influenza viruses.

1. Virus may grow with the development of cytopathic changes. Few strains produce cytopathic effects in early passages, but may do so on adaptation. Cyto-pathic effects are most readily produced in kidney tissue cultures ($H_1$, $M_{110}$, $H_2$, $B_{143}$, $D_{79}$, $N_{10}$, $M_{107}$, $M_{52}$, $G_{80}$). Neurotropic strains of virus such as NWS may grow and produce cytopathic changes in cells in which other strains fail to grow or produce no obvious effects ($K_{50}$).

2. Virus may grow without producing obvious cell changes, but with the release of virus into the culture fluid in which it can be detected by haemag-glutinin titrations. This is the usual result in lung tissue and chorioallantoic membrane tissue cultures.

3. An incomplete growth cycle may occur, complement fixing antigen and haemagglutinin being produced in the cells but no release of infective or haemag-

Table 1. *Tissue Culture Studies with Influenza Viruses*

---

*Human Cells*

Embryonic lung — : $F_2$, $F_1$, $F_4$, $R_{32}$, $R_{37}$, $M_{110}$, $B_{143}$, $H_{122}$, $H_{123}$
Embryonic kidney — : $M_{105}$, $M_{110}$, $B_{143}$, $H_{58}$   Embryonic liver — : $C_{60}$   Amnion — :
   $B_{143}$, $T_{10}$
Embryonic skin, conjunctiva, fibroblasts, KB cells — : $G_{72}$
HeLa cells — : $H_{65}$, $G_{33}$, $G_{72}$, $R_{68}$, $B_{143}$   Keratoacanthoma cells — : $P_{86}$
Conjunctival (Chang) cells — : $W_{58}$, $T_{60}$   Diploid cells (Wyeth 1024, Wyeth 3) — : $K_{50}$

*Monkey Cells*

Rhesus monkey kidney — : $T_{11}$, $M_{106}$, $G_{72}$, $R_{68}$, $B_{143}$, $H_{58}$, $G_9$, $S_{233}$, $S_{25}$, $B_{113}$, $M_{107}$,
   $M_{104}$, $S_{263}$
Monkey heart — : $B_{143}$, $S_{233}$   Testicle — : $B_{143}$
Cynomolgus monkey heart (SOT) cells — : $S_{232}$

*Chicken Embryo Cells*

Chorioallantoic membrane — : $F_{163}$, $G_1$, $H_{174}$, $F_{53}$, $S_{343}$, $S_{347}$, $Z_3$, $A_{53}$
Lung — : $S_{295}$, $T_{74}$, $L_{51}$, $B_{143}$, $S_{150}$, $S_{265}$, $T_{60}$, $S_{264}$   Amnion — : $W_{32}$
Kidney — : $M_{35-37}$, $H_{58}$, $B_{113}$, $L_{95}$   Fibroblasts — : $S_{295}$, $G_{69}$, $S_{265}$, $P_{67}$, $S_{264}$
Skin-muscle — : $S_{150}$, $C_{19}$   Liver — : $T_{74}$   Cornea — : $S_{295}$   Brain — : $W_{32}$

*Mouse Cells*

Kidney — : $S_{233}$, $H_{58}$   Fibroblast (L cells) — : $S_{160}$   Ehrlich ascites tumour cells — :
   $A_{12}$, $O_9$
Mouse embryo lung — : $G_{48}$

*Other Types of Kidney Cell*

Calf kidney — : $H_2$, $G_{72}$, $N_{10}$, $H_{58}$, $H_{157}$, $N_{29}$, $L_{56}$
Guinea pig, mouse, rat, rabbit — : $S_{233}$, $H_{58}$   Ferret — : $H_1$   Hamster — : $H_{58}$, $G_{80}$
Adult fowl — : $M_2$, $B_{113}$

*Other Types of Lung Cell*

Pig lung — : $H_{122}$, $H_{123}$   Rabbit lung — : $H_{123}$   Rat embryo lung — : $B_{143}$

---

glutinating virus into the culture fluid. In these cases virus production may be detected by the haemadsorption technique. Incomplete growth cycles occur in HeLa cells ($H_{65}$, $G_{33}$).

4. Virus may fail to grow. Total failure has been recorded with rabbit lung ($H_{123}$), human amnion ($T_{10}$), monkey heart ($B_{143}$, $S_{232}$), chick fibroblasts ($S_{295}$, $S_{265}$), human fibroblasts and human conjunctiva ($G_{72}$). It is possible that in some of these cases an incomplete growth cycle may have occurred and not been recognized ($S_{264}$). There are considerable strain variations, thus swine strains have been reported to fail to grow in human embryo lung while human strains grow ($H_{123}$).

*Culture of influenza viruses in calf kidney cells:* Kidney cells are very suitable for the cultivation of influenza viruses and calf kidney cells introduced by LEHMANN-GRUBE ($L_{55}$) are very satisfactory. Kidneys were collected from 1—5 months old calves within 3 hours of slaughter. Trypsinized cells washed with BSS were suspended in a growth medium consisting of medium 199 containing 0.04% $NaHCO_3$ and 10%

calf serum. Tubes were seeded with $6-8 \times 10^5$ cells in 1 ml and incubated at an angle of $5°C$. After 2 days at $37°C$ the medium was changed, the $NaHCO_3$ being increased to $0.17\%$ and gassed with $CO_2$. Cultures were used $1-2$ days later. Tubes were drained of medium, the cells washed with BSS and inoculated with 0.1 ml of virus suspension. After 15 minutes at room temperature 0.9 ml of medium 199 containing $0.17\%$ $NaHCO_3$ but without calf serum was added and the tubes incubated at $35°C$. Good growth of virus was obtained in $3-5$ days. After a brief period in which haemadsorption can be demonstrated the cells show cytopathogenic effects and haemagglutinin appears in the fluid.

### 1. Plaque Methods with Influenza Viruses

Attempts to produce plaques by the tissue culture methods of DULBECCO and VOGT ($D_{93}$, $D_{94}$) have been less successful than with many other viruses. GRANOFF ($G_{69}$) using chick embryo fibroblast monolayers only obtained plaques with neurotropic strains, with chick embryo lung plaques were obtained with PR8 and MEL but not with LEE. The plaque titre was 800 times less than the egg $ID_{50}$ titre and there was evidence that non-plaque forming particles interfered with the growth of plaque formers.

Plaques were obtained by LEDINKO with chick embryo lung ($L_{48}$), by CHOPPIN ($C_{41}$) and VONKA ($V_{29}$) with monkey kidney cells, and by GROSSBERG ($G_{81}$) with hamster kidney, but in all these cases the plating efficiency was very low. Better results were obtained by LEHMANN-GRUBE ($L_{53}$) using calf kidney cells, and after serial passage in monkey kidney cells CHOPPIN ($C_{41}$) obtained plaque formation with a ratio of p.f.u. to egg $ID_{50}$ of 0.7 with one strain RI/5$^+$ of influenza A2 virus.

SUGIURA and KILBOURNE ($S_{327}$) used plaque cultures with a variant of the Chang human conjunctival cell line in recombination studies, and STAIGER ($S_{252}$) used plaque type cultures in chick embryo fibroblasts in recombination studies with fowl plague virus.

### 2. Organ Cultures

WAGNER ($W_3$) transplanted embryonic mouse tissues infected with virus into the subcutaneous space of homologous adult hosts. Infection was localised to the transplanted donor tissue and the draining lymph glands. Non-neurotropic strains only grew in transplanted lung tissue, but the NWS strain grew in lung, intestine, kidney, heart, bladder and spleen but not in skin.

TYRRELL and HOORN ($T_{77}$) used organ cultures of ferret and human tracheal epithelium in medium 199 at $33°C$. Egg adapted strains of virus grew, even influenza C, but there was no detectable damage to the epithelium except with influenza B.

# VII. Purification of the Influenza Virus

Crude preparations of influenza virus such as infected tissue extracts or infected allantoic fluid do not usually contain more than about 1 part in 10,000 of virus, while they may contain from $1-10\%$ of non-viral protein, so that the separation of virus from the large amount of contaminating material is not easy. It is essential to start with a sufficiently large amount of material and in general three steps are necessary to obtain a satisfactory result. The usual first step is a clarification procedure designed to eliminate as much as possible of the non-viral material without serious loss of virus. This should be followed by a concentration procedure which is necessary partly to reduce the bulk of the material to an amount convenient to handle, and also to reduce losses of virus by denaturation at interfaces, adsorption to glassware, oxidation etc., which are much more serious with dilute than with more concentrated preparations. Finally some highly selective procedure must be used to produce the final product. Such selective procedures may be based on differences in particle size between virus and contaminating material, differences in density in electric charge, or in chemical properties. Experience has shown that it is best to employ two different types of selective procedure rather than to repeat the same process many times as this merely results in loss of virus without significant gain in purity.

## 1. Purification by Centrifugation Methods

### a) Differential Centrifugation

In early work on influenza virus purification the procedure of differential centrifugation was much used. The material was first centrifuged at low speeds to remove particles of larger size than the virus. The supernatant was then centrifuged at a speed sufficient to deposit the virus. The deposit was then resuspended, again centrifuged at low speed and any deposited material discarded. The method was first used in influenza virus work by ELFORD and ANDREWES ($E_{35}$) in 1936. In 1937 HOYLE and FAIRBROTHER ($H_{203}$) using mouse lung extracts and a belt-driven centrifuge separated influenza virus from soluble antigen and used the semi-purified virus as a vaccine in experimental animals.

In 1944 STANLEY ($S_{258}$) using differential centrifugation at 24,000 and 3000 r.p.m. in a Bauer and Pickels air turbine centrifuge obtained very satisfactory concentration and purification of virus from infected allantoic fluid. It was found that 97% of the haemagglutinating activity of the original material appeared in the first 24,000 r.p.m. deposit and the final product contained 22,500 haemagglutinin units per mg N, a result superior to that attained by red cell adsorption-elution methods. STANLEY ($S_{260}$) also used a Sharpless centrifuge to handle large volumes and obtained a 70% recovery of virus in the final product. In later work the Spinco L centrifuge has usually been used. SCHÄFER and SCHRAMM ($S_{39}$) using fowl plague virus obtained results similar to those of STANLEY.

Although many workers attempted to improve the purity by repeating the differential centrifugation procedure several times it is impossible by the use of differential centrifugation alone to separate virus from non-viral material of the same particle size. A considerable improvement could, however, be attained by the use of density gradient centrifugation.

### b) Density Gradient Centrifugation

Two methods are used. In the rate zonal method semi-purified and concentrated virus is layered on top of a sucrose density gradient prepared by layering sucrose solutions ranging from 40% to 10% in a centrifuge tube. A swinging-bucket type rotor such as the SW 39 is used and on centrifugation virus concentrates in a band at a level depending partly on the density of the particles, partly on the size and partly on the duration of centrifugation.

In the equilibrium technique virus is mixed with an equal volume of saturated caesium chloride solution and centrifuged for 24—36 hours at 35,000 $g$ until equilibrium is attained. The slow sedimentation of the heavy caesium chloride produces a density gradient in the tube and virus collects in a band dependent on its density. Alternatively, virus may be mixed in during the formation of a caesium chloride or sucrose density gradient when equilibrium is attained with a shorter period of centrifugation. In both methods samples are collected from a pin-hole made in the bottom of the tube, or by siphoning serial samples from the top.

STANLEY ($S_{259}$) employed the rate-zonal method using a sucrose gradient. Recently REIMER and others ($R_{18}$) have described the use of a B IV centrifuge rotor capable of handling large volumes of material and using both the rate-zonal and equilibrium methods with a sucrose gradient. After preliminary differential centrifugation of allantoic fluid virus a highly purified product was obtained, containing 22,000 haemagglutinin units per mg of protein.

The density gradient technique is of most value when used at the final step of a purification procedure.

## 2. Purification by Filtration Methods

It is possible to purify virus by passing crude preparations through a filter with a pore size which retains large particles but allows passage of virus, and then to allow virus to silt up against a membrane of smaller pore size which will not allow the virus to pass, but such methods result in large losses of virus and are generally impractical. In 1942 SCHAEFER ($S_{26}$) used a method of this kind with infected mouse lung extracts, but this work is chiefly of interest in that it describes the first use of enzymes to destroy non-viral material. By treatment with trypsin and ribonuclease he produced a 99% reduction in nitrogen content of the final product without loss of infectivity.

## 3. Purification by Adsorption-Elution from Red Blood Cells

The discovery of the adsorption and elution of virus from red cells ($H_{125}$, $M_{54}$) was soon followed by the use of this method for purification and concentration of virus especially for vaccine production ($H_{42}$, $F_{116}$, $K_{16}$, $S_{216}$). Allantoic fluid virus is readily concentrated up to 20 times and a high degree of purity can be attained. The method may also be used with mouse-lung extracts ($K_{11}$). By the use of two successive cycles of adsorption-elution virus can be almost entirely freed of contamination by material of host-cell origin. SHEFFIELD, SMITH and BELYAVIN ($S_{98}$) used four cycles but this resulted in great losses of virus.

Successful purification of virus by the adsorption-elution technique depends on the use of correct concentration of red cells. Crude infected allantoic fluid should be adsorbed at room temperature with $1\frac{1}{2}$—2% red cells and the cells immediately centrifuged out. Alternatively, a more prolonged adsorption period may be used at 4°C. Fowl, guinea pig or human cells may be used. Some 80 - 90% of the virus is adsorbed. The use of larger amounts of red cells may lead to difficulty in elution. The deposited cells are washed with ice-cold saline and eluted at 37°C into 5 times their volume of saline. Some haemolysis usually occurs at this first cycle, which is primarily a concentrating cycle. A second cycle is then carried out using the same amount of red cells as in the first cycle. There is little or no loss of virus in the second cycle and with most strains of virus haemolysis is minimal. Best results are obtained with strains which elute readily — in 30 minutes to 1 hour.

To overcome the difficulty of haemolysis on elution YEREMEYEV ($Y_5$) recommended adsorption with washed erythrocyte stroma followed by elution into 5% NaCl, and Russian workers have also used formolised erythrocytes ($Y_8$, $D_{76}$). TAKATSY ($T_3$) found that virus purified by red cell adsorption-elution precipitated on dialysis and the deposit contained less than 0.7% of non-viral nitrogen. Precipitation of different strains of virus occurred at different salt concentrations, B virus requiring a lower salt concentration than A viruses.

## 4. Combined Use of Adsorption-Elution from Red Cells and Differential Centrifugation

This method has been used by the author for many years ($F_{154}$). Virus from infected allantoic fluid is purified by two cycles of adsorption-elution from guinea pig red cells as described above and the second cycle eluate is clarified by centrifugation at low speed, the virus deposited by centrifugation at 26,000 $g$ for 1 hour, resuspended to a concentration of about 1% by volume and centrifuged at 5000 $g$ for 10 minutes. The resulting virus suspension proved satisfactory for electron microscopy and chemical analysis and contained 100,000 haemagglutinin units per mg of dry virus. This result is superior to any obtained by centrifugation methods and is equivalent to the best results obtained by adsorption-elution from aluminium phosphate columns (see below).

## 5. Adsorption-Elution from Inhibitors

CURTAIN ($C_{95}$) found that if powdered cellulose was converted to p-aminobenzyl cellulose it could be coupled to haemagglutinin inhibitors from urine or salivary gland by diazotisation. The product adsorbed influenza virus which could be recovered by elution. The adsorbing power was destroyed by RDE or by virus.

## 6. Non-specific Adsorption Methods

If infected allantoic fluid is frozen and then allowed to slowly thaw a precipitate forms which carries down 75—95% of the virus ($H_{150}$, $H_{44}$). The precipitate can be redissolved at 37°C or can be eluted in buffer pH 5.5. Concentration of virus is possible but the procedure has no value as a purification method. CHAMBERS and HENLE ($C_{28}$) found that virus could be precipitated from allantoic fluid by protamine. HARE and CURL ($H_{41}$) studied the adsorption of virus by silica earths, kaolin, FULLERS earth and charcoal. Satisfactory elution was possible only from silica earths and by eluants containing protein (1% isinglass). Recovery of virus was only 50%. TOVARNITSKY ($T_{52}$) used kaolin and eluted with ammonia at pH 7.6.

### a) Adsorption with Calcium Salts

SALK ($S_3$) partially purified virus from mouse lung by adsorption on calcium phosphate at alkaline reaction followed by elution with citrate buffer pH 6.5.

SALK and STANLEY ($S_5$, $S_{261}$) used calcium phosphate adsorbed virus for vaccination. VOORSPUIJ ($V_{30}$) adsorbed virus from allantoic fluid on calcium carbonate, resuspended the precipitate in saline at pH 7, added sodium dusarit (zeocarb) and dissolved the precipitated $CaCO_3$ by passing $CO_2$ into the suspension. On centrifugation the supernatant contained all the original virus and there was no loss of infectivity. TAVERNE, MARSHALL and FULTON ($T_{32}$) adsorbed virus on calcium phosphate columns and eluted with increasing concentrations of phosphate buffer pH 7. Elution with 0.2 M phosphate removed most of the protein but little or no haemagglutinin. Some 50—80% of the haemagglutinin was eluted by 0.5 M buffer and infectivity was unaffected.

### b) Adsorption with Aluminium Salts

BODILY, COREY and EATON ($B_{94}$) precipitated virus from allantoic fluid with alum and redissolved in sodium citrate without loss of properties. Formolised alum precipitated virus gave good results when used as a vaccine.

MILLER and SCHLESINGER ($M_{91}$) prepared columns of silica gel coated with aluminium phosphate. To 100 ml of grade 912 silica gel was added 80 ml of 16% $Na_3PO_4$ and 80 ml of 10% $AlCl_3$. The aluminium phosphate coated gel was then thoroughly washed and formed into a $30 \times 200$ mm column. At suitable pH and salt concentration the column would adsorb all the virus from 1500 ml of infected allantoic fluid. Adsorption conditions varied with different strains of virus but in general was best with 0.063 M buffer at pH 6—6.5. Virus was eluted with 0.25 M buffer at pH 7—8. It was found useful to carry out a preliminary concentration of virus by adsorption and elution from bulk aluminium phosphate gel before passing through the column. Virus was not inactivated and the final product contained $10^{15.2}$ $ID_{50}$ per gram of nitrogen. By varying the conditions of adsorption-elution it was possible to separate LEE and PR8 viruses. FROMMHAGEN and KNIGHT ($F_{157}$) also used the method and considered it superior to red cell adsorption-elution methods.

VEERARAGHAVAN and SREIVALSAN ($V_{17}$) compared various methods of purification and found adsorption-elution from aluminium phosphate gels to give the greatest purity.

DRESCHER ($D_{71}$) made a study of the adsorption of influenza B on aluminium oxide and hydroxide. With gamma aluminium oxide adsorption of virus was uniform and predictable with the aid of Freundlich's adsorption isotherm. Aluminium hydroxide had a greater adsorption capacity but uptake of virus was not uniform.

### c) Adsorption with Zinc Salts

METCALF ($M_{82}$) precipitated virus from allantoic fluid with 0.5 M zinc acetate and recovered by serial extraction with 1 M glycine. There was 90% recovery as measured by haemagglutination and complement fixation tests but the product was non-infective. HAUSLER and DICK ($H_{56}$) modified the method by removal of the zinc from the glycine extract by passage through an Amberlite IR 120 resin column, followed by differential centrifugation.

### d) Adsorption with Magnetic Iron Oxide

WARREN, NEAL and REYNOLDS ($W_{20}$) adsorbed the virus from infected allantoic fluid by addition of 20 mg/ml of finely divided $Fe_2O_3$. The virus was eluted without loss of infectivity by 10% phosphate buffer pH 8.0. The method could be adapted to a continuous flow process by passing allantoic fluid over a layer of $Fe_2O_3$ held in place by a magnet.

### e) Adsorption on Lipids

NOLL and YOUNGER ($N_{26}$) found that PR8 virus could be adsorbed by fatty acids, alcohols, amines and an amide with a chain of 13—21 C atoms. Palmitic acid adsorbed 55,000 haemagglutinin units of B virus per gram, cholesterol 15,000 and hexadecylamine 200,000 units/gram. Concentration was easily achieved on hexadecylamine but infectivity was lost.

## 7. Purification by Alcohol Precipitation

TOVARNITSKY and CHALKINA ($T_{54}$) clarified mouse lung suspensions by addition of saturated $CaSO_4$ solution and then precipitated the virus by addition of 30% cold ethanol. The virus was then redissolved in Ringer's solution.

Cox and others ($C_{85}$) found that methyl alcohol produced less denaturation than ethanol. Infected allantoic fluid cooled to —5°C was precipitated by addition of 25% of ice cooled methanol. The precipitate was centrifuged off in the cold and resuspended in 0.1 M phosphate buffer pH 7. After clarification at 2000 r.p.m. the virus was deposited in a Sharples centrifuge and the deposit washed with buffer in the centrifuge. The final product contained 24,500 haemagglutinin units per mg N. Cold methanol precipitation was also used by LYTLE and WOOLDRIDGE ($L_{135}$).

In the author's experience methanol precipitation has not been very successful. Influenza virus is easily disintegrated by fat solvents, and it is possible that the finding by Cox et al. that the final yield of virus as measured by haemagglutination was sometimes greater than 100% indicates the occurrence of such disintegration.

## 8. Purification by the Use of Ion Exchangers

MULLER and ROSE ($M_{147}$) found that the Amberlite exchange resin XE 64 converted into the sodium form and washed with N/1 acetate buffer pH 5.6 adsorbed 90% of virus from infected allantoic fluid. The virus could be eluted with 10% NaCl.

MATHEKA and ARMBRUSTER ($M_{45}$) adsorbed virus on Amberlite IRA 400 in the carboxyl form at pH 6.0 in 0.01 N citrate and eluted with increasing concentrations from 0.01—2 N citrate at pH 6.0.

LAVER ($L_{36}$) found that if LEE virus partially purified by red cell adsorption-elution was passed through the cellulose phosphate cation exchanger Celex P, impurities remained adsorbed on the column while the virus passed through. A virus strains were adsorbed on the column and could not be eluted.

The author has found that virus partially purified by red cell adsorption-elution and dialysed to reduce the salt concentration is adsorbed by DEAE cellulose columns and can be eluted as a single peak by buffered saline solutions of increasing concentration, but the recovery of virus was less than 50%.

## 9. Evaluation of Methods and Assessment of Purity

Two methods of purification have been most used, adsorption on aluminium phosphate followed by controlled elution, and two cycles of adsorption-elution from red cells followed by differential centrifugation. The two methods give closely comparable results and are superior to methods based on centrifugation alone. In the case of the influenza virus there are considerable difficulties in assessing the purity of the final product.

In general highly purified virus preparations appear uniform on electron microscopy, have a high infectivity and high serological titres in relation to protein or nitrogen content, give a single sharply defined band on density gradient centrifugation, move as a single band on electrophoresis, and on inoculation into animals do not provoke the appearance of antibody to host materials. With the influenza virus none of these tests is completely satisfactory. Influenza particles are not uniform in size and morphology. Highly purified preparations tend to rapidly lose infectivity, probably as a result of oxidation. The particle may disintegrate with an increase in haemagglutinin titre. Highly concentrated preparations may become aggregated with reduction in complement fixing antigen titre. Disintegration may release sub-units with different densities or electrophoretic mobilities. Thus RUTTKAY-NEDECKY and IVANICOVA ($R_{70}$) found that highly purified preparations contained some 10% of a virus-specific material moving as a separate component in moving boundary electrophoresis tests. And finally influenza virus particles contain host-cell material as an integral part of their structure.

# VIII. Fractionation of Influenza Virus Particles

Studies of the release of influenza virus from the infected cell ($H_{195}$, $W_{69}$) showed that the virus acquired an outer membrane derived from the wall of the infected cell, and this led to attempts to disrupt the virus particle by the use of agents which disintegrate cell membranes. The most satisfactory results were obtained by the use of ethyl ether.

## 1. Ether Fractionation

In 1949 ANDREWES and HORSTMANN ($A_{76}$) found that viruses could be divided into ether-sensitive and ether resistant groups but the mechanism of action of ether was not studied. In 1952 HOYLE ($H_{193}$) described the fractionation of the influenza virus particle by ether treatment. Virus from infected allantoic fluid was partially purified and concentrated by adsorption-elution from red blood cells and the virus suspension shaken with half its volume of ethyl ether. On centrifugation a layer of membranous material collected at the ether-water interface, while in the aqueous phase two separate sub-units could be detected; a haemagglutinating component, and a complement fixing antigenic component which was indistinguishable from the soluble S antigen extractable from infected cells. The two components could be separated by red cell adsorption. In 1953, SMITH, COHEN and others ($S_{172}$) studied the effects of ether on virus haemagglutination but did not observe the release of S antigen. In 1954 SCHÄFER and ZILLIG ($S_{41}$) fractionated fowl plague virus and obtained results similar to those of HOYLE, and the results with influenza virus were confirmed by LIEF and HENLE ($L_{75}$), PAUCKER, BIRCH-ANDERSEN and VON MAGNUS ($P_{17}$) and by DAVENPORT, ROTT and SCHÄFER ($D_{29}$).

Successful fractionation of influenza virus by ether treatment depends chiefly on the use of virus preparations of suitable concentration and purity. Best results are obtained with preparations containing some 10,000 haemagglutinin units per ml. More concentrated preparations may disintegrate less completely, while with weak preparations there may be losses by denaturation. Very satisfactory results are obtained by the use of virus preparations made from infected allantoic fluid by two cycles of adsorption-elution from red cells, concentrating from 10 to 20 times. More highly purified preparations require minimal treatment with ether — it is easy to denature such preparations by excessive shaking. Ether treatment may be carried out at 37°C, room temperature, or in the cold, but pure redistilled ether free from peroxide should be used. Three successive periods of shaking for 30 seconds at 30 minutes intervals is usually satisfactory, but some workers have used more prolonged periods of treatment ($L_{75}$, $D_{29}$). After treatment the ether-virus mixture is centrifuged when three separate virus fractions can be produced. The membranous material which collects at the ether-water interface contains protein, lipid and carbohydrate and appears to represent the outer membrane of the virus particle. If it is resuspended in saline it agglutinates red blood cells. Only small amounts of lipid material are found in the ether layer. The haemagglutinin and complement fixing "soluble antigen" components present in the aqueous phase can be separated by red cell adsorption. The haemagglutinin is adsorbed and can be subsequently eluted. Both components can then be deposited by centrifugation at $80-100,000\ g$ for 5 hours. The complement fixing component can be completely freed from haemagglutinin, but the haemagglutinating component usually contains a small amount of S antigen. This S antigen can be removed by precipitation with lanthanum acetate ($M_{101}$).

The ether fractionation technique has been much used in studies of the chemical structure of the virus particle ($F_{154}$, $Z_{43}$, $W_{30}$, $P_{21}$, $F_{153}$, $L_{36}$) and for the production of non-toxic sub-unit vaccines ($M_{101}$, $W_{27}$).

Fat solvents other than ether also disrupt influenza virus particles but are less satisfactory than ether.

## 2. Fractionation by the Use of Detergents

Influenza virus particles are disrupted by detergents ($B_{182}$), which have been used in chemical studies of the virus structure by LAVER ($L_{36}$, $L_{40}$).

Highly purified virus suspensions have usually been used. Suspensions containing 100,000 haemagglutinin units per ml readily disintegrate on treatment with $1-2\%$ sodium deoxycholate at 37°C, the opalescence clearing in $15-30$ minutes. Infectivity is greatly reduced but the serological properties are retained. Neuraminidase activity is unaffected. Haemagglutinin titres may be increased or may be reduced, the results depending on the type of red cells used. The haemagglutination is less vigorous and end points may be irregular. If the deoxycholate is removed by dialysis haemagglutinin and complement fixing S antigen can be separated by red cell adsorption.

Virus preparations disintegrated by deoxycholate are very satisfactory for use in studies of the antigenic structure of the virus by the use of gel-diffusion methods ($S_{308}$, $H_{17}$) and are superior for this purpose to ether-treated preparations.

Sodium dodecyl sulphate produces a more complete disintegration of virus than deoxycholate ($L_{40}$) and dodecyl sulphate treated virus is less effective as an immunising agent and less satisfactory in gel-diffusion studies.

### 3. Combined Use of Detergents and Ether

Ether treatment of virus usually results in partial or complete loss of the strain-specific complement fixing power ($H_{193}$). This is not due to destruction of the specific antigen as complement fixation may occur if the fixation time is prolonged. The antigen appears to be modified so that it no longer gives complement fixation in 1 hour at 37°C, and in fact may prevent fixation by intact virus. Prolonged ether treatment also produces a reduction in haemagglutinin titre ($C_{46}$) apparently as a result of disintegration into small particles which are possibly monovalent and unable to link red blood cells. Detergents may produce similar effects. HOBSON ($H_{153}$) found that if deoxycholate treated preparations are extracted with ether all haemagglutinating activity is lost, but specific antigen was still present and could be detected by its ability to block haemagglutinin inhibiting antibody in haemagglutination-inhibition tests, and by its ability to provoke antibody production in animals.

### 4. Other Methods of Disruption

TYRRELL and HORSFALL ($T_{79}$) found that exposure of virus to 50 cycles of freezing and thawing destroyed the haemagglutinating power, enzymic activity and infectivity, but a highly strain-specific antigen was released which could be detected by its ability to block haemagglutinin-inhibiting and neutralising antibody.

LIEF and HENLE ($L_{76}$) found that sonic vibration released S antigen from virus particles but was inferior to ether for this purpose.

DORMAN ($D_{62}$) found that treatment of A2 influenza virus for 1 minute at 0°C with 7.2 M urea caused disruption with release of S antigen which could be separated from haemagglutinin and V antigen by density gradient centrifugation.

NOLL, AOYAJI and ORLANDO ($N_{27}$) found that treatment of LEE virus by trypsin did not disrupt the virus particle but released 90% of the neuraminidase as material with a sedimentation constant of 9.0 S, corresponding to a molecular weight of 190,000. The sedimentation constant of the virus particle was reduced from 891 to 797 S.

The haemagglutinin of some strains of influenza virus is destroyed by trypsin with loss of the surface spikes ($B_{106}$).

REGINSTER ($R_{16}$) found that caseinase C released neuraminidase from purified PR8 virus and also material which blocked haemagglutinin-inhibiting antibody. Most of the surface projections of the virus were removed. Similar results were obtained with pronase ($R_{17}$). MIZUTANI and MIZUTANI ($M_{102}$) showed that disintegration of the virus phospholipid by phospholipase C did not result in loss of infectivity but the virus particle became susceptible to the action of proteinases with reduction in infectivity.

# IX. Chemical Composition

Early studies of the chemical composition of influenza virus particles ($T_{33}$, $K_{67}$) were carried out with insufficiently purified preparations, and in addition to protein, lipid, and carbohydrate both types of nucleic acid were found ($M_{90}$). More recent work with highly purified virus has shown that only ribonucleic acid is present. The virus does, however, contain large amounts of lipid and carbohydrate, and the nitrogen content of dried virus is only about 10% ($M_{91}$).

## 1. Fatty and Lipid Constituents of the Virus

The total amount of material extractable from influenza virus by fat solvents has been found to range from 18—36% ($T_{33}$, $G_{64}$, $U_1$, $F_{154}$, $F_{158}$), and to be 24% in fowl plague virus ($Z_{43}$). The variable results may be due partly to different methods of extraction. It is very difficult to extract all the lipid material from dried virus, while extraction of hydrated virus by ethanol may remove some non-lipid material. There are, however, variations in lipid content which are not due to technical difficulty. UHLER and GARD ($U_1$) found that incomplete virus produced by the VON MAGNUS technique contained about twice as much lipid as standard virus. The most complete studies of the lipid constituents of the virus have been made by FROMMHAGEN, KNIGHT and FREEMAN ($F_{158}$) and KATES, ALLISON, TYRRELL and JAMES ($K_{18}$, $K_{19}$). They found a total lipid content of 18—19%, consisting of 11—12% phospholipid, 6—6.5% cholesterol, and 1% glycolipid. There was practically no triglyceride, free fatty acid, or neutral fat. The phospholipid component consisted of 49% cephalin, 36% spingomyelin, and 15% lecithin.

### Origin of the Virus Lipid

The chemical analysis of the virus lipids shows them to be remarkably similar to those of the host cell ($F_{158}$). KATES et al. ($K_{18}$, $K_{19}$) made a very detailed study of the composition of the lipids of virus grown in fertile eggs and in calf kidney tissue cultures and compared the composition with that of lipids from the respective host cells. The two types of virus had different lipid compositions and in each case the lipid composition was almost identical to that of the host cell lipid. In experiments on the labelling of influenza virus with radiophosphorus, HOYLE and FRISCH-NIGGEMEYER ($H_{207}$) found that if virus and $P^{32}$ were simultaneously introduced into fertile eggs then most of the $P^{32}$ in the resulting labelled virus was found in the nucleic acid and only about 20—30% in the phospholipid. If $P^{32}$ was introduced into the eggs 24 hours before inoculation of virus then the lipid $P^{32}$ became greater than the nucleic acid $P^{32}$, indicating the incorporation of preformed host lipid into the virus particle. Similar results were obtained with fowl plague virus by WECKER ($W_{99}$).

## 2. Polysaccharide Constituents of the Virus

TAYLOR ($T_{33}$) showed that the influenza virus contained carbohydrate in amounts greater than could be accounted for by the nucleic acid present. By extraction with formamide KNIGHT ($K_{67}$) isolated a mucoprotein which gave precipitin reactions with anti-PR8 serum and also with antisera to the sedimentable protein of normal allantoic fluid. By the use of the carbazole reaction the carbohydrate was shown to contain galactose, mannose, and glucosamine. ADA and GOTTSCHALK ($A_{19}$) showed that in addition to 0.28% ribose the virus particle contained galactose, mannose, fucose and glucosamine and the same sugars were found in the mucoproteins of allantoic fluid. The total carbohydrate content of the virus particle was estimated to be 4.6 $\pm 1.0$%. FRISCH-NIGGEMEYER and HOYLE ($F_{154}$) found that when the virus particle was fractionated by ether treatment the haemagglutinin component contained 4.2% of carbohydrate while the soluble antigen component contained no sugar other than ribose.

By the use of two dimensional paper chromatography with saturated collidine and butanol-water-acetic acid applied to acid hydrolysates of purified virus FROMMHAGEN, KNIGHT and FREEMAN ($F_{156}$, $F_{158}$) found that several strains of virus A and one of virus B contained 6% of polysaccharide made up of galactose, mannose, fucose and amino sugar in amounts similar to those found in the sedimentable component of normal allantoic fluid and in polysaccharide from normal chorioallantoic membrane. The results suggest that, like the lipid, the polysaccharide component of the virus particle may be derived from the host cell.

### 3. Protein Components of the Virus

Protein accounts for some 70% of the dry weight of the influenza virus particle. KNIGHT ($K_{68}$) identified 17 amino acids in hydrolysates of purified virus and found significant differences in the amino acid composition of A and B viruses. HOYLE and DAVIES ($H_{201}$) fractionated influenza virus particles by ether treatment with the production of three separate protein constituents, the ribonucleoprotein "soluble antigen", haemagglutinin, and a "membrane protein" which collected at the ether-water interface. The amino acid composition of hydrolysates of the three proteins was determined and compared with those of proteins derived from cytoplasmic particles released from normal chorioallantoic cells. These normal cell particles could be separated by ether treatment to yield a "membrane protein" which collected at the interface and a "cytoplasmic protein" which was found in the aqueous phase. In all the five proteins analysed 18 amino acids were detected. Results are shown in Table 2. The soluble antigen protein contained 9.8% arginine, an amount double that found in the other proteins. The virus haemagglutinin and virus membrane proteins appeared to have the same composition, values did not differ by more than the experimental error ($\pm 10$%) of the method, and this composition differed from that of the normal membrane protein only in having a higher content of histidine and a lower content of cystine.

LAVER ($L_{35}$, $L_{36}$) coupled the N-terminal amino acids of the virus protein to $S^{35}$ labelled phenylisothiocyanate and on hydrolysis was able to identify two N-terminal amino acids, aspartic acid and glycine. Disintegration of the virus with deoxycholate showed that the N-terminal aspartic acid and glycine were associated with the haemagglutinin. In later work, however, LAVER ($L_{37}$) was only able to find one N-terminal amino acid, aspartic acid, in LEE virus and in 4 strains of virus A. No N-terminal amino acid was found in the internal protein component of the virus. LAVER ($L_{38}$) also studied the C-terminal amino acids by treatment of cuprammonium sulphite-treated LEE virus with carboxypeptidase. Leucine and tyrosine were released, and the results suggested that LEE

virus contained 3 different polypeptide chains. The C-terminal amino acids of
A viruses could not be certainly determined.

By disrupting virus with sodium dodecylsulphate LAVER obtained three
proteins separable by electrophoresis. One fraction which appeared to represent
the haemagglutinin contained N-terminal aspartic acid and smaller amounts
of N-terminal glycine. Peptide mapping of tryptic digests of this protein indicated
differences between the A viruses MEL and BEL ($L_{40}$). Another fraction possibly
representing the internal antigen gave identical peptide maps in the two strains.

Table 2. *Percentage of Amino Acid Residues in the Protein Components of the DSP
Strain of Influenza Virus A Grown in the Allantoic Sac of Hens' Eggs, and in the Protein
Components of Cytoplasmic Particles Released from Normal Chorioallantoic Membranes*

| Amino acid | Normal cell proteins | | Virus proteins | | |
|---|---|---|---|---|---|
| | Normal cytoplasmic protein | Normal membrane protein | Virus membrane protein | Haemagglutinin | Soluble antigen protein |
| Alanine | 7.55 | 9.0 | 8.5 | 7.45 | 7.75 |
| Arginine | 5.45 | 4.4 | 4.35 | 5.3 | 9.8 |
| Aspartic acid | 8.3 | 8.15 | 8.25 | 8.9 | 9.0 |
| Cystine | 3.65 | 4.4 | 1.9 | 2.55 | 3.35 |
| Glutamic acid | 8.65 | 12.4 | 9.5 | 10.45 | 10.8 |
| Glycine | 5.85 | 5.15 | 5.0 | 5.3 | 5.45 |
| Histidine | 3.5 | 2.6 | 5.25 | 4.45 | 2.5 |
| Isoleucine | 4.0 | 4.1 | 4.8 | 4.55 | 4.7 |
| Leucine | 7.95 | 8.0 | 9.55 | 9.2 | 8.3 |
| Lysine | 7.55 | 7.05 | 6.9 | 7.65 | 6.0 |
| Methionine | 1.8 | 1.6 | 1.8 | 1.5 | 1.65 |
| Phenylalanine | 7.35 | 6.85 | 8.25 | 7.0 | 7.6 |
| Proline | 4.45 | 4.55 | 4.4 | 4.35 | 3.55 |
| Serine | 5.25 | 5.35 | 4.6 | 4.8 | 4.9 |
| Threonine | 7.7 | 5.45 | 5.95 | 5.7 | 5.75 |
| Tryptophan | 1.6 | 2.2 | 2.15 | 2.3 | 2.25 |
| Tyrosine | 2.75 | 2.45 | 2.5 | 2.55 | 2.3 |
| Valine | 6.6 | 6.5 | 6.3 | 5.8 | 4.6 |

By treatment of virus and virus fractions with a range of chemical reagents
reacting with different amino acids HOYLE and HANA ($H_{208}$) were able to show
that the major antigenic determinant of the internal S antigen was the amino-
acid tyrosine, though tryptophane and histidine were also involved. Four amino
acids, histidine, tyrosine, cystine and an incompletely identified amino acid with
an amide group were shown to be involved in the haemagglutinating and neuraminidase activities of the viruses, and there were differences in the active centres
of the proteins in virus A and virus B.

Several workers have found that neuraminidase can be split off the virus particle by treatment with proteinases ($N_{27}$, $R_{16}$, $R_{17}$, $L_{39}$), the results suggesting
that the enzyme is present on the virus surface either as part of the projecting
spikes or stacked in between them.

*Electrophoretic Properties — Isoelectric Point*

TOVARNITSKY and SHISHKINA ($T_{56}$) and MILLER, LAUFFER and STANLEY
($M_{88}$) studied the electrophoretic properties of the virus. The virus carries a

negative charge and the isoelectric point is pH 5.3. Tovarnitsky and Karlina ($T_{55}$) found that the virus was precipitated by papain, protamine, globin, cyto-chrome C and clupeine which have isoelectric points at a more alkaline pH than the virus. If virus is disintegrated by ether the released haemagglutinin and soluble antigen can be partially separated by electrophoresis or by passage through DEAE cellulose columns. Schmidt, Hartmann and Grossgebauer ($S_{57}$) found that after ether treatment separate haemagglutinin fractions could be eluted from DEAE cellulose columns.

## 4. The Virus Nucleic Acid

Highly purified preparations of influenza virus contain only ribonucleic acid, the amount present ranging from $0.7-1.1\%$. Estimates were made by Ada and Perry ($A_{24}$, $A_{25}$), Frisch-Niggemeyer and Hoyle ($F_{154}$) and Frommhagen, Knight and Freeman ($F_{158}$). The purified virus was defatted with alcohol or chloroform-methanol, the nucleic acid extracted with hot $10\%$ sodium chloride and precipitated with alcohol. The amount was estimated by ultraviolet absorption spectrophotometry, by ribose analysis, or by phosphorus analysis. Influenza virus particles tend to adsorb mineral phosphate and estimates based on phosphorus analysis gave values $10\%$ higher than by other methods.

By fractionating virus labelled with $P^{32}$ Hoyle, Jolles and Mitchell ($H_{211}$) showed that the inner "soluble antigen" component of the virus carried the whole of the nucleic acid and similar results were obtained by Paucker, Lief and Henle ($P_{21}$). The inner component was shown to contain $5.3\%$ of RNA by Frisch-Niggemeyer and Hoyle ($F_{154}$) and by Ada ($A_{17}$). In similar experiments with fowl plague virus Schäfer and Zillig ($S_{41}$, $Z_{43}$) estimated the RNA content of the inner component at $15\%$, but later experiments gave lower values ($S_{32}$).

The RNA content of incomplete virus was shown by Ada and Perry ($A_{25}$) to be lower than that of fully infective virus. Frisch-Niggemeyer ($F_{153}$) showed that fractionation of virus by ether under conditions of increased ionic strength yielded soluble antigen in the form of long threads and suggested that the whole of the RNA of the intact virus is present as a single macromolecule of molecular weight about 2 millions.

Davies and Barry ($D_{31}$) extracted RNA from purified $P^{32}$-labelled A2 virus with sodium dodecyl sulphate and phenol and studied its sedimentation properties. They found that most of the RNA sedimented as a sharp band with a sedimentation constant of 18S, corresponding to a molecular weight of 500,000. A similar procedure applied to tobacco mosaic virus yielded an RNA of sedimentation constant 31S equivalent to a molecular weight of 2 millions, while treatment of the influenza virus RNA with ribonuclease reduced the S value to about 4. They concluded that the 18S RNA extracted from A2 virus was undegraded, and that in the intact virus the RNA was present in separate pieces of molecular weight 500,000. Pons ($XP_1$), using the WSN strain labelled with tritiated uridine extracted two types of RNA with sedimentation values of 38S and 20S, and considered that the 38S RNA broke down into 20S RNA. He also isolated a ribonuclease-resistant RNA from infected cells with a sedimentation value of $11-12S$ and suggested that this might be a replicative form acting as a template for $18-20S$ single-stranded RNA.

### Base Composition of the Virus Ribonucleic Acid

ADA and PERRY ($A_{26}$, $A_{27}$) estimated the base composition of the RNA of several strains of A and B viruses by acid hydrolysis followed by separation of the components by paper chromatography and their estimation by UV spectrophotometry. Significant differences were found between A and B viruses. BURKE, ISAACS and WALKER ($B_{151}$) made similar analyses but their results differed from those of ADA and PERRY, their analyses for virus A being very similar to those of ADA and PERRY for virus B. HOYLE developed a method of analysis using $P^{32}$-labelled virus which required much smaller amounts of material ($H_{197}$, $H_{198}$). The $P^{32}$-labelled nucleic acid was hydrolysed by alkali and the resulting nucleotides separated by paper chromatography with phenol-tertiary butanol-formic acid. They were located by scanning with a Geiger counter, eluted, and estimated by their radioactivity. Results of most analyses were very similar to those of ADA and PERRY. The base ratio of the virus RNA was very different from that of host cell RNA and mean values obtained by HOYLE and by ADA and PERRY are shown in Table 3.

Table 3. *Base Composition of Influenza Virus Ribonucleic Acid and Ribonucleic Acid from Normal Chorioallantoic Membranes*

|  | Molar precentage | | | | $\dfrac{A + U}{G + C}$ |
|---|---|---|---|---|---|
|  | Adenine | Guanine | Cytosine | Uracil |  |
| Virus A (ADA and PERRY) | 22.8 | 20.1 | 24.5 | 32.6 | 1.24 |
| Virus A (HOYLE, winter results) | 22.1 | 20.3 | 24.7 | 32.9 | 1.22 |
| Virus B (ADA and PERRY) | 22.9 | 18.0 | 23.3 | 35.8 | 1.42 |
| Virus B (HOYLE) | 21.5 | 16.9 | 24.1 | 37.5 | 1.44 |
| Normal CAM (ADA and PERRY) | 19.6 | 32.6 | 27.5 | 20.2 | 0.67 |
| Normal CAM (HOYLE) | 18.6 | 33.8 | 29.5 | 18.1 | 0.58 |

It is, however, by no means certain that the base ratio is constant. Thus, ADA ($A_{17}$) found that the nucleotide composition of the RNA of virus grown in chick embryo lung differed considerably from that of allantoic fluid virus. HOYLE ($H_{197}$, $H_{198}$) found that there were considerable differences in the base ratio when analyses were made from virus grown in eggs at different times of the year. In the winter months results agreed with those of ADA and PERRY, but in the summer months the RNA content of cytosine increased and uracil diminished. These results have not been confirmed and in later work HOYLE found much smaller seasonal variations. It seems possible that these differences may be due to the fact that the method of release of virus from the host cell makes possible the inclusion of host cell RNA in the particle. ADA and PERRY ($A_{29}$) suggested that the RNA of a filamentous strain of virus was a mixture of cell and virus RNA. Another possible source of error is the contamination of preparations by ribonuclease. However, the properties of influenza viruses are so variable that if the variations are genetically controlled some variation in the composition of the RNA may well occur.

### 5. Radioactive Labelling of Influenza Virus

Studies of the chemical structure and growth cycle of the influenza virus have been greatly assisted by the use of virus labelled with radioactive tracers.

Influenza and fowl plague viruses are easily labelled with radio-phosphorus by cultivation in fertile eggs in which radioactive sodium phosphate has been introduced into the allantoic fluid ($G_{67}$, $G_{64}$, $G_{68}$, $H_{211}$, $L_{100}$, $P_{21}$, $H_{207}$, $W_{30}$, $O_{19}$). The 10—12 day fertile egg will tolerate up to 1 millicurie of $P^{32}$ and satisfactory labelling is achieved with doses of 100—200 microcurie per egg. Very heavy labelling can be attained by the use of de-embryonated eggs. The labelled virus is purified by two cycles of adsorption-elution from red cells, the cells carrying the adsorbed virus being very thoroughly washed with ice-cold saline at each cycle to remove non-viral $P^{32}$. As a result of the tendency of influenza virus to adsorb mineral phosphate final preparations usually contain some non-viral $P^{32}$ but it is easy to produce preparations in which over 95% of the radioactivity is carried by the virus. Influenza virus can be labelled with $S^{35}$ by growth in eggs inoculated with radioactive methionine and the virus is easily completely freed from non-viral $S^{35}$ ($H_{206}$, $H_{196}$). The virus proteins are labelled and by ether fractionation HOYLE and FINTER found that most of the $S^{35}$ appeared in the interface protein and in the haemagglutinin, the virus S antigen component containing only 15—20% of the label.

Virus can also be labelled with $I^{131}$ by simply mixing it with radioactive iodine, but iodine is extremely viricidal so that the uses of iodine-labelled virus are limited. All the components of the virus take up iodine but much of it is not firmly fixed and may be removed by washing or by treatment with thiosulphate.

# X. Antigenic Structure

Serological studies of the influenza virus have shown the presence of two types of virus-specific antigen, one showing type specificity and distinguishing influenza A and B viruses, and one showing strain-specificity. In addition antigenic components are present with serological properties indicating them to be derived from the host cell.

## 1. The S Antigen

In early work on the influenza viruses it was found that while the human and swine viruses were easily distinguished by the neutralisation test they appeared to share a common antigen which could be detected by complement fixation methods. HOYLE and FAIRBROTHER ($H_{203}$) showed that the common antigen present in extracts of mouse lung infected with human and swine viruses had a smaller particle size than the infective virus and it came to be known as soluble or S antigen to distinguish it from the strain-specific V antigen associated with the infective virus particle. However, in 1945 HOYLE ($H_{186}$) in an analysis of the complement fixation reaction found evidence that the virus particle contained S antigen, and this conclusion was verified in 1952 when it was shown that disintegration of purified virus by ether treatment released a complement fixing antigen identical in every respect with S antigen ($H_{193}$, $L_{75}$, $D_{29}$). The S antigens of all strains of virus A appear to be identical ($H_{186}$, $H_{96}$, $L_{58}$, $W_{51}$, $L_{74}$, $B_{75}$, $B_{83}$) and to differ from the S antigen of virus B ($F_{91}$, $H_{186}$). Recently, however, DAVENPORT, ROTT and SCHÄFER ($D_{29}$) have suggested that there may be slight differences in the S antigens of human and animal strains of virus A, and HANA and HOYLE ($H_{17}$) have found that S antigen can be disintegrated with the production of fractions giving separate lines in precipitin tests by the immunodiffusion technique. However, all strains of virus A appeared to give the same fractions.

Highly purified preparations of influenza virus do not react with antibody to S antigen ($H_{90}$) and S antibody does not inhibit haemagglutination or neutralize virus. It appears that the S antigen is an internal component of the virus particle and by the ether fractionation of $P^{32}$ labelled virus it was shown ($H_{211}$, $P_{21}$) to be the protein component of the ribonucleoprotein. Similar conclusions were arrived at by SCHÄFER ($S_{32}$) in studies of fowl plague virus.

## 2. The V Antigen Complex

The surface V antigen of the virus particle appears to be a complex of a number of different antigens, and a very large number of studies have been made of the variations of the V antigen in different strains of virus. A variety of serological methods are available for the study of the V antigen. The antigen is responsible for the production of virus neutralizing and protecting antibody in the sera of convalescents or vaccinated animals. SMITH and ANDREWES ($S_{166}$) used the neutralization test with convalescent ferret sera and demonstrated antigenic differences in different strains of virus. MAGILL and FRANCIS ($M_{12}$) obtained similar results with the sera of vaccinated rabbits. FRANCIS and MAGILL ($F_{107}$) used cross-protection tests in mice, mice being immunised by intraperitoneal injection of virus and then challenged with intranasal homologous and heterologous virus one week later. MAGILL and SUGG ($M_{16}$) tested the mouse protective properties of human convalescent sera with different strains of virus.

The V antigen complex is responsible for the agglutination of red blood cells by virus and very numerous investigations of antigenic structure have been made using the haemagglutination-inhibition test ($H_{131}$, $H_{113}$, $V_9$, $S_{17}$, $M_{137}$, $J_{15}$). The complement fixation test ($H_{186}$, $K_1$, $G_{40}$, $H_{89}$, $L_{78}$) and antibody absorption tests ($S_{200}$, $L_{128}$, $F_{144}$, $J_{12}$) have also been used.

Although the different types of test do not always give precisely comparable results ($H_{117}$), in general the results are very similar, similar strain relationships being indicated by different tests. All the methods show that the V antigen is a complex of several different components the proportion of which differ in different strains of virus A. A similar situation exists with virus B ($B_{118}$, $B_{121}$, $H_{97}$, $M_{136}$, $H_{104}$, $J_{21}$). The results suggest that the outer surface of the virus particle is a mosaic of a number of different antigens. In some strains of virus one antigen appears to be dominant and others present in only small amounts, while in other strains several antigens may appear to be equally prominent. There is, however, an alternative to the surface mosaic concept. Influenza convalescents commonly produce antibodies to V antigen components not demonstrable in the infecting strain by in vitro methods ($L_{79}$). This suggests that some of the components of the V antigen complex may not be present on the surface of the virus but in the interior. The results of ether fractionation of influenza virus show that the haemagglutinin component is partly present on the surface and partly in the interior, and there is some evidence that vaccination with ether-split virus gives a broader antibody response than vaccination with intact virus

($D_{22}$). The serum-blocking antigen produced by disintegrating virus with ether and detergent ($H_{153}$) produces on injection a wider range of antibodies than would be expected from its in vitro blocking action. ISAACS, DEPOUX and FISET ($I_{15}$) have also produced evidence suggesting that some of the V antigen components may be in the interior, and MAGILL ($M_{10}$) in experiments on changes in antigenic properties resulting from passage of virus in partially immune mice suggested that a spatial rearrangement of antigens occurred, antigens formerly present deep in the particle appearing on the surface.

### 3. The Host Antigen Components

Observations of the release of influenza virus from the infected cell have shown that the virus acquires a membrane derived from the wall of the host cell, and many studies have indicated the presence in the virus particle of antigenic material characteristic of the host.

#### a) The Lipid Component

Influenza virus particles contain about 12% of phospholipid in combination with protein. Lipoproteins are very good antigens, but the sera of animals convalescent from influenza virus infection do not contain antibody reacting with alcoholic extracts of influenza virus derived from the same host. However, sera containing heterophile antibody of the Forssman type may react strongly in complement fixation tests with antigens prepared by mixing alcoholic extracts of purified virus grown in eggs with saline. The author has found that normal ferret sera react with alcoholic extracts of virus grown in eggs or in mice. BURNET and McCREA ($B_{214}$) found that normal ferret sera gave a non-specific inhibition of haemagglutination by influenza virus grown in eggs. BORECKY and RATOVA ($B_{107}$) found that rat antisera to influenza virus of egg origin lysed sheep red blood cells. SMITH, BELYAVIN and SHEFFIELD ($S_{170}$) found that sera prepared by immunising rabbits with heat-degraded virus in which the V antigens had been destroyed gave strong cross-reaction between A and B viruses in complement fixation tests. ROTT and DRZENIEK ($R_{60}$) showed that heterophile antigens of the Forssman type could be demonstrated in purified virus only if these substances were also present in the host cells.

These results support the conclusions arrived at by chemical analysis that the virus lipids are of host cell origin.

#### b) The Polysaccharide Component

KNIGHT ($K_{63}$, $K_{66}$) found that antisera to the sedimentable component of normal allantoic fluid give precipitin reactions with highly purified PR8 and LEE viruses and also inhibit haemagglutination by these viruses. An intensive study of this phenomenon has been made by HARBOE and his collaborators ($H_{34}$, $H_{35}$, $H_{53}$, $H_{54}$). Antibody prepared in rabbits against normal chick allantoic fluid inhibited haemagglutination by A and B viruses, and rabbits immunised with purified PR8 virus from infected allantoic fluid produced an antibody which reacted with LEE virus. Chemical studies of the antigen in normal allantoic fluid showed it to be a high molecular weight carbohydrate associated with protein. Purified preparations contained 27% galactose, 24% N-acetyl-glucosamine, 11% N-acetylgalactosamine, 12% ester sulphate, 4.3% fucose and 5.8% peptide. Chemical analyses of the antigen have also been made by other workers ($S_{280}$, $L_{43}$). The antibody inhibits haemagglutination by A and B viruses grown in the allantoic sac of chick, turkey or duck eggs but not virus grown in other tissues. Influenza C virus is not inhibited. The antibody inhibits haemagglutination both by intact virus and by haemagglutinin produced by ether disintegration. The results suggest that the polysaccharide component of the virus particle is of host cell origin. LAVER ($L_{43}$) estimated that the host antigen amounts to some 5% of the dry weight of the virus. It is strongly bound to the virus haemagglutinin. KOSYAKOV and ROVNOVA ($K_{96}$) showed that the host component was no longer detected after a single passage of virus in a different host.

### c) Possible Presence of Host Protein in the Virus Particle

Although both the lipid and carbohydrate of the influenza virus are closely linked to protein it is improbable that this protein is of host cell origin. KROEGER ($K_{108}$) studied the anaphylactogenic properties of PR8 influenza virus suspensions by the Schultz-Dale reaction with uterine muscle from guinea pigs sensitized to host protein. It was found that crude preparations of virus produced anaphylactic reactions but virus purified by two cycles of red cell adsorption-elution followed by differential centrifugation did not, and it was concluded that purified virus contained no detectable host protein. LAVER and WEBSTER ($L_{43}$) showed that peptide maps of virus proteins grown in eggs and in calf kidney cultures were identical. HOYLE and DAVIES ($H_{201}$) showed that the amino acid composition of the virus protein which is linked to lipid "membrane protein" is the same as that of the haemagglutinin component.

# XI. Morphology and Physical Structure

Influenza virus particles are too small to be seen by ordinary light microscopy but may be rendered visible if fixed films are heavily overstained, as in the MOROSOV silver impregnation technique ($M_{124}$). LINDENMANN ($L_{87}$) was able to see particles after fixation with formalin followed by treatment with permanganate and Victoria blue. Filaments with terminal bodies could be seen in infected allantoic fluid.

By the use of dark-field method with a powerful light source the particles are easily seen and some degree of resolution is attained, filamentous particles found in many strains of virus have a characteristic appearance, rigid rods often with sharp angular bends and sometimes showing terminal spherical swellings (Figs. 3, 4).

### 1. Size of the Influenza Viruses

The first measurements of the size of influenza virus particles were made in 1936 by ELFORD, ANDREWES and TANG ($E_{36}$) using filtration through Gradocol membranes. They estimated the size at 80—120 mμ. By similar methods FRIEDWALD and PICKELS ($F_{149}$) obtained a value of 80—135 mμ.

Many estimates of the size of PR8 and LEE viruses have been made by centrifugation methods. The sedimentation constant of PR8 virus has been

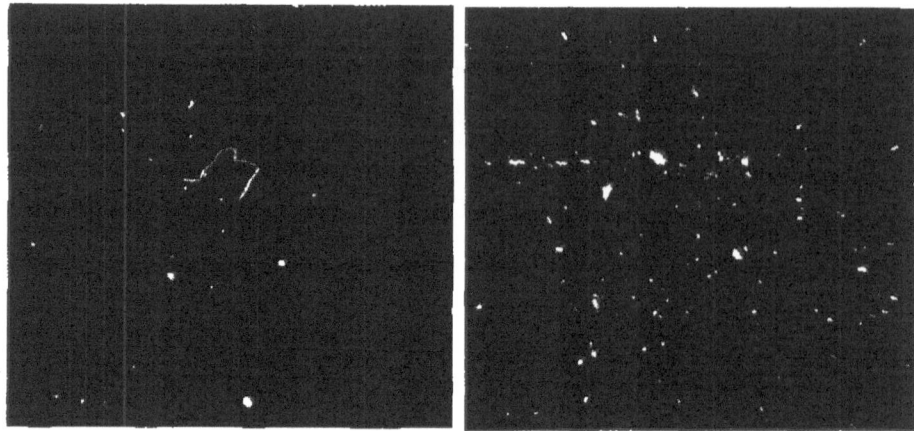

Figs. 3 —5. Filamentous forms of influenza virus.
Figs. 3 and 4. Dark field microscope photographs of a purified preparation of A1 virus. ×1250.

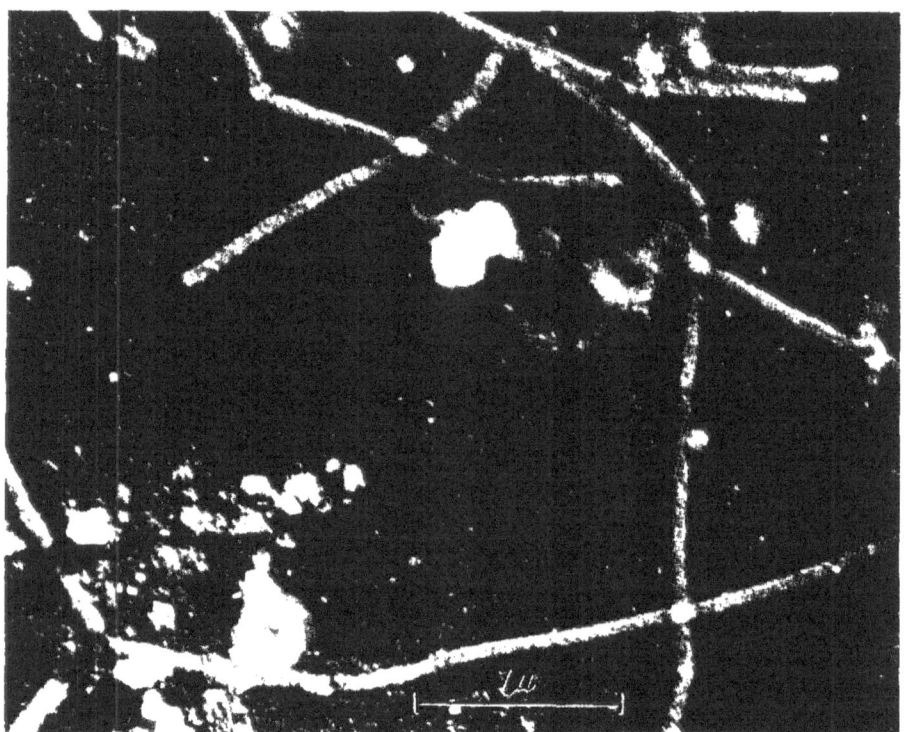

Fig. 5. Electron microscope picture of the same material as in Figs. 3 and 4. Chromium shadowed.

Figs. 6—12. Dark field microscope studies of the formation of cytoplasmic protrusions and their conversion into virus filaments.
Fig. 6. Production of large spherical cytoplasmic particles from normal uninfected chorioallantoic membrane.
Figs. 7 and 8. Production of spherical, tubular, and filamentous protrusions from chorioallantoic membrane infected with A1 virus.
Fig. 9. Entodermal cell in the allantoic fluid of an egg infected with A1 virus showing filamentous protrusions.
Fig. 10. Filamentous particle of complex morphology seen just after release from an entodermal cell (A1 virus).
Fig. 11. Flexible virus filament with two spherical swellings in the allantoic fluid of an egg infected with A1 virus.
Fig. 12. Agglutination of virus filaments in allantoic fluid by homologous immune serum (A1 virus).
(All Figures ×1250.)

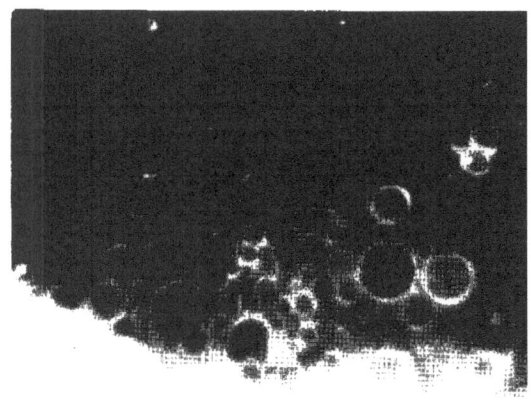

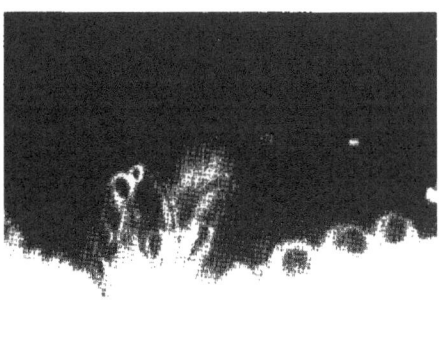

Fig. 7.

Fig. 6.

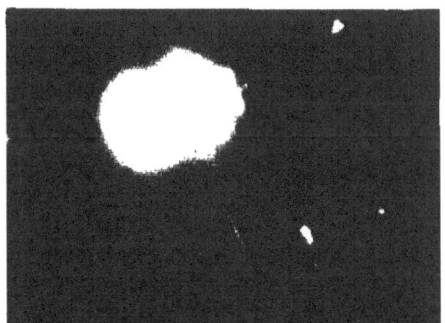

Fig. 9.

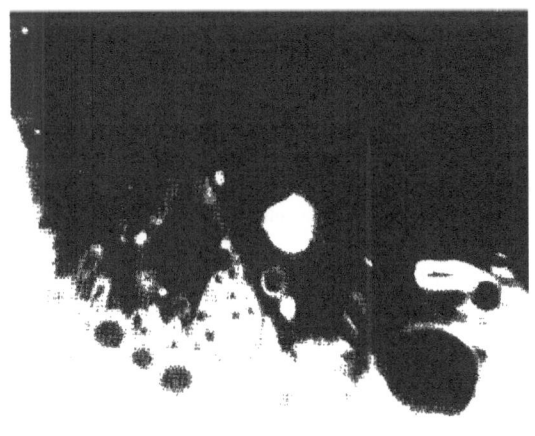

Fig. 8.

Fig. 11.

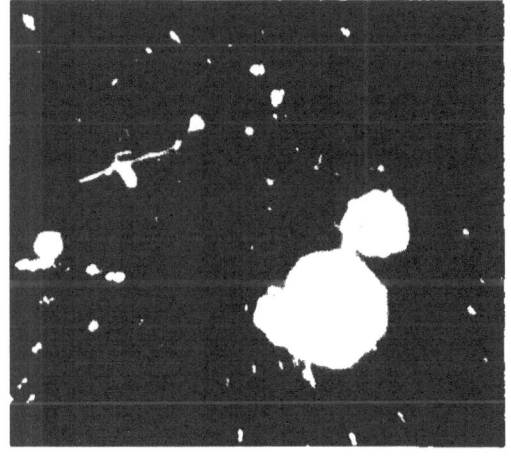

Fig. 10.

Fig 12.

4*

variously estimated at from 600—800 S and LEE virus from 800—894 S ($T_{34}$, $S_{95}$, $F_{149}$, $B_{30}$, $S_{253}$, $G_7$, $H_{27}$, $G_8$, $P_{88}$, $F_{159}$). These results indicate a particle diameter of about 80 m$\mu$ for PR 8 and 100 m$\mu$ for LEE virus. Many workers have found in addition to the major virus component in infected allantoic fluid, smaller amounts of virus-specific material sedimenting more slowly ($F_{149}$, $G_7$, $G_8$, $P_{88}$, $F_{159}$, $H_{27}$). These components probably represent incomplete virus.

## 2. Electron Microscopy — Shadow Casting Techniques

Early electron microscope pictures revealed little more than the fact that the particles were predominantly spherical or bean-shaped of a diameter of 80—100 m$\mu$ ($T_{34}$, $S_{95}$). More satisfactory results were obtained following the introduction by WILLIAMS and WYCKOFF ($W_{53}$) of the shadow-casting technique, and with purified virus preparations many excellent pictures were obtained by shadowing particles deposited on Formvar grids with chromium, palladium or gold ($M_{130}$). Laboratory strains of virus such as PR 8 or LEE were found to be predominantly spherical, LEE virus being slightly larger than PR 8. Occasional short rod-like forms were also seen, and rod-like and filamentous forms were more frequently found when freshly isolated strains were examined (Fig. 5). Very numerous filaments were shown in A 2 viruses ($R_{58}$). Most of these studies were made with virus preparations which had been purified by differential centrifugation, a procedure which tends to select particles of a particular size. With the introduction of the technique of adsorption of virus on red cell ghosts by HEINMETZ ($H_{59}$) and DAWSON and ELFORD ($D_{36}$, $D_{37}$, $C_{56}$) it became possible to examine crude infected fluids without the need to employ differential centrifugation, and it was found that both spherical and filamentous forms were adsorbed by the cells, and also spherical particles of larger size ($V_{33}$). Excellent pictures of virus particles adsorbed on red cells were secured using both shadowed and unshadowed preparations, and replica techniques have also been used ($F_{62}$, $R_{22}$).

## 3. Globular, Filamentous and Bizarre Forms of Virus

### a) Dark-field Microscopy

HOYLE ($H_{191}$, $H_{195}$) examined detached fragments of chorioallantoic membrane from infected eggs in the dark-field microscope at a time when virus was being released from the cells, and a complete understanding of the origin and varied morphology of the virus particles was obtained. It was found that the entodermal cells of the normal chorioallantoic membrane possessed the property of detaching fragments of cytoplasm enclosed in cell membrane into the allantoic fluid. In the normal membrane these detached fragments were almost all in the form of large spherical globules 2—10 m$\mu$ in diameter, but in the infected membrane while large globules were produced, in many cases they did not break away but tended to become drawn out into a tubular form (Figs. 6—8) usually with a terminal spherical swelling which often separated from the tubule. These tubular protrusions became longer and narrower and when the tubule diameter became reduced to about 0.5 $\mu$ it tended to break suddenly into a chain of small spherical particles. In cells infected with filamentous strains of virus this tubule fragmentation often did not occur. All types of protrusion ultimately broke away from

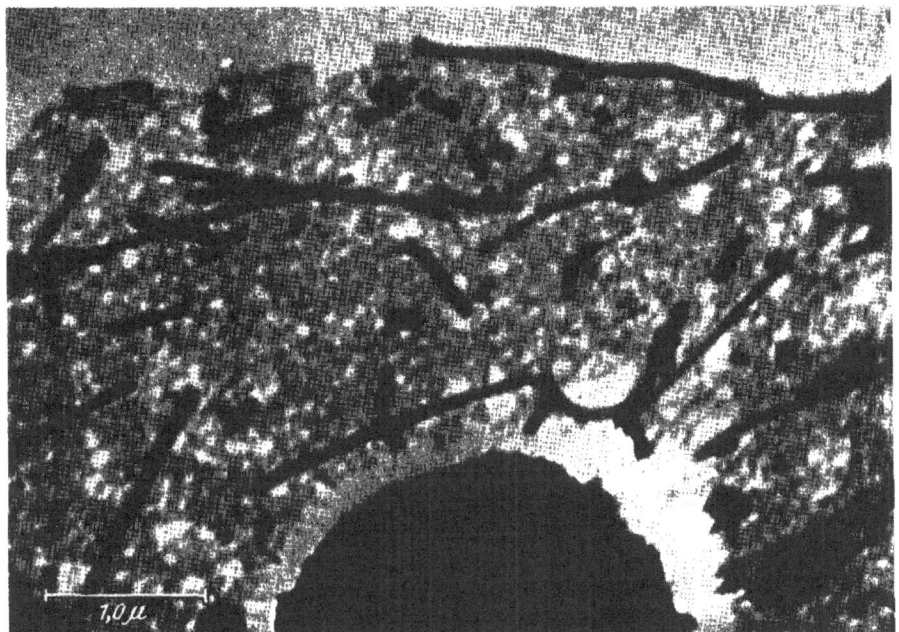

Figs. 13 and 14. Virus particles absorbed on fowl red cell ghosts.
Allantoic fluid from eggs infected with A 1 virus adsorbed with laked fowl red cells.
Fig. 13. Unshadowed preparation. Note beaded appearance of one virus filament.

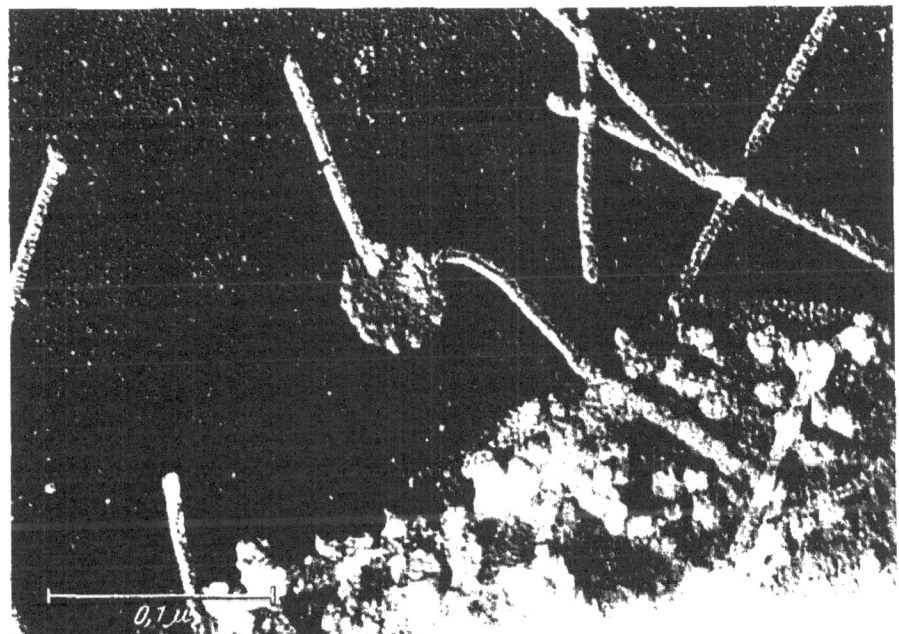

Fig. 14. Metal shadowed preparation showing spherical and filamentous forms. Reproduced by permission of Dr. I. ARCHETTI from ARCHETTI ($A_{95}$). The magnification index should be: 1.0 μ.

0,5 μ

Fig. 15. Formation of virus particles at the cell surface.
Electron microscope photograph of a thin section of chorioallantoic membrane infected with PR8 virus, showing the formation of virus particles at the surface of an entodermal cell. Reproduced by permission of Dr. C. Morgan from Morgan et al. (M₁₁₀).

the cells and on detachment appeared to lose water and become much smaller in size. The result was the appearance in the allantoic fluid of large spherical particles of diameter 200—2000 mμ derived from large globules, small spherical particles of diameter around 100 mμ derived from fragmented tubules, and filaments of diameter 100 mμ derived from unfragmented tubules. As the fragmentation of a single tubule might produce 40—50 small spherical particles

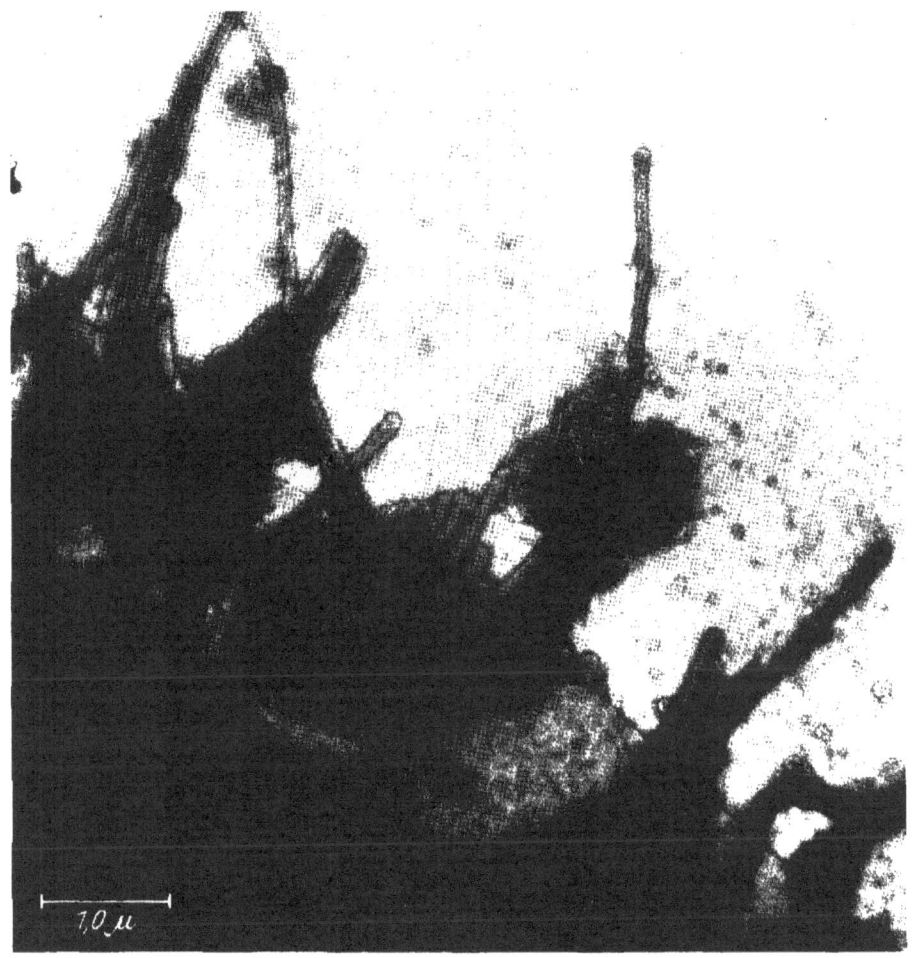

Fig. 16. Release of virus from the cell.
Whole cell mount of monkey kidney cell infected with A 2 virus. Negative stained with phosphotungstate. Reproduced by permission of Drs. P. W. CHOPPIN and S. DALES from CHOPPIN (C₄₂).

the majority of particles released from the cell had a diameter of 100 mμ. In addition to the spherical and filamentous particles a number of bizarre forms were encountered, filaments with large terminal spherical swellings, large spherical forms with two or more filaments attached, branching filaments etc. (Figs. 10, 11). In the allantoic fluid of infected eggs desquamated cells showing all types of protrusion could be found (Fig. 9).

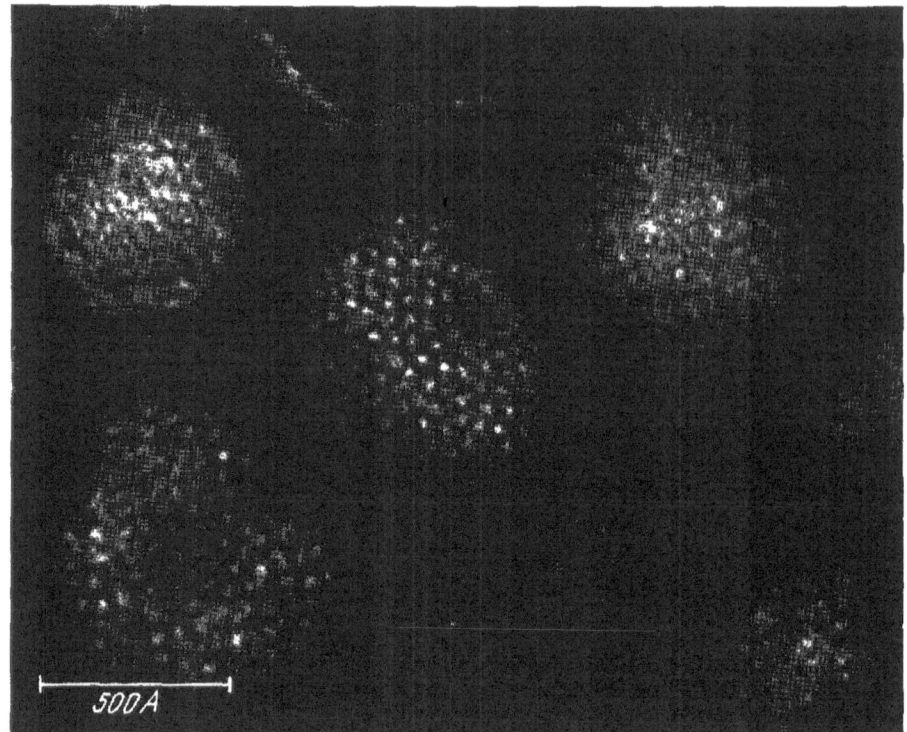

Fig. 17.

Figs. 17—19. Electron microscope pictures of influenza virus particles obtained by the negative staining technique.

Fig. 17. Purified DSP virus showing surface projections.

Fig. 18. Preparation of DSP virus in which some penetration of phosphotungstate has occurred, revealing that the particles have a membrane, and that the surface projections pass through holes in the membrane.

Fig. 19. Filamentous particles of an A 1 strain. The whole surface of the filament is covered by projections.

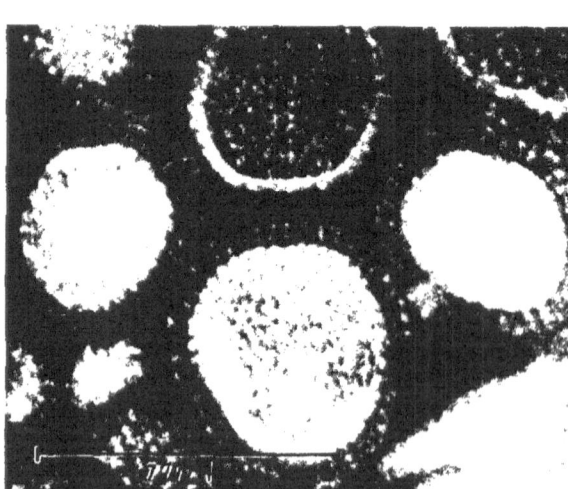

Fig. 18.

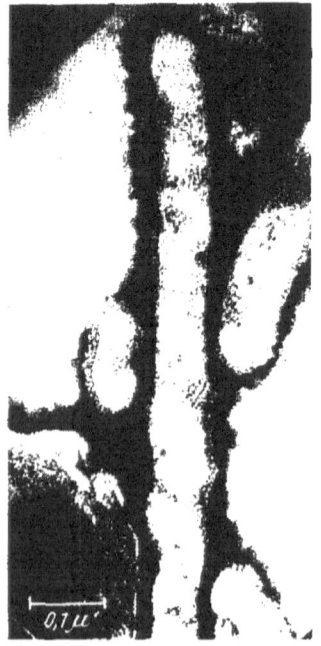

Fig. 19.

The dark-field method was also used by BURNET and LIND ($B_{208}$, $B_{182}$) in studies of virus filaments. They found that filaments were only seen in infected allantoic fluids were adsorbed and eluted from red cells, and were agglutinated by immune sera (Fig. 12). They were destroyed by heat, hypotonic solutions, ether, chloroform, detergents, cobra venom, saponin and digitonin. Adaptation of virus by serial passage resulted in a reduction in the number of filaments produced.

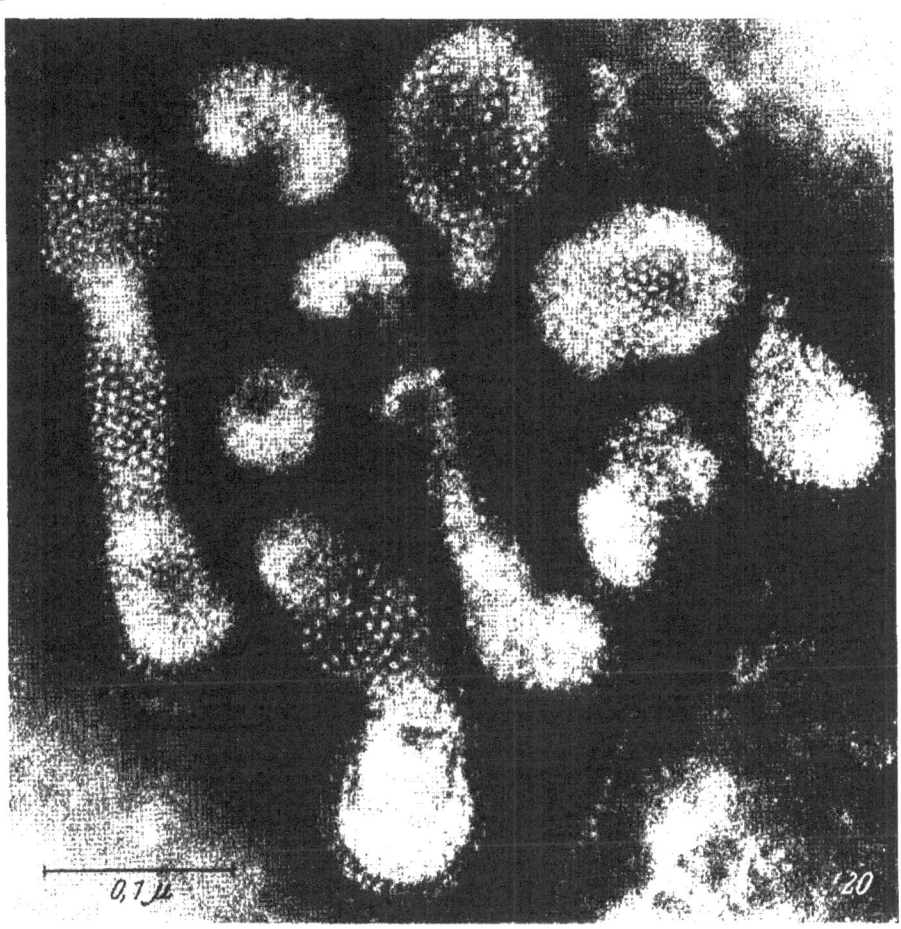

Fig. 20. Pleomorphic virus particles (A 2 strain).
Showing irregular size and shape of the particles in a recently isolated strain of virus. Photograph by Drs. J. ALMEIDA and A. P. WATERSON.

### b) Electron Microscopy

Many workers have obtained excellent electron microscopic pictures of virus filaments by adsorption on red cell ghosts. ARCHETTI ($A_{95}$) produced pictures showing complex globular and filamentous forms and forms segmenting into bead-like chains (Figs. 13, 14). VALENTINE and ISAACS ($V_9$) showed that filaments treated with acid broke up into chains of spherical particles. Remarkable pictures of the more bizarre forms of virus were obtained by DRAGANOV

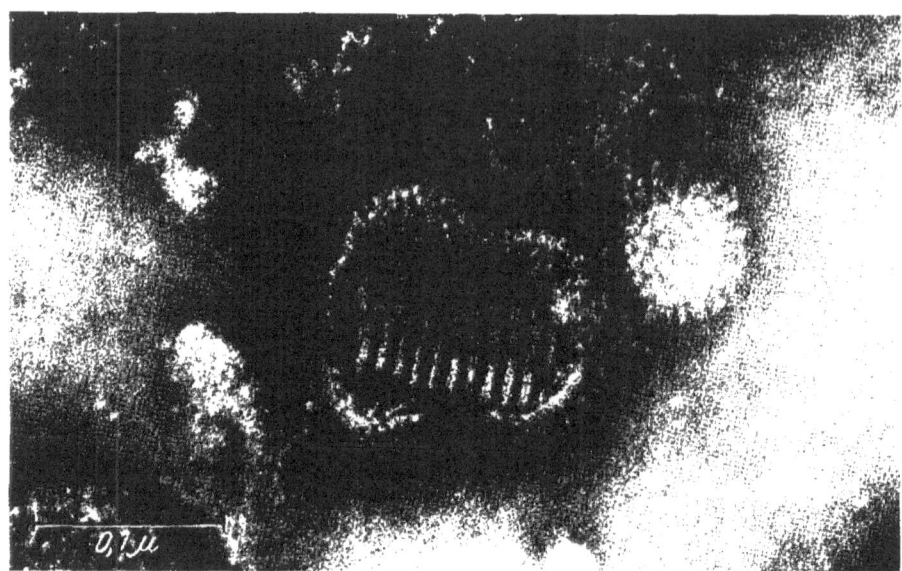

Fig. 21. Virus particle showing the internal coiled component.
Partially disrupted particle of A2 virus in which penetration of phosphotungstate has revealed the
internal coiled helical component. Photograph by Drs. J. ALMEIDA and A. P. WATERSON.

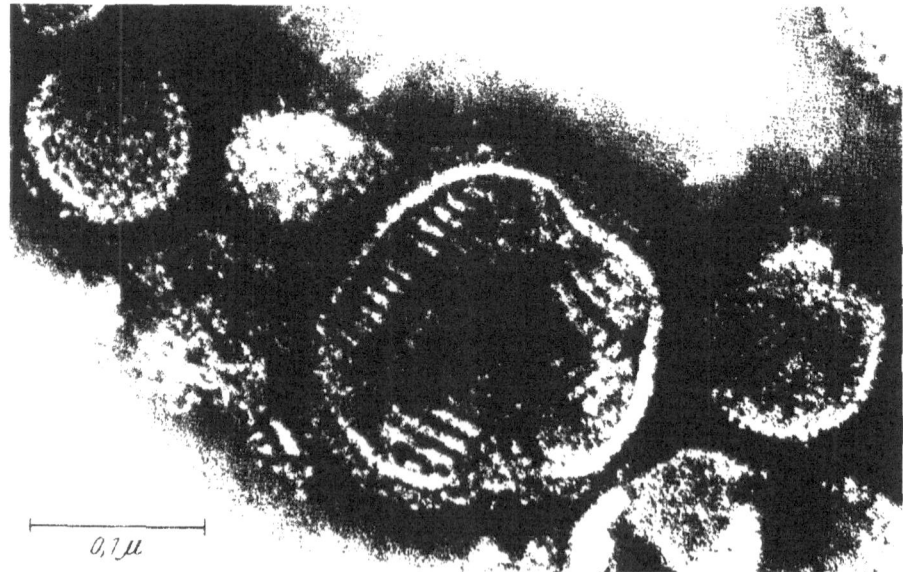

Fig. 22. Virus particle with multiple internal components.
Large particle of A2 virus containing three or possibly four separate internal coiled components.
Photograph by Drs. J. ALMEIDA and A. P. WATERSON.

($D_{64\cdot 66}$). Although ANGULO ($A_{86}$) considered the filaments to be artifacts, all other workers are agreed that the filaments are virus material and ADA, PERRY and EDNEY ($A_{30}$) found that filaments were rather more infective than spherical particles.

The release of virus from the cell has been studied by electron microscope techniques by EDDY and WYCKOFF ($E_{24}$, $W_{69}$); MURPHY, KARZON and BANG ($M_{150}$); MORGAN, ROSE and MOORE ($M_{117}$, $M_{118}$); CHOPPIN ($C_{42}$); and by SCHÄFER

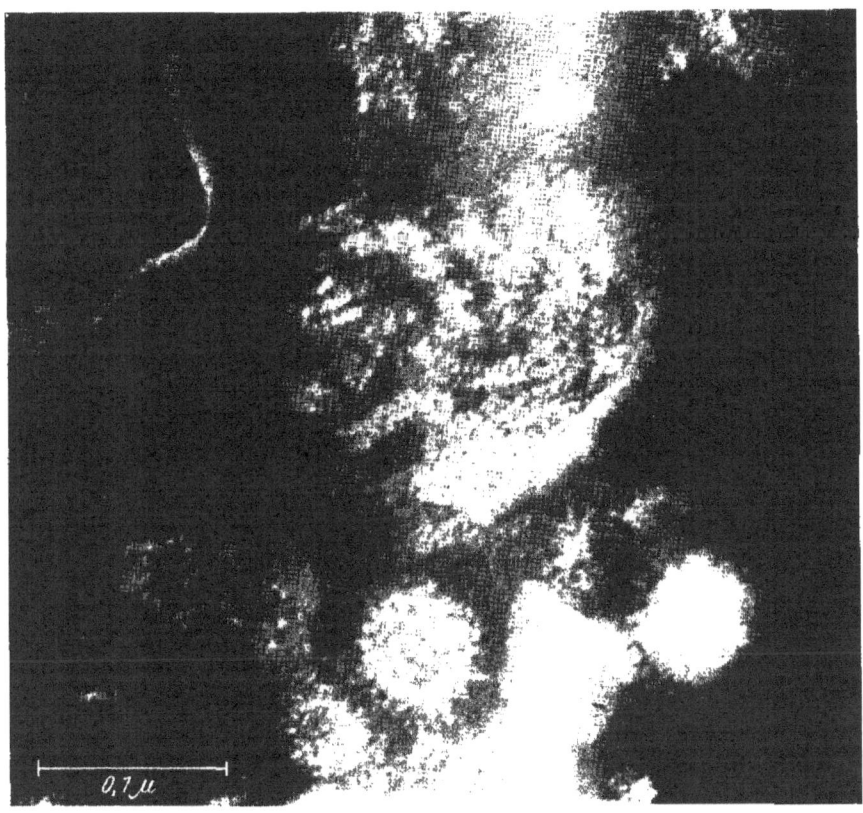

Fig. 23. Virus particle partially disrupted by ether treatment.
Particle of DSP virus disrupting as a result of a brief period of ether treatment. The internal coiled helix breaks into short lengths, and the diameter of the helix is apparently increased as compared with preparations not treated with ether (Figs. 21, 22). Photograph by Dr. R. W. HORNE.

($S_{31}$) using fowl plague virus (Figs. 15, 16). Results were in general similar to those obtained by HOYLE using the dark field method, but also indicated that virus particles might be released from cells, enclosed in cell membrane, without the formation of cell protrusions. Such a release would not be detected by dark field methods and would result only in the formation of small spherical particles.

### 4. Electron Microscopy by the Negative Staining Technique

In 1959 BRENNER and HORNE ($B_{124}$) introduced the technique of negative staining into electron microscopy. Virus suspensions in water or 1% ammonium acetate were mixed with 2% potassium phosphotungstate and sprayed on carbon

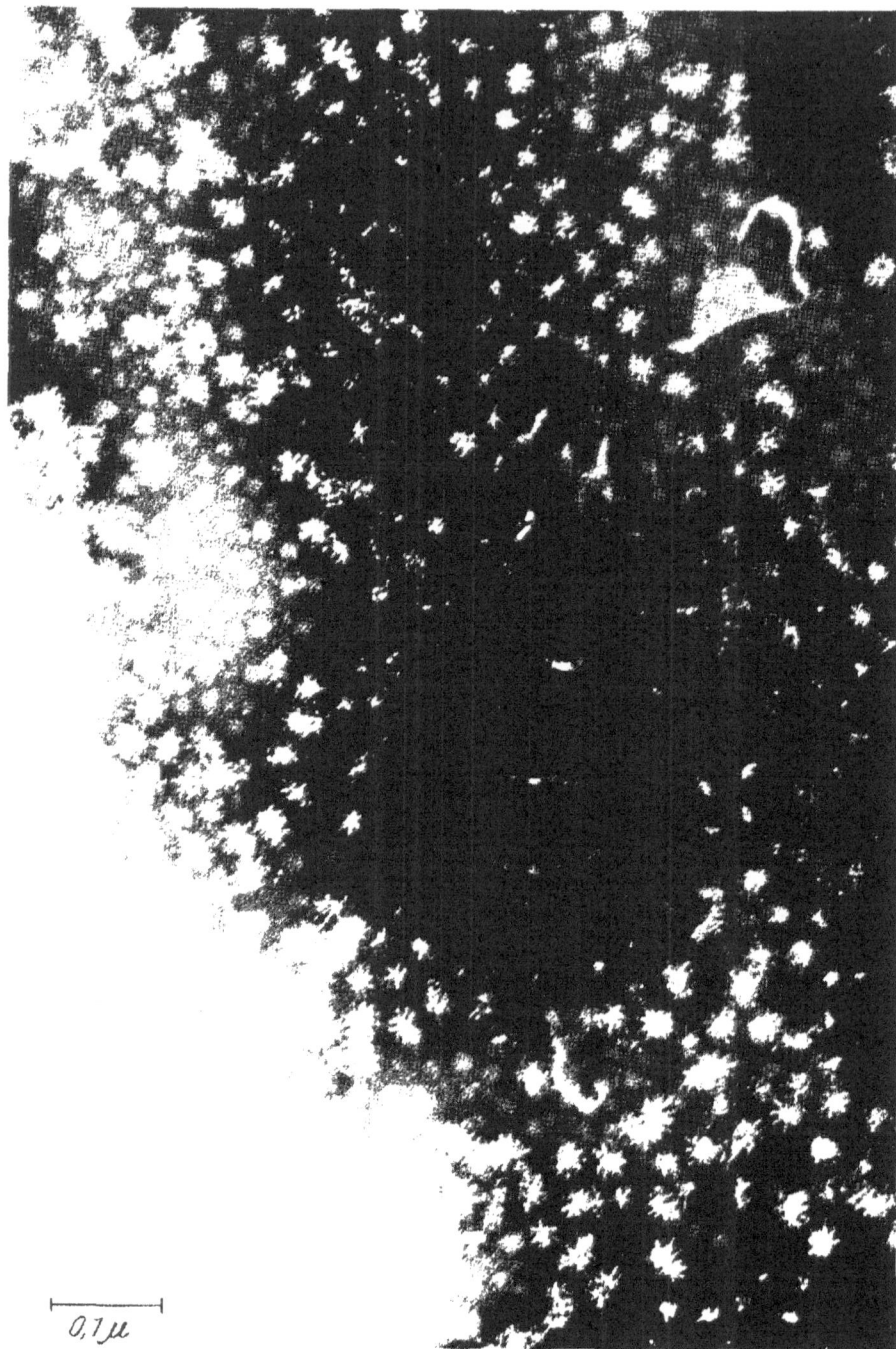

Fig. 24. Haemagglutinin of DSP virus in the sol form.
Purified DSP virus disintegrated by ether, the aqueous phase adsorbed with red cells, and the haemagglutinin eluted and deposited by centrifugation. Preparation made from the upper layer of the deposit. Predominantly in the sol form, star-like clusters of rods.

Fig. 25. Haemagglutinin of DSP virus in the gel form.

Purified DSP virus disintegrated by ether, the aqueous phase adsorbed with red cells, and the haemagglutinin eluted and deposited by centrifugation. Preparation made from the lower more tightly packed part of the deposit. Amorphous masses of irregular size and shape.

grids. Virus particles were seen against a black background produced by the electron-dense phosphotungstate and an enormous increase in resolution was attained. The technique was applied to influenza virus by HORNE, WATERSON. WILDY and FARNHAM (XH$_3$); HOYLE, HORNE and WATERSON (H$_{209}$); ARCHETTI and BOCCIARELLI (A$_{97}$, A$_{98}$); and CHOPPIN, MURPHY and STOEKENIUS (C$_{44}$). With most preparations only the outer surface of the virus particle was revealed, and was found to be covered with evenly spaced rod-like projections about 100 Å in length, the overall diameter of the particle being 800 – 1000 Å. Filamentous strains of virus were also found to be covered over the whole length with projections (Figs. 17, 19, 20).

In some cases some degree of penetration of phosphotungstate into the interior of the particle occurred, and in these cases there was revealed an outer membrane 60—90 Å in thickness. The projecting spikes appeared to pass through holes in this membrane (Fig. 18). In a few cases where particles ruptured on the electron microscope grid an internal coiled helical structure was revealed (H$_{209}$, A$_{91}$). (Figs. 21, 22.)

## 5. Morphology of Virus Fractions

Studies of the morphology of the haemagglutinin and S antigen components which are produced from influenza and fowl plague particles by ether treatment were made using shadow-casting techniques (H$_{212}$, S$_{32}$), but the resolution was insufficient to yield satisfactory results. VALENTINE and ISAACS (V$_1$, V$_2$, V$_3$) treated influenza virus particles with acid followed by trypsin and obtained ring-like structures which appeared to be nucleoprotein, but the treatment was so drastic that no conclusions could be drawn as to the morphology of the nucleoprotein component.

Satisfactory pictures were obtained by the use of the negative staining technique (H$_{209}$, S$_{34}$, B$_{32}$) with influenza, fowl plague and the tern virus.

### a) The Haemagglutinating Component

Electron microscope pictures made by the negative staining method of the haemagglutinating component recovered from the aqueous phase of ether-treated virus suspensions by adsorption-elution from red cells followed by centrifugation show particles with an overall diameter of 300—400 Å (H$_{209}$). These particles, sometimes called "rosettes" appear to be made up of star-like aggregates of rods (Fig. 24). The size and morphology of these rods resembles that of the surface projections seen in intact virus, they appear to have a drumstick form with a spherical end of larger diameter than the rod. The rods aggregate with the large spherical ends in contact. The outer end of the rods may also appear to be swollen or possibly forked. Prolonged centrifugation of the haemagglutinin causes it to pack into a transparent gel which on electron microscopy appears as irregular amorphous masses of variable size (Fig. 25). Some pictures show the component partly in the gel form and partly in the star-like sol form (Fig. 26).

If the material which accumulates at the interface when virus preparations are shaken with ether, and which agglutinates red cells, is examined it shows appearances suggesting that it is composed of fragments of the outer membrane of the particle. "Bottle-brush" forms which appear to consist of haemagglutinin rods attached to a more dense material may be seen.

## b) The Nucleoprotein "Soluble Antigen" Component

The S antigen component obtained by ether fractionation of virus appears in the form of elongated structures about $90-100$ Å in diameter and of irregular length, the average length being about 600 Å (Fig. 27). FRISCH-NIGGEMEYER ($F_{153}$) found that if purified virus was mixed with a six times isotonic solution of Na, K, Ca and Mg chlorides subsequent ether treatment resulted in the release of S antigen in threads of up to 10,000 Å in length, and suggested that in the intact virus the soluble antigen component is present as a single structure of length $8-10,000$ Å. Partially disrupted influenza virus particles sometimes show an internal component in the form of a central coil of $6-12$ turns (Fig. 21), and large particles may show two or more separate internal components (Fig. 22). APOSTELOV and FLEWETT ($A_{91}$) suggest that there is a fundamental length of six turns and that particles showing an internal coil of 12 turns are doublets. Coiled inner components can only be seen in a comparative small number of virus particles, even when the phosphotungstate has penetrated to the interior of the particle in most cases the inner component appears amorphous in structure. It is probable that in most particles the nucleoprotein component is embedded in an amorphous haemagglutinin gel, and that it is only in large particles with relatively less haemagglutinin than normal that the coiled form can be seen. In most virus preparations it is probable that there are some particles which contain no internal helix at all (incomplete virus).

The coiled structure itself appears to be made up of small sub-units arranged in the form of a double helix, two intertwined spirals of the same directional sense (Fig. 28). When seen end on there appear to be 5 or 6 subunits to each turn of each helix. The inner component of fowl plague and the tern virus shows similar appearances ($B_{32}$, $S_{34}$). The helical particles released from virus by ether treatment have a diameter of $90-100$ Å, but in the intact particle the helix appears to have a smaller diameter of some $50-60$ Å ($A_{91}$). Probably ether treatment causes some relaxation of the helix as well as its fragmentation. The differences are seen by comparison of Figs. $21-23$.

## 6. Physical Structure of the Virus Particle

The dark-field and electron microscope studies show that on release from the cell the virus acquires a membrane derived from the cell wall. This membrane, composed of protein, lipid and carbohydrate has certain important physical properties. It forms an osmotic barrier, retaining protein molecules but allowing the passage of smaller molecules. Dark-field observations show it to be extremely elastic — it can be blown out to form large spherical protrusions which on breaking away from the cell contract to smaller size. It is probable that the membrane can exist in two alternative physical states — a closed form in which the lipoprotein is closely aggregated to form an impenetrable barrier, and an open form in which it is perforated by pores ($K_{25}$). There is probably a balance between the elasticity of the membrane and the internal osmotic pressure. Chemical modifications of the membrane may result in a weakening of its mechanical strength and this might be expected to result in a "blow-out" at the weak point, and it is possible that the release of virus from the cell is a result of such an action by the virus on the cell membrane. The resulting conversion of the membrane

Fig. 26. Haemagglutinin of DSP virus.
Preparation partly in the sol and partly in the gel form. Note apparent transition between the two forms.

Fig. 27. Ribonucleoprotein (S antigen) component of DSP virus.
Preparation of purified S antigen of DSP virus produced by ether disintegration of virus. Negative staining technique.

into the open form would allow the passage of the haemagglutinin projections through the pores. BLOUGH ($B_{85-89}$) showed that modification of the cell lipoprotein membrane by vitamin A and other agents acting on lipoproteins resulted in the production in eggs of pleomorphic and filamentous particles by strains of virus such as PR8 which normally give predominantly small spherical particles.

ALMEIDA and WATERSON ($A_{52}$) studying the distribution of the projecting spikes on the envelope of fowl plague virus noted that in some cases one spike was surrounded by 6 others, while in other cases it was surrounded by 5, and drew analogies with the capsid structure of icosahedral viruses.

It is a curious fact that no matter what the shape of the virus particle the surface projections always appear to be more or less evently spaced. In particles which fall into flattened forms on the electron microscope grid the projections appear to be hexagonal close packed, each spike being surrounded by 6 others. In more solid spherical units the projections show a variety of arrangements. As during the course of formation a virus protrusion may change in morphology

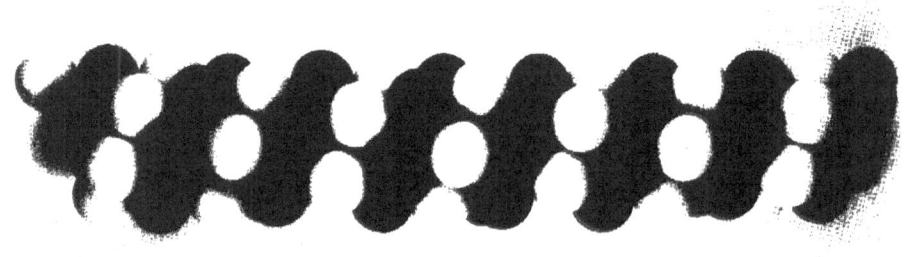

Fig. 28. Model showing the probable arrangement of sub-units in the ribonucleoprotein double helix.

from a large spherical form, through a tubular form which then fragments into small spherical forms, the surface spikes must be capable of some degree of redistribution in relative positions. The lipoprotein membrane must have a fluid or semi-fluid consistency to permit these rearrangements. It almost appears as if the projections were floating in an oily film. As they probably carry a fairly high electric charge their even distribution under all conditions might be due to mutual electric repulsion. This might also explain the star-like arrangement of haemagglutinin rods in ether-split virus haemagglutinin.

When first released from the cell the contents of cytoplasmic protrusions appear to be fluid. Large globules may show internal particles in rapid Brownian movement ($H_{195}$). On detachment the globules become smaller and filamentous particles which are at first flexible become converted into rigid rods. All types of virus particle, large globules, filaments, and small "typical" particles are covered with projecting spikes ($M_{115}$). The results of ether fractionation indicate that the volume of haemagglutinin released is greater than can be accounted for by the projections and it is probable that there is a large amount of haemagglutinin inside the particle. This internal haemagglutinin must be in the gel form since the water content of virus particles is only of the order of 50%. Probably the development of rigidity in filamentous particles after release from the cell is due to the conversion of internal haemagglutinin into the gel form as water is lost.

The final shape of the virus particle appears to result from the development of certain stable states of the lipoprotein membrane when in the closed form. These stable states are the spherical, tubular, and a flattened shape resembling that of a human red blood cell. All virus particles appear to adopt one or other of these forms or combinations of them. Once in the closed state no further contraction of the lipoprotein is possible. If in this condition a spherical particle does not contain enough haemagglutinin or internal ribonucleoprotein to completely fill it then on electron microscopy the particle will collapse on the grid to give a flattened, "poached" egg, or bean-shaped form.

The internal ribonucleoprotein helix appears to adapt itself to the shape of the particle determined by the lipoprotein ($B_{87}$), being coiled up in spherical particles and arranged in parallel lines in filamentous particles. Treatment of spherical particles with acid and trypsin ($V_1$) results in their developing a swollen flattened shape resembling that of a red cell, and the nucleoprotein is then found in the form of peripheral rings.

## 7. Weight, Density, Volume and Water Content of the Virus Particle

The variability in size and shape of influenza virus particles makes attempts to determine their weight and volume almost meaningless. However, studies have been made using highly egg-adapted strains such as PR8 or DSP which are almost entirely composed of small spherical particles.

LAUFFER and STANLEY ($L_{33}$) determined the density of PR8 virus by centrifugation in sucrose gradients as 1.18, and in the absence of sucrose as 1.1. They calculated the average diameter of anhydrous particles to be 80 m$\mu$. Assuming the particles to contain 60% water they calculated the hydrated particle to measure 100 m$\mu$. TOVARNITSKY ($T_{53}$) states that the virus contains 52% water. VALENTINE ($V_1$) estimated the density of freeze-dried virus particles at 0.9 from electron density measurements and assuming the density of the hydrated particle to be 1.2 concluded that the particles contained 25% water. ADA, PERRY and ABBOT ($A_{28}$) estimated the dry weight of PR8 spherical particles at $6 \times 10^{-16}$ g. This is equivalent to a molecular weight of $3.6 \times 10^8$ which is higher than would be expected from sedimentation data. Also if the particles have a diameter of 100 m$\mu$, possibly an over-estimate, the volume would be $5.23 \times 10^{-16}$ ml and with a density of 1.18 the weight would be $6.1 \times 10^{-16}$ g. This means that according to the results of ADA et al. the particles contain no water at all. REIMER, BAKER, NEWTON and HAVENS ($R_{18}$) determined the density at 1.185 and found the particles to contain $4.2 \times 10^{-16}$ g of protein. They assumed the protein content to be 75% which gives a total weight of $5.25 \times 10^{-16}$ g.

The author attempted to directly measure the water content of DSP virus. A large volume of concentrated purified virus was centrifuged and the volume of the deposited pellet measured. The virus was then resuspended, precipitated with trichloracetic acid, washed with distilled water, dried and weighed. The actual volume occupied by the virus particles in the deposited pellet was calcu-

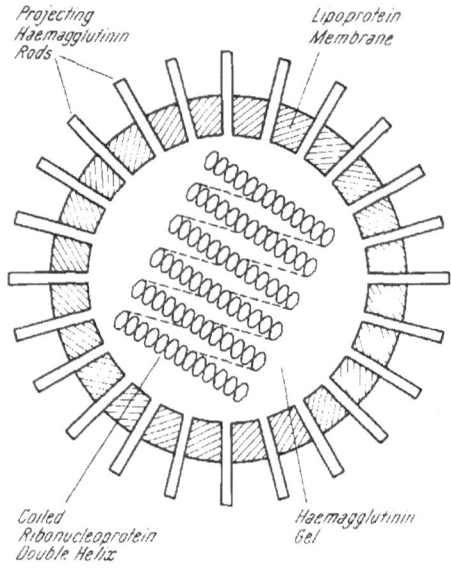

Fig. 29. Structure of a spherical particle of influenza virus.

lated by assuming the pellet to consist of closely packed spheres, an arrangement found by experiment with glass beads to include 34.5% of non-particle water. From these measurements the water content of the particles was found to be 51%.

## 8. Conclusions

The structure of a typical spherical particle of influenza virus is shown diagrammatically in Fig. 29. The particle has an outer lipoprotein membrane some 60 Å in thickness and covered with evenly distributed projecting rods some 100 Å in length which carry the haemagglutinin and enzyme properties of the virus. The projecting rods pass through holes in the lipoprotein membrane. The interior of the particle is occupied by a coiled ribonucleoprotein double helix which is embedded in haemagglutinin in the gel form.

A model of the probable arrangement of subunits in the ribonucleoprotein double helix is shown in Fig. 28.

# XII. Resistance to Physical and Chemical Agents

In general the influenza viruses show a resistance to destructive agents comparable to that of vegetative bacteria. The environment of the virus particle has a very great influence on its resistance. Differences in pH, ionic strength and ionic composition of the surrounding fluid have a marked effect on the resistance to most agents, and non-viral protein in the fluid exerts a considerable protective effect. For these reasons it is not possible to give absolute values for the viricidal concentrations of chemicals, but the resistance of different strains of virus may be compared if tested under parallel conditions, and it is possible to compare the relative sensitivities of the virus components and the different virus properties. The serological properties are usually found to be more resistant to destructive agents than the infective and toxic properties of the virus. Highly purified virus preparations are very unstable, infectivity is rapidly lost, the major destructive factors being thermal inactivation and oxidation. The infective property is best preserved at low temperatures, neutral pH, in media with a balanced salt composition and containing protective protein. The low water content of the virus particle and the elasticity of its outer membrane make it relatively resistant to osmotic effects, and in the presence of protective protein it may be preserved for long periods in glycerol.

## 1. Heat Resistance

The influenza viruses are relatively sensitive to thermal inactivation. HORS-FALL ($H_{166}$) found that in allantoic fluid at 35°C the infective property of LEE virus had a half-life of only 85 minutes. The haemagglutinin was much more resistant, being unaffected in 12 hours.

At temperatures between 50°C and 65°C the virus properties are rapidly destroyed, the infectivity and enzymic properties being more sensitive than the haemagglutinating and antigenic properties ($C_{52}$, $Z_{39}$, $O_{20}$, $H_{126}$). Different strains of virus show differences in heat resistance ($S_6$). By heating virus in allantoic fluid at 56°C for 30 minutes it is possible to destroy the infectivity and enzymic activity of many strains and produce virus preparations which agglutinate red blood cells but do not elute and are very sensitive to non-specific inhibitors of haemagglutination ($F_{94}$, $C_{52}$, $P_{51}$).

LAUFFER and WHEATLEY ($L_{34}$) showed that heat denaturation of virus is a first order reaction, and corresponds to the rate of haemagglutinin inactivation. Infectivity is more sensitive. There appeared to be two types of haemagglutinin centre in PR8 virus, one destroyed at the same rate as denaturation and one destroyed more rapidly.

## 2. Resistance to Freezing and Thawing

The results of freezing and thawing influenza virus preparations are extremely variable, depending on the rapidity and depth of freezing and on the purity of the preparations. GRIEF, BLUMENTHAL, CHIGA and PINKERTON ($G_{77}$) found that the infectivity of PR8 virus in allantoic fluid fell from $10^{8.6}$ $ID_{50}$ to $10^{0.8}$ as a result of five cycles of freezing and thawing, the second, third and fourth cycles causing a greater loss than the first and fifth cycles. TYRRELL and HORS-FALL ($T_{79}$) found that infectivity, enzymic activity, and haemagglutinating power were all destroyed by 50 cycles of freezing and thawing, but the virus still combined with antibody.

Very rapid freezing in $CO_2$ alcohol or liquid nitrogen produces much less damage. Freezing and storage at $-10$ to $-20$°C produces considerable losses in haemagglutinin possibly because of the concentration of solutes in intercrystallic films at this temperature. Such films may freeze at lower temperatures ($P_{29}$). Highly purified virus is very susceptible to damage by freezing, but considerable protection is afforded by non-viral protein. Infected allantoic fluid containing 5% of heated horse serum and frozen and stored at $-60$°C retains infectivity unchanged for long periods. Peptone also gives considerable protection ($P_{11}$).

*Lyophilisation:* The procedure of drying from the frozen state is much used for the preservation of virus strains. Best results are obtained by pre-cooling to $-80$°C and subliming at 0°C or pre-cooling at $-190$°C and subliming at $-80$°C ($G_{77}$). Once dried the survival of virus depends on the degree of dehydration attained and on the exlusion of oxygen ($G_{77}$, $P_{11}$).

The author has obtained very satisfactory preservation of virus strains by freeze-drying pieces of infected chorioallantoic membrane or infected mouse lung and storing in sealed evacuated tubes containing magnesium perchlorate at 4°C. Under these conditions virus has survived for 10 years.

### 3. Resistance to Drying

EDWARD ($E_{30}$) dried virus preparations on fabrics, glass, and house-hold dust. From 1—10% of the virus survived under ordinary conditions, but only 0.01% under humid conditions. Once dried there was little further loss in 3 days if kept in the dark, but more rapid loss of infectivity occurred in the light. In the dark dried virus survived for several weeks. Similar results have been obtained by other workers ($M_{79}$, $P_{12}$, $P_{13}$, $S_{223}$). Addition of mucin greatly increased survival times ($P_{12}$). Virus survived drying on the hands very badly, allantoic fluid virus being dead in 50 minutes ($P_{13}$).

### 4. Resistance to Ultrasonic Vibration

RHEINS and FINLAY ($R_{21}$) found that infectivity and haemagglutinin were unaffected by exposure to ultrasonics at 450 kilocycles if the temperature was kept below 30°C. Virus was rapidly liberated from infected chorioallantoic membrane and antigenicity was unaffected.

### 5. Resistance to pH Changes

TOVARNITSKY and SHISHKINA ($T_{58}$) found influenza virus to be stable between pH 5.2 and 7.8. MILLER ($M_{86}$) states that stability is maximal at pH 7—8, infectivity and haemagglutinating power being lost more rapidly on the acid than the alkaline side. CHAGNON, GILKER and BOUDREAULT ($C_{21}$) found that there was a marked fall of infectivity of B virus at pH 6—7, A virus at pH 5—6 and equine strains at pH 4—5. All strains showed a fall in haemagglutinin titre preceding loss of infectivity.

The author has found that the haemagglutinin of most strains is relatively stable between pH 5 and 10 but that the neuraminidase is more alkali-labile. The S antigen is more acid resistant and less alkali resistant then the haemagglutinin.

### 6. Effect of Ions on Resistance

KNIGHT ($K_{64}$) studied the persistance of infectivity of purified PR8 virus suspended in water, saline, phosphate, borate and veronal buffers at pH 7. Survival was best in 0.1 M buffers. EDNEY and BURNET ($E_{28}$) observed the effect of various ions on the thermal inactivation of dialysed virus at 56°C. Haemagglutinin and neuraminidase were destroyed in 30 minutes at pH below 6.9 or over 10.6. Additions of 0.001 Molar Cl, $SO_4$, citrate and hexametaphosphate ions protected against inactivation at pH 6.9.

### 7. Resistance to Ionising Radiations

#### a) Ultraviolet Light

Many workers have studied the effects of ultraviolet light on the virus properties ($S_{211}$, $S_{212}$, $H_{81}$, $T_{16}$, $Z_{39}$, $P_{83}$). HENLE and HENLE ($H_{81}$) found that infectivity, toxicity, interfering power, haemagglutinin and neuraminidase, and complement fixing antigen are inactivated in that order. By the use of monochromatic UV light TAMM and FLUKE ($T_{16}$) showed that infectivity was more sensitive to radiation at 2652 Å than at 2803 Å indicating probable action on the nucleic acid, while the haemagglutinin was more sensitive at 2803 Å suggesting action on protein. Several hundred times more energy was needed to inactivate haemagglutinin than infectivity. The appearance of the particles in the electron microscope was unchanged. POWELL and SETLOW ($P_{83}$) found that the interfering property of the virus was inactivated at 2600—2800 Å, the action spectrum being different from that for loss of infectivity. UV treated virus has been much used in interference experiments and in studies of the virus growth cycle.

### b) Photodynamic Inactivation

In the presence of toluidine blue the infectivity of influenza virus in allantoic fluid is destroyed by visible light without effect on haemagglutinin or antigenic properties ($W_{14}$). If the virus is purified the haemagglutinin becomes sensitive to photodynamic inactivation.

### c) Resistance to X Rays

LACORTE, MONTEIRO and LOURES ($L_{4-6}$, $L_8$) found that if suspensions of mouse lung infected with PR8 or LEE virus were irradiated with X-rays in a dose of 200 r the preparations became more virulent for mice than unirradiated controls. Brain tissue suspensions from mice infected with a neurotropic strain of A1 virus showed a similar increase of intracerebral virulence after irradiation with 200 r. Allantoic fluid preparations of PR8 virus retained mouse virulence after a total dose of 96,000 r delivered in fractionated doses of 600r over a period of several weeks.

BUZZEL, BRANDON and LAUFFER ($B_{294}$) found that the infectivity of PR8 and LEE viruses was inactivated by X-rays as if associated with a sensitive volume of about one eigth of the total volume of the virus particle. Irradiation of chick chorioallantoic membrane indicated that the ability to support virus growth was associated with a single sensitive volume of the same size as that associated with virus infectivity. X-ray inactivation of the haemagglutinin was found to be a multi-hit process, either requiring a single hit on three targets, or three hits on a single target.

KROEGER and KEMPF ($K_{109}$) found that X-ray inactivation of the infectivity of PR8 virus followed an exponential curve with a 37% survival after irradiation with 65,000r, from which a target diameter of 42 mμ was calculated. Haemagglutinin of purified virus showed a 37% survival with a dose of $6.2 \times 10^5$ r indicating a target diameter of 17 mμ, while haemagglutinin of ether-split virus had a target diameter of 23 mμ.

### d) Resistance to Gamma Rays

POLLEY ($P_{68}$) exposed semi-purified PR8 virus in saline to a cobalt 60 source emitting $1.25 \times 10^6$ rads/hour. Virus haemagglutination was often destroyed before total loss of infectivity, apparently because of oxidation as the haemagglutinin could be selectively protected by sulphur containing amino acids. Ascorbic acid, histidine, tryptophane and tyrosine also protected. In the presence of 0.2% histidine infectivity could be destroyed without loss of haemagglutinin.

### e) Resistance to Fast Charged Particles

JAGGER and POLLARD ($J_1$) exposed MEL and DSP viruses dried in thin layers to bombardment in a vacuum by deuterons of 3.9 Mev, α particles of 7.8 Mev, and electrons of 2.0 Mev energy. It was possible to determine the size, shape and multiplicity of the structural entities responsible for various properties of the virus. In all cases infectivity was inactivated as a single-hit process. Results with MEL virus suggested the presence of two regions, one of high sensitivity with a radius of 24.6 Å and one of lower sensitivity with a radius of 346 Å. With both strains the haemagglutinin was relatively radiation-resistant, requiring at least 3 ion pairs per particle for inactivation. The haemagglutinin appeared to be a flat target with a molecular weight of 190,000. In similar experiments POWELL and POLLARD ($P_{82}$) found the interfering property of MEL virus to have a radiation-sensitive volume of $2 \times 10^6$ cubic Å corresponding to a molecular weight of 1,600,000.

## 8. Resistance to Chemical Agents

KNIGHT and STANLEY ($K_{69}$) and DUNHAM and MacNEAL ($D_{96}$) studied the resistance of virus to a variety of antiseptics. Virus was inactivated by oxidising agents, heavy metals, detergents, alcohols and formalin. Apart from ascorbic acid reducing agents had little effect. Protein protected the virus against many antiseptics.

### a) Oxidising Agents

Influenza virus is very sensitive to oxidising agents, but protein may afford considerable protection. Although extensive studies of the effects of oxidising agents on influenza viruses do not appear to have been made, it would be expected from analogy with other viruses that the sensitivity would depend on the oxidation potential developed. The inactivating effect of halogens at alkaline reactions is probably due to oxidation. KLEIN ($K_{52}$) showed that the inactivating effect of ascorbic acid was due to the production of hydrogen peroxide as it was prevented by catalase.

### b) Halogens

STONE and BURNET ($S_{279}$) found that haemagglutinin was destroyed by $6 \times 10^{-5}$ N iodine, and chlorine and bromine had similar effects. HOYLE ($H_{200}$) found that even at pH 6.0 iodine was extremely viricidal, being 800 times as active as formaldehyde. Iodine kills influenza virus almost instantly and this was attributed to an ability to pass through the lipid of the outer membrane of the virus particle.

### c) Heavy Metal Salts

Heavy metal ions destroy infectivity and precipitate the virus by combination with COOH groups. The author found silver nitrate to be the most viricidal, but oxidation may be involved in this case. With zinc and copper salts subsequent removal of the metal may restore some of the properties such as haemagglutination.

Many mercury salts are not highly ionised and do not precipitate virus. They react with sulphydryl and disulphide groups. KLEIN and PEREZ ($K_{55}$, $K_{53}$) found that the inactivation of virus by 1:10,000 mercuric chloride was reversible, infectivity being restored by treatment with BAL (2,3 dimercaptopropanol) either in vitro or in vivo. ALLISON, BUCKLAND and ANDREWES ($A_{51}$) found that infectivity of influenza and fowl plague viruses was greatly reduced by p-chlormercuribenzoate while haemagglutinin was unaffected. ALLISON ($A_{50}$)

showed that virus treated with mercuric chloride or p-chlormercuribenzoate was adsorbed by the cells of the chorioallantoic membrane but was unable to penetrate. The effects of mercurials were reversed by cysteine or glutathione.

LACORTE, MONTEIRO and LOURES (L₉) found that influenza viruses were resistant to contact with metallic silver, while most bacteria were destroyed in 24—48 hours.

### d) Organic Solvents

Many organic solvents destroy influenza virus by disrupting the particles. Virus suspensions are disrupted by shaking with ethyl ether ($A_{76}$, $H_{193}$, $S_{172}$). Infectivity is lost but haemagglutinin and antigenic properties are retained. Chloroform produces similar effects ($F_{61}$) but may cause some protein denaturation. Petroleum ether has little effect. Water-miscible solvents such as alcohol and acetone at 0°C precipitate virus without much destruction of its properties, but at higher temperatures the virus protein is denatured. The author found that concentrations of alcohol or acetone of 36% destroyed all virus properties in 2 hours at 20°C. GINSBERG and WILSON ($G_{31}$) found that the infectivity of influenza viruses was completely destroyed by 1% ethylene oxide at 35°C in 24 hours, even in the presence of 99% horse serum.

### e) Phenolic Antiseptics

Influenza viruses are more resistant to phenolic antiseptics than are most vegetative bacteria. In the early years of influenza virus research it was found useful, when passing influenza virus serially in mice, to add 0.5% phenol to mouse lung suspensions and hold them at 4°C overnight before inoculating intranasally, this procedure preventing secondary infection by bacteria of the Pasteurella group. Virus is destroyed by 0.5% phenol in 7—10 days.

### f) Detergents

Soap quickly destroys influenza virus on the hands ($P_{13}$). Solutions containing 1% of soap inactivate 10% infected mouse lung suspensions in 40 minutes, and 0.1% suspensions in 15 minutes, B virus being more sensitive than A ($K_{110}$).

Most synthetic detergents destroy virus infectivity ($S_{268}$, $K_{56}$). In contrast to vegetative bacteria which are much more sensitive to cationic than anionic detergents, the author found that influenza virus was equally sensitive to both types of detergent, the virus being destroyed by concentrations similar to those which lysed red blood cells. Similar observations were made by BURNET ($B_{182}$) in studies of the destruction of virus filaments by detergents.

Purified virus preparations are rapidly disintegrated by sodium deoxycholate with results similar to those produced by ether. Serological properties are retained. Sodium dodecyl sulphate is more destructive than deoxycholate.

### g) Agents Acting on Ribonucleic Acid

Influenza and fowl plague viruses are inactivated by hydroxylamine ($F_{126}$, $S_{38}$) nitrous acid ($R_{64}$), and an ethylene iminoquinone Bayer A 139 ($S_{67}$). These agents appear to act on the virus nucleic acid and have been used in studies of the role of the nucleic acid in the synthesis of virus components.

### h) Enzymes

Trypsin reduces the infectivity of influenza virus and destroys the neuraminidase activity ($S_{270}$, $H_{190}$). Neuraminidase activity may be lost as a result of separation from the virus particle ($N_{27}$). The surface projections of the virus particle may be distorted or destroyed. Prolonged trypsin treatment may reduce haemagglutinin titre and release serologically active material from the particle. Caseinase C and pronase may produce similar effects ($R_{15-17}$). Different strains of virus vary considerably in susceptibility to trypsin and CLEELAND and SUGG ($C_{62-64}$, $S_{322}$) obtained trypsin sensitive and trypsin resistant lines from the same strain of virus A.

Phospholipase C disintegrates influenza virus probably by action on the phospholipid ($S_{140}$). Influenza virus particles are resistant to nucleases.

### i) Other Chemical Viricidal Agents

The action of sulphur mustard derivatives was studied by FONG and others ($F_{80-83}$). Treatment of virus with $5 \times 10^{-3}$ M sulphur mustard destroys infectivity without affecting haemagglutinating, interfering and antigenic properties. Cyclo and chloromustard derivatives destroyed virus properties in the order of infectivity, toxicity, interfering power, haemagglutination-elution, and antigenicity.

CARSON and FRISCH ($C_{15}$) found that PR8 virus was inactivated by neutralized tannic and digallic acids. Gallic, pyrogallic and gentisic acid had only slight action.

UTZ ($U_5$) found that a lecithin-like substance derived from serum and liver destroys infectivity at $37°C$ without affecting haemagglutinin. No such material was found in egg yolk or in brain.

## 9. Survival and Resistance of Influenza Virus in Air

Because of the possible importance of such studies in the control of the spread of influenza many observations have been made of the survival and resistance of influenza virus when distributed as an aerosol ($E_{31}$, $L_{115}$, $L_{63}$, $M_{79}$). Mice are readily infected by exposure in closed chambers to an aerosol of infected mouse lung suspension and lung lesions are more regularly produced by this method than by intranasal instillation. All workers are agreed that the major factor controlling the survival of virus in air is the relative humidity. Virus survives much longer in dry than in humid atmospheres ($E_{31}$). LOOSLI, LEMON, ROBERTSON and APPEL ($L_{115}$) found that virus survived for 30 minutes at 80—90% humidity, 6 hours at 45—55%, and 24 hours or more at 17—24% humidity. LESTER ($L_{63}$) obtained rather different results. Exposure of mice for 15 minutes to a virus aerosol produced by atomising 0.82—0.85 g of 10% infected mouse lung suspension into 640 cubic feet of air produced 100% death rate at humidities below 30%, but only 22.5% death rate at humidities of 45—60%. With higher humidities the death rate increased and reached 100% at humidities over 80%. Probably various factors are involved, survival of virus in the air, variations in susceptibility of mice at different humidities, and variation in stability of the aerosols.

## Inactivation of Virus Aerosols

EDWARD, LUSH and BOURDILLON ($E_{32}$) found that 99% of the virus in an aerosol was killed by 6 seconds exposure at 2 cm distance from a Hanovia lamp or GEC sterilamp emitting UV light at 2537 Å. SMORODINTSEVA ($S_{212}$) found that virus aerosols were killed by UV light and also by ozone, which in a concentration of 0.25 mg/cubic metre in air killed virus in 30—60 minutes.

Virus aerosols are rendered non-infective by propylene glycol vapour in a concentration of 1 in 2 million ($R_{40}$). STONE and BURNET ($S_{279}$) found iodine vapour effective at 1 in 10 million and suggested the use of iodine-impregnated gauze masks for the protection of medical and nursing staff during an epidemic. VASHKOV and ASTAF'YEVA ($V_{10-12}$) recommended lactic acid at 5 mg/cubic metre and glycerine at 10 mg/cubic metre. They found pyruvic, laevulinic, glyceric and mono-, di-, and trichloracetic acids rather less effective, while acetic, citric, isovaleric, butyric, formic and caproic acids were only slightly active. Other viricidal vapours include resorcin, hexylresorcin, pyrogallol ($V_{19}$), isocyanates and p-nitrobenzylchloride ($G_{88}$). Oxyquinoline, thiourea, nutmeg oil and mustard oil have slight action.

## 10. Inactivation of Vaccines

The viricidal effects of formaldehyde, $\beta$-propiolactone, and hydroxylamine are of special importance because of the use of these substances in the preparation of killed vaccines.

The effects of formaldehyde have been particularly studied by KOCH and CSONKA ($K_{72}$, $K_{73}$, $C_{89}$). They found the reaction to follow first order kinetics. Inactivation increased with increasing pH, temperature, and concentration of formaldehyde. Infectivity was destroyed without loss of antigenicity by 0.01% formaldehyde in 30 minutes. At 37°C 1% formaldehyde destroyed 99% of the complement fixing power, but some antibody was still produced on inoculation. They suggest that destruction of infectivity results from partial reaction of the ε-amino groups with one molecule of HCHO, inactivation of haemagglutinin from reaction of all ε-amino groups, and loss of antigenicity to "rough damage to surface structure".

The effects of $\beta$-propiolactone were studied by POLLEY and GUERIN ($P_{70}$, $P_{71}$). Infectivity was destroyed more rapidly at pH 8 than at pH 6. Haemagglutinin was best preserved at pH 7. At pH 7 PR8 virus was inactivated in 2 hours at 37°C by 0.5% $\beta$-propiolactone, and haemagglutinin, antigenic, and protective capacities were retained.

SCHÄFER and ROTT ($S_{38}$) mixed influenza and fowl plague virus suspensions with equal volume of hydroxylamine at 20°C. At intervals samples were diluted 1:10, dialysed against saline and infectivity tested. The viruses were inactivated exponentially. Hydroxylamine reacts with the nucleic acid and haemagglutinin, antigenicity, and protective capacity were unaffected.

# XIII. The Infective Property of the Influenza Viruses

Influenza viruses are capable of infecting a wide variety of animals, and a considerable range of cell types. A suitably chosen infectivity test is usually the most sensitive method of detecting the presence of virus. Infectivity tests do not, however, measure any absolute characteristic of the virus particle. The result of an infectivity test depends on a combination of many factors; the number and type of virus particles present, the amount of nucleic acid in the particle, the type of host cell from which the virus was derived, the host cell in which infectivity is tested, and the method used to detect infectivity. Some of the difficulties are illustrated by the following examples.

HIRST ($H_{133}$) showed that if egg adapted viruses are inoculated intranasally to mice they appear to be non-infective if the development of lung lesions is used as the test of infection. If, however, extracts of the mouse lungs are titrated in eggs it can be shown that considerable multiplication of virus in the mouse lungs has occurred. But even if this mouse-egg titration method is used GINSBERG ($G_{24}$) found that it required 100 times as much egg adapted virus to infect mice as to infect chick embryos. With mouse adapted strains the infective dose was about the same in either host.

BURNET ($B_{158}$) found that influenza virus in human nasopharyngeal washings was much more infective when introduced into the amniotic sac of developing eggs than if inoculated to the allantoic sac.

VON MAGNUS ($M_{20-23}$) compared virus produced by serial passage in eggs using small inocula with that produced by the use of large inocula. Although the yield of virus as measured by haemagglutinin titration was the same in both cases, the virus produced by serial passage of large inocula appeared to be very much less infective than if small inocula were used. ADA and PERRY ($A_{25}$) showed that the VON MAGNUS phenomenon was related to the presence of a smaller amount of nucleic acid in the less infective form of the virus.

BEALE ($B_{28}$) compared the infectivity of intracellular virus prepared by extracting chorioallantoic membranes in the early stages of virus multiplication with that of extracellular virus. The intracellular virus appeared to be relatively more infective if tested by determining the amount of S antigen produced at 6 hours in the chorioallantoic membrane of eggs receiving large inocula, than if tested by determining the minimal dose required to produce haemagglutinin in the allantoic sac at 48 hours. The differences were attributed to "multiplicity reactivation" ($B_{29}$).

ADA, PERRY and ABBOT ($A_{28}$) were able to partially separate filamentous and spherical particles of the Ryan strain of virus and showed that, in terms of the number of particles required to initiate infection, the filaments were more infective than the spheres.

The great lability of the infecting property also causes difficulties. Virus preparations rapidly lose infectivity, especially if highly diluted or highly purified. Considerable dilution is necessary in most types of quantitative infectivity tests and it is essential to make dilutions in medium containing some non-viral protein which exerts a protective effect.

The presence of inactivated virus may interfere with infection by active virus. As a result of the rapid loss of infectivity on purification and the interfering effect of inactivated virus it is not unusual to find that the infectivity of a highly concentrated purified virus preparation may be less than that of the original crude unconcentrated preparation from which it was made.

Infectivity titration is, therefore, a procedure of somewhat limited value in influenza virus research. Although infectivity titrations have been widely used as a measure of the amount of virus in a given material, they are singularly ill-adapted for this purpose, and their use has often led to erroneous conclusions. The most satisfactory results are obtained when virus strains highly adapted to growth in a particular host tissue are titrated in the same tissue, and under these conditions replicate titrations give consistent results which bear a close relation to the amount of virus present ($D_7$).

## 1. Infectivity Titration in Mice

Groups of 5—10 mice are inoculated intranasally under light ether anaesthesia with 0.05 ml of a range of decimal dilutions of the material to be tested. Dilutions may be made in buffered saline pH 7, or in broth-saline, and it may by useful to add 600 units of penicillin and 1.25 mg streptomycin per ml to protect against bacterial infection. DAVENPORT ($D_7$) recommends addition of 10% horse serum to the diluent, but horse serum may be inhibitory to some strains of virus. The 50% death rate, $LD_{50}$, calculated by the method of REED and MUENCH ($R_{14}$) may be used as a measure of the end point, or alternatively the end point may be estimated by an assessment of the extent or incidence of lung lesions in mice killed at 4 days. LAUFFER and MILLER ($L_{32}$) used a composite method taking account both of the extent of lung consolidation and death.

## 2. Infectivity Titration in Fertile Eggs

HIRST ($H_{127}$) inoculated 0.1 ml of 10-fold dilutions of virus in 10% horse serum broth to the allantoic sac of 11—12 day old eggs, six eggs being used to each dilution. The presence or absence of haemagglutinin in the allantoic fluid was observed after 37—48 hours incubation and 50% end points calculated. Similar methods have been used by many other workers ($XK_1$, $H_{87}$, $M_{20}$). Virus diluents may include penicillin and streptomycin and a variety of protective proteins have been used. Broth or peptone are very satisfactory and the use of animal sera is probably best avoided.

### Titration in Chorioallantoic Membrane Tissue Suspensions

The orthodox egg titration methods described above are laborious and expensive of materials.

FULTON and ARMITAGE ($F_{163}$) developed a technique in which small pieces of chorioallantoic membrane are suspended in 1 ml amounts of a balanced salt solution of pH 6.8 containing 10 units of penicillin, 40 µg of streptomycin, and 0.05% glucose, distributed in a series of cups in a plastic tray. Dilutions of virus in 1% egg white saline are added to the various cups, five cups to each dilution. The tray is then covered and placed in a rocking machine at 37°C for 60 hours. The pieces of membrane are then removed and 40 cmm of 4% chick red cells added to each cup and the haemagglutination pattern read after 1 hour. The method is somewhat less sensitive than the in ovo titration method.

FAZEKAS DE ST GROTH, and WHITE ($F_{53}$, $F_{54}$) improved the method by the use of pieces of membrane still attached to the egg shell. This prevented union of virus with the cells of the chorionic layer of the membrane, in which an incomplete growth cycle occurs, and improved the buffering of the solution. Hanks solution was used, buffered with Tris to pH 7.2 and containing 0.5% gelatine and 0.03% glucose. The sensitivity of the method varied with different strains of virus but was comparable to that of *in ovo* titrations ($F_{55}$).

Infectivity has also been measured in tissue cultures of minced chorioallantoic membrane ($G_1$, $H_{171}$).

### The Six Hour Soluble Antigen Production Test

This test was studied by BEALE and FINTER ($B_{28}$, $B_{29}$, $F_{69}$). Groups of fertile eggs are inoculated to the allantoic sac with 1 ml amounts of decimal dilutions of virus. After 6 hours incubation the chorioallantoic membranes are removed, suspended in 1 ml of saline, frozen and thawed three times and centrifuged. The content of soluble antigen in the supernatant fluid is then titrated with a suitable antiserum. The amount of antigen produced is a function of the number of cells infected and hence the amount of infective virus in the inoculum. The test is only applicable to preparations containing large amounts of virus, and its main use has been in its ability to detect multiplicity reactivation. The orthodox *in ovo* test measures the number of virus particles which are of such infectivity that the entry of one particle into a cell results in infection. The six hour test measures the number of cells infected whether the infection results from the entry of one or more than one particle. If a virus preparation appears more infective by the six hour test than would be expected from its *in ovo* titration then multiplicity reactivation is occurring.

### 3. Infectivity Titration in Tissue Cultures

In addition to the use of chorioallantoic membrane tissue, infectivity may also be measured in a variety of other types of tissue culture.

MOGABGAB, SIMPSON and GREEN ($M_{110}$) used human embryo lung and kidney tissue cultures and compared the results with titration in the amniotic cavity of fertile eggs. End points were determined by the detection of haemagglutinin in the culture fluids. Kidney cultures were suitable for both A and B viruses, but lung cultures were less susceptible to virus A than to virus B. Tissue culture lines of virus A were more infective to kidney cultures than to the amnion of fertile eggs, but with B virus the titre was higher in the amnion than in lung tissue cultures.

ROZEE, WILLIAMS and VAN ROOYEN ($R_{68}$) titrated Asian virus in HeLa cell and monkey kidney cell cultures. Titration was based on the haemadsorption technique. The infected cell sheets were washed with a suspension of chick red cells. The number of red cell clumps adhering to the cell sheet was linearly related to the inoculum dilution. LEHMANN-GRUBE ($L_{55}$) used a variety of tissue cultures and found calf kidney cultures most useful.

Tube cultures in medium 199 were used, and 0.1 ml of serial virus dilution was added to the cell sheets. After 15 minutes adsorption 0.9 ml of medium 199 containing 0.17% $NaHCO_3$ and gassed with $CO_2$ was added. Tubes were stoppered and incubated for 3 days. Titre was determined by observation of cytopathogenic effect and also by haemadsorption. It was considered to be important to avoid the use of additives such as animal sera, amniotic fluid, or embryo extracts to the basal maintenance medium. Sensitivity of the test was only slightly lower than *in ovo* methods.

Plaque assay methods have in general proved unsatisfactory with influenza viruses, but LEHMANN-GRUBE ($L_{53}$) found that most virus strains form plaques on calf kidney cells. Trypsinized calf kidney cells were grown in medium 199 with 10% calf serum.

Monolayers were formed in 5 cm Petri dishes by suspending $3-4 \times 10^6$ cells in 5 ml of medium and incubating under $CO_2$ for $4-5$ days. Monolayers were washed with Hanks' solution and inoculated with 0.5 ml of virus dilution. After an adsorption period of 90 minutes they were washed and overlaid with 5 ml of Eagle's basal medium containing bicarbonate and agar. After 4 days incubation under $CO_2$ a second overlay of medium containing neutral red was added and plaque counts made $12-14$ hours later.

## 4. Infectivity of Influenza Virus Ribonucleic Acid

The work of ADA and PERRY ($A_{25}$) indicated that the infectivity of influenza virus particles was related to their content of ribonucleic acid.

Evidence that influenza virus infection might be transmitted by pure virus nucleic acid was first advanced by PORTOCALA, BOERU and SAMUEL ($P_{74-78}$). Purified influenza virus was extracted three times with ether and the RNA extracted by the phenol technique of GIERER and SCHRAMM ($G_{22}$, $G_{23}$). Inoculation of the RNA into the allantoic sac of fertile eggs resulted in the production of virus with a haemagglutinin titre of $1280-2560$ in 38 hours. The original strain of virus was of A1 type, but the derived strain did not contain A1 antigen and also differed from the original in having a filamentous morphology and in failing to infect mice. Infective RNA was isolated from 5 strains of A1 virus but negative results were obtained with an A2 strain ($P_{79}$). The RNA solutions were very labile, losing infectivity on storage and on treatment with ribonuclease.

Other workers were unable to confirm these results. ADA, LIND, LARKIN and BURNET ($A_{21}$) were unable to infect chick embryo or mouse brain with phenol-treated preparations of three strains of virus A and one of virus B, and no recombinants were obtained by mixed infection with RNA from WSE virus and intact MEL or WS viruses. SOKOL and SZURMAN ($S_{215}$), SZURMAN ($S_{349}$), and SOKOL and SCHRAMEK ($S_{214}$) were unable to isolate infectious RNA and SCHÄFER ($S_{33}$) also had negative results. PORTOCALA and his colleagues ($P_{74}$, $P_{80}$) suggested that the negative results might be due to the action of ribonuclease in allantoic fluid or to the presence of lead in the phenol used.

MAASSAB ($M_1$, $M_3$) was unable to isolate infectious RNA from purified virus but obtained an infectious ribonucleic acid from infected chorioallantoic membrane extracts by phenol treatment. The material was only pathogenic for chick kidney cell tissue cultures, and failed to infect the amniotic cavity of eggs or mouse lungs. Infectivity was destroyed by ribonuclease or by incubation at 37°C for 3 hours. The RNA derived strain of virus was antigenically identical to the original strain, but differed from it in being less infective for chick kidney cells, and in being unable to grow in the allantoic sac, though it would grow in the amniotic sac of fertile eggs.

Several groups of Russian workers have claimed successful isolation of infectious RNA. ORLOVA and DISKINA ($O_{17}$) isolated an infectious RNA from A2 virus. DUBROVINA ($D_{88}$) isolated RNA from purified virus by treatment with a mixture of sodium dodecyl sulphate and phenol. The RNA was inoculated to chick embryo cell cultures and apparently produced only one cycle of multiplication, but on subsequent passages in eggs virus of the original serological type emerged. The infectivity of the RNA was destroyed by ribonuclease. Dubrovina, Polyak and Smorodintsev ($S_{193}$) obtained RNA preparations from purified virus by treatment with phenol, chloroform and ether followed by alcohol precipitation. The product infected eggs but virus was detectable only after several passages. They found that if the RNA was inoculated to monolayers of chick fibroblasts in which intact virus does not grow, infection apparently occurred and virus could be recovered by subsequent passages in eggs.

Zhumatov and Isaeva ($Z_{36}$) found that RNA from purified virus usually failed to infect eggs but infection occurred if the preparations were diluted 1:2 or 1:4 with 0.1 M magnesium chloride or with 0.15 M sodium chloride +0.007 M phosphate buffer pH 7.2. The infectious RNA was resistant to ribonuclease and to alcohol. Mice inoculated intranasally survived, but on passage lung lesions and death occurred. Infective RNA could also be obtained from extracts of infected chorioallantoic membrane. They state that RNA preparations are only infective if they contain at least 0.1 µg/ml. Arkangelsky, Savchenko and Zhumatov ($A_{102}$) found that infection of mice by intranasal RNA could be revealed by staining cells of the respiratory tract with fluorescent antibody.

### 5. Some Reflections on the Nature of the Infecting Property of Influenza Viruses

The most outstanding characteristic of the infecting property is its unpredictability. Successful transmission of infection from one cell to another or from one animal to another depends on a large number of extremely variable factors. All that can be said is that infected cells may often discharge particles into the surrounding environment which under suitable conditions may be able to unite with another cell and may sometimes induce the production in that cell of certain substances which are characteristic of influenzal infection. Ribonucleic acid derived from the infected cell or from extracellular particles may be able to induce infection, but here the results are even more irregular and unpredictable than with more complex extracellular particles. It is very difficult to define an influenza virus in terms of infectivity, and it is unjustifiable to assume that particles which fail to infect are defective. Such an assumption really depends on the concept that the virus is a parasite with an independent existence apart from the cell from which it originated, and that the purpose of virus reproduction is to produce such a parasite. If so the process is often quite remarkably inefficient.

The author has found it useful to regard influenza as a cell disorder associated with the synthesis of abnormal ribonucleoprotein and enzymically active protein, and the production of cell fragments able to transmit the disorder as a more or less irrelevant accident. Such a concept leads to the study of influenza as an exercise in cell chemistry, an approach which has not been unprofitable. It may be that the production of infective particles is a method of transmitting biochemical information from one cell to another, and that influenzal infection may be a detectable example of a much more widespread phenomenon which does not normally produce any obvious effects.

The application of very precise mathematical methods to the assessment of infectivity is in most instances unjustified. The mathematical approach depends on the basic assumption that the entities considered are all identical. This assumption is not true for influenza virus particles.

# XIV. Haemagglutination

The agglutination of red blood cells by influenza virus was discovered by HIRST ($H_{125}$) and McCLELLAND and HARE ($M_{54}$). The haemagglutination was inhibited by antibody and the test could be used to detect and measure the amount of virus in a suspension and also the antibody content of sera. The essential features of the reaction were described by HIRST ($H_{126}$, $H_{129}$). The haemagglutinin titre was shown to be proportional to the mouse lethal dose, and the haemagglutinin-inhibiting titre of antiserum to be closely related to the virus neutralisation titre. Adsorption of virus suspensions with red blood cells produced a parallel reduction in infectivity and haemagglutinin titre. Adsorbed virus was eluted from the red cells on incubation at 37°C, and could be again serially adsorbed and eluted from fresh red cells without significant loss. Cells from which virus had been eluted could not be agglutinated by fresh virus. The results indicated an enzymic destruction of a receptor substance in the red cells.

For the simple detection of virus in infected fluids a few drops may be mixed with an equal volume of 5% chick or guinea pig red cells in the concavity of a porcelain tile and the mixture rocked. Haemagglutination can be detected by the naked eye in 30 seconds to 1 minute. By using a range of virus dilutions a rough estimate of the amount of virus can be made, but for most purposes a more accurate method is necessary.

## 1. The Salk Test

This method ($S_4$) is the most widely used type of haemagglutinin titration.

A series of two-fold dilutions of the material in 0.5 ml of saline is placed in round-bottomed tubes of internal diameter 10 mm. To each tube is added 0.5 ml of an 0.25% suspension of chick red blood cells. The mixtures are shaken and then allowed to stand at room temperature for $1\frac{1}{2}-2$ hours. At the end of this time control tubes with no virus show a compact button of red cells in the centre of the bottom of the tube. Haemagglutination is shown by the sedimentation of the red cells to form a flat film covering the whole area of the bottom of the tubes. Tubes of any convenient size may be used, and plastic plates with a series of round-bottomed holes are commonly used.

The end point is mainly dependent on the concentration of red cells used. The 0.25% suspension recommended by SALK is the lowest giving satisfactory results and most workers have preferred to use 0.5% cells, as recommended by the Committee on standard serological procedures in influenza studies ($C_{83}$). If guinea pig or human red cells are used an 0.5% suspension is necessary.

The haemagglutinin titre is inversely related to the red cell concentration over a wide range, The titre is also affected by the temperature, higher titres being given at 4°C when the cells sediment more slowly. At 37°C results are unsatisfactory because of the rapid elution of virus from the cells.

## 2. Densitometric Methods

In these methods, introduced by HIRST and PICKELS ($H_{148}$) mixtures of virus dilutions with red cell suspension are set up in optically standardized tubes and after a suitable period of standing the optical density at the centre of the column is measured photoelectrically. Calibration curves are made relating the cell concentration to the optical density and the end point is read in tubes showing 30—50% sedimentation of the cells. More precise end points can be determined than with the SALK technique.

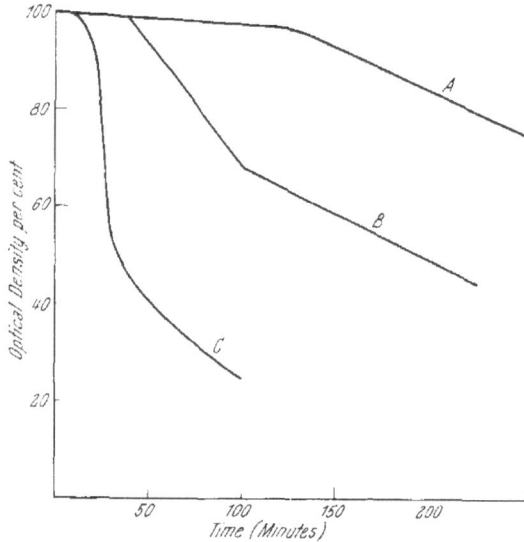

Fig. 30. Absolute assay by densitometric methods.
Optical density of the central part of a tube containing virus-red cell mixtures. Adapted from LEVINE, PUCK and SAGIK ($L_{67}$). A: Control, red cells with no virus. B: Excess of cells over virus. Initial lag followed by linear sedimentation of dimers and later by sedimentation at the same rate as the control when all dimers have passed the scanning level. C: Excess virus over cells, showing the "avalanching" effect.

### a) Absolute Assay by Densitometric Methods

LEVINE, PUCK and SAGIK ($L_{67}$) introduced an absolute assay method depending on counting the number of dimers (two red cells linked by virus) in a mixture containing excess of cells. Optically standard tubes $13 \times 100$ mm containing 6.0 ml of virus dilution and $5.8 \times 10^7$ chick red cells per ml are allowed to stand and the optical density measured at 10 minute intervals, the light beam of the densitometer scanning the central part of the tubes.

It was found that, with large amounts of virus relative to red cells, after an initial lag representing the time taken for collisions to occur there was a very rapid fall in optical density due to the agglutinated cells descending with increasing velocity, "avalanching", followed by a slower fall as the tubes became denuded of cells. With amounts of virus particles less than half the number of cells the initial lag was followed by a linear decrease in optical density as the dimers sedimented more rapidly than single cells. After a time this linear fall ceased when all the dimers had passed through the illuminated part of the tube and from this time on the fall in optical density became the same as in tubes with no virus (Fig. 30). By comparing the optical density at this point with that in control tubes without virus it was possible to calculate the number of dimerized cells and the number of virus particles was taken to be half this number. Results agreed closely with counts made by electron microscopy of mixtures of virus and latex spheres. A similar method has been used by HORSFALL ($H_{165}$).

### b) The Drescher Technique

DRESCHER ($D_{67}$) using the photoelectric method found with PR8 virus and chick red cells that in the absence of virus the optical extinction was linearly related to the concentration of red cells over the range $0-1\%$. If all the cells were aggregated by addition of large amounts of virus the extinction was linearly related to the concentration of aggregates. From the observed gradients it was possible to derive an equation by means of which the concentration of aggregated cells could be calculated from the observed value of the extinction. A linear relation was obtained when the logarithm of the aggregate concentration was plotted against the reciprocal of the virus dilution, and it was possible to determine the virus concentration independently of the sedimentation time and in spite of variations in the properties of different red cell suspensions. The mathematics is somewhat complicated and the original papers should be consulted by those interested in the method. MILLER ($M_{87}$) confirmed DRESCHER's work and improved the method by the use of a longer period of incubation of the virus-cell mixtures to minimise interference due to inhibitors and by addition of gelatin to stabilise haemagglutinin. SCHMIDT ($S_{54}$) confirmed the practical use of the method but criticised the theoretical aspects. POTEL, HLAWATSCH and DEGEN ($P_{81}$) found that haemagglutinin could be measured to an accuracy of $\pm5\%$ by the DRESCHER technique but found the method complicated, difficult and time consuming and not suitable for routine work.

## 3. Other Methods of Haemagglutinin Titration

SCHMIDT and GROSSGEBAUER ($S_{55}$) found a linear relation between the haemagglutinin titre of allantoic fluids and the logarithm of the time at which the onset of haemagglutination can be observed in mixtures of virus and red cells drawn up into sedimentation tubes. Haemagglutinin titres could be determined in a few minutes without carrying out serial dilutions.

KOHN and DANON ($K_{80}$) have recently described the use of the fragiligraph with an automatically registered photo-electric system. A mixture of virus and cells is placed in the instrument and the changes in light transmission observed. The concentration of virus can be calculated from the time it takes for the inflection of the curve to start and its angle.

COHEN and BELYAVIN ($C_{72}$) have also developed an automatic titration method by the adaptation of a component of the "Technicon" auto analyser which will separate agglutinated from non-agglutinated cells. The agglutinated cells are decanted and the non-agglutinated cells lysed by sodium hydroxide and the haemoglobin estimated colorimetrically and recorded.

## 4. Effect of pH, Temperature and Salt Concentration on Haemagglutination

HIRST found that at 4°C adsorption of virus by red cells was rapid and complete. At higher temperatures the elution effect led to progressively less adsorption. Maximum adsorption at 4°C occurred in 5 hours, at 27°C in 25 minutes, and at 37°C in 3—5 minutes ($H_{129}$). The haemagglutinin was inactivated at 60°C but was more heat resistant than infectivity. The cell receptors were comparatively heat stable, resisting 100°C for 5 minutes.

MAKOVER and SKURSKA ($M_{29}$) showed that if PR8 virus was titrated in parallel at 4°C and 20°C the titres were the same, but if the cells were resuspended and the temperatures reversed a higher titre was found with the test originally carried out at 4°C. OVERMAN and FRIEDEWALD ($O_{21}$) found that sheep and rabbit cells were strongly agglutinated by PR8 and LEE viruses at 4°C but not at 20°C or 37°C. The sheep cell receptors were destroyed by virus at 37°C. MAGILL and SUGG (M₁₇) found that sheep cells were agglutinated at pH 5.8.

The effects of electrolytes were studied by DAVENPORT and HORSFALL (D₂₈), YEREMEYEV (Y₇) and FLICK, SANDFORD and MUDD (F₇₈). Adsorption of virus is eliminated at low electrolyte concentrations. BARUA and FAUCONNIER (B₁₆) using formolised red cells showed that both active and heated virus elute rapidly from red cells at 4°C if the cells are suspended in distilled water. DAVENPORT and HORSFALL dialysed virus against 0.001M phosphate buffer pH 7.2 and then tested haemagglutination with virus dilutions in isotonic dextrose, sucrose,

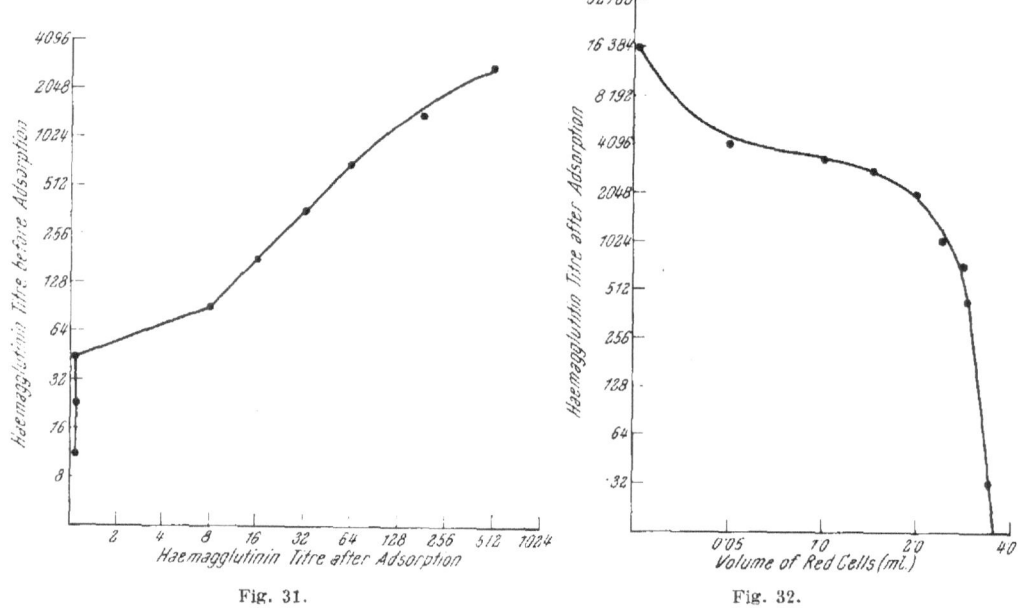

Fig. 31.                                    Fig. 32.

Fig. 31. Adsorption of influenza virus by red cells.
Adsorption of influenza virus (DSP strain) by guinea pig red cells (10⁸ cells/ml).
Fig. 32. Adsorption of virus by red cells (DSP virus).
1.0 ml samples of virus concentrate adsorbed in the cold with varying amounts of guinea pig RBC.

and a range of alkali metal salts. No haemagglutination occurred in isotonic sugar solutions, but agglutination occurred in all the salt solutions at pH 7.2. Some strains did not agglutinate at pH 5.0.

DEREVICI (D₄₈) found that small amounts of nickel, manganese, iron and zinc salts interfere with the haemagglutination reaction and stressed the importance of using pure materials in the tests.

## 5. Mechanism of the Haemagglutination Reaction

HIRST showed that increasing the number of red cells increases the rate of adsorption and the amount adsorbed, but the time of maximum adsorption was independent of the cell concentration.

The work of MAGILL (M₈) on the adsorption of influenza virus by red cells shows this to be a somewhat complex phenomenon. He found that if a given quantity of red cells adsorbed a certain proportion of virus from a suspension, increasing the amount of red cells did not produce a proportional increase in the virus adsorbed. Successive adsorption of a virus preparation with red cells removed successively lower and lower percentages of the available virus. If a sufficiently large amount of cells was used all haemagglutination could be removed, but some virus still remained unadsorbed and could be detected by infectivity tests. The observed data did not fit the Freundlich adsorption isotherm, nor did they fit the law of mass action. It was found that for a given mass of red cells the ratio of adsorbed to unadsorbed virus was a parabolic function of the total quantity of virus. MAGILL's work was done with relatively small amounts of fowl cells and an adsorption period of 30 minutes at 0°C.

The author has studied the adsorption of virus by guinea pig red cells using larger amounts of cells and an adsorption period of 10 minutes. The DSP strain of virus A was used, the virus being partially purified by two cycles of adsorption-elution from red cells. Serial dilutions of virus were made and the haemagglutinin titres measured before and after adsorption (Fig. 31). The curve shows a central region in which there is a linear relation between the titres before and after adsorption, indicating the adsorption of a constant percentage (92%) of

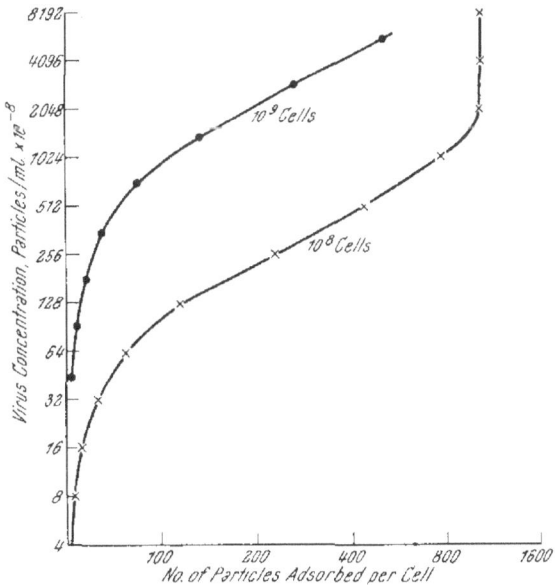

Fig. 33. Adsorption of virus by red cells. Relation between concentration of virus and number of virus particles adsorbed by guinea pig red cells.

the available virus. With large amounts of virus the adsorption percentage is reduced, probably as a result of exhaustion of the available cell receptors. With small amounts of virus the whole of the virus appeared to be adsorbed.

If a constant amount of virus is adsorbed with varying amounts of red cells the result obtained is shown in Fig. 32. Addition of 0.05 ml of cells adsorbed 75% of the virus, but increasing the amounts of cells to 0.2 ml only increased the amount of virus adsorbed to 87.5%. A further increase to 0.4 ml, however, removed all detectable virus. MAGILL obtained similar results.

It appears that the red cells do not adsorb a constant amount of virus. With excess virus the adsorption powers of the red cell become exhausted. With lower virus-cell ratios there is an equilibrium in which the amount of virus adsorbed is mainly dependent on the virus concentration, but this equilibrium is only set up after the cells have adsorbed a minimal amount of virus, so that with large amounts of cells all the virus is apparently adsorbed. The fact that, as shown by MAGILL, some unadsorbed virus can be detected by infectivity tests probably means that the collision rate between virus and cells never quite reaches 100%.

By dividing the volume of the virus sediment deposited on centrifugation from a large volume of purified DSP virus by the calculated volume of one virus particle the author estimated the Salk unit to represent $3.9 \times 10^7$ particles. From this it was possible to determine the number of virus particles adsorbed per red cell under different conditions of cell-virus ratio (Fig. 33). With cell-virus ratio of 1:1 or more all the virus is adsorbed. With lower cell-virus ratios the number of particles adsorbed per cell increases with the virus concentration until a maximal value is reached at which no further adsorption occurs. This saturation dose is about 1100 particles per cell.

WERNER and SCHLESINGER ($W_{37}$) using fowl cells found a maximum adsorption of 1000—1400 particles per cell. BATEMAN, DAVIS and McCAFFNEY ($B_{20}$) showed that active and heated virus are adsorbed by the same receptors and using heated virus they found that red cells were saturated by 10,000 Salk units per ml of 1% cells; assuming the Salk unit to be $5 \times 10^7$ particles they found the saturation dose for heated virus to be 8000 particles per cell. It seems probable, therefore, that with active virus the maximum adsorption depends on a balance between the rate of adsorption and the rate of elution, and BATEMAN et al. calculated velocity constants for the initial reaction of formation of haemagglutinin-receptor complex, and for the secondary breakdown of the complex into free haemagglutinin and receptor decomposition product.

### 6. Effect of Virus on the Red Cells

#### a) The Receptor Gradient

BURNET and his collaborators ($B_{191}$, $B_{215}$) found that while red cells which had been treated with one strain of virus could no longer be agglutinated by the same strain, they might still be agglutinated by a different strain. They suggested that there was a graded arrangement of receptors on the red cell surface. Virus strains could be placed in a series in which treatment of cells with any one virus removed all receptors for that virus and for all viruses preceding it in the series, but not for viruses later in the series. Using human O type red cells they found the following series (O and D indicate original and derived strains in the O-D variation). Mumps, Newcastle disease virus, MEL (A), MIL (AO), WS (A), BEL (AO), IAN (AO), M12 (AD), LEE (B), BEL (AD), IAN (AD), SWINE (A), HUT (B), MIL (B).

HIRST ($H_{139}$) found that all the receptors of fowl red cells could be removed by any strain of virus or by RDE, but some receptors were removed more rapidly than others. For fowl cells treated with RDE the sequence of receptor destruction was PR8, Newcastle disease virus, MEL, WS, mumps, SWINE, LEE.

#### b) Effects on Electrophoretic Mobility

ADA and STONE ($A_{32}$, $S_{278}$) found that the electrophoretic mobility of human O cells was reduced by treatment with influenza viruses, and that viruses could be arranged in a series according to the extent of the effect. The series corresponded closely to the receptor gradient series except that swine and Newcastle disease viruses had a greater effect than would be expected from their positions in the gradient. The results suggested that there were two substrates for the virus enzymes, one being attacked by all the viruses, while the second was attacked only by SWINE, Newcastle disease virus and RDE.

SEREDA ($S_{81}$) found that viruses reduced the electrokinetic potential of guinea pig, human, and chick cells, but not of horse cells which were not agglutinated. The decrease in potential was produced by a dose of virus 3—4 times less than the minimum haemagglutinating dose.

BATEMAN, ZELLNER, DAVIS and McCAFFREY ($B_{21}$) treated guinea pig red cells with PR8 virus. The cells were then washed and suspended in buffered glucose solutions of different pH and ionic strength and electrophoretic mobilities and surface charge densities determined. Mobilities were less negative by some $4 \times 10^{-5}$ sq.cm/V/sec. than controls. The decrease of charge was independent of pH. It was calculated that each receptor consists of one molecule of mucoprotein.

CHESBRO and HADRICK ($C_{39}$) studied the tendency of red cells to stick to glass beads before and after virus action. They suggested that haemagglutination is not due to bridge formation but to the hydrophobic character of the virus-cell complexes which tend to aggregate and to stick to glass under the influence of surface tension.

### c) Other Effects of Virus on the Red Cell

COX, ZERSCHLING and MELROY ($C_{86}$) found that chick red cells treated at $4°C$ with PR8 virus were haemolysed at room temperature by rabbit anti-viral serum and complement. Virus heated at $56°C$ was also effective.

DANON, JERUSHALMY, KOHN and DE VRIES ($D_4$) found that the human red cells carrying adsorbed FM1 virus were phagocyted by human leucocytes in the presence of human serum.

SVET-MOLDAVSKIY and LOZOVSKAYA ($S_{342}$) inoculated chickens and guinea pigs parenterally with virus and found that their red cells showed a decreased agglutinability by virus 2—10 days later.

### 7. Nature of the Red Cell Receptors

HIRST ($H_{135}$) found that the receptors of fowl red cells were stable at high temperatures and in solutions of pH as high as 10. They were resistant to several oxidising agents; ferricyanide, permanganate, dichromate, iodine and hydrogen peroxide; but were destroyed by periodate and by trypsin. He suggested that the receptors were mucoproteins.

FAZEKAS DE ST GROTH ($F_{24}$) found that if red cells were treated with an equal volume of M/800 periodate they were still agglutinated by virus but the virus did not elute. If the residual receptors on such cells were destroyed by treatment with RDE they were no longer agglutinated. Such "PVR" cells still carried firmly adsorbed virus. They were not agglutinated by virus but were agglutinated by specific antisera to the virus strain used. They also agglutinated normal red cells and destroyed the cell receptors. There are probably at least two types of receptor in the red cell ($S_{278}$, $M_{122}$), some viruses attacking both types and some only one.

TAKATSY, BARB and FARKAS ($T_6$) found that the red cell receptors for avid A2 strains were different from those reacting with non avid A2, A1, A, and B viruses in that they were resistant to the action of RDE, periodate and other influenza viruses.

## 8. Agglutination of Red Cells of Different Animal Species

Extensive studies of the agglutination of the red cells of different animal species have been made by SHUBLADZE and SOLOV'YEV ($S_{128}$), CLARK and NAGLER ($C_{59}$), CHU ($C_{50}$), SELIMOV ($S_{76}$), WANG and LIN ($W_{17}$) and SCHMIDT and GROSSGEBAUER ($S_{66}$,), and observations have been made by many other workers ($T_{72}$, $T_{14}$, $O_{21}$, $B_{101}$, $Y_2$). The results of different workers are not in complete agreement, and it appears that there may be differences in behaviour of different batches of cells from the same animal species, differences between different strains of virus, and especially differences due to the use of different temperatures in the haemagglutination test ($O_{21}$, $T_{14}$). HORVATH, SZOLLOSY and IVANOVICS ($H_{175}$) found that the agglutination of red cells of sixteen vertebrate species by influenza viruses at 2°C was parallel to their ability to adsorb virus at this temperature but the agglutination at 20°C had no direct relation to the ability to adsorb virus in the cold.

The red cells of man, monkey, guinea pig, dog, ferret, hedgehog, squirrel, rat, mouse, frog, fowl, duck and most other birds are agglutinated at both 4°C and 20°C by almost all A and B viruses. Some strains of virus A may fail to agglutinate monkey cells ($W_{17}$) and difficulties may be experienced with rat and mouse cells which may give unsatisfactory patterns in Salk tests ($M_{29}$).

In general horse, ox, sheep, goat, pig, rabbit, cat and hamster cells are agglutinated by A and B viruses at 4°C but not at 20°C, and by A2 viruses at both 4°C and 20°C. Failure to agglutinate at 20°C is probably due to very rapid elution, and virus heated at 56°C for 30 minutes usually agglutinates these types of cell ($C_{50}$).

The red cells of the ground squirrel are agglutinated by B and C viruses but not by A ($B_{101}$). Influenza C viruses have very restricted powers of haemagglutination. Only fowl, mouse, rat, frog and ground squirrel cells are agglutinated and these only in the cold ($W_{17}$).

Different results may be obtained with freshly isolated strains of virus as compared with the same strains after adaptation. Strains of virus A on first isolation from man agglutinate human and guinea pig cells to much higher titre than fowl cells, but on adaptation to the egg the differences disappear. Strains of virus B agglutinate fowl and guinea pig cells equally even on first isolation.

It has often been found that widely different results may be obtained in haemagglutinin and haemagglutinin-inhibition tests with the red cells of different fowls ($S_{284}$, $M_{89}$, $D_{44}$, $P_{31}$). Sheep cells also vary, some being totally inagglutinable, some only agglutinate at 4°C, and some at both 4°C and 20°C ($M_{29}$). Rabbit cells vary considerably in adsorptive powers and agglutinability ($H_{175}$). Human cells, however, give the same titre whether the cells are of groups O, A, B or AB, and after virus treatment the cells still agglutinate normally with their appropriate antisera ($M_{39}$). The red cells of guinea pig in which an experimental haemolytic anaemia has been produced by inoculation of specific rabbit haemolytic serum adsorb more influenza virus than normal cells ($XC_2$).

## 9. Relation between the Number of Virus Particles Present in a Virus Suspension and the Haemagglutinin Titre

Levine, Puck and Sagik ($L_{67}$) and Horsfall ($H_{165}$) claimed that the number of virus particles present in a suspension could be directly determined from the haemagglutinin titre obtained in the "absolute assay" densitometric method. The claim is based on the assumptions that at the end point all the virus is adsorbed on the cells, and that each particle acts as a bridge linking two red cells to form a dimer. The first assumption is approximately true, as with excess cells the adsorption of virus approaches 100%. The second assumption is probably not true. It involves a collision rate between the red cells which is so high that in the duration of the test any cell carrying a virus particle will collide with another cell at exactly the point at which the virus particle is present, and that any bridge so formed is strong enough to resist mechanical forces tending to separate the cells, and to overcome the electric repulsion between the cells. It is much more probable that a stable dimer is formed only when several particles have formed bridges between the cells and when the electric charge on the cells has been sufficiently reduced. Tyrrell and Valentine ($T_{81}$) compared direct virus particle counts made by electron microscopy using both the red cell adsorption method ($D_{36}$, $D_{60}$) and the spray method using mixtures of virus and latex spheres, with the "absolute assay" haemagglutinin count and found that the haemagglutination method underestimated the number of particles by a factor of 6 to 20 fold.

In the Salk or pattern type of haemagglutination test the end point is certainly not entirely determined by the bridging effect, another factor is the reduction of electric charge on the red cell leading to a tendency to adhere to glass. Fazekas de St Groth and Cairns ($F_{33}$) found that at the end point in the Salk test only 15% of the cells were present as dimers. Gard ($G_5$) found the partial haemagglutination end point in the Salk test to correspond to one particle per cell, and by comparing electron microscope counts with the results of Salk tests Donald and Isaacs ($D_{61}$) and Tyrrell and Valentine ($T_{81}$) found values ranging from 0.5 to 2 particles per cell. By comparing the results of Salk tests with an assessment of the number of particles made by measurement of the volume of virus deposited by centrifugation from purified preparations the author estimated the 50% end point in the Salk test to represent 1.5 particles per cell.

Great caution is necessary in accepting the results of these quantitative assessments, and while results obtained with highly adapted strains of virus, in which the great majority of the particles are spherical, are probably reasonably accurate the results should never be applied to virus strains of filamentous morphology, or to virus-cell systems other than those in which the measurements were made.

# XV. Enzymic Activity of Influenza Viruses

The original studies of Hirst ($H_{125}$, $H_{126}$, $H_{129}$) demonstrated the enzymic nature of the action of influenza viruses on red blood cells and indicated the presence of a receptor substance on the red cell surface which was destroyed by virus action. The receptor was shown to be heat resistant and to be destroyed by trypsin and by periodate and HIRST suggested that it was probably a muco-protein ($H_{135}$, $H_{136}$).

Following the discovery by FRANCIS ($F_{94}$) of a haemagglutinin inhibitor in normal serum a very large range of mucoproteins from many sources were found to possess the ability to inhibit haemagglutination and these "Francis inhibitors" were inactivated by the enzymic action of the virus.

## 1. Nature of the Virus Receptor

GOTTSCHALK and LIND ($G_{61}$, $G_{62}$) found a haemagglutinin-inhibitor in the ovomucin fraction of egg white. After treatment with MEL virus dialysis yielded a carbohydrate-peptide complex. The carbohydrate appeared to be an oligo-saccharide containing one or more N-acylhexosamine residues linked by an alkali-labile glycosidic linkage to a non-amino sugar. Later work ($G_{53}$) with a mucoprotein from urine suggested that the virus substrate was a polysaccharide containing hexoses, methylpentose, hexosamines and 2-carboxy-pyrrole, and that the compound released by the virus enzyme consisted of 2-carboxy-pyrrole and a single sugar linked by an N-glycosidic bond.

KLENK, FAILLARD and LEMPFRID ($K_{59}$), treating urinary mucin with LEE virus, obtained the split product in crystalline form and identified it as N-acetyl-neuraminic acid. FAILLARD ($F_7$, $F_8$) isolated N-acetyl-neuraminic acid from uri-nary mucin and from bovine submaxillary gland mucin after treatment with the receptor-destroying enzyme of *V. cholerae*, and KLENK and LEMPFRID ($K_{60}$) obtained N-acetyl-neuraminic acid from human red cells treated with RDE.

GOTTSCHALK and GRAHAM ($G_{55}$, $G_{59}$, $G_{60}$) showed that the neuraminidase susceptible prosthetic group of bovine salivary mucoprotein was a disaccharide, 6-a-D-sialyl-N-acetylgalactosamine, and that the virus enzyme and RDE split the disaccharide into N-acetyl-neuraminic acid and N-acetyl-galactosamine.

LAUCICOVA ($L_{30}$) showed that N-acetyl-neuraminic acid was released from a haemagglutinin inhibitor in chorioallantoic membrane by the action of virus or RDE.

CARUBELLI, BHAVANANDAN and GOTTSCHALK ($C_{16}$) obtained sialoglycopeptides from ovine submaxillary glycoprotein by digestion with pronase, and after removal of N-acetyl-neuraminic acid with neuraminidase, were able to show that the N-acetyl-galactosamine component of the prosthetic group was linked to the hydroxyl groups of serine and threonine.

## 2. The Receptor Destroying Enzyme of V. cholerae, RDE

BURNET and STONE ($B_{220}$) found that filtrates of *V. cholerae* cultures contained an enzyme which destroyed the virus receptors of red blood cells, and this material, RDE, has been much used in influenza virus research. A good yield of the enzyme is obtained by growth of the vibrio on soft agar plates or in the allantoic sac of fertile eggs. It is thermolabile, being destroyed in the absence of calcium ions at 52.5°C in 30 minutes, but in the presence of M/100 calcium it is only inactivated at 60°C. It is, therefore, possible to sterilise the material without filtration by heating in the presence of calcium. The enzyme can be purified by adsorption to red cells in the cold followed by elution at 37°C.

Neuraminidases are produced by other bacteria, pneumococci, proteus and *C. welchii* ($C_{51}$, $B_{103}$, $B_{104}$, $B_{97}$). LAVER, PYE and ADA ($L_{42}$) purified the neuraminidase of the 42 strain of *V. cholerae* and estimated it to have a molecular weight of 90,000.

## 3. Estimation of Neuraminidase Activity

The neuraminidase activity of a virus strain may be estimated by allowing the strain to act on a suitable substrate for a selected period and estimating the amount of neuraminic acid released. Any suitable mucoprotein may be used as substrate, but sialyl lactose is often used. Neuraminic acid is usually estimated by the method of AMINOFF ($A_{56}$).

The sample (0.5 ml) is incubated at 37°C for 30 minutes with 0.2 ml of a 25 mM solution of periodate in 0.125N $H_2SO_4$. Then 0.2 ml of 2% sodium arsenite in 0.5N HCl is added, followed by 2 ml of 0.1 M thiobarbituric acid adjusted to pH 9.0 with NaOH. The mixture is heated at 100°C for $7\frac{1}{2}$ minutes, cooled in ice and the colour extracted with 5 ml of butanol containing 5% of 12N HCl. The colour is measured at 590 mμ and compared with standard N-acetyl-neuraminic acid controls.

### a) Estimation by Determination of the Rate of Elution from Red Cells

The neuraminidase activity of virus strains may be compared by determining the rate at which they elute from red cells. Preparations with an equal haemagglutinin titre are adsorbed with a suitable amount of red cells, the mixture incubated at 37°C and samples removed at intervals, centrifuged, and the haemagglutinin titre of the supernatant measured ($S_{171}$). A more satisfactory method, used by HOYLE and HANA ($H_{208}$) is to adsorb the virus preparations at 0°C under conditions in which the virus is present in large excess so that the red cells are saturated with virus. The cells are then centrifuged out and washed with ice-cold saline, resuspended and incubated at 37°C and the haemagglutinin titre of the supernatant measured at intervals.

In most cases elution is an indication of neuraminidase activity, but the two are not always in complete correlation ($S_{87}$). Some strains may fail to elute even though they release neuraminic acid from the cells and strains of equal ability to release neuraminic acid may elute at different rates ($T_{65}$). The elution of virus from cells which occurs at $0°C$ in the absence of ions is not associated with receptor destruction ($B_{16}$).

Some chemical agents such as phloridzin ($K_{57}$) and phenylisothiocyanate ($H_{208}$) may prevent elution while having no effect on neuraminic acid release.

CARLINFANTI ($C_{13}$) noted that in Salk tests the time taken for the dissociation of agglutinated cells on the bottom of the tubes to occur varied with different viruses and with red cells of different species.

### b) Other Methods of Estimating Enzymic Activity

McCREA ($M_{60}$) found that solutions of sheep salivary gland mucoproteins developed a turbidity when mixed with acidified horse serum. Previous incubation of the mucoproteins with virus or RDE prevented this effect, and the turbidity-reducing activity could be used as a measure of enzymic activity. STEWART and QUILLIGAN ($S_{267}$) found that human O cells treated with virus or RDE developed antigenic changes and the "cell modifying activity" of virus fluids could be estimated.

### 4. Mechanism of Neuraminidase Action

LANNI and LANNI ($L_{20}$) studied the reaction of egg adapted SWINE virus with egg white mucoprotein. They concluded that nearly every collision between virus and mucoprotein molecule results in combination. The mucoprotein appeared to consist of a protein core with possibly 200 polysaccharide groups distributed over its surface, and dissociation of virus from the inhibitor did not occur until most of these groups had been destroyed. The progressive loss of the terminal groups of the polysaccharide produced a progressive decrease in affinity of the mucoprotein molecule for virus.

MATHIEU, COLOBERT, CREACH and FONTANGER ($M_{46}$) studied the kinetics of hydrolysis of purified 3 lactose lactaminic acid from cow colostrum by LEE virus. There was a linear relation between the initial velocity of reaction and the concentration of virus over the range $0-128$ haemagglutinin units.

The optimum pH was 6.8 and the Michaelis constant $2.75 \times 10^{-4}$ M. The reaction velocity was reduced by N-acetyl-neuraminic acid and by copper ions.

ISHIDA and TOZAWA ($I_{40}$) found the optimum pH to vary with different viruses, the highest activity being usually between pH 4.5 and 6.0.

NIHOUL ($N_{25}$) found the rate of destruction of ovomucin inhibitor by virus to be proportional to the concentration of both components and to the temperature between $4°C$ and $37°C$.

Virus neuraminidase activity is usually destroyed by heating at $56°C$ for 30 minutes, while the haemagglutinin is more resistant ($H_{136}$), but the neuraminidase of A2 and swine viruses is said to be more heat-resistant than the haemagglutinin ($S_{87}$).

The neuraminidase activity of RDE of *V. cholerae* requires the presence of calcium ions ($E_{26}$, $P_{73}$, $N_{23}$) but the influenza virus enzyme is unaffected by the absence of Ca ions.

### 5. Purification of Virus Neuraminidase

LAVER ($L_{39}$) found that disruption of LEE virus with sodium dodecyl sulphate released neuraminidase which could be separated as a minor component on electrophoresis. The neuraminidase had a sedimentation coefficient of 9S. LAVER suggested that the neuraminidase and haemagglutinin activities of the virus were present in separate co-valently bonded structures. The enzyme can also be separated from the virus particle by trypsin or chymotrypsin treatment ($M_{51}$, $N_{27}$).

SETO, DRZENIEK and ROTT ($S_{86}$) separated neuraminidase from A2 virus by treatment with ether and pronase and isolated it on a Sephadex G 200 column. The sedimentation coefficient was 8.8S. The material appeared to be electrophoretically heterogeneous when tested by disc electrophoresis and disc immunodiffusion methods.

## 6. Variations in the Neuraminidase Activity of Different Strains of Virus

The work of BURNET and his colleagues on the receptor gradient indicated that there were great differences in the enzymic activity of different strains of virus. STONE ($S_{275}$) showed that virus strains could be arranged in a gradient according to their action on various mucoprotein inhibitors, and that the order varied with different mucoproteins. Differences occurred in the same strain of virus in the course of adaptation to eggs ($S_{276}$). SMITH and COHEN ($S_{171}$) showed that the rate of elution of different strains from red cells showed great variation, and that even in a single allantoic fluid the particles present were heterogeneous in respect to eluting ability and that they could be fractionated into rapidly eluting, slow eluting and non-eluting forms. SETO and others ($S_{89}$) correlated neuraminidase activity with particle counts and found strains to differ by as much as 100 fold. A2 strains were more active than A or A1 strains, and neurotropic strains were apparently devoid of enzymic activity ($S_{87}$). JAMESON and LEVINE ($J_3$) showed that neurotropic WS strains were unable to attack pig submaxillary gland mucin while WS strains were able to do so. This mucin contains 90% glycolyl neuraminic acid and only 10% of acetyl neuraminic acid, and using other mucins it was shown that the neurotropic strains did not attack the glycolyl form.

BOLDESOV ($B_{98}$) found neuraminidase activity greater in inhibitor resistant than in inhibitor sensitive strains though both were able to destroy ovomucin inhibitor. Strains of A2 virus which were avid or non avid to antibody had the same enzymic activity.

STONE and ADA ($S_{277}$) found that if red cells were treated successively with two viruses the electrophoretic mobility was reduced to the value characteristic of the more active virus alone. They suggested that most influenza A strains have one enzyme, that swine and Newcastle disease viruses have 2 and RDE consists of 3 enzymes.

## 7. Relation of Neuraminidase to Haemagglutinin

There can be no doubt that the virus neuraminidase is an integral part of the virus particle ($G_{63}$) and the results of ether fractionation indicate that it is part of the haemagglutinating subunit. However, by various methods neuraminidase can be separated from haemagglutinin in some strains of virus ($M_{51}$, $N_{27}$, $R_{16}$, $L_{39}$).

HOWE, ROSE and LEE ($H_{183}$) tested various haemagglutinin inhibitors and found no relation between the inhibition and the content of N-acetyl-neuraminic acid. Some mucoids such as amniotic mucoid and serum ovosomucoid were rich in N-acetyl-neuraminic acid and susceptible to the virus enzyme but had no haem-agglutinin-inhibiting activity. They concluded that haemagglutinin and neuramin-idase were distinct independent attributes of the virus particle.

Antisera to neuraminidase can be produced by immunisation of rabbits especi-ally with the use of oily adjuvants ($B_{171}$, $M_7$, $S_{85}$, $A_{22}$). Bacterial and virus neura-minidases are serologically distinct. SETO ($S_{85}$) found that antisera to purified virus neuraminidases contained some haemagglutinin-inhibiting antibody which was attributed to the presence of some residual V antigen in the purified prepara-tions. FAZEKAS DE ST GROTH ($F_{32}$) found that specific antibody to the V antigen inhibited neuraminidase activity when tested against macromolecular substrates of molecular weight 100,000 or more, was somewhat less active against muco-proteins of molecular weight 10,000 and almost ineffective against sialyl lactose of molecular weight 633. He concluded that the effect was due to steric inhibition and was dependent on the distance between antigenic determinants and enzymic-ally active centres on the virus particle and on the shape and size of the substrate and antibody molecules.

HOYLE and HANA ($H_{208}$) attempted to identify the amino acids present in the active centres of the haemagglutinin and neuraminidase of A and B virus by treatment of virus preparations with a range of chemical reagents acting on the chemically reactive groups of protein molecules. They found that haemagglu-tinin and neuraminidase were unaffected by agents acting on sulphydryl or amino groups, but that agents acting on disulphide bridges, the phenolic hydroxyl group, the imino group of histidine, the phenol or imidazole rings, and the amide group, produced reduction or destruction of one or both the activities. The results suggested the presence in virus A of two different active centres, both of which possessed haemagglutinating activity, but only one having neuraminidase activity. Virus B also contained two centres, one similar to the neuraminidase-active centre of virus A, and a second centre which possessed neuraminidase activity but did not haemagglutinate. Virus A thus contained two haemagglutinating centres but only one neuraminidase, while virus B contained two neuraminidases but only one haemagglutinin centre. BLOUGH ($B_{90}$) found that glutaraldehyde in-activates both the haemagglutinin and neuraminidase of PR8 virus, but with LEE and MEL viruses the haemagglutinin was destroyed, but 50% of the neuramin-idase activity was retained.

LAVER and KILBOURNE ($L_{41}$) carried out a recombination experiment using the NWS strain of virus A and an A2 virus, A2/RI/5$^-$. They isolated a recombinant strain X7 which was shown by immunologic and by peptide mapping methods to possess the haemagglutinin of the original A strain and the neuraminidase of the A2 strain. The recombinant eluted much more rapidly than either parent strain.

## 8. Relation of Neuraminidase to the Virus Growth Cycle

The possibility that the virus neuraminidase plays some part in the process of entry of virus into the cell has often been considered ($G_{56}$). However, removal of the surface receptors of the cell by RDE does not apparently damage the cells so that it is unlikely that penetration of virus into the cell is a result of neuramini-dase action. It is possible that a highly active neuraminidase might result in virus particles becoming released from the cell after the initial adsorption before penetration could occur.

Another possibility is that neuraminidase plays some role in the release of virus from the infected cell. NOLL, AOYAGI and ORLANDO ($N_{26}$) found that in eggs inoculated with LEE virus no neuraminidase could be detected in membrane extracts 4 hours after inoculation, but that the enzyme appeared at 6 hours and rapidly increased from that time onwards. Release of virus from the cell occurred suddenly between 8 and $9\frac{1}{2}$ hours. KELLY and GRIEF ($K_{28}$) found that if mice were inoculated intracerebrally with a non-neurotropic strain of virus there occurred an increase in viral neuraminidase reaching a peak at 36 hours. No infective virus was released from the cells but the mice died from toxic effects which it was thought might be a result of action of the enzyme.

ADA and LIND ($A_{20}$) made the important observation that neuraminidase could be detected in small quantities in normal chorioallantoic membranes, and suggested that virus infection might stimulate a change in cell metabolism leading to a great increase in neuraminidase production.

# XVI. Inhibitors of Haemagglutination

In 1946 McCREA ($M_{56}$) described a heat-labile haemagglutinin inhibitor present in normal rabbit serum, and in 1947 FRANCIS discovered a component of normal human serum which strongly inhibited haemagglutination by heated LEE virus while having little or no effect on unheated virus ($F_{94}$). BOVARNICK and DE BURGH ($B_{116}$) found a haemagglutinin inhibitor in lipid extracts of human red blood cells. Later studies by HIRST ($H_{135-137}$), BURNET and his colleagues ($B_{214}$, $A_{61}$, $M_{57}$, $B_{167-169}$, $A_{63}$, $B_{172}$, $B_{173}$, $B_{177}$), GINSBERG and HORSFALL ($G_{29}$), SMITH and WESTWOOD ($S_{174}$, $S_{175}$), CHU ($C_{53}$), and SAMPAIO and ISAACS ($S_{18}$, $S_{16}$) showed that haemagglutinin inhibitors were present in most animal sera and in mucoproteins from many sources. These inhibitors are of two types, the Francis or $\alpha$ inhibitor present in the mucoprotein components of serum and in other mucoproteins; and a heat labile protein, the Chu or $\beta$ inhibitor, present in most animal sera.

## 1. The Francis or $\alpha$ Inhibitors

Francis type inhibitors have the following characteristics. They are powerfully inhibitory to haemagglutination by heated influenza viruses but have only slight action against unheated viruses. They are destroyed by live influenza viruses, by crude *V. cholerae* filtrates and by purified receptor destroying enzyme (RDE). They are usually heat stable, resisting 100° C for variable periods. They are destroyed by periodate. Resistance to trypsin and other proteolytic enzymes is variable. They do not neutralize virus.

All Francis inhibitors are mucoproteins containing N-acetyl-neuraminic acid and the inhibiting effect is due to competition for the virus haemagglutinin between the inhibitor and the red cell receptors.

Francis inhibitor can be detected and measured by determining its ability to prevent haemagglutination by LEE virus which has been heated at 56° C for 30 minutes. Such virus is called "indicator virus" and conversion of virus by heat into the indicator state is usually believed to be a result of the destruction of the virus neuraminidase ($S_{274}$). This concept has been criticised by SMITH and WESTWOOD ($S_{175}$) and it is probable that other factors may be involved in the conversion to the indicator state.

Strains of virus differ in the ease with which they are converted to the indicator state by heating. In addition to their presence in human and animal sera, inhibitory mucoproteins have been obtained from many sources and have been extensively studied (Table 4).

Table 4. *Sources of Mucoproteins with Haemagglutinin-inhibiting Properties of the Francis Type*

Egg allantoic fluid — : $S_{333-339}$, $H_{39}$, $L_{22}$, $N_{16}$

Egg chorioallantoic membrane — : $S_{51}$, $S_{52}$, $C_6$, $L_{28}$, $L_{29}$, $A_{42}$, $K_{104}$

Egg white (ovomucin) — : $L_{26}$, $E_{22}$, $G_{61}$, $S_{274}$, $C_{88}$, $L_{23-25}$, $S_{94}$, $F_{40}$, $G_{63}$, $K_{99}$, $N_{24}$, $N_{25}$, $L_{20}$.

Human red cells — : $H_{135}$, $D_{38}$, $M_{59}$, $H_{184}$, $R_{49}$

Urine — : $C_{93}$, $C_{94}$, $XL_2$

Salivary glands of sheep and ox — : $M_{58}$, $C_{98}$, $G_{60}$, $G_{57}$, $C_{16}$

Saliva, sputum, tears and nasal secretions — : $S_{15}$, $C_{97}$, $M_{43}$, $A_{78}$, $B_{192}$

Lungs and other animal tissues — : $F_{147}$, $M_{47}$, $S_{348}$, $L_{109}$

Milk — : $L_{21}$, $S_{137}$

Bile — : $T_1$

Bovine amniotic fluid — : $L_{134}$

Ovarian cyst — : $S_{274}$, $B_{167}$, $B_{169}$

Human meconium — : $C_{96}$, $Z_{44}$

Vaginal mucus — : $B_{45}$

Seaweed (carrageenin) — : $G_5$, $K_{20}$

Edible bird nest — : $B_{67}$

### Chemistry of the Francis Inhibitors

All Francis inhibitors are mucoproteins, but in many cases they appear to be combined with lipid. Lipomucoprotein inhibitors are present in human and fowl red blood cells ($B_{116}$, $H_{160}$, $M_{59}$) in chorioallantoic membrane ($L_{28}$), mouse lung ($L_{109}$) and ox brain ($R_{59}$). HANA and STYK ($H_{20}$, $H_{21}$) found that the inhibitory activity of guinea pig serum against A2 viruses, and of rat serum against C viruses was increased by removal of the lipid, probably as a result of an increase in reacting groups following the separation of glycoprotein from combination with lipid.

Most inhibitors, however, appear to be mucoproteins, usually of high molecular weight. The inhibitor of normal egg allantoic fluid has a sedimentation constant of 200 S and is highly viscous ($S_{339}$). The egg-white inhibitor is present in the ovomucin fraction ($N_{24}$) and is a highly viscous mucoprotein with a sedimentation constant of 31—37 S ($L_{26}$). It appears as a fibrous material on electron microscopy ($S_{94}$). It contains 26% polysaccharide and 73% protein ($K_{99}$) and LANNI and LANNI ($L_{20}$) suggest that it consists of about 200 polysaccharide groups attached to a protein core. Treatment with virus greatly reduces the viscosity ($E_{22}$).

Ovine submaxillary gland mucoprotein has a molecular weight of 1 million and contains 41% carbohydrate and 59% protein. The carbohydrate is present in the form of about 800 disaccharide units ($G_{57}$). Sheep salivary gland mucoprotein is relatively less viscous and has a molecular weight of 87,000.

The inhibitory power may depend on the molecular weight of the inhibitor. Thus orozomucoid from the urine of nephrotic children is only a weak inhibitor, but its polymer with a molecular weight of 1 million inhibits strongly ($XL_2$). FAZEKAS DE ST GROTH and GOTTSCHALK ($F_{41}$) consider that the inhibitory potency of glycoproteins depends on the number of sialic acid residues attached simultaneously to complementary groups on the virus surface, depending in turn on the number of residues per molecule of glycoprotein and the steric fit to the virus surface.

Mixtures of influenza virus with inhibitor may give precipitates ($S_{336}$, $S_{338}$) which with active virus redissolve on incubation. With heated virus an equilibrium exists between the free reactants and the virus-glycoprotein complex ($F_{41}$).

The whole of the N-acetyl-neuraminic acid groups of mucoproteins are not removed by virus action, and the activity of different strains of virus varies, so that they can be arranged in a linear gradient of increasing activity. The gradient differs with different inhibitors ($B_{173}$, $L_{25}$, $C_{98}$, $Z_{44}$, $G_{58}$).

The inhibitor of allantoic fluid can be deproteinised with phenol with the release of an active polysaccharide ($N_{16}$), but the egg-white inhibitor is destroyed by phenol. Resistance to trypsin is variable. Inhibitors of serum and sheep salivary gland are destroyed by trypsin, while the inhibitors of human meconium, of mouse lung, and human red cells are resistant. Allantoic fluid inhibitor is only slightly affected by trypsin unless the fluid is first dialysed, when it becomes sensitive ($H_{39}$).

Periodate in minimal amounts increases the activity of inhibitors, but larger amounts inactivate them ($B_{172}$).

## 2. The Chu or β Inhibitors

Chu inhibitors are highly active against unadapted strains of virus, but have little action against mouse-adapted lines. They are equally active against live and heated viruses, and they neutralize the virus. They are heat labile, being destroyed by heating at 65°C for 30 minutes. They are destroyed by trypsin, but are unaffected by periodate or purified RDE, and are not destroyed by virus. They are inactivated by crude cholera filtrate owing to the presence of a trypsin-like enzyme.

The presence of heat-labile haemagglutinin inhibitors in serum was first noted by HIRST ($H_{126}$), and McCREA described a heat labile inhibitor in normal rabbit serum. CHU ($C_{53}$) described the properties of the heat-labile inhibitors of a variety of animal sera, noting the resistance to periodate and RDE, and the relative insusceptibility of mouse-adapted strains. The inhibitor was adsorbed by kaolin. Calcium ions are necessary for the action of CHU inhibitors. Inhibitory titres are usually much lower than with Francis-type inhibitors ($A_{61}$, $C_{29}$). GINSBERG and HORSFALL noted that neutral mixtures of virus and inhibitor did not become infective on dilution ($G_{29}$). SMORODINTSEV and SHISHKINA ($S_{207}$) found great differences in the virus neutralizing power of different types of normal animal sera. GINSBERG and HORSFALL showed that the heat labile inhibitor was not complement as it was not reduced by adsorption with an antigen-antibody complex. HENRY and YOUNGER ($H_{106}$) found that the inhibitor would neutralize the cytopathogenic effects of influenza strains in monkey kidney cultures. SAMPAIO ($S_{18}$, $S_{16}$) found that the serum content of Chu inhibitor could be titrated by using the A/England/1/51 strains and the same strain after adaptation to mice, while Francis inhibitor could be titrated with heated influenza B strains. FURESZ ($F_{167}$) found that Chu inhibitor was adsorbed by A1 strains but not by B, while Francis inhibitor was adsorbed by B but not by A1 strains. SMITH, WESTWOOD and BELYAVIN ($S_{176}$) noted the great susceptibility of A1 strains to $\beta$ inhibitor of normal rabbit serum.

### 3. Other Types of Haemagglutinin Inhibitors in Serum

Several groups of workers have observed the presence in horse and guinea pig sera of a non-specific inhibitor active almost exclusively against Asian A2 viruses ($S_{100}$, $T_4$, $C_{70}$, $V_{26}$, $C_{69}$, $C_{75}$). This is usually referred to as $\gamma$ inhibitor.

#### a) Properties of $\gamma$ Inhibitor

The inhibitor is present in normal horse serum and is of extremely high titre. It is active against both live and heated A2 viruses but has no inhibitory action against heated LEE virus. It is heat stable and resistant to RDE but is inactivated by trypsin and periodate. It neutralises A2 virus, and intranasal inoculation of horse serum protects mice for 6 hours against subsequent inoculation of live A2 virus ($C_{69}$, $L_{93}$).

The $\gamma$ inhibitor of horse serum causes flocculation of A2 viruses ($B_{69}$). Flocculation may also be given by strains of virus not sensitive to $\gamma$ inhibitor, and such strains adsorb the inhibitor from serum.

The reaction of $\gamma$ inhibitor with virus is very similar to that of specific antibody, the virus-inhibitor complex dissociating on dilution ($S_{344}$). Purified $\gamma$ inhibitor appears to be a glycoprotein containing sialic acid, and RDE removes $40-50\%$ of the sialic acid present without any loss of inhibitory or neutralizing power ($K_{105}$). The inhibitor contains $12\%$ of polysaccharide, consisting of galactose, mannose, glucosamine and sialic acid ($B_{111}$).

Human sera apparently do not contain $\gamma$ inhibitor, but inhibitors of this type are present in horse, guinea pig, rabbit and pig sera ($C_{76}$, $C_{71}$, $B_{42}$, $K_{105}$, $L_{96}$).

#### b) The Influenza C Inhibitor of Rat Serum

STYK ($S_{297}$) discovered in rat serum a powerful inhibitor of haemagglutination by influenza C virus, only active if chick red cells were used. A and B viruses were unaffected but there was some inhibition of mumps and NDV viruses.

The C inhibitor is a heat stable glycoprotein, resistant to periodate and RDE, but destroyed by trypsin ($S_{303}$, $K_{105}$, $L_{96}$).

### 4. Electrophoretic Analysis of Serum Inhibitors

By starch zone electrophoresis TYRRELL ($T_{73}$) showed that Francis inhibitor is present in the $\alpha$ globulin fraction of the serum, $\beta$ inhibitor in the slow-moving

$\beta$ globulin, and specific antibody in the $\gamma$ globulin fraction. HARBOE, REENAAS and OPPEDEL ($H_{37}$) by similar methods found the $\alpha$ inhibitor to be electrophoretically heterogeneous, appearing in several serum fractions. By fractional precipitation methods COHEN and BELYAVIN ($C_{71}$) found the Francis inhibitor to have three components, and to be present mainly in the pseudoglobulin and albumin fractions.

BURNET and McCREA ($B_{214}$) found the $\beta$ inhibitor of ferret serum to be present in the gamma globulin fraction. ACKERMANN and DINKA ($A_9$) studying the $\beta$ inhibitor of bovine serum found no activity in the 7S gamma globulin fraction, but activity was present in macroglobulin fractions in which early or primary antibodies are usually found. KRIZANOVA and SOKOL ($K_{103}$) found that purified $\beta$ inhibitor migrated between the gamma and beta globulins and had a sedimentation constant of 4S.

COHEN and BELYAVIN ($C_{71}$) found that the $\gamma$ inhibitor of rabbit serum was found in the $\gamma$ globulin fraction. The $\gamma$ inhibitor of horse serum consists of two components, a glycoprotein with a sedimentation constant of 3.66S migrating

Table 5. *Distribution of Inhibitors to Egg Adapted Viruses in Animal Sera*

| Francis inhibitor | | Chu inhibitor | | $\gamma$ inhibitor | | C inhibitor |
|---|---|---|---|---|---|---|
| Ferret | $+++$ | Rabbit | $++$ | Horse | $+++$ | Rat $+++$ |
| Fowl | $++$ | Guinea pig | $++$ | Rabbit | $++$ | |
| Rabbit | $++$ | Ox | $++$ | Guinea pig | $++$ | |
| Guinea pig | $++$ | Sheep | $++$ | Pig | $++$ | |
| Human | $++$ | Human | $++$ | Human | $-$ | |
| Mouse | $+$ | Mouse | $++$ | | | |
| Cat | $-$ | Ferret | $++$ | | | |
| Hamster | $-$ | Cotton rat | $++$ | | | |
| Rat | $-$ | Fowl | $--$ | | | |
| | | Hamster | $+-$ | | | |
| | | Rat | $-$ | | | |

with the $\alpha$ globulins, and a component migrating between the $\beta$ and $\gamma$ globulins. The second component was separated by sedimentation into a major fraction of 14.7S and a minor 5.9S component ($K_{107}$). By separation in carboxymethyl cellulose three fractions could be obtained from the $\gamma$ inhibitor of horse serum ($K_{102}$).

## 5. Nature of the Non-specific Inhibitors of Normal Sera

In addition to the studies described above many other extensive observations have been made of the incidence and properties of serum inhibitors of haemagglutination ($B_{15}$, $B_{102}$, $B_{138}$, $F_{17}$, $J_2$, $J_4$, $XT_1$, $V_{26}$, $A_{49}$). The results indicate an extraordinary complexity in the reactions. There would appear, however, to be two major types of inhibitor. The first type are mucoproteins or lipomucoproteins which possess receptors for the virus haemagglutinin and therefore compete with the red cell for attachment to virus. If all the receptors are susceptible to the virus neuraminidase the inhibitor will be destroyed by virus and will not neutralise it (Francis inhibitor). If, however, the haemagglutinin receptors are not all sensitive to the virus neuraminidase, possibly because of unsatisfactory steric relations, then the inhibitor-virus complex will not dissociate and virus will be neutralized ($\gamma$ and C inhibitors).

The second type of inhibitor, Chu inhibitor, does not depend for its activity on the presence of polysaccharide and is probably best regarded as a natural antibody, possibly related to the heterophile antibodies and reacting with some relatively non-specific components of the virus particle. This is suggested by the fact that the activity of Chu inhibitor does not only depend on the strain of virus but on the host cell in which the virus has been grown.

The distribution of the various types of serum inhibitor is shown in Table 5.

# XVII. Serological Reactions

Influenza virus particles are very good antigens, antibodies of high titre are found in convalescent sera and can be produced by immunisation. These antibodies are of two types. Antibodies directed against the internal ribonucleoprotein component of the virus (S antigen) are group or type specific, distinguishing between viruses A, B and C but not between individual strains. Antibodies directed against the outer component of the virus particle (V antigen) are strain-specific.

Convalescent sera usually contain both types of antibody, but sera produced by artificial immunisation usually contain little or no antibody to the inner component ($W_{51}$, $M_{129}$, $L_{71}$).

Antibody to the host component of the virus particle is not found in convalescent sera but may be produced if animals are vaccinated with virus grown in a different type of host.

Antibody to the outer virus component protects against infection, neutralizes virus in vitro, neutralizes the toxic property of the virus, inhibits haemagglutination, and can be detected by agglutination, precipitation and complement fixation tests.

Antibody to the inner component does not prevent infection, and does not neutralize virus or inhibit haemagglutination. It can be detected by complement fixation and precipitation tests.

## A. Serological Reactions of the Inner Ribonucleoprotein Component
### 1. The Group Complement Fixation Test

FAIRBROTHER and HOYLE ($F_9$, $H_{202-205}$, $F_{11}$, $H_{186}$, $H_{187}$, $H_{205}$) found that extracts of the lungs of mice infected with influenza virus contained a complement fixing antigen reacting with antibody in human convalescent sera, and developed the test as a method of serological diagnosis. They showed that the effective antigen in the extracts was not the infective virus but an agent of smaller size, the soluble or S antigen. A similar antigen was found in infected chorio-allantoic membrane of fertile eggs. The S antigens of the human and swine viruses were found to be identical. These results were soon confirmed ($L_{126}$, $L_{57}$, $B_{37}$, $L_{58}$, $E_3$, $N_{20}$, $F_{142}$, $H_{96}$, $W_{51}$, $R_{23}$, $N_5$, $H_{84}$, $Y_9$), and the test became accepted as a reliable method of serological diagnosis.

Some of the earlier workers used infected allantoic fluid as antigen, but this material was unsatisfactory as it contained variable amounts of S antigen and infective virus and some degree of strain specificity was found. The work of FRIEDE-

WALD ($F_{142}$) and HENLE and WIENER ($H_{96}$, $W_{51}$) showed that infected allantoic fluid contained two complement fixing antigens, the infected virus with a sedimentation constant of 600S which gave strain-specific complement fixation and a smaller agent of sedimentation constant 30S which was not adsorbed by red cells and was apparently identical with the S antigen present in chorioallantoic membrane extracts ($K_{51}$).

The S antigens of all strains of virus A were identical, and completely different from the S antigen of virus B ($F_{91}$)

An analysis of the complement fixation reaction was made by HOYLE ($H_{186}$) using both S antigen and virus particle (V antigens) from different strains of virus and a range of human convalescent sera. Chessboard titration methods were used,

a range of antigen dilutions being tested against a range of serum dilutions. The results of such a titration with concentrated S antigen from mouse lung and a human convalescent serum is shown in Fig. 34. The curve is obtained by drawing a line through all mixtures giving 50% haemolysis. Where no tube in a series gives exactly 50% haemolysis the 50% point is obtained by interpolation. All mixtures inside the curve give complete fixation and all mixtures outside the curve give complete haemolysis. Three quantitative measurements can be made. The line AB gives a measure of the maximal antigen titre. The point C gives a measure of the maximal serum titre and the optimal antigen titre, i.e. the antigen concentration giving fixation with the smallest amount of serum. With small amounts

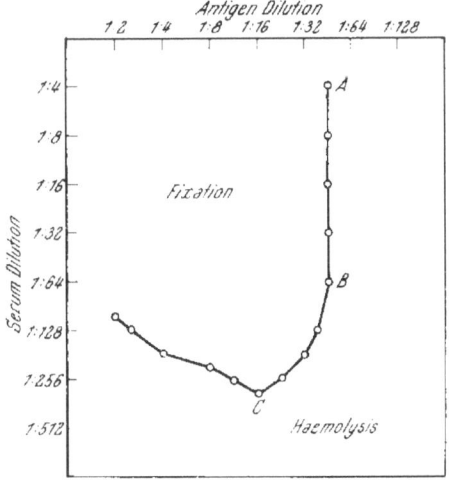

Fig. 34. Chessboard titration of group complement fixing antigen.

Chessboard complement fixation test showing the 50% haemolysis curve obtained with concentrated S antigen from mouse lung infected with WS virus and a human convalescent serum.

of serum fixation is reduced if the antigen is in excess and a zone phenomenon occurs, but there is no optimal serum titre, as with human sera no zone phenomenon occurs with serum excess. The maximal antigen titre is usually three times the optimal titre. It was found that the maximal and optimal antigen titres were the same when the antigen was titrated with a number of different convalescent sera, and the maximal serum titre was the same when the serum was titrated with S antigen from different strains of virus or from virus grown in different host cells.

## 2. Technique and Use of the Group Complement Fixation Reaction

Complement fixation with the S antigen is a very reliable test for use in the serological diagnosis of influenza, and the demonstration of S antigen in infected tissues is the most certain criterion for the identification of an influenza virus on isolation.

### a) Preparation of Antigens

Mouse lung antigens are prepared by inoculating mice intranasally with a mouse-adapted strain of virus. Mice are killed on the 3rd or 4th day, the lungs cut into small pieces and dried in vacuo. Dried mouse lung will retain its antigenic potency for many years. Liquid antigen is prepared by grinding the dried lungs with sand and saline to make a 1% suspension. After centrifugation a few drops of chloroform are added to the supernatant which is allowed to stand overnight at 4°C. This results in the sedimentation of finely particulate matter, Forssman antigen is removed, and the fluid becomes less anti-complementary. The liquid is again centrifuged and 0.08% sodium azide added to the supernatant. The fluid antigen will keep at 4°C for several weeks. A more concentrated antigen can be prepared by precipitating the antigen by adjustment of the fluid to pH 5.3 with acetic acid, followed by extraction of the precipitate with phosphate buffered saline pH 7.3.

Egg chorioallantoic membrane antigens are more commonly used. Eggs may be inoculated with a small dose of virus and the membranes collected after 24—36 hours, or a large dose of virus may be used and the eggs harvested at 6 hours. Membranes are suspended in azide saline, 1 ml per membrane, and frozen and thawed three times. After centrifugation the supernatant fluid is adsorbed with 5% chick red blood cells to remove haemagglutinin. The cells are then centrifuged off and the supernatant fluid used as antigen.

Highly purified antigen can be prepared from infective virus particles by ether disintegration. Virus is purified and concentrated by two cycles of adsorption-elution from red cells. The fluid is then shaken with half its volume of ether until a precipitate appears at the ether-water interface. The aqueous phase is then adsorbed with red cells to remove haemagglutinin. After removal of the cells 0.08% sodium azide is added. Further purification and concentration can be attained by centrifuging the antigen at 80,000—100,000 $g$ and resuspending the deposited material in a smaller volume.

### b) Preparation of Antisera

The most satisfactory antisera are human convalescent sera. It is usually possible to find sera which contain antibody to only one type of virus, and it is sometimes possible to find sera reacting only with S antigen and containing little or no V antibody. For most purposes the presence of V antibody is unimportant. V antibody can be selectively adsorbed from sera with purified homologous virus or with virus adsorbed on periodate-treated red cells.

Antisera to the S antigen are not easily produced by immunisation. Guinea pigs or rabbits may be immunised by repeated doses of purified S antigen but antibody response is often poor. Antigen adsorbed on aluminium phosphate gives rather better response. LIEF, FABIYI and HENLE ($L_{74}$) found that if guinea pigs were inoculated intranasally with a small dose of active virus and intraperitoneal inoculations of purified S antigen given later, high titre S antibody was produced, and LIEF ($L_{71}$) improved the method by using an initial stimulation by intranasal inoculation of live swine virus followed by a booster inoculation of S antigen from an equine strain of virus. By this means pure anti-S sera were obtained.

*Technique of the complement fixation test.* For the detection and estimation of anti-body in serum the antigen is first titrated to determine the optimal titre. Sera are heated at 56°C for 30 minutes and unit volumes of a range of dilutions from 1:2 to 1:128 set up. Dilutions may be made in physiological saline or buffered saline pH 7.3. A unit volume of guinea pig complement containing 2.5 MHD is added to each tube, followed by a unit volume of antigen diluted to the optimal titre. Control tubes are included with antigen and serum replaced by saline. Tubes are incubated in a water bath at 37°C for 1 hour and 2 volumes of 3% sensitised sheep cells added. After a further incubation of 30 minutes the cells are allowed to settle and readings made, the end point being taken as that serum dilution giving 50% haemolysis ($H_{187}$).

Similar methods may be used for detection and estimation of antigen in infected tissues, using a range of antigen dilutions and a positive serum which is first titrated to determine the maximal serum titre, four times this amount being used in the test.

Satisfactory results are obtained using a unit volume of 0.05 ml in small glass tubes. For very accurate work a larger unit volume, 0.2 ml, is preferable. Plastic plates with cup-like depressions may be used in place of glass tubes but are less satisfactory as they do not conduct heat readily, taking longer to reach 37°C on incubation in a dry incubator, and the fluids are not easily mixed by shaking.

Sensitivity of the test may be increased by the use of long fixation methods (overnight at 4°C followed by 1 hour at 37°C) and such methods have been used by Russian workers to detect antigen in nasopharyngeal washings, blood, and autopsy material ($P_{48}$, $P_{49}$, $S_{184}$). In general long fixation methods gain sensitivity at the expense of a loss of accuracy and reproducibility.

## 3. Precipitin Reactions with the S Antigen

A typical precipitin reaction is given when concentrated preparations of purified S antigen are mixed with convalescent sera, but the sensitivity of the reaction is very much less than with complement fixation so that the test has little practical value. Precipitation lines are given in immuno-diffusion tests ($S_{308}$, $H_{17}$).

## 4. Specificity of the S Antigen

The analysis of the complement fixation reaction by HOYLE ($H_{186}$) showed that the S antigens of different human strains of virus A were serologically identical. BLASKOVIC ($B_{75}$, $B_{83}$) showed that the S antigens of human, duck, swine and fowl plague strains of virus A were qualitatively and quantitatively the same.

DAVENPORT, ROTT and SCHÄFER ($D_{29}$) using the FULTON and DUMBELL complement fixation test ($F_{164}$) claimed that slight differences could be distinguished between different human strains of virus A and somewhat larger differences between the S antigens of human and swine and fowl plague viruses. These results have been criticised by LIEF ($L_{71}$) who considered that the differences were due to the presence of non-haemagglutinating V antigen in the S antigen preparations. The Fulton-Dumbell test is a very unorthodox procedure in which a range of serum dilutions is titrated in a chessboard experiment against a series of increasing doses of antigen and complement. The final result is assessed by measuring the area of fixation in the chessboard. As the serum is in effect being titrated in both directions this area is a function of the square of the serum titre, and small differences in serum titre are being mathematically magnified. The validity of this procedure is dubious.

HANA and HOYLE ($H_{17}$) carried out immunodiffusion tests with purified S antigen from human A, A1 and A2 viruses and found that if the antigens were stored for some time or were treated with deoxycholate several precipitation lines were obtained. The antigen appeared to disintegrate into a number of serologically distinct components, but all the components were present in S antigen from all the strains of human virus. There was no evidence of any qualitative difference in the various S antigens. If, however, there are quantitative differences the results of DAVENPORT, ROTT and SCHÄFER might be explained.

## B. Serological Reactions of the Outer Virus Component
## 1. The Neutralization Test

SMITH, ANDREWES and LAIDLAW ($S_{167}$) showed that human or ferret convalescent sera would neutralize virus and prevent the development of febrile illness in ferrets, and with the successful adaptation of the virus to mice it became possible to study the antibody content of sera by neutralization tests on mice.

Early work was concerned with the detection of antibody in normal human sera ($A_{80}$, $F_{105}$, $F_9$). It was found that a large proportion of adult normal human sera was able to neutralize both the human and swine viruses. The incidence of antibody in the sera of children under 10 years was very low, but antibody was present in new-born children up to the age of 6 months. Some sera neutralized one virus but not the other. FRANCIS and SHOPE ($F_{120}$) showed that convalescent animal sera neutralized the homologous virus better than the heterologous. Neutralizing antibody in human sera is considerably increased as a result of infection, but the levels return to normal in about 1 year ($F_{11}$).

## 2. Quantitative Neutralization Tests in Mice

Mouse-adapted strains of virus must be used. Serial dilutions of serum heated at 56°C for 30 minutes are mixed with virus suspension and 0.05 ml of the mixtures inoculated intranasally under light ether anaesthesia, using 6 mice for each dilution. The infectivity of the virus suspension is first titrated and a dose of from $100-1000$ $LD_{50}$ used in the test. With large doses of virus the end point is assessed by determining the 50% lethal dose, or smaller doses of virus may be used and the end point assessed by the incidence of lung lesions in mice killed at 4 days.

Comparisons of the neutralizing antibody in serum against different strains of virus is rendered difficult by the fact that the results depend on the mouse pathogenicity of the various strains ($G_{38}$, $Z_{34}$).

### The Mouse-egg Neutralization Technique

KILBOURNE and HORSFALL ($K_{48}$) developed a method for measuring neutralizing antibody in mice which could be applied to non-mouse adapted strains. Virus serum mixtures were inoculated intranasally to mice and after 2 days the mice were killed and the presence or absence of virus in the lungs determined by inoculating lung extracts into the allantoic sac of fertile eggs. The method is claimed to be superior to direct *in ovo* neutralization tests because in the mouse the quantitative relation between virus and antibody is very constant and not affected by the nature of the virus or the type of serum, and the slope of the line relating the variables is much less steep than in eggs. In the mouse a 5 fold dilution of serum results in a 10 fold decrease in the virus neutralized ($H_{163}$).

### 3. Quantitative Neutralization Tests in Eggs

BURNET ($B_{160}$) found that the method of amniotic inoculation in fertile eggs could not be used for serum neutralization tests because of extreme sensitiveness, mixtures heavily over-neutralized as tested by other methods will still infect by the amniotic route, though the death of the embryo is delayed.

Neutralization tests in the yolk sac have been used ($M_{121}$), the end point being assessed by the death of the embryo. Certain strains of virus produce pocks on the external surface of the chorioallantoic membrane and in these cases neutralization tests can be carried out by pock-counting methods.

All these methods are inferior to neutralization tests carried out by inoculation of the allantoic sac and this method is almost always used.

In the original methods of HIRST ($H_{127}$) and BURNET and BEVERIDGE ($B_{188}$) a range of virus dilutions is made in 10% horse serum broth or horse serum saline and mixed with undiluted serum. After standing at 4°C for 1 hour the mixtures are inoculated to the allantoic sac of 10—12 day fertile eggs and after 36—48 hours incubation the allantoic fluids are tested for the presence of haemagglutinin. From 6 to 9 eggs are used for each dilution. Control tests are made with virus alone and the results expressed as the number of infective doses neutralized by undiluted serum. Diluents containing horse serum may cause non-specific inactivation of virus and DICK and TAYLOR ($D_{53}$) recommended dilutions in 0.2% bovine albumin in buffered saline. The results showed considerable strain-specificity, neutralization titres of convalescent sera being higher against the infecting virus than against other strains. In these methods, using a constant amount of serum and varying amounts of virus there are apparently enormous increases in antibody content between acute and convalescent sera, and it is difficult to relate the results to antibody titrations by other methods. WALKER and HORSFALL ($W_{11}$) found that a ten-fold increase in serum concentration results in a 500,000 fold increase in the amount of virus neutralized, and they recommended the use of a constant amount of virus with a range of serum dilutions and this method is now usually used. The results of antibody titrations are only slightly affected by variations in the amount of virus used in the tests and results are much more comparable to other methods.

There are, however, many difficulties inherent in the *in ovo* neutralization test ($G_{38}$, $Z_{34}$). Because of the great sensitivity of the fertile egg only comparatively small amounts of virus can be used in the tests and the results may be affected by the presence of non-specific inhibitors in the sera and by variation in the susceptibility of the egg to different strains of virus. The presence of inactive virus may cause difficulty because of the interference phenomenon.

Neutralization tests in de-embryonated eggs have been used ($S_{251}$) but have no special advantages.

### 4. Neutralization in Tissue Cultures

FULTON ($F_{162}$) adapted his infectivity testing technique using small pieces of egg chorioallantoic membrane to the determination of neutralizing antibody.

Virus-serum mixtures are set up, using 100 $ID_{50}$ of virus and a range of serum dilutions, distributed in cups in a plastic plate. Pieces of chorioallantoic membrane are added and the plate rocked for 18 hours, during which time un-neutralized virus is able to infect the cells. The membrane pieces are then washed and transferred to fresh glucosol and rocked at 37°C for 24 hours. Penicillin and streptomycin are used throughout to prevent bacterial contamination. The membrane pieces are then removed and red cells added to the fluids to detect the presence of haemagglutinin. A similar method was used by ZAKSTELSKAYA ($Z_3$).

FAZEKAS DE ST GROTH, WITHELL and LAFFERTY ($F_{56}$) improved the method, using membrane pieces still attached to the egg-shell. Non-specific inhibitors in the sera were destroyed by treatment with RDE and heating at 65°C. An extensive analysis of the method was made and it was considered to be superior to other methods of neutralization in both sensitivity and accuracy.

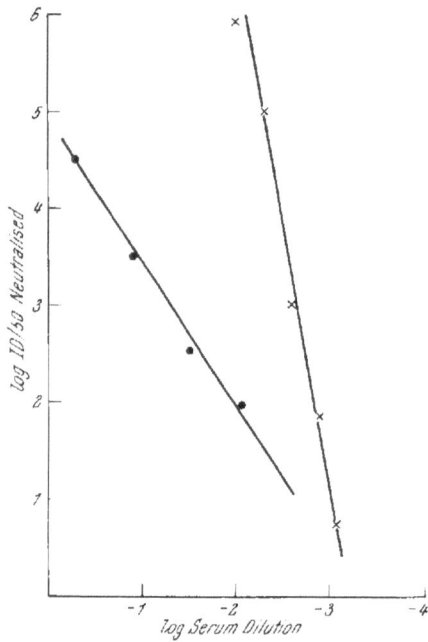

Fig. 35. Relation between serum dilution and amount of virus neutralised in eggs and in mice. Neutralisation of mouse-adapted DSP virus by homologous serum. Neutralisation in mice ●——●; neutralisation in the allantoic sac ×——×.

CLARKE and TYRRELL ($C_{60}$) carried out neutralization tests in roller tube cultures of trypsinized chick embryo lung, human and monkey kidney, hamster kidney and human liver.

WEINBERGER, BUESCHER and EDWARDS ($W_{31}$) used neutralization tests in monkey kidney cultures. A series of doubling dilutions of periodate treated human sera were tested against 300—500 haemadsorption doses ($HAD_{50}$) of virus, four culture tubes being used for each dilution. The virus-serum mixtures were allowed to stand 1 hour before inoculation. The test was superior to haemagglutination-inhibiting methods for the study of antigenic variation.

## 5. Mechanism of the Neutralization Reaction

The initial union of influenza virus with neutralizing antibody is a reversible reaction, and neutral mixtures become infective on dilution ($H_{163}$, $B_{165}$, $T_{78}$, $C_{60}$). The initial reversible reaction appears, however, to be followed by a more stable union. BURNET ($B_{165}$) found that a mixture reactivated by dilution was less infective than if virus and serum were first diluted and then mixed. SMORODINTSEV and SHISHKINA ($S_{205}$) found that active virus could be recovered from over-neutralized mixtures by washing centrifuged deposits or by washing virus-serum complexes silted up against Gradocol membranes. Recovery of active virus by simple dilution was possible only if the antibody excess was small.

There is a linear relation between the serum titre and the amount of virus used in neutralization tests ($H_{163}$, $W_{11}$, $T_{78}$, $C_{60}$), but the slope of the line varies according to the host cell system used, and in the chick embryo by the route of inoculation used ($T_{78}$). In mice a 5 fold dilution of serum results in a 10 fold decrease in amount of virus neutralized ($H_{163}$), but in the allantoic sac the slope of the line is much steeper ($B_{165}$, $W_{11}$), and a 10 fold increase in serum may result in a 500,000 fold increase in virus neutralized. Fig. 35 illustrates results of neutralization tests of a mouse-adapted virus in mice and in the allantoic sac. The steep slope of the neutralization line in the allantoic sac explains the great differences in the results of neutralization tests using the constant virus varying serum technique as against the constant serum varying virus method. Using roller tube tissue cultures CLARKE and TYRRELL ($C_{60}$) found that if the serum virus mixtures were removed after a period of adsorption there was a reduction in the slope of the neutralization line, and suggested that the steeper slopes obtained if the mixtures were not removed was due to the action of the antibody being exerted in successive cycles of reproduction.

An extensive analysis of the neutralization reaction was made by FAZEKAS DE ST GROTH and WEBSTER ($F_{31}$, $F_{48}$, $F_{49}$). They found that the virus-serum equilibrium was the same whether approached from the side of virus or antibody excess, and suggested that the law of mass action was applicable. They calculated that the number of antigen sites on virus particles was of the order of 2000. AMELUNXEN and WERDER ($A_{54}$) using immune rabbit serum found that in mixtures containing traces of un-neutralized virus the number of antibody molecules per particle was 1200, but with complete neutralization the number was 6040.

Kinetic studies by LAFFERTY ($L_{11}$, $L_{12}$) indicated that the union of virus with antibody occurred in two stages. The first was a freely reversible combination, followed by a reaction leading to a more stable complex which did not dissociate on dilution. There appeared to be two types of antibody involved, a non-avid type which was unable to form a stable complex with virus, and could prevent union with a more avid type able to produce an irreversible union. A monovalent antibody produced by digestion of immune gamma globulin with papain combined reversibly with virus and could protect it against irreversible interaction with divalent antibody.

McKEE and HALE ($M_{62}$) found that about 10% of the virus in over-neutralized mixtures could be reactivated by addition of concentrated heat inactivated virus of homologous type but not of heterologous type. A similar reactivation *in ovo* could be obtained up to 20 hours after inoculation of over neutralized mixtures ($H_{215}$).

## 6. The Haemagglutination-inhibition Test

HIRST ($H_{126}$) showed that the agglutination of red blood cells by influenza viruses was inhibited by antibody in convalescent sera, and HIRST and PICKELS ($H_{148}$), SALK ($S_4$) and others developed methods of estimating the antibody content of sera by haemagglutination-inhibition. The simplicity of the method led to its wide use both for the determination of serum antibody titres and for the identification and classification of virus strains. Any type of haemagglutination test may be adapted to the measurement of serum antibody, but methods based on the Salk test are almost always used.

*Technique:* Any virus-containing fluid free of particulate matter may be used as antigen, and crude infected allantoic fluids are commonly used, but preparations partially purified by adsorption-elution from red cells are more satisfactory. The haemagglutinin content of the fluid is first titrated by the Salk test. Sera are heated at

$56°C$ for 30 minutes and 0.25 ml of a range of two-fold dilutions from $1:16-1:16384$ placed in round-bottomed tubes. To each tube 0.25 ml of antigen containing 4 haemagglutinin units is added. After mixing 0.5 ml of a $1:200$ dilution of washed fowl red cells is added to each tube. Tubes are shaken, allowed to stand at room temperature for $1-1\frac{1}{2}$ hours when readings can be made ($C_{83}$). Human or guinea pig cells may be used, and if desired the serum dilutions may be made in red cell suspension and the virus added later ($S_4$). The serum titre is expressed as the highest initial dilution which completely prevents haemagglutination.

Certain difficulties may be experienced with HI tests. Red cells from different fowls may give different results ($S_{287}$). RODRIGUEZ DA SILVA and FELDMAN ($R_{42}$) found that formolized chick red cells may be used and can be stored for 3 months, so that the same batch of cells could be used for many tests. Formolized human cells were not satisfactory. The major difficulty with HI tests is due to the presence of non-specific haemagglutinin inhibitors in serum. These may completely invalidate the results and it is essential to treat the sera by some procedure which destroys non-specific inhibitors without damage to specific antibody.

## 7. Destruction of Non-specific Inhibitors in Serum

Many methods of inhibitor destruction have been used, none are completely satisfactory. Some procedures destroy Francis type inhibitor but have no action on heat-labile inhibitors, while others have the reverse action. Methods destroying both types of inhibitor may have some action on specific antibody.

### a) Inhibitor Destruction by Periodate

HIRST ($H_{138}$) mixed the serum with 2 volumes of saturated $K10_4$ and heated for 15 minutes at $56°C$. Periodate was then neutralized by addition of 2 volumes of 7% glucose. JAMES and FISET ($J_2$) recommended the use of M/90 periodate at room temperature. Periodate destroys Francis inhibitor but has little effect on heat-labile inhibitors.

### b) Inhibitor Destruction by RDE and Cholera Filtrates

Purified RDE destroys Francis inhibitor only, but crude cholera filtrates destroy both Francis and Chu inhibitors because of the presence of a trypsin-like enzyme in the filtrate, and crude cholera filtrates have been extensively used ($M_{142}$, $V_9$, $A_{93}$, $I_{11}$, $B_{122}$, $B_{59}$, $L_{121}$).

Sera are mixed with 5 volumes of cholera filtrate and incubated overnight at $37°C$. The mixture is then heated at $58°C$ for 50 minutes to destroy the enzymes. The cholera filtrate is prepared by growing the 4Z strain of *V. cholerae* on 0.8% nutrient agar pH 6.9 for 16 hours. The soft agar is then pressed through layers of sterile gauze and the resulting liquid sterilised by filtration. The activity is tested by determining the amount which will destroy the receptors of an 0.5% red cell suspension in 30 minutes at $37°C$. Satisfactory filtrates have receptor-destroying titres of $256-1024$ ($V_9$).

### c) Inhibitor Destruction by Trypsin

Both Francis and Chu inhibitors are destroyed by trypsin ($S_{18}$, $J_4$). To each 1 ml of serum is added 4 mg of commercial trypsin dissolved in 0.5 ml of 0.1 M phosphate buffer pH 8.2. The mixture is heated at $56°C$ for 30 minutes when the trypsin first digests the inhibitors and is then heat-inactivated. Specific antibody titre is unaffected.

### d) Other Methods of Inhibitor Destruction

Filtrates of broth culture of *Pseudomonas fluorescens* were used by GOREV ($G_{47}$), inhibitors being destroyed by a trypsin-like action.

Heat-labile inhibitors may be destroyed by heating the serum at $56°C$ for 1 hour, but Francis inhibitors are not affected.

Both Francis and Chu inhibitors are destroyed by saturation of the serum with $CO_2$ ($S_{253}$, $C_{32}$), but the results are not always satisfactory ($S_{255}$).

Adsorption of the serum with kaolin was recommended by SPENCE ($S_{248}$) especially for use with Asian strains of virus.

ANANTHANARAYAN and PANIKER ($A_{59}$) found no single procedure adequate for all strains of virus or for all types of inhibition, but a combination of trypsin and periodate gave best results.

The effects of non-specific inhibitors may be prevented by titrating sera in the presence of citrate, oxalate, fluoride or metaphosphate solutions which remove calcium salts. Titrations in the presence of physiological oxalate solution gave the best results ($V_9$).

### e) Use of the Drescher Technique

DRESCHER and LANGE ($D_{74}$) described a method of antibody titrations in which it is possible to distinguish between the effects of antibody and non-specific inhibitors and to estimate both without the use of cholera filtrate. The method is complicated and unsuitable for routine use.

## 8. Antigenic Analysis by the Use of HI Tests

The haemagglutinin-inhibition test reveals antigenic differences between different strains of both virus A and virus B, and is, therefore, unsuitable for the identification of virus on first isolation. It has, however, been much used in studies of antigenic variation in the viruses. In general strains isolated in the same epidemic are serologically identical, but considerable differences occur in strains isolated in different outbreaks. These differences are revealed by the use of cross HI tests ($H_{131}$, $V_9$, $B_{118}$, $B_{121}$, $H_{141}$, $H_{117}$). Specific antisera are prepared in animals and after suitable treatment to destroy non-specific inhibitors are used in HI tests with homologous and heterologous viruses, all the viruses being tested against all the sera. The sera of ferrets convalescent after intranasal inoculation of virus were first used by HIRST ($H_{131}$) and have been much employed. They have the disadvantage of a high content of non-specific inhibitors. HIRST ($H_{141}$) also used rabbit antisera but such sera may contain antibody to the host component of the virus and require to be adsorbed with heterologous virus to render them specific. HILLEMAN and HORSFALL ($H_{115}$, $H_{117}$) used fowl sera prepared by intravenous and intraperitoneal inoculation of allantoic fluid virus.

The antigenic relationship between different strains of virus revealed in HI tests is indicated by the ratio of the heterologous titre to the homologous titre, often expressed as a percentage. With complete cross HI tests the extent of immunological relationship between two strains may be represented as an R value when R is equal to $\frac{1}{2} r_1 + r_2$, $r_1$ and $r_2$ being the percentage ratio of heterologous to homologous titres with each virus ($H_{117}$).

## 9. The Specific Complement Fixation Test

In 1937 Lush and Burnet observed that, in complement fixation tests using egg membrane antigens of human and swine viruses and immune ferret sera, complement fixation was greater with homologous than with heterologous sera, and Eaton ($E_3$) obtained similar results with mouse lung antigens and immune mouse sera. Later work showed that two complement fixing antigens were present in influenza virus preparations, the infective virus particle and the soluble S antigen, and that the strain-specific phenomena were due to the infective particle ($H_{84}$, $W_{51}$, $H_{96}$, $H_{186}$). Hoyle ($H_{186}$) and Wiener, Henle and Henle ($W_{51}$) showed that infective virus preparations contained S antigen,

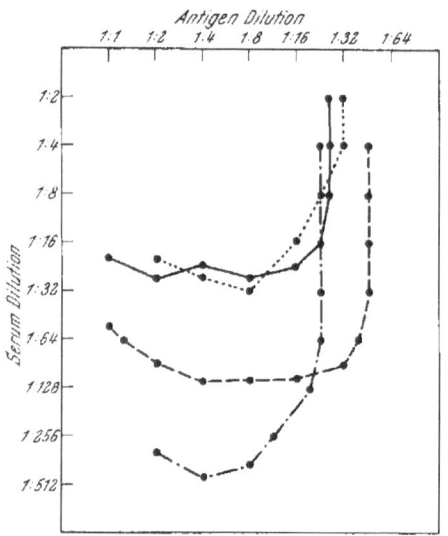

Fig. 36. Chessboard titration of specific complement fixing antigen.

Chessboard complement fixation test with V antigen of DSP virus and four different human convalescent sera.

but Kirber and Henle ($K_{51}$) considered that the S antigen was probably superficially adsorbed on the virus particle. By the use of virus particle (V antigen) preparations made by adsorption-elution from red cells Fulton and Dumbell ($F_{164}$) were able to demonstrate serological differences between different strains of virus A, and Kirber and Henle ($K_{51}$) obtained results very similar to those obtained in haemagglutinin-inhibition tests. Russian workers obtained similar results ($K_1$, $G_{40}$). The development of the specific complement fixation test is, however, mainly due to the work of Henle, Lief and their associates ($H_{90}$, $L_{78-81}$, $H_{89}$). They found that purified virus particle antigens prepared by two cycles of adsorption-elution from red cells were free of extraneous S antigen and did not react with anti-S sera.

Such preparations could be used for identification and serological classification of virus strains, and the results were superior to the use of haemagglutination-inhibition tests for the detection of serum antibodies as they were not affected by the presence of non-specific inhibitors in the sera. The preparations were also very useful in studies of the antibody response to vaccination and in the diagnosis of infection in young children in whom there is little production of antibody to the S antigen.

## 10. Technique of the Specific Complement Fixation Test

### a) Preparation of Specific V Antigens

This consists of the preparation of virus particle suspensions free of external S antigen. Preparations produced by a single cycle of adsorption-elution from red cells usually react with S antibody. Kirber and Henle ($K_{51}$) believed that this S antigen was adsorbed on the virus particle, but it is more probably produced by disintegration of some of the larger virus particles, which when first released from the infected cell are somewhat delicate in structure and easily burst. On standing for some time the particles become more rigid and stable. Second cycle eluates are usually free of detectable external S antigen and are suitable for use. Virus preparations made by two cycles of adsorption-elution from red cells followed by differential centrifugation are completely free from external S antigen, and are relatively stable. On standing some disintegration may occur with the release of internal S antigen, so that V antigen preparations should be used within a few days. The V antigen is a complex of several different components so that the determination of the optimal amount for use in serum titrations presents some difficulty. Chessboard titrations give different types of curve with different sera (Fig. 36) and there may be no well defined optimal titre such as is seen in S antigen titrations. It is probably best to determine the maximal antigen titre and use from 4—6 times this amount in serum titrations.

### b) Preparation of Specific V Antisera

LIEF and HENLE ($L_{78}$) recommend the intraperitoneal inoculation of guinea pigs with the haemagglutinin fraction obtained from the virus particle by ether disintegration. The preparations should have a haemagglutinin titre of 2000— 5000 and three doses of 1 ml are given, the first two doses at an interval of 7 days and the third dose 3 weeks later. Sera are collected 1 week after the last dose, inactivated at 56°C for 30 minutes, and adsorbed with sheep red cells. They may be stored in the frozen state.

### c) Technique of the Test

Complement fixation with V antigens may be carried out by the same methods as are used with S antigen. Short fixation periods (1 hour at 37°C) are usually used.

## 11. Agglutination, Precipitation and Flocculation Tests

Concentrated preparations of influenza virus particles give precipitation reactions with immune sera ($K_{66}$, $B_{39}$), and with filamentous strains of virus the agglutination of filaments by immune sera can readily be observed by dark-field microscopy (Fig. 12). Fowl plague virus preparations also give precipitates with antiserum ($M_{148}$). GROSSGEBAUER ($G_{83}$) carried out cross agglutination tests with filamentous forms of 17 strains of virus and their respective antisera. In most cases agglutination with homologous serum was observed, but there was a considerable degree of cross-reaction.

Flocculation reactions have been particularly studied by BELYAVIN ($B_{39}$, $B_{41}$). Crude infected allantoic fluid could be used as antigen but better results were given by virus preparations partially purified and concentrated by red cell adsorption-elution or differential centrifugation. Mixtures of virus and dilutions of rabbit immune sera were mixed in Dreyer tubes and incubated at 37°C in a water-bath with half-immersion of the columns. Typical precipitation reactions were observed, with the usual zone phenomenon, precipitation failing to occur with large excess of antigen or antibody. Serological relations between strains could be demonstrated by cross-flocculation reactions, and A, A1 and B strains could be readily distinguished. Scandinavian and Liverpool strains could be distinguished from FM1 but not from each other. The flocculation reaction was inhibited by thermolabile inhibitors present in human, horse, guinea pig, rabbit and fowl sera, the inhibitor being destroyed by trypsin but not by RDE ($B_{40}$).

Certain inhibitors in normal sera may cause flocculation of virus ($B_{41}$) but this effect is destroyed by pre-treatment of the sera with RDE. SVEDMYR ($S_{338}$) found that purified influenza virus was precipitated by normal allantoic fluid inhibitor but that the precipitates dissolved on incubation. Stable precipitates were given with heated virus. Influenza A2 strains give a flocculation reaction with normal horse serum due to the presence of gamma inhibitor. This effect is abolished by treatment of the serum with periodate ($B_{69}$). Owing to the presence of a host cell component in the virus particle influenza virus preparations are precipitated by antisera to the host cell ($K_{66}$).

## 12. Precipitin Reactions by the Gel-diffusion Technique

JENSEN and FRANCIS ($J_{11}$) showed that precipitin lines could be produced in solid medium by layering virus preparations on top of serum-impregnated agar in small tubes. SVET-MOLDAVSKY and GHENDON ($S_{341}$) inserted serum impregnated agar tubes into human embryo lung tissue cultures. Virus antigen diffused through the agar and gave specific precipitation lines; A, A1, and B viruses could be identified.

STYK and HANA ($S_{308}$) employed the Ouchterlony immuno-diffusion methods. Intact virus particles diffuse very slowly through agar gels and are unsuitable for use, but very satisfactory results were obtained by using purified virus preparations disintegrated by treatment with 2% deoxycholate. Numerous precipitation lines were seen when deoxycholate-treated virus was allowed to interact with human convalescent sera, the results indicating the antigenic complexity of the virus particle. The method has considerable promise for the study of antigenic relations between virus strains.

## 13. Antibody Absorption Techniques

Neutralising, haemagglutinin inhibiting and specific complement fixing antibodies can all be removed from sera by absorption with virus preparations. Antibody absorption appears to have first used by SMORODINTSEV ($S_{200}$) to demonstrate the serological identity of the 1936 strains of virus. Convalescent ferret sera were absorbed with mouse lung suspensions and virus + absorbed antibody removed by filtration through porcelain filters. Neutralization tests were then carried out with the absorbed sera. Absorption of neutralising antibody has not proved of much value in studies of antigenic structure ($S_{291}$) and the method has not been much used, but absorption of haemagglutinin-inhibiting antibody has been more successful ($F_{144}$, $L_{128}$, $H_{141}$, $J_{12}$).

HIRST ($H_{141}$) using an absorption technique developed by WALKER and HORSFALL ($W_{11}$) was able to classify strains of virus A into seven groups, using rabbit antisera.

Sera were treated with RDE and diluted 1:3 with saline. They were then absorbed with allantoic fluid virus concentrated 100-fold by centrifugation, one part of virus being used to absorb 2 parts of diluted serum. After standing overnight at 4°C the virus was removed by centrifugation at 13,000 r.p.m. for 90 minutes. Usually several serial absorptions were carried out. After addition of 2% sodium citrate the sera were heated at 65°C for 30 minutes to destroy any residual haemagglutinin. By absorbing sera in this way with heterologous virus strains, highly strain-specific absorbed sera were produced.

Russian workers have used a technique developed by SMORODINTSEV (L$_{128}$). Virus was purified by adsorption-elution from red cells and 4 ml with an HA titre of at least 1:640 was added to 5 ml of serum diluted 1:16. Haemagglutination tests were made to demonstrate the presence of excess virus. Mixtures were then left overnight at 4°C and then free virus was removed by absorption with 1 ml of packed red blood cells. There are many technical difficulties in the preparation and use of antibody-absorbed sera (Z$_{34}$).

## 14. Relation between Neutralizing, Haemagglutinin-inhibiting, Specific Complement Fixing and Precipitating Antibodies

HIRST (H$_{126}$) found a close correlation between the neutralising and haemagglutinin-inhibiting antibodies in sera, and in general the results of antigenic analysis of virus strains give similar results whatever type of serological test is used. However, many workers have found striking differences in the results of antibody titrations by the use of haemagglutinin-inhibition tests and by neutralization (B$_{188}$, F$_{144}$, S$_{293}$, W$_{11}$, H$_{117}$, H$_{158}$). Many of these discrepancies were undoubtedly due to the effects of non-specific haemagglutinin inhibitors in the sera, the importance of which was not realised in the early work. Differences in the results of complement fixation and precipitation tests are probably always a result of the action of inhibitors which may produce precipitates which do not fix complement.

Differences between haemagglutinin-inhibition tests and neutralization tests have, however, certain features which suggest that the two tests are measuring different antibodies. FRIEDEWALD (F$_{144}$) and WALKER and HORSFALL (W$_{11}$) found that when sera are absorbed with virus the neutralizing antibody is preferentially absorbed. HOLTERMANN and PETERSEN (H$_{158}$) studied the development and persistence of specific complement fixing and neutralizing antibodies in convalescent sera and found that whereas the CF antibody declined to pre-epidemic levels within 2 years of infection the neutralizing antibody remained relatively stationary.

In spite of these results the author finds it difficult to believe that the neutralizing antibody is reacting with a different antigen to that involved in the other tests. It is possible that the discrepancies are caused by the great differences in the technique of the various tests. In HI and CF tests large amounts of virus are used and the reaction with antibody is tested immediately after mixing, whereas in neutralization tests only very small amounts of virus are used and the virus-serum mixtures are allowed to stand for some time before testing. In the neutralization test the result of the test depends on the pathogenicity of the residual virus and on the susceptibility of the host employed, factors which are not involved in the other tests.

Immunisation with antigens of large molecular weight results in the appearance of a range of antibody molecules of smaller size which therefore react

with only a part of the antigen molecule. Haemagglutination is probably a
function of the presence in the antigen molecule of an active centre of small
dimension, and only antibody molecules reacting with this small region will
prevent haemagglutination. Antibody molecules reacting with non-haemagglutinating regions of the antigen molecule might be active in other types of
test such as complement fixation or neutralization. The occurrence of differences
in the results of the various tests does not, therefore, necessarily indicate that
different antigens are involved.

HILLEMAN and HORSFALL ($H_{117}$) found that the ratio of *in ovo* neutralization
to HI titres of fowl antisera against homologous virus varied widely, but that
when strains of virus were classified by complete cross neutralization tests the
results were closely similar to those obtained by cross-HI tests.

## XVIII. Growth Cycle of the Influenza Virus — General Description

Following the discovery that influenza viruses could be readily cultivated
in the allantoic sac of fertile eggs with a high yield of relatively clean virus,
many workers studied the conditions of growth with the object of determining
the optimal conditions for maximal yield of virus. Observations of this kind
were stimulated by the discovery by VON MAGNUS that the inoculation of large
doses of virus led to the production of virus with a much lower infectivity titre
in relation to haemagglutinin titre than if small inocula were used.

Studies of the growth cycle, however, were primarily initiated by the work
of HENLE, HENLE and ROSENBERG ($H_{87}$, $H_{95}$). They inoculated eggs to the
allantoic sac with PR8 or LEE virus and one hour later introduced a large
dose of the heterologous virus which had been rendered non-infective by irradiation with ultra-violet light. The irradiated virus blocked the cell receptors
and prevented any subsequent infection of the cells. By this means the growth
of the original inoculum was restricted to a single cycle. The infectivity of the
allantoic fluid was measured at intervals. It was found that the infectivity remained stationary for a period of 6 hours with virus A and 9 hours with virus B.
Then there occurred a sharp rise in infectivity over a period of 2—4 hours after
which the level again remained constant. For each infective dose of virus adsorbed
by the cells of the chorioallantoic membrane some 30—50 infective doses were
released with virus B, and 60—100 doses with virus A. The duration of the
stationary period and the yield of virus in the one-step growth were unaffected
by the inoculum dose over a wide range.

If single eggs are inoculated with virus and the haemagglutinin or infectivity
titre of the allantoic fluid measured at intervals it is quite often found that
step-like increases of titre occur (Fig. 37). Similar results may be obtained in
tissue culture preparations of chorioallantoic membrane ($H_{172}$, $H_{173}$). These
results show that the multiplication of virus in the egg is a cyclic process, cycles
of intracellular growth being followed by the release of virus in relatively short
bursts. Other workers, however, have found that the increase of virus in the
allantoic fluid follows an exponential curve, a hundred-fold increase occurring
every 5 hours ($Z_{38}$, $H_{165}$, $H_{166}$, $C_7$), and that the release of virus from the cells

may extend over several hours ($C_3$, $F_{71}$). These different results may be explained in the following way. If a small dose of virus is inoculated into the allantoic sac infection of the cells occurs only after a variable delay which is due to some of the particles combining with inhibitor in the allantoic fluid and only reaching the cells after destruction of inhibitor by the virus enzyme. As a result there is a lack of synchrony in the infection of the cells, the release of virus from the cells is spread over a wide period, and the growth cycles tend to partly overlap in time ($P_{10}$). If large doses of virus are inoculated inhibitor destruction is very rapid and cells tend to become infected nearly simultaneously. In such cases the step-like increase will be observed. The effects of allantoic fluid inhibitor probably account for the finding that the duration of the first cycle of multiplication is longer than subsequent cycles ($C_3$, $C_7$, $H_{172}$, $W_{45}$, $F_{33}$).

A further variable delay occurs in the release of virus from the cells ($C_{10}$). This delay is abolished if the cell receptors are destroyed by RDE. If eggs are inoculated with relatively large doses of virus the residual virus in the allantoic fluid will destroy the cell receptors and lead to a more rapid release of virus. Infection with large doses of virus has been observed to reduce the latent period ($H_{166}$).

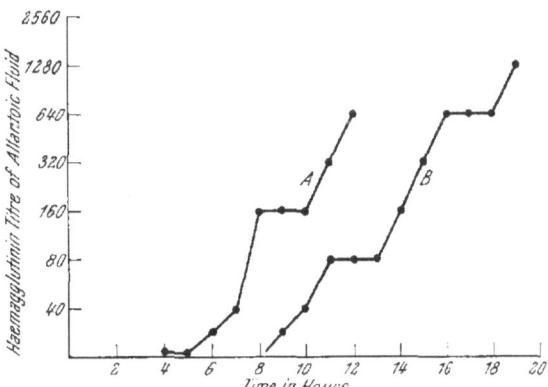

Fig. 37. Step-like increase in haemagglutinin titre in allantoic fluid.

Examples of step-like increases of haemagglutinin titre in allantoic fluid in individual eggs. Egg A: Inoculated with 100 million infecting doses of DSP virus. Egg B: Inoculated with 1 million infecting doses of DSP virus.

SCHLESINGER and KARR ($S_{52}$) estimated the amount of extractable inhibitor in the cells of the chorioallantoic membrane at intervals during the growth of virus and found that there was a step-like breakdown followed by partial restoration of cell-associated inhibitor, the destruction steps occurring just before the release of virus in the first and second cycles. They concluded that the infective process in eggs is a cyclic and not a continuous process.

These effects of inhibitors explain the finding that in infectivity titrations where eggs are inoculated with very small doses haemagglutinin may only become detectable in some eggs after 3—4 days incubation ($L_{104}$).

## 1. Some Quantitative Considerations

When virus is introduced into the allantoic sac the whole of the inoculum is not adsorbed by the cells of the chorioallantoic membrane. With small doses a constant proportion of the virus is adsorbed, the proportion being found by different workers to vary from 70—90% ($H_{66}$, $H_{166}$, $H_{191}$). Differences are probably due to the use of different strains of virus and the use of eggs at different stages of development. With large doses of virus the percentage adsorption is reduced to 50% or less.

The number of cells lining the allantoic sac has been the subject of some dispute, values ranging from 20 million to 500 million having been obtained by different workers using different methods ($H_{69}$, $H_{189}$, $C_8$, $F_{33}$, $T_{80}$, $C_9$). By direct counting under the microscope of cells in measured pieces of chorioallantoic membrane TYRRELL, TAMM, FORSSMAN and HORSFALL ($T_{80}$) estimated the number in 10 day eggs at 20 million. CAIRNS and FAZEKAS DE ST GROTH ($C_9$) found the number to be 30 million in 10 day eggs and 47 million in 12 day eggs, but noted that the entodermal layer was more than one cell thick, and if the deeper cells were included the count was 76 million in 12 day eggs. These direct counting methods almost certainly underestimate the number of cells involved in the growth process, since between the 10th and 12th day the cells are actively multiplying, and during infection many cells are desquamated into the allantoic fluid to be replaced by fresh cells. An estimate of the number of cells may be obtained by dividing the total yield of virus by the cyclic increment determined in growth cycle experiments, and this method gives values between 100 and 200 million. By the use of a somewhat complicated analysis of the results of interference experiments FAZEKAS DE ST GROTH and EDNEY ($F_{39}$) estimated the number of cells at 300—500 million, but the methods involved some assumptions of uncertain validity.

It is clear that the number of cells involved in the growth of virus in eggs cannot be estimated with certainty, but the original estimate of 100 million by HENLE is probably about right.

The yield of virus from one cell has been variously estimated to range from 50—120 $ID_{50}$ in the case of virus A ($F_{33}$), but results are affected by the inoculum dose ($H_{172}$). The rate of growth of virus in calf kidney tissue cultures is much slower than in eggs but the yield per cell was estimated at 100 infective units by LEHMANN-GRUBE and FAZEKAS DE ST GROTH ($L_{56}$). The electron microscope studies of infected cells by MORGAN, ROSE and MOORE ($M_{118}$) show that the number of virus particles released from the cell must be much greater than 100. It has usually been found that the infective dose represents about 10 particles, so that the number of virus particles released from one cell may be about 1000. HORSFALL ($H_{166}$) found that the yield of haemagglutinating particles from one cell was 900—1400 with LEE virus and 500—900 with PR 8.

## 2. Intracellular Reproduction of the Influenza Viruses

Studies of the intracellular phase of growth were first made by HOYLE ($H_{188}$, $H_{192}$). Fertile eggs were inoculated to the allantoic fluid with a dose of virus large enough to infect all the cells lining the sac. At hourly intervals after infection chorioallantoic membranes were removed, saline extracts prepared, and the amount of complement fixing S antigen, haemagglutinin and infective virus in the extracts measured. No increase of infective virus could be demonstrated until 5—6 hours after infection, but S antigen could be detected by complement fixation at 2 hours and attained its maximal level in 5—6 hours. Haemagglutinin was detected at 3 hours and rapidly increased between 3 and 5 hours. Not only was no increase in infective virus demonstrable in the early stages, but the amount detectable in membrane extracts was very considerably less than the amount originally entering the cells. The inoculated virus appeared to be almost entirely destroyed. It was suggested that on entry into the cell the infecting virus was disrupted with the release of sub-units which were then synthesized in the cell and later aggregated to form new virus.

The results of a typical growth cycle experiment are shown in Fig. 38. Twelve day old fertile eggs were inoculated with 0.3 ml of 18 hours infected allantoic fluid of known haemagglutinin titre. After 30 minutes the residual haemagglutinin in the allantoic fluid was measured and the amount of virus adsorbed by the cells calculated. The amount of haemagglutinin, specific complement fixing antigen and group S antigen present in the virus entering the cells was

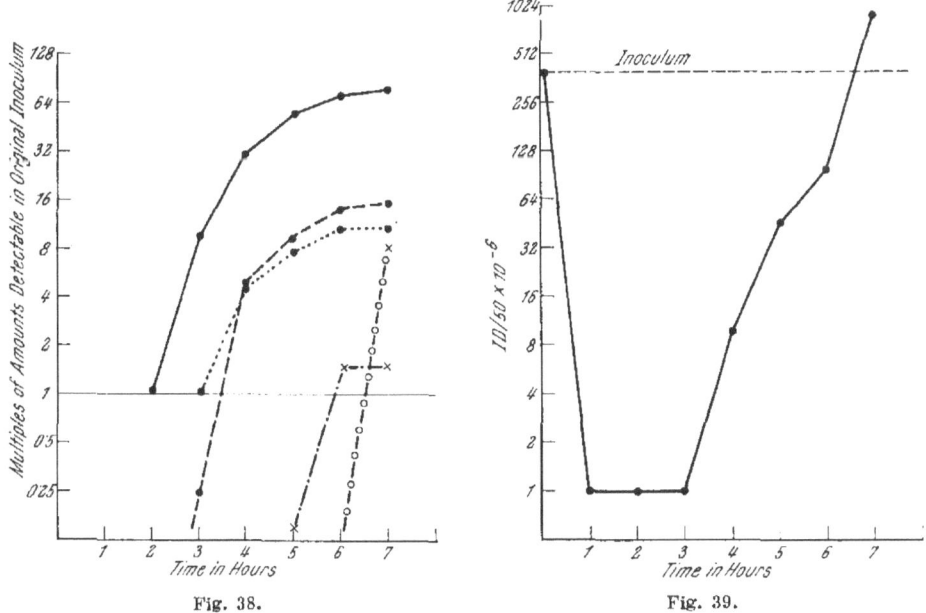

Fig. 38.                                                Fig. 39.

Fig. 38. Growth cycle of virus A, intracellular synthesis of virus components.
Growth cycle of virus A, DSP strain. Soluble antigen in membrane ●————●; specific CF antigen in membrane ○·····○; haemagglutinin in membrane ●— — —●; infective virus in membrane ×—·—·—×; infective virus in allantoic fluid ×—○—○—×.

Fig. 39. The eclipse phase.
Recovery of infective virus from chorioallantoic membranes of eggs inoculated with a large dose of DSP virus.

calculated, haemagglutinin and specific CF antigen being measured in the intact inoculated virus and the S antigen being measured after disintegration of the virus by ether treatment. Chorioallantoic membrane extracts were made at hourly intervals, the membranes being suspended in 1 ml of saline, frozen and thawed three times and then incubated for 6 hours to allow of elution of virus from inhibitors. The suspensions were then centrifuged and the content of CF antigen, haemagglutinin and infective virus measured. Infective virus was also measured in the allantoic fluids. The results were recorded in terms of multiples of the amounts of the various agents present in the virus originally entering the cells. It is seen that the S antigen is synthesized first, commencing at 2 hours, the V antigen and haemagglutinin are synthesized about an hour later, while increase of infective virus only occurs after 6 hours and only small amounts are produced in the membrane, the virus being immediately excreted to appear in the allantoic fluid.

The "eclipse" of the infecting virus is shown in Fig. 39. In this experiment eggs were inoculated with a large dose of virus and the amount entering the cells measured. Chorioallantoic membranes were collected at hourly intervals, washed, extracts prepared and the content of infective virus measured by titration in the allantoic sac. The amount of infective virus recovered from the cells 1 hour after infection was only 1/400th of the amount originally entering the cells, and remained at this low level for 3 hours. Increase then commenced but the amount only became greater than that originally inoculated at $6\frac{1}{2}$ hours.

The original work was confirmed and extended in subsequent studies ($H_{188-192}$, $H_{194}$). Essentially similar results were obtained by HENLE and his associates ($H_{66-68}$, $L_{102}$) and by SCHÄFER and MUNK ($S_{36}$) with fowl plague virus. The cyclic multiplication of virus has also been demonstrated in mice ($K_8$) and in mouse lung S antigen production precedes release of infective virus ($F_{25}$).

WATSON and COONS ($W_{24}$) studied the growth of virus in the amniotic sac using fluorescent antibody to detect intracellular virus antigen. Virus growth occurred in the cells lining the amnion and in the epidermal and pharyngeal epithelium. Staining with fluorescent antibody was first detected in the cell nuclei and later in the cytoplasm. LIU ($L_{98}$, $L_{99}$) working with infected nasal epithelium in ferrets showed that nuclear staining was due to S antigen, while the V antigen was found only in the cytoplasm. Similar observations were made with fowl plague virus ($B_{123}$, $R_{65}$). The S antigen appears to be synthesized in the cell nucleus and later spreads to the cytoplasm, while V antigen is only found in the cytoplasm and later accumulates near the cell surface ($H_{157}$, $T_{60}$).

The disintegration of virus on entry into the cell was confirmed by the use of virus labelled with radioactive tracers ($H_{206}$, $H_{207}$). It was found that free nucleic acid was released and appeared to become associated with the nuclear material of the cell.

The growth cycle of the influenza virus appears to be similar in many different types of host cell and to consist of a number of well-defined stages, adsorption, penetration, eclipse, nuclear synthesis, cytoplasmic synthesis, maturation and release. These stages will now be considered in greater detail.

# XIX. Stages in the Growth Cycle of Influenza Viruses

## A. Initial Union of Virus with the Cell: Adsorption Stage

The presence of virus receptors in the cells of the respiratory tract was first demonstrated by HIRST ($H_{130}$). Excised ferret lungs were perfused with virus suspension by means of a cannula inserted into the trachea. It was found that 99% of the virus was adsorbed. By perfusion of the lungs with saline at intervals after introduction of virus it was found that 25% of the adsorbed virus was eluted in 5 hours in the case of PR8 virus, and 75% with LEE virus. If living ferrets were inoculated intranasally and then killed at intervals and the lungs perfused with saline, it was found that some virus could be recovered in animals killed after 10 minutes, but more was recovered in ferrets killed after 2 hours. Large amounts of virus were recovered from animals killed at 24 hours indicating multiplication of the virus. Similar experiments were carried out by SZOLLOSY, IVANOVICS and HORVATH ($S_{348}$) using excised lung preparations from mice, rats, hamsters, guinea pigs, rabbits and chickens. Adsorption rates of from 90—99% were found and most of the adsorbed virus was not eluted. They noted that the amount of receptor substance in the respiratory tract was very great and was not exhausted by 10 successive administrations of highly potent virus.

Essentially similar results were obtained in fertile eggs ($H_{188}$, $H_{191}$, $H_{66}$). In *in vitro* experiments with washed chorioallantoic membranes suspended in saline containing varying amounts of virus it was found that the percentage of virus

Table 6. *Recovery of Virus from Chorio-allantoic Membrane in vitro*

1 ml amounts of dilutions of DSP allantoic fluid were adsorbed with the washed chorio-allantoic membrane of one egg. After 15 minutes at 37°C the membranes were removed, washed, frozen and thawed three times, and eluted into azide saline at 37°C for 6 hours. Haemagglutination tests were made on the original, adsorbed, and eluted fluids

| Amount of virus HA units | Amount taken up of membrane | | Amount recovered in eluate | |
|---|---|---|---|---|
| 16 | 16 | 100% | 0 | 0% |
| 32 | 30 | 96% | 0 | 0% |
| 64 | 56 | 87% | 0 | 0% |
| 128 | 96 | 75% | 2 | 2% |
| 256 | 192 | 75% | 16 | 9.3% |
| 512 | 384 | 75% | 64 | 16.6% |
| 1024 | 768 | 75% | 256 | 33.3% |
| 2048 | 1536 | 75% | 768 | 50% |

Table 7. *Adsorption and Destruction of Virus by the Living Chorio-allantoic Membrane*

Eggs inoculated with varying amounts of DSP allantoic fluid, incubated 1 hour, and the allantoic fluids and chorio-allantoic membranes collected. Membranes eluted into azide saline for 24 hours at 37°C

| Amount of virus inoculated HA units | Amount recovered in allantoic fluid | | Amount recovered in membrane eluate | |
|---|---|---|---|---|
| 50 | 0 | | 0 | |
| 100 | 0 | | 0 | |
| 200 | 0 | | 0 | |
| 400 | 30 | 8% | 0 | |
| 800 | 80 | 10% | 12 | 1.5% |

adsorbed was constant over a fairly wide range, but with very small amounts of virus the percentage adsorption increased ($H_{191}$). The results were very similar to those obtained by adsorption of virus by red blood cells, but whereas in the case of red blood cells the whole of the adsorbed virus could be recovered by elution, with chorioallantoic membrane only a small part of the adsorbed virus could be recovered either by simple elution for 6 hours at 37°C or by elution after disintegrating the cells by freezing and thawing (Table 6). In the intact living egg inoculated to the allantoic sac HOYLE found that 90% of the inoculated virus was adsorbed in 1 hour, but no haemagglutinating virus could be recovered in chorioallantoic membrane extracts except when very large doses were used, and even then only about 1% of the adsorbed virus could be recovered (Table 7). In the living egg HENLE ($H_{66}$) found that 70% of the inoculated virus was adsorbed and only 1—5% could be recovered from the membranes, and this was believed to represent superficially adsorbed virus.

In tissue culture studies with chorioallantoic membrane TAMM and TYRRELL ($T_{25}$) found that less than 10% of the adsorbed virus could be recovered and 90% of this was eliminated by immune serum. LIPPELT and MANNWEILER ($L_{95}$) found that chick embryo kidney monolayers take up 80% of the inoculated virus over a 1000-fold range. ALTUCCI, CORREGIO, MARGHERITA and GALEOTA ($A_{53}$) using chorioallantoic membrane tissue cultures found that the amount of adsorbed virus ranged from 50% with inocula of 1000 $EID_{50}$ per cell to 98% with 0.1 $EID_{50}$ per cell.

BURNET ($B_{170}$) found that in living eggs virus introduced into the allantoic sac is adsorbed but not eluted. If the embryos were killed by washing out the allantoic sac with formalin adsorption-elution occurred as with red blood cells.

## 1. Effects of Receptor Destruction by RDE

Extensive studies of the effects of destruction of the cell receptors by the *V. cholera* enzymes were made by Australian workers ($F_{21}, F_{22}, S_{271}, S_{272}, B_{170}, B_{175}$).

FAZEKAS DE ST GROTH ($F_{21}, F_{22}$) in perfusion experiments with mouse lung found that both active and heated virus was adsorbed by the respiratory surfaces. Some active virus was eluted 50 minutes after addition, but heated virus remained adsorbed. Preliminary treatment with RDE prevented adsorption and treatment with RDE caused elution of heated virus.

The 50% elution time of active virus could be used as an indirect measure of the lung receptors and it was found that a single exposure of mice to an RDE aerosol caused a progressive reduction in receptors for about 8 hours. Regeneration of receptors began after 30 hours and reached normal levels in six days. Regeneration occurred even after 14 consecutive exposures to RDE, and RDE produced no changes in the mouse lungs.

STONE ($S_{271}$) found that injection of RDE into the allantoic sac 30 minutes before inoculation of virus prevented infection. RDE had no direct effect on the virus. Mice exposed to RDE mists or inoculated intranasally with RDE were protected against virus given intranasally shortly afterwards. After 24 hours the protective effect wore off, presumably a result of receptor regeneration. Different strains of virus varied in susceptibility to the inhibition by RDE.

The author found that if eggs were inoculated with a large dose of virus after pre-treatment with RDE infection did occur, but the yield of S antigen in the first cycle was only 25% of that in control eggs. Treatment with RDE 2 hours after infection had no effect on the yield of S antigen at 7 hours.

CAIRNS ($C_1$) found that mice inoculated intracerebrally with 12000 units of RDE were partially protected against subsequent inoculation of neurotropic WS virus. Protection was maximal 4 days after inoculation of enzyme at which time the brain content of Francis inhibitor reached a minimal value. FINTER, LIU, LIEBERMANN and HENLE ($F_{71}$) used the de-embryonated egg technique to demonstrate the protective effect of RDE, and BORECKY ($B_{104}$) showed in de-embryonated eggs that the receptor-destroying factor in pneumococcal cultures caused a prolongation of the eclipse phase in influenza virus growth and limited multiplication to a single cycle. The factor also protected mice and caused temporary loss of red cell receptors on intravenous inoculation of guinea pigs. HAFF and STEWART ($H_6$) found that treatment of chick embryo cell monolayers with RDE produced a 2.5—4.7 fold reduction of plaque formation by various myxoviruses. They considered that some of the cell receptors were resistant to the enzyme.

ACKERMANN, ISHIDA and MAASSAB ($I_{37}$, $A_{10}$) in studies of the binding of influenza virus to chorioallantoic membrane pieces in tissue culture found that some infectious virus was bound to the tissue but was unable to initiate infection. This bound infectious virus, BIV, could not be removed by repeated washing with saline which removes superficial virus, but could be recovered by disruption of the cells. The amount of BIV was reduced by treatment with RDE and also by pre-treatment with x-amino-p-methoxyphenylmethane sulphonic acid (AMPS), and was entirely abolished by the use of both RDE and AMPS. It was suggested that active virus is held by two types of binding at the same site, one binding being sensitive to RDE and the other to AMPS.

## 2. Effects of Immune Serum on Virus Adsorption

Homologous anti-viral sera neutralize virus by preventing adsorption to the cells. Union with the cells is not completely prevented, HULTON and McKEE ($H_{215}$) showed that over-neutralized virus slowly united with the cells and after 24 hours in fertile eggs more active virus could be recovered from membrane extracts by addition of concentrated heat-inactivated virus than could be recovered from the allantoic fluid.

The fact that when virus is introduced into the allantoic sac a constant percentage is adsorbed by the cells over a wide range indicates that the initial union of virus and cell is to some extent reversible, and if adsorption is carried out with chorioallantoic membrane *in vitro* in the cold much of the adsorbed virus can be recovered by repeated washing of the cells with saline. ISHIDA and ACKERMANN ($I_{37}$) showed that this initial reversible union is followed by a more stable union in which virus cannot be removed by washing or by treatment with RDE, but in which it is still sensitive to neutralization by antiserum. This serum-sensitive stable virus-cell complex is formed both at 3°C and 37°C. The complex is stable at 3°C but at 37°C sensitivity to antiserum is rapidly lost, presumably as a result of penetration of virus into the interior of the cell. However, even after 2 hours at 37°C some reduction in the infectious process was produced by addition of neutralizing serum.

## 3. Effects of Haemagglutinin Inhibitors

Haemagglutinin inhibitors of the Francis type which are destroyed by the virus neuraminidase do not prevent infection by virus but may retard the growth of virus in the allantoic sac ($T_1$). Chu-type inhibitors which are not destroyed by the virus enzyme show virus neutralizing properties, preventing union of virus with the cells in a manner similar to immune serum ($C_{53}$).

## 4. Effects of Receptor-blocking Agents

SZANTO, LAUCICOVA, KOCISKOVA and RATHOVA ($S_{345}$, $S_{346}$) prepared an antiserum against the haemagglutinin inhibitor of chorioallantoic membrane. The antibody was adsorbed by chorioallantoic membrane and treated chorioallantoic membranes showed reduced ability to adsorb virus. Virus multiplication in chorioallantoic membrane tissue cultures pre-treated with the serum was greatly reduced as compared with controls. Treatment with the serum after virus adsorption was without effect.

The growth of influenza virus in fertile eggs is greatly reduced by the previous inoculation of homologous or heterologous virus inactivated by heat or ultraviolet light ($H_{75}$). This is the interference phenomenon. The effects are not, however, entirely due to receptor blocking, as interference is produced by doses of UV inactivated virus which are too small to block all the cell receptors.

### 5. Summary of the Observations on Initial Union of Virus with the Cell

The work described above shows that the union of virus with the cells is a two stage process. The first stage is comparable to the union of virus with red cells and is dependent on the union of the virus haemagglutinin with mucoprotein cell receptor substance. The reaction is reversible and is prevented by immune serum or by previous receptor destruction by RDE. This is followed by a second stage in which the union with the cell is more stable and is unaffected by RDE but the virus is still accessible to antibody. It is probable that at this stage a second type of bonding occurs which does not involve RDE sensitive receptors. Both stages are independent of temperature, occurring at both 3°C and 37°C. At low temperatures nothing further occurs, but at 37°C the virus enters a new stage in which it is no longer sensitive to antibody.

### B. Stage of Penetration and Eclipse

Following the original discovery of the enzymic action of influenza viruses on the red blood cell ($H_{129}$) it was thought that the penetration of virus into susceptible cells might be a result of the action of the virus enzyme. Later work in which it was found that the only effect of the virus enzyme was to split off the terminal N-acetylneuraminic acid from the surface mucoprotein of the cell, and that the cell receptor could be completely destroyed by RDE without apparent damage to the cell, made it seem improbable that penetration could be a result of the neuraminidase action.

Fazekas de St Groth ($F_{24}$) showed that cell receptors could be modified by treatment with periodate so that while they still adsorbed influenza virus the virus was not eluted, nor could it be removed by treatment with RDE. Fazekas de St Groth and Graham ($F_{42}$) modified the receptors of the respiratory passages of mice by treatment with periodate and showed that infection could occur through the periodate-treated cells, the infectivity end points being identical in both periodate-treated and normal lungs. It was also shown that the protective effect of RDE was greatly reduced in the mice treated with periodate indicating that infection could occur as a result of union of virus with periodate-treated receptors. Similar results were obtained in the allantoic sac of fertile eggs. It was concluded that penetration of virus into the cells was not a result of the activity of the virus enzyme, but was a passive process described by

FAZEKAS DE ST GROTH ($F_{23}$) as "viropexis", the cell ingesting the virus particle in a manner similar to the taking up of colloidal dyes.

INGLOT and DAVENPORT ($I_1$) showed that polymorph leukocytes actively ingest influenza virus but the intracellular virus was not inactivated. By contrast HeLa cells were found to adsorb virus but on staining with fluorescent antibody only the surface of the cells become stained.

## 1. Electron Microscope Studies of the Penetration of Virus into the Cell

DALES and CHOPPIN ($D_2$) made thin sections of the pellet produced by mixing chorioallantoic membrane or monkey kidney cells with the RI/5⁺ and RI/5⁻ strains of A2 virus. After 20 minutes virus particles were seen closely attached to the cell surface and also enclosed in intracellular vesicles. The virus particles appeared to be actively phagocyted by the cells. Later the particles disintegrated within the vesicles.

HOYLE showed that the cells of the chorioallantoic membrane in normal eggs detach small fragments of cytoplasm enclosed in cell membrane into the allantoic fluid in the form of spherical globules of diameter ranging from 1—5 μ, and by incubating membranes in saline large numbers of these cytoplasmic particles can be produced and can be concentrated and partially purified by centrifugation methods. These particles are easily distinguished from virus particles by their variation in size and by the absence of any layer of surface spikes.

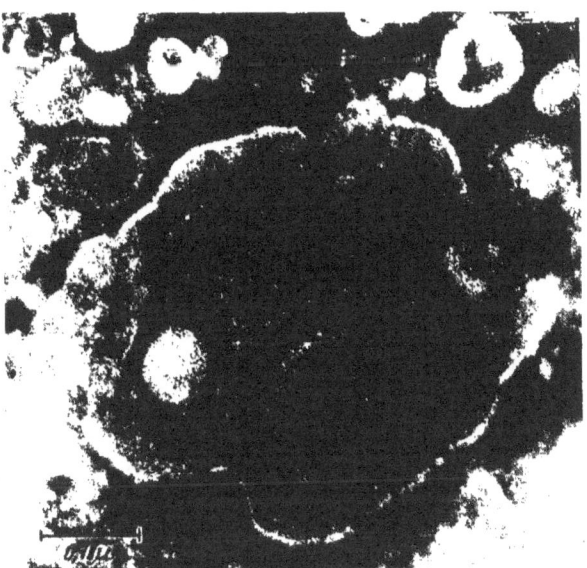

Figs. 40—42. Interaction of virus and normal cytoplasmic particles. Fig. 40. Virus particles adsorbed by a large normal cytoplasmic particle and becoming engulfed. Note the breakdown of the normal particle membrane at the points of contact with virus.

HOYLE, HORNE and WATERSON ($H_{210}$, $H_{199}$) studied the interaction between these normal cytoplasmic particles and influenza virus in the electron microscope. If concentrated preparations of virus and normal cell particles are mixed an agglutination reaction visible to the naked eye occurs in a few seconds. The agglutination rapidly disappears on incubation, and is prevented by pre treatment of the cell particles with RDE. If mixtures are immediately examined in the electron microscope virus particles are seen adherent to the normal cell particles and becoming engulfed by them (Figs. 40, 41). The virus membrane shows a patchy disintegration and the particles burst and release their inner components. The same disintegration is seen in virus particles which do not appear to be inside normal cell particles. The membrane of the normal cell particles shows a patchy disintegration at points where it is in contact with virus (Fig. 40). Many particles are seen which seem to be a result

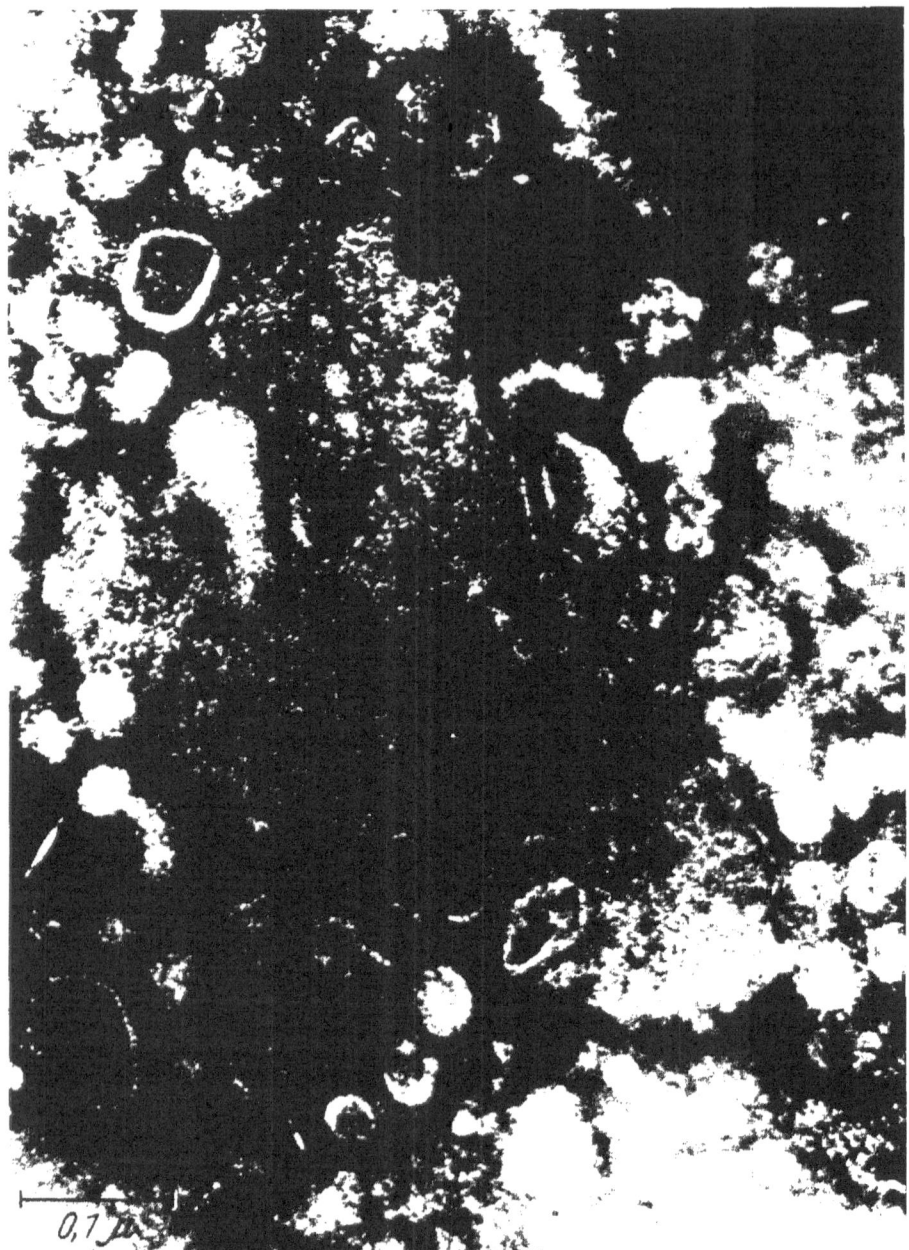

0,1 µ

Fig. 41. Further stage of interaction. Almost total destruction of normal particle membrane. Virus particles disrupting and releasing their internal contents.

of fusion of normal cell and virus. Preparations examined 20 minutes after mixing show almost total disintegration of the virus particles (Fig. 42).

These results suggested that the penetration of virus into the cell was a result of some interaction between the lipoprotein membranes of virus and cell leading to a partial destruction of both membranes.

## 2. Studies of the Penetration Stage by the Use of Radioactive Tracers

The studies of interaction between normal cytoplasmic particles and virus showed a number of curious features. In spite of the almost total disintegration of virus seen in the electron microscope there was no measurable loss of infectivity, and the material released from the particles appeared amorphous in structure, there being no sign of the helical ribonucleoprotein and star-like haemagglutinin particles which appear in virus disintegrated by ether. If virus labelled with $S^{35}$ was used and after interaction with normal cell particles the

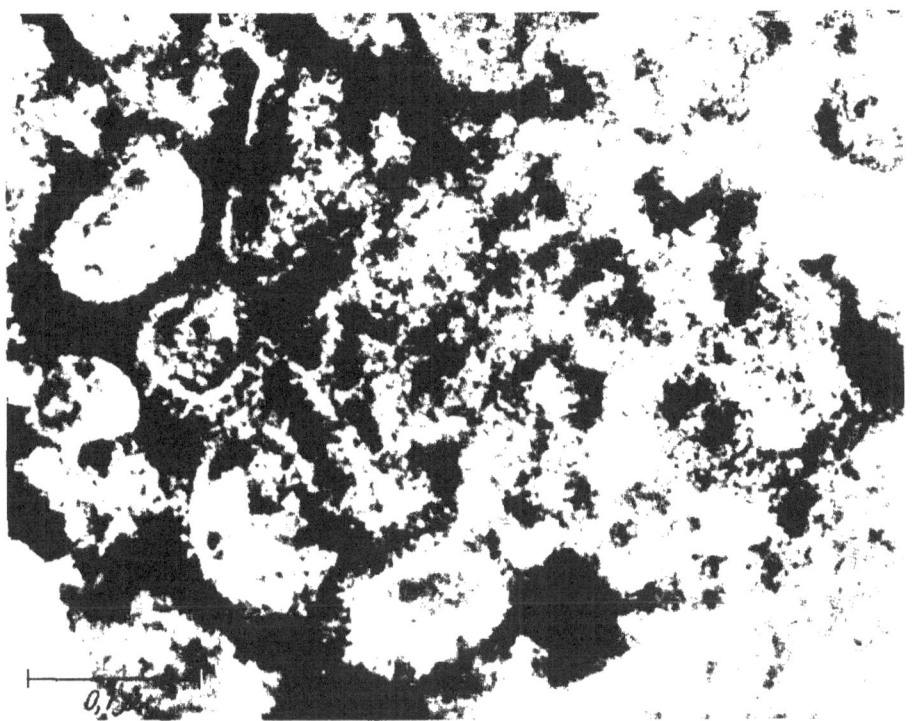

Fig. 12. Late stage of interaction. Almost total disintegration of virus particles, membranes broken down and internal contents released.

mixture was adsorbed with red blood cells, it was found that $99\%$ of the $S^{35}$ was adsorbed, there being no evidence either of the breakdown of virus protein or release of free S antigen ($H_{210}$).

Evidence of the interaction of virus and cell lipoproteins was obtained by labelling the materials with $P^{32}$ ($H_{199}$). In both virus and normal cell particle membranes the lipid is present in two forms: "free" lipid extractable with ether, and "combined" lipid which can only be extracted by denaturing the protein with alcohol. When virus was labelled with $P^{32}$ it was found that $12.6\%$ of the lipid $P^{32}$ was extractable by ether. After interaction with normal cell particles the ether-extractable lipid increased to $36\%$, indicating a breakdown of virus lipoprotein. If the normal cell particles were labelled with $P^{32}$ it was found that $55\%$ of the lipid $P^{32}$ was extractable by ether. After interaction with virus only $40\%$ of the lipid $P^{32}$ was ether soluble, and after incubation of the mixture to destroy haemagglutinin inhibitors it was found that about one third of the lipid $P^{32}$ was adsorbed by the red cells. The results suggested that labelled "free" lipid from the normal cell particle displaced unlabelled virus lipid from combination with protein.

There are several observations which indicate that the lipid constituents of the virus envelope are somewhat labile and may interchange with other lipids. NOLL and YOUNGER ($N_{28}$) found that influenza viruses were adsorbed by a variety of water-insoluble polar lipids, and in the case of hexadecylamine the adsorption was associated with loss of infectivity. UTZ found that virus infectivity was destroyed by a lecithin-like substance from serum or liver at $37°C$ but not at $24°C$ ($U_5$). SIMPSON and HAUSER ($S_{140}$) found that influenza strains grown in chick embryo fibroblasts were very sensitive to phospholipase C but allantoic sac cultures were resistant. Addition of sphingomyelin to culture medium greatly increased the resistance of virus to phospholipase C apparently because the added phospholipid became incorporated in the virus envelope. HOYLE ($H_{199}$) found that influenza viruses were partially disintegrated by shaking with arachis oil and that arachis oil was able to extract $P^{32}$ from the lipoprotein of $P^{32}$ labelled virus.

## Studies with Radioactive Virus Introduced into the Living Cell

HOYLE and FRISCH-NIGGEMEYER ($H_{207}$, $H_{196}$) inoculated both intact and de-embryonated eggs with virus labelled with radiophosphorus and studied the physical and chemical state of the $P^{32}$ present in the chorioallantoic membranes $1\frac{1}{2}$ hours after inoculation. It was found that 77% of the virus phospholipid $P^{32}$ entering the cells was apparently destroyed, as only 23% could be recovered by extraction with alcohol. Saline extracts of the membranes contained water-soluble $P^{32}$ of small molecular weight in amounts comparable to the missing lipid $P^{32}$ and it was concluded that much of the virus phospholipid was broken down on entry of virus into the cell. Saline extracts of the membranes contained small amounts of nucleoprotein $P^{32}$ and much larger amounts of $P^{32}$ which was precipitated by alcohol but not by ammonium sulphate and appeared to be free nucleic acid.

Most of the original nucleoprotein $P^{32}$ of the virus, however, was not extractable from the membranes by physiological saline, and appeared to be associated with the insoluble nuclear material of the cell, since it could be extracted by molar NaCl. When the molar NaCl extracts were diluted with water the precipitated deoxyribonucleoprotein carried 30% of the radioactivity. It was concluded that on entry into the cell the virus nucleoprotein was broken down with the release of free nucleic acid which became associated with the cell nucleus. SCHÄFER ($S_{32}$) in similar experiments with fowl plague virus also demonstrated the release of free nucleic acid on entry of virus into the cell.

HOYLE and FINTER ($H_{206}$) carried out similar experiments with virus labelled with radiosulphur. It was found that chorioallantoic membrane extracts made $1\frac{1}{2}$ hours after inoculation of virus labelled with $S^{35}$ contained radioactive material which could not be sedimented at 100,000 $g$, and was not precipitated by ammonium sulphate, trichloracetic acid, or alcohol. It appeared that on entry into the cell some of the virus protein was broken down, probably into amino acid. The amount of protein $S^{35}$ broken down corresponded fairly closely to the amount of $S^{35}$ present in the original virus nucleoprotein. Most of the $S^{35}$ present in the membranes was, however, present in the form of large molecular weight material of size comparable to that of the original virus, but which was almost entirely non-infective. This large molecular weight protein was extractable only with difficulty, the results suggesting that it was adsorbed to some insoluble cell constituent. There was, however, no evidence of the presence of $S^{35}$ in the nuclear material of the cell, as extraction with molar NaCl did not increase the release, and it was suggested that the material was associated with the cell membrane.

### 3. Effect of Ribonuclease on Virus Production

Le Clerc ($L_{46}$, $L_{47}$) found that the yield of influenza virus in both intact eggs and in chorioallantoic membrane cultures was greatly reduced by ribonuclease added before or shortly after inoculation of virus. No reduction of growth occurred if the enzyme was added 2 hours after virus inoculation. It was shown that the ribonuclease had no action on intact virus and did not prevent adsorption or penetration. It was suggested that on entry into the cell virus disintegrated with the release of free ribonucleic acid which was susceptible to the enzyme.

### 4. Chemical Inhibition of Penetration and Eclipse

A number of inhibitory substances are known which suppress growth of virus by interfering with penetration into the cell. The action of α-amino-p-methoxyphenylmethane sulphonic acid, AMPS, has been described above. The anti-viral substance adamantanamine ($D_{32}$, $D_{33}$, $XF_4$) appears to act in this way. Ammonium chloride and a variety of aliphatic amines similarly do not prevent union of virus with the cell but prevent penetration. The virus remains on the cell surface and is neutralisable by antiserum ($XF_4$).

Allison ($A_{50}$) found that fowl plague virus was inactivated by treatment with mercuric chloride but the inactivation was reversed by cysteine or glutathione. Mercury treated virus was adsorbed by cells and penetrated to the interior but eclipse did not occur, and virus could be recovered from the cells 3 hours after inoculation at a time when untreated virus was eclipsed. It was suggested that the mercury salt prevented uncoating of the nucleic acid by union with disulphide bonds.

Ackermann and Maassab ($A_{15}$, $A_{16}$) found that virus production was inhibited by methoxinine. Methoxinine had no effect on adsorption or penetration but was effective at an early stage in the latent period, preceding the time at which p-fluorophenylalanine inhibited virus production.

Zimmermann and Schäfer ($Z_{45}$) working with fowl plague virus and chick embryo cell tissue cultures showed that addition of p-fluorophenylalanine, FPA, before infection and up to 1 hour after infection prevents all production of virus material. The effect was reversed by addition of phenylalanine. They showed that FPA does not prevent adsorption or penetration of virus, the virus becoming inaccessible to antibody in FPA treated cultures in the same way as in normal cultures. Eclipse of virus also occurred in the presence of FPA since the amount of virus recoverable in cell extracts was less than one hundredth of the number of cells capable of producing virus on further incubation as estimated by distributing intact cells in chick embryo cell monolayers and counting the number of plaques developing on incubation. It was concluded that FPA prevents the synthesis of an "early protein" which is necessary before synthesis of viral components can occur. SCHOLTISSEK and ROTT ($S_{84}$) showed that FPA added early in the cycle not only prevents all production of virus antigens but also prevents synthesis of virus RNA.

## 5. Analysis of the Penetration and Eclipse Stages

Although in some cases it may be preceded by a process of engulfment by the cell resembling phagocytosis, the actual penetration of virus material through the cell membrane is a result of a complex interaction between the lipoprotein membranes of virus and cell leading to partial breakdown of both membranes at the point of contact. The process involves some interchange of lipids between the two membranes and also some lipid destruction with the release of water soluble phosphorus. The penetration is temperature dependent and possibly involves enzyme action. Penetration is inhibited by ammonium salts, aliphatic amines, adamantanamine and AMPS, but the mode of action of these agents is unknown.

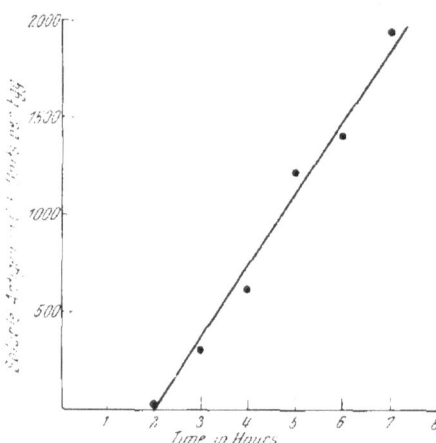

Fig. 43. Linear increase of soluble antigen in the cell. Production of soluble antigen in chorioallantoic membranes of eggs inoculated with $8 \cdot 10^8$ ID$_{50}$ of DSP virus. Mean result of seven experiments.

Eclipse of virus is not a simple result of penetration. Virus treated with mercuric salts penetrates the cell but does not become eclipsed. Eclipse appears to be a result of disintegration of virus material with the release of nucleic acid. The haemagglutinin component of the virus probably remains attached to the cell wall by RDE resistant bonds as it cannot be recovered in cell extracts. The nucleic acid of the ribonucleoprotein component is "uncoated" probably as a result of digestion of the protein by intracellular protease. It is possibly the uncoating stage which is inhibited by methoxinine, though if the uncoating involves synthesis of a new enzyme it is not easy to understand why this stage should not also be inhibited by p-fluorophenylalanine.

Following the release of nucleic acid the further development of the growth cycle is inhibited by FPA which possibly prevents the synthesis of an "early protein" of unknown function.

## C. The Intranuclear Phase

Following the eclipse of infective virus on entry into the cell the first detectable evidence of virus reproduction is the appearance of complement fixing S antigen in cell extracts. In eggs inoculated to the allantoic sac with doses of virus sufficient to infect all the cells the antigen is first detectable in membrane extracts 2 hours after infection. Its subsequent increase is linear and not logarithmic, equal amounts being produced in equal times (Fig. 43).

## 1. Studies with Fluorescent Antibody

WATSON and COONS ($W_{24}$) by staining infected cells with fluorescent antibody at intervals after infection showed that virus antigen first appeared in the cell nuclei, and LIU ($L_{99}$) showed that the nuclear staining was due to S antigen. BREITENFELD and SCHÄFER ($B_{123}$) found in chick fibroblast cultures infected with fowl plague virus that nuclear staining due to S antigen was detected 3 hours after inoculation, while V antigen was only detected in the cytoplasm 2 hours later. Similar results were obtained by many other workers ($W_{40}$, $T_{60}$, $H_{157}$, $S_{68}$, $N_{29}$, $R_{13}$) using a variety of infected cells. All workers are agreed that S antigen is produced in the nucleus and later transported to the cytoplasm, while V antigen is only produced in the cytoplasm. The nuclear synthesis of S antigen differentiates influenza and fowl plague viruses from the NDV, mumps and Sendai group of viruses in which antigen is only produced in the cytoplasm ($W_{40}$, $T_{60}$, $R_{13}$).

WHEELOCK and TAMM ($W_{40}$) noted that fluorescent cells might be encountered in all stages of mitosis, so that virus infection did not prevent cell division.

MORGAN and others ($M_{114}$, $M_{116}$) stained infected chorioallantoic membrane cells with ferritin-conjugated antibody and on electron microscopy found dense aggregates of viral antigen in the cell nucleus.

## 2. Nuclear Synthesis of Virus RNA

By inoculation of cells with influenza virus labelled with $P^{32}$ HOYLE and FRISCH-NIGGEMEYER ($H_{207}$, $H_{196}$) showed that virus nucleic acid entered the cell nucleus and apparently became associated with cell DNA.

SCHOLTISSEK and ROTT ($S_{63}$, $S_{64}$, $S_{68}$) developed a method of distinguishing between different types of RNA. RNA labelled with $P^{32}$ was digested with pancreatic ribonuclease which resulted in the release of pyrimidine nucleotides and oligonucleotides consisting of purine nucleotides with a terminal pyrimidine nucleotide. These were then separated on Dowex 1 columns and the radioactivity of the fractions measured. There was a striking difference in the resulting pattern between virus RNA and cell RNA. By applying pulses of $P^{32}$ to label the newly synthesized RNA in chick embryo fibroblasts infected with fowl plague virus and then determining the oligonucleotide pattern it was possible to determine how much of the new RNA was virus RNA and how much was cell RNA. It was found that at all stages of infection the oligonucleotide pattern was intermediate between that of virus RNA and cell RNA, and it was possible to follow the synthesis of viral RNA in the cells. It was found that virus RNA synthesis commenced after 1 hour and continued throughout the rest of the cycle. By growth of virus in the presence of $C^{14}$ leucine it was shown that S antigen synthesis commenced at the same time as RNA synthesis, and by precipitation of cell extracts with antiserum to S antigen it was shown that almost the whole of the virus RNA synthesised was combined with S antigen. After removal of the S antigen the remaining cell RNA newly synthesized during the first 3 hours had an oligonucleotide pattern similar to that of non-infected cells.

ZHANTIEVA and STAKHANOVA ($Z_{15}$) grew fowl plague virus in human embryo skin-muscle monolayers and at various times of incubation a medium containing 1 microcurie per ml of tritiated uridine was added. After washing the cells were fixed, covered with autoradiographic emulsion, and exposed for 14 days in the dark. Preparations were then developed, fixed and stained. It was found that synthesis of RNA began in the nuclei between 1 and 2 hours after infection, reaching a maximum at 3 hours. Later labelled RNA was found in the cytoplasm. It was concluded that the RNA of fowl plague virus was synthesised in the nucleus. Similar results were obtained by ZHDANOV ($Z_{23}$) with PR8 virus. Autoradiographic studies show that the normal pattern of RNA synthesis in cells is not greatly modified by infection with fowl plague virus ($S_{68}$, $H_{55}$).

SPRECHER-GOLDBERGER ($S_{249}$) suggests that there are two different metabolic patterns for the synthesis of RNA viruses. Ethidium chloride is said to inhibit the utilisation of pre-formed nucleotides in nucleic acid synthesis. It inhibits fowl plague virus much more than NDV, while NDV synthesis is more susceptible to inhibitors of *de novo* synthesis of purines or pyrimidines. Viruses such as fowl plague which are susceptible to ethidium chloride are also more susceptible to protein synthesis inhibitors such as FPA and puromycin, and to inhibitors of DNA function (UV light, actinomycin and histone). It is suggested that the DNA-dependent event in the synthesis of fowl plague virus is the synthesis of an RNA involving the preferential use of pre-formed bases rather than *de novo* synthesis.

### 3. The DNA-dependence of Virus Synthesis

The growth of influenza virus is not inhibited by agents such as aminopterin and fluorodeoxyuridine which inhibit DNA synthesis, but BARRY, IVES and CRUICKSHANK ($B_{14}$) found that actinomycin D, which inhibits DNA controlled RNA synthesis, in concentrations of $1-10$ µg/ml inhibited the growth of influenza virus in chorioallantoic membrane cells. The antibiotic acts by forming a complex with DNA.

ROTT and SCHOLTISSEK ($R_{67}$) found that the synthesis of fowl plague virus RNA, S antigen, haemagglutinin, neuraminidase and infective particles was suppressed by 5 µg/ml of actinomycin D in tissue cultures, while much larger amounts (40 µg/ml) had no effect on multiplication of NDV. If actinomycin was added to fowl plague virus cultures at $1\frac{1}{2}$ to 2 hours after virus RNA synthesis had begun, then further synthesis of virus RNA ceased, but the S antigen titre continued to rise. BARRY ($B_{13}$) noted that fowl plague virus grown in chick embryo cell cultures was much more sensitive to actinomycin than influenza virus grown in chorioallantoic membrane. BUKRINSKAYA ($B_{145}$, $B_{147}$) also noted the resistance of influenza virus to low concentrations of actinomycin D.

Actinomycin inhibits influenza virus reproduction only if added very early in the cycle, once production of virus protein has commenced the antibiotic is completely ineffective. This suggests that no cell-coded messenger RNA synthesis is required for the synthesis of virus protein ($W_{45}$).

The relative resistance of influenza virus to low concentrations of actinomycin has suggested to some workers that it may be acting in some way other than on DNA ($R_{62}$). There is, however, other evidence that DNA is involved in influenza virus synthesis.

Irradiation of cells with ultraviolet light causes severe damage to DNA, multiplication of the cells is stopped and there is a decrease in RNA synthesis.

BARRY ($B_{13}$) exposed chick embryo cell monolayers to ultraviolet light for varying periods and then infected them with either fowl plague virus or NDV. The multiplication of fowl plague virus was progressively inhibited while there was no effect on the growth of NDV. WHITE and CHEYNE ($W_{44}$) obtained similar results with influenza virus and showed that virus production could be stopped by irradiation of the cell-virus complex up to $1\frac{1}{2}$—2 hours after infection but irradiation at late stages had no effect on the yield of virus. ROTT, SABER and SCHOLTISSEK ($R_{62}$) also found that irradiation of cells before infection prevented the synthesis of all components of fowl plague virus but had no effect on NDV.

The antibiotic mitomycin C which acts on DNA also suppresses the growth of fowl plague virus if added early in the cycle but has no effect on the growth of NDV ($R_{62}$, $N_7$). Mitomycin C has no effect on penetration or eclipse but suppresses all RNA synthesis in the cell. The DNA analogues bromo- and iodo-deoxyuridine and cytosine arabinoside had no inhibitory effect so that the block in RNA synthesis must be due to inactivation of existing cellular DNA or of the replicating form of viral RNA ($N_7$).

## 4. The Phenomenon of "Multiplicity-dependent Delay"

CAIRNS ($C_7$) showed that there is a variable delay in the release of virus from infected cells indicating a lack of synchrony in the infective process in different cells. Increase of the inoculum dose reduces this delay and increases the synchrony of virus release. The effect is partly due to variation in the time of adsorption of virus to the cells, but mainly occurs at some stage after infection since it still occurs if the cell receptors are destroyed shortly after addition of virus by treatment with RDE. The phenomenon was studied by WHITE and CHEYNE ($W_{44}$, $W_{45}$). They found that if the haemagglutinin yield was titrated at different times in cultures infected at different multiplicities the production of haemagglutinin occurred more rapidly in cells infected at high multiplicity than at low multiplicity. If the cultures were treated with small doses of fluorophenyl-alanine, which inhibits synthesis of virus protein but not synthesis of "early protein", and after some hours the FPA block was removed by addition of phenylalanine the multiplicity delay phenomenon was abolished. If however large doses of FPA were used so as to block the "early protein" synthesis, then the delay phenomenon still occurred on removal of the FPA block. The delay phenomenon therefore occurred at some stage between the production of "early protein" and the synthesis of virus protein. It was similarly shown that the delay occurred before the stage of synthesis inhibited by actinomycin D.

It is possible that the delay occurs in the interaction of the virus RNA entering the cell nucleus with nuclear DNA, such interaction might occur more rapidly if the amount of virus RNA is increased by increasing the multiplicity of infection.

## 5. Infection with Damaged RNA

The ethylene-iminoquinone Bayer A 139 reacts with the phosphate groups of nucleic acids, breaking the sugar-phosphate "backbone". Hydroxylamine attacks nucleic acid pyrimidines without breaking the strand. Fowl plague and

NDV viruses are inactivated by both agents. SCHOLTISSEK and ROTT ($S_{65}$, $S_{67}$) partially inactivated fowl plague virus by treatment with Bayer A 139 or hydroxylamine for varying periods and then inoculated the virus to chick embryo fibroblasts and studied the yield of infective virus, haemagglutinin, neuraminidase, S antigen and virus RNA, as compared with the yield produced by untreated virus. It was found that while the yield of infectious virus was reduced by short periods of exposure of the virus to Bayer A 139 or hydroxylamine, a longer period of inactivation was required to reduce the yield of haemagglutinin, and still longer periods to reduce the yields of neuraminidase, S antigen and virus RNA (Fig. 44). The results suggested that more chemical "hits" on the virus RNA were required to prevent subsequent synthesis of some of the virus components than others. The results could be explained by supposing that the "target size" for haemagglutinin was one half that for infectivity, for neuraminidase one quarter, and for S antigen and RNA one eighth to one tenth. This type of result was not obtained with NDV, in this case production of all the virus components was equally susceptible to prior treatment with Bayer A 139 or hydroxylamine.

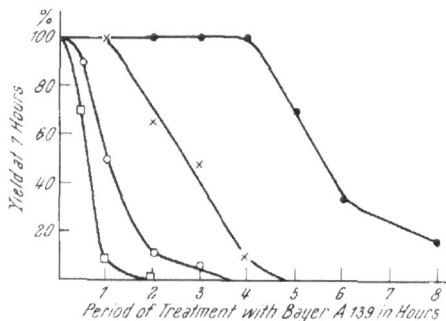

Fig. 44. Effect of treatment of fowl plague virus with Bayer A 139 on ability to synthesize virus components.

Effect of treatment of fowl plague virus with 1% Bayer A 139 on ability to synthesize virus components in chick embryo tissue culture. Plaque-forming units □————□; haemagglutinin ○————○; neuraminidase ×————×; S antigen and RNA ●————●. Modified from SCHOLTISSEK and ROTT ($S_{67}$).

SCHOLTISSEK, DRZENIEK and ROTT ($S_{61}$) prepared VON MAGNUS incomplete virus by serial passage of fowl plague virus in eggs using undiluted allantoic fluid inocula. At each passage chick embryo fibroblast cultures were infected and the production of virus components measured at 6 hours. With increasing undiluted passages there was a loss of capacity to produce infectious virus, haemagglutinin+neuraminidase, and S antigen in that order. In contrast to the results obtained with hydroxylamine and Bayer A 139 there was no uncoupling of capacity to synthesize haemagglutinin and enzyme and it was suggested that the loss of RNA in VON MAGNUS incomplete virus starts at one end of the molecule and is not random as with the chemicals.

## 6. Size of the Infecting RNA Molecule

Virus particles are not infective if they contain less than a certain minimal amount of RNA, corresponding to a molecular weight of $2-2\frac{1}{2}$ million, and it is commonly held that the difficulty in isolating an infectious RNA from influenza virus is due to a difficulty in extracting undegraded RNA. However, the phenomenon of multiplicity reactivation and genetic recombination with influenza viruses suggests that it may not be necessary for the infecting RNA to enter the cell in one piece. DAVIES and BARRY ($D_{31}$) extracted RNA from purified virus with sodium dodecyl sulphate and phenol and centrifuged the RNA in a sucrose gradient. Almost all the RNA sedimented in a sharp band with a sedimentation constant of 18 S. RNA degraded by ribonuclease sedimented much more slowly. The 18 S peak of the virus RNA corresponded exactly with the central 18 S ribosomal RNA peak obtained when chick embryo fibroblast RNA was centrifuged in a sucrose gradient. Ribosomal RNA has a molecular weight of about 500,000. It was suggested that the RNA of the influenza virus particle exists in 4 or 5 separate pieces with molecular weight of 500,000, and that it may be a variety of ribosomal RNA. This idea is supported by the fact that the synthesis of both viral RNA and ribosomal RNA occurs in the nucleus and both are inhibited by actinomycin D.

## 7. Possible Synthesis of V Antigen Precursors in the Cell Nucleus

Although the results of staining infected cells with fluorescent antibody indicates that the V antigen is produced in the cell cytoplasm at a later stage than the production of S antigen in the nucleus, there is some evidence that peptide precursors of the V antigen may be synthesized in the nucleus. SCHOLTISSEK and ROTT ($S_{64}$) grew fowl plague virus in tissue culture in the presence of $C_{14}$ leucine and replaced the $C_{14}$ leucine at intervals with unlabelled leucine. It was found that even when the radioactive leucine was replaced only 2 hours after infection the resulting haemagglutinin protein was labelled. The results suggested that both S antigen protein and haemagglutinin protein synthesis commenced at about 1 hour after infection. FAZEKAS DE ST GROTH ($B_{13}$ discussion) describes similar results.

## D. The Cytoplasmic Phase

The original studies of HOYLE and HENLE ($H_{188}$, $H_{66, 68}$, $H_{190}$) showed that in eggs inoculated to

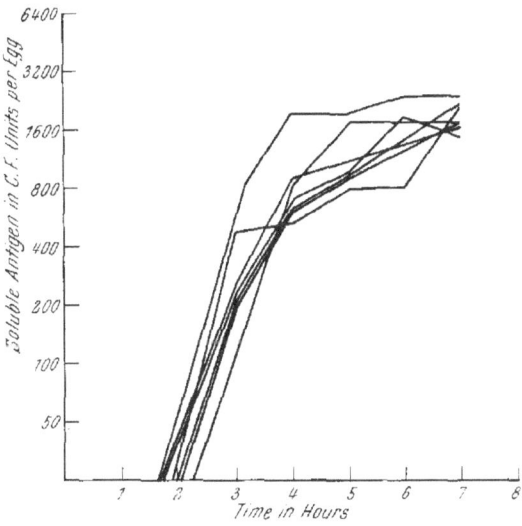

Fig. 45. Intracellular production of soluble antigen. Results of seven experiments on intracellular production of S antigen. Eggs inoculated with $8 \times 10^8$ infecting doses of DSP virus and chorioallantoic membrane extracts made at intervals. The CF unit = Volume of extract × 20 (0.05 ml of extract used as antigen in the CF test).

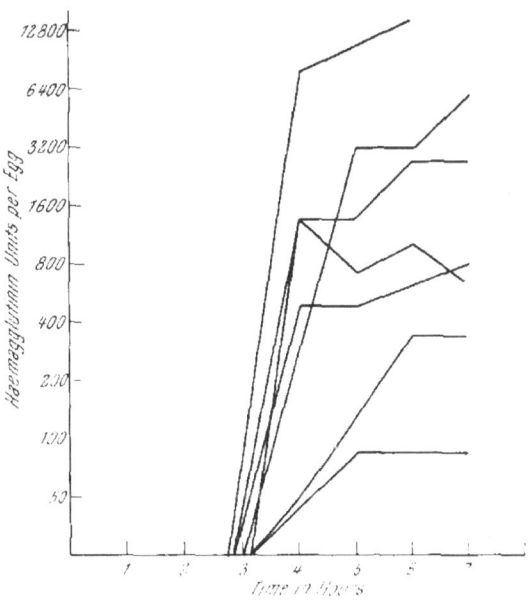

Fig. 46. Intracellular production of haemagglutinin. Results of seven experiments on production of haemagglutinin in the chorioallantoic membrane of eggs inoculated with a large dose of virus.

the allantoic sac with large doses of virus, haemagglutinin and specific complement fixing V antigen were detectable in chorioallantoic membrane extracts 3—4 hours after inoculation, about 1—2 hours after the first detection of S antigen.

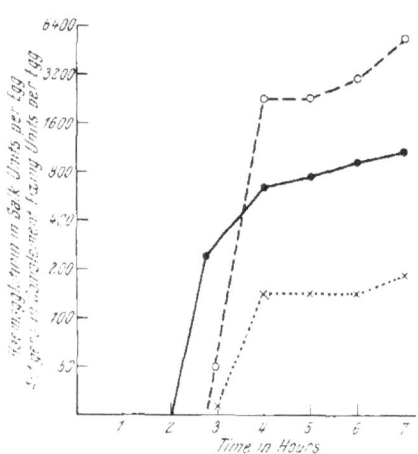

Fig. 47. Production of haemagglutinin, specific complement fixing antigen, and soluble antigen in chorioallantoic membrane.

Production of haemagglutinin, specific CF antigen and soluble antigen in the chorioallantoic membrane of eggs inoculated with 1000 HA units of DSP virus. Haemagglutinin ,— ; specific antigen ⨯ - - - - - ⨯ ; soluble antigen ● — ●.

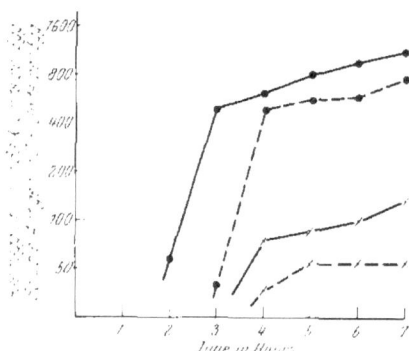

Fig. 48. Effect of dahlia violet on production of S antigen and haemagglutinin.

Effect of 1 mg of Dahlia violet added 30 minutes after inoculation of 300 million ID₅₀ of DSP virus on the production of S antigen and haemagglutinin in the chorioallantoic membrane. Controls (No dye) S antigen ● ●; haemagglutinin ● - - - -●. Dahlia violet S antigen ; Haemagglutinin - - - - - .

The production of haemagglutinin in different experiments is more variable than that of S antigen (Figs. 45, 46). The yield of haemagglutinin at first rises sharply for about 1 hour and then levels off as a result of its release into the allantoic fluid. Although V antigen is sometimes detectable slightly before the detection of haemagglutinin the production of the two agents is closely parallel (Fig. 47) and both are removed by red cell adsorption. The haemagglutinin appears to be a part of the V antigen complex.

HENLE, LUI, PAUCKER and LIEF (H₉₃) studied the hourly production of haemagglutinin and V antigen in de-embryonated eggs and showed that the production could be arrested at any time by addition of 0.01—0.001 M KCN. On removal of the cyanide production was resumed. There appears to be a close relation between the production of haemagglutinin and the prior production of S antigen. If S antigen production is retarded by proflavine or triphenylmethane dyes (H₁₉₁, H₁₈₉) haemagglutinin production is also delayed. As the effect of the dyes wears off and S antigen begins to be produced haemagglutinin production follows after the usual 1—2 hours delay (Fig. 48).

Studies with fluorescent antibody (T₆₀, B₁₂₃, H₁₅₇, S₆₈, S₃₃) show that after its first appearance in the cell nucleus S antigen spreads to the cytoplasm and at this time V antigen becomes detectable in the cytoplasm, at first near the nucleus, later throughout the cytoplasm and finally accumulates at the cell surface.

ZIMMERMANN and SCHÄFER (Z₄₅) showed with fowl plague virus that addition of large doses of fluorophenylalanine, FPA, one hour after infection prevents all production of virus material, but the addition of small doses of FPA 2 hours after infection has no effect on S antigen production but suppresses production of haemagglutinin and infective virus. Under these conditions FPA prevents the escape of S antigen from the nucleus. The effects are reversed by addition of phenylalanine. Proflavine has similar effects on the transport of virus antigen ($A_{108}$). SCHOLTISSEK and ROTT ($S_{62}$) showed that the S antigen produced in the nucleus in the presence of small doses of FPA and labelled with $C_{14}$ leucine did not become incorporated into infective virus particles on subsequent removal of the FPA block with phenylalanine, suggesting that it was defective as a result of incorporation of FPA. ROTT and SCHOLTISSEK ($R_{66}$, $R_{65}$) found that when incomplete fowl plague virus

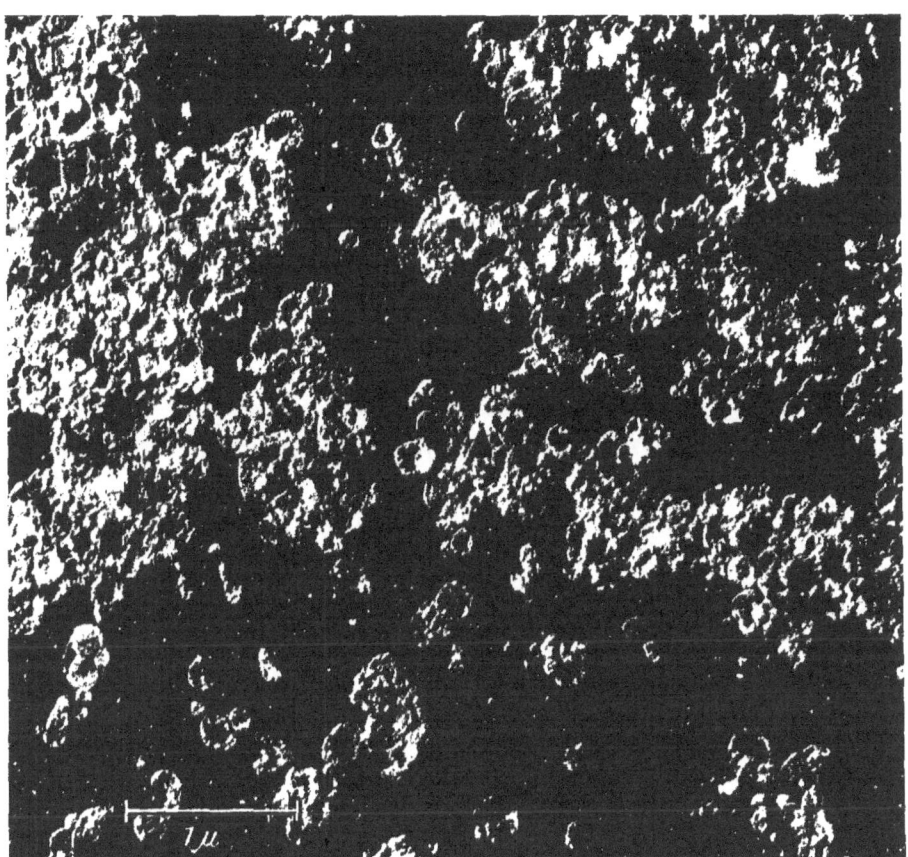

Fig. 49. Intracellular haemagglutinin particles, "viromicrosomes".

was produced by inoculation of large doses of seed virus, the S antigen produced was retained within the cell nucleus and hence did not appear in the final incomplete virus and similar findings were made by FRANKLIN and BREITENFELD ($F_{125}$) studying the incomplete growth cycle of fowl plague virus in Earle's L cells.

## 1. Morphology and Properties of Intracellular Haemagglutinin

The morphology of the intracellular haemagglutinin of fowl plague virus was studied by SCHÄFER and his colleagues ($S_{42}$, $S_{34}$, $R_{65}$). By adsorption and elution from red blood cells they obtained balloon-like particles of variable size resembling microsomes of normal cells. They possessed laminated membranes and contained about 60% of lipid. They carried haemagglutinin and neuraminidase and also glucose-6-phosphatase activity. They were given the name "viromicrosomes" (Fig. 49). By shaking with fluorocarbon ribosome-like particles containing 50% RNA could be isolated from them (Fig. 50). In serological tests the viromicrosomes reacted with anti-V sera but only slightly or not at all with anti-S sera. They reacted strongly with antiserum to normal cell components.

Similar particles were found in cells infected with influenza virus ($R_{63}$). The incomplete influenza virus particles from mouse brain obtained by WERNER and SCHLESINGER ($W_{37}$) had a similar morphology.

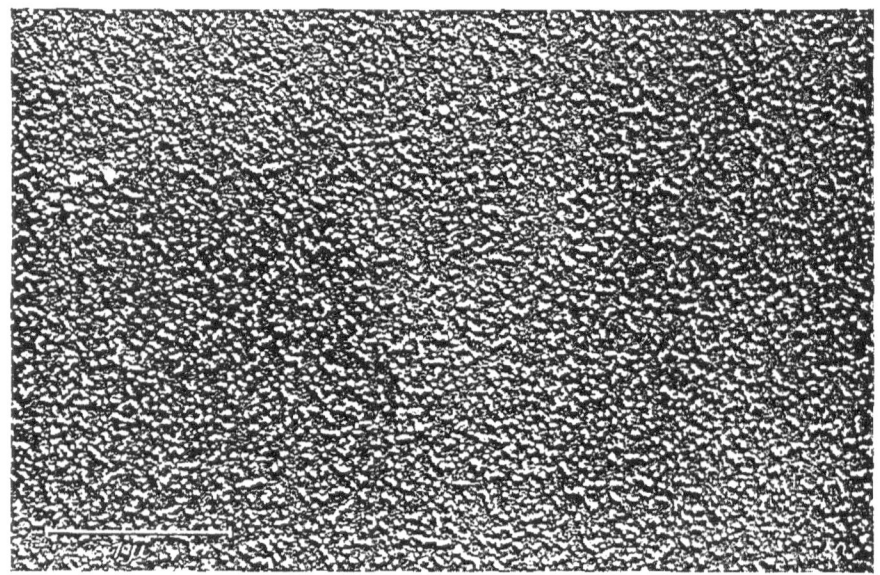

Fig. 50. Ribosome-like particles obtained from viromicrosomes by treatment with fluorocarbon. Reproduced by permission of Dr. W. SCHÄFER from SCHÄFER et al. ($S_{42}$).

FRANKLIN ($F_{124}$) found that the inhibition of haemagglutinin production by proflavine is a multi-hit process and suggested that the haemagglutinating structures are multivalent sites of haemagglutinin synthesis.

## 2. Intracellular Synthesis of Neuraminidase

HENLE, LIU, PAUCKER and LIEF ($H_{93}$) found that the content of haemagglutinin-inhibitor in the chorioallantoic membrane decreased sharply between 3 and 4 hours after infection, coinciding with the appearance of intracellular haemagglutinin. NOLL, AOYAGI and ORLANDO ($N_{26}$) found that normal chorioallantoic membrane extracts do not contain detectable neuraminidase but that in membranes inoculated with LEE virus it appeared and rapidly increased between 6 and $9\frac{1}{2}$ hours after infection. Release of virus from the cell occurred between 8 and $9\frac{1}{2}$ hours.

### 3. State of the Cytoplasmic S Antigen

It is a remarkable fact that in none of the many studies made by electron microscopic methods on cells infected with influenza virus has any helical structure resembling the S antigen component been seen in the cytoplasm ($R_{65}$, $P_{84}$). That such a structure could be seen if it was there is suggested by the finding by PROSE, BALK, LIEBHABER and KRUGMAN ($P_{84}$) that masses of helical material can be seen in thin sections of the cytoplasm of cells infected with a myxovirus of unknown nature isolated from patients with infective hepatitis.

HANA and HOYLE ($H_{17}$) studied the appearance of viral antigens in the course of the growth cycle by gel-diffusion methods. Extracts of infected chorioallantoic membranes were made at hourly intervals after infection and tested against S antibody in gel-diffusion tests. No precipitin lines were seen in 1 hour and 2 hours extracts, but from 3 hours onwards several precipitin lines were obtained indicating the presence of separate S antigen components. It may be that the S antigen disintegrated during the preparation of the extracts or during the course of the gel-diffusion, but it is also possible that S antigen in the cytoplasm is present as units of comparatively small molecular weight.

### 4. Cytoplasmic Inclusion Bodies

In general the usual staining techniques do not reveal the presence of inclusion bodies in cells infected with influenza viruses though there may be some increase in pyronin staining in the cytoplasm. However, inclusion bodies of various types have been described in some types of cell. BURNET ($B_{159}$) found that the tracheal fluid of the chick embryo in eggs inoculated to the amnion contained large vacuolated phagocytic cells in which finely granular eosinophilic material could be seen. PANTHIER, CATEIGNE and HANNOUN ($P_7$) made tracheal washings of mice infected with PR8 virus intranasally and on staining by Machiavello's method found many distorted vacuolated epithelial cells containing minute red-staining bodies on the third day. HARFORD, HAMLIN and PARKER ($H_{47}$) found, on electron microscopic examination of thin sections of the bronchi of mice infected by virus aerosols, cytoplasmic inclusions containing many small virus particles. PIGAREVSKY ($P_{61}$) found that the bronchial cells of infected mice "wall off" the virus forming eosinophilic inclusions which may be eliminated from the cell. Russian workers have also found inclusion bodies in the columnar cells of the human respiratory tract in influenzal infection ($U_4$, $P_{62}$). Mouse fibroblasts infected with A2 virus may show basophilic, eosinophilic or pyroninophilic inclusions ($K_{92}$).

These results may indicate that in some types of cell virus material is excreted into intracellular vacuoles producing stainable inclusions.

## E. Stage of Maturation and Release

### 1. Microscopical Observations on the Release of Influenza Virus from the Cell

The first observations on the release of influenza virus from the infected cell were made by HOYLE ($H_{191}$, $H_{194}$, $H_{195}$) using dark-field microscopy. It was shown that virus material was released from the cell by a process of detachment from the cell of cytoplasmic protrusions enclosed in cell membrane. These protrusions might be spherical or filamentous. The process has been described in a previous chapter. These observations were soon confirmed by the use of electron microscopy ($W_{69}$, $W_{70}$, $M_{150}$). MORGAN, ROSE and MOORE ($M_{118}$) examined thin sections of infected chorioallantoic membranes in the electron microscope and their observations indicate clearly that infective particles are not found in the cytoplasm but are produced at the surface of the cell and are almost immediately released (Fig. 15.) HOTZ and SCHÄFER ($H_{180}$) obtained similar results with fowl plague virus. Some process of concentration and aggregation of virus protein appears to occur at the cell surface and can be demonstrated by the use of ferritin-labelled antibody ($R_{28}$, $M_{113}$). In the earlier stages ferritin-labelled antibody is found in close association with virus particles undergoing release from the cell (Figs. 51, 52), but in later stages of infection antibody accumulates over the whole of the cell surface, suggesting that virus antigen is distributed over the whole of the cell membrane. CHOPPIN ($C_{42}$) used the negative staining technique to study the release of a filamentous strain of virus from monkey kidney cells and found that the surface spikes characteristic of virus particles were present over the whole surface of the filamentous cell protrusions. DUC-NGUYEN, ROSE and MORGAN ($D_{89}$) studied the process using ferritin-labelled antibody against both the V antigen of the virus and against host cell antigen. It was found that as the virus antigen accumulated at the cell surface the host antigen diminished in amount and it was concluded that normal host protein did not constitute an integral part of the virus surface.

## 2. Assembly of Virus Components

The first stage in the maturation of virus at the cell surface would appear to be a replacement of the normal membrane protein by the V antigen complex of the virus. At this stage infected cells show the property of haemadsorption which precedes release of virus into the surrounding medium. At the same time the S antigen component appears to become incorporated into a complex with the V antigen. This process is probably not a simple mechanical enclosure of S antigen in the virus particle, since some methods of disintegration of virus particles do not result in separation of free S antigen. Thus the disintegration which results from interaction of virus particles with normal cell particles ($H_{199}$) results in the release of amorphous material containing both haemagglutinin and S antigen, and the same result occurs if virus particles are disintegrated by treatment with arachis oil. It seems probable that the S antigen in the cytoplasm is present in an incompletely assembled form and that some kind of selective process is involved in its incorporation into the infective particle. Thus, if S antigen is produced in cells in the presence of fluorophenylalanine and the FPA block later removed by addition of phenylalanine the abnormal FPA containing S antigen does not become incorporated ($S_{62}$). Linkage between S antigen and haemagglutinin may be effected by lipid as it is broken by ether treatment. Large amounts of lipid and polysaccharide are incorporated in the final particle, serologically these materials appear to be of host cell origin, and the chemical composition of the virus lipid reflects that of the host cell.

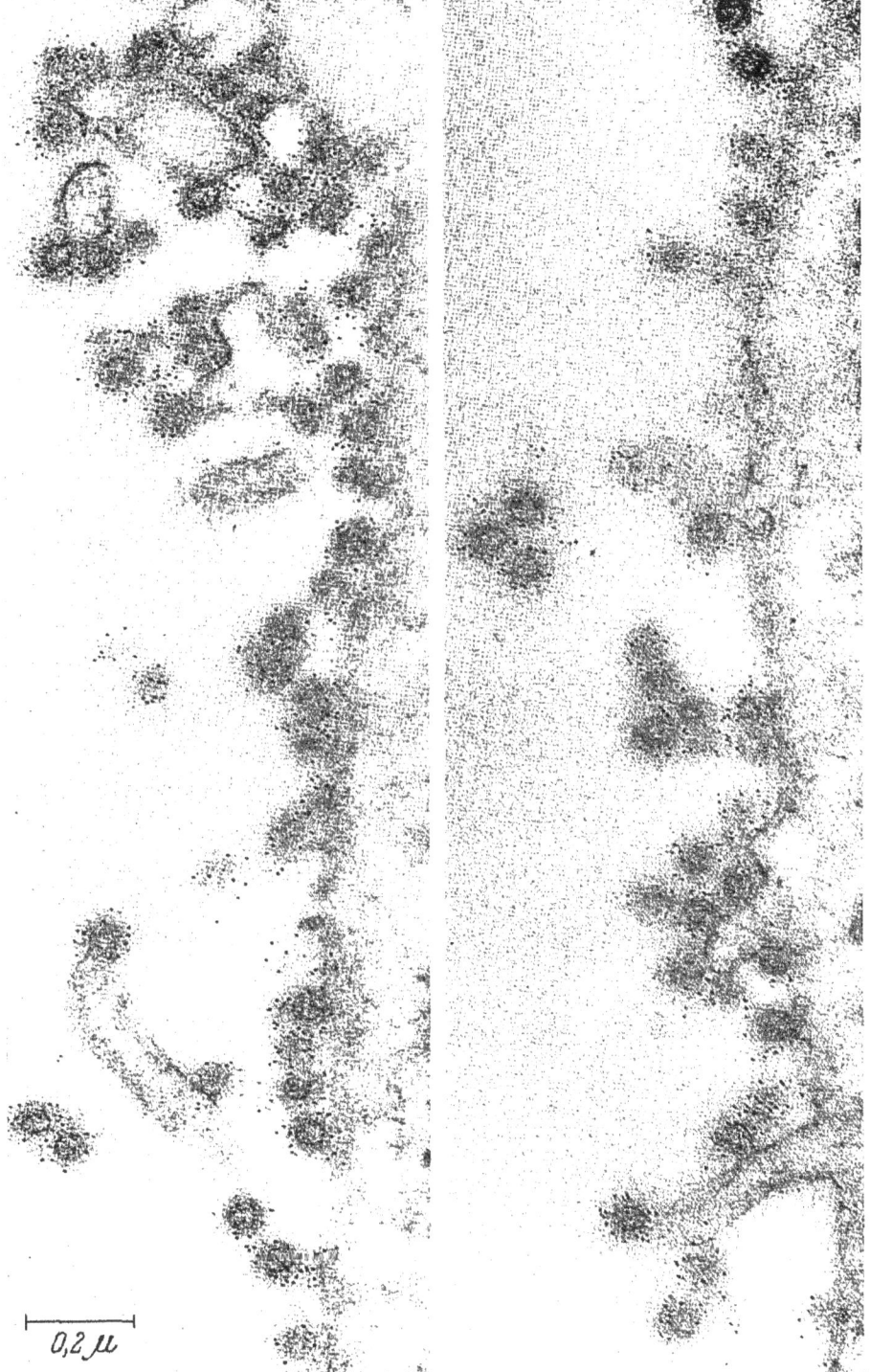

0,2 μ

Figs. 51 and 52. Release of virus from the cell surface.
Thin sections of chorioallantoic membrane infected with PR8 virus and immersed in ferritin-conjugated antibody before fixation. Emerging virus particles tagged with ferritin. Reproduced by permission of Dr. C. MORGAN from MORGAN et al. (M113).

In the process of labelling influenza virus with $P^{32}$ by cultivation in eggs in the presence of radiophosphorus it is found that the intensity of labelling of the virus nucleic acid is almost unaffected by the length of time of exposure of cells to $P^{32}$ before inoculation, but the labelling of the phospholipid is greatly increased if $P^{32}$ is added to the eggs for some time before inoculation of virus ($S_{33}$). This indicates that while the virus nucleic acid is entirely synthesized after inoculation of virus, the virus incorporates pre-formed cell lipids into its structure. MOGAB-GAB and HORSFALL ($M_{108}$) showed that sodium fluoroacetate which blocks the metabolism of carbohydrate and fat at the tricarboxylic acid stage does not inhibit virus production, again suggesting that pre-formed lipid and carbohydrate is utilised in the formation of infective virus.

DAWSON, EPSTEIN and HUMMELER ($D_{35}$) found, in HeLa cells infected with fowl plague virus, that virus particles budding from areas of cell membrane in which adenosine triphosphatase activity could be demonstrated by cytochemical methods incorporated the enzyme in their outer membranes, while particles associated with cell surfaces without enzyme activity were themselves free of it.

### 3. Relation of Neuraminidase to Virus Release

SCHLESINGER and KARR ($S_{52}$) showed that release of virus from the infected cell is immediately preceded by a destruction of the haemagglutinin inhibitor in the cells, and other workers have confirmed the finding ($H_{93}$, $N_{26}$). CAIRNS and MASON ($C_{10}$) found that the delay of about 1 hour which occurs between the formation of haemagglutinin and the release of virus from the cell is almost entirely abolished by pre-treatment of the cells with RDE. SETO and ROTT ($S_{90}$) showed that an antiserum active against fowl plague neuraminidase completely prevented all release of virus from infected cells. It seems possible, therefore, that neuraminidase plays an important role in the release of virus from the cell. This may explain the finding that the duration of the first cycle of virus multiplication is longer than succeeding cycles ($C_3$). Virus released in the first cycle may cause the destruction of cell receptor from outside and thus accelerate release in later cycles. Similarly receptor destruction from outside in the case of eggs inoculated with large doses of virus may result in a more rapid release of virus leading to a "burst" effect, while removal of extracellular virus at intervals as in the experiments of HENLE, LIU and FINTER ($H_{92}$) with de-embryonated eggs may lead to a prolongation of the period of virus release.

### F. The Biochemical Events in Influenza Virus Reproduction

In spite of the large amount of work which has been done on the virus growth cycle we are still unable to give a complete account of the biochemical processes involved. The influenza viruses have certain properties such as multiplicity reactivation, genetic recombination, serological variation and formation of incomplete virus which differentiate them sharply from other RNA-containing viruses, even of closely related types. Their reproduction must involve processes which are different from those occurring with other RNA viruses, and it may be unwise to assume that events known to occur in the reproduction of these viruses also occur in influenza.

## 1. Zhdanov's Theory

ZHDANOV ($Z_{22}$) suggested that six proteins are formed during the multiplication of influenza virus, three structural proteins, the S antigen, haemagglutinin, and neuraminidase; and three "early" proteins, an inhibitor of synthesis of cell messenger RNA, a protein which disintegrates cell RNA in the ribosomes, and a viral RNA polymerase. These are all supposed to be formed along the RNA chain of the virus, and possible arrangements of the six cistrons involved are suggested.

Later ZHDANOV ($Z_{23}$) proposed the following sequence of events. (1) Viral RNA becomes associated with a pre-existing cell enzyme which synthesises a complementary strand forming a double-stranded replicating form of RNA. (2) Cellular RNA polymerase uses the complementary strand to synthesise messenger RNA for the virus RNA polymerase. (3) This polymerase then synthesises numerous molecules of virus RNA. (4) Cellular RNA polymerase transcribes cistrons of virus RNA into messenger RNA for the S and V antigens. (5) S antigen becomes bound to free virus RNA and is transported to the sites of synthesis of V antigen. Three enzymes are needed, one to synthesise the double-stranded RNA, one to synthesise messenger RNA, and one to cause replication of virus RNA.

The concept is extremely speculative and largely unsupported by experimental evidence. It rejects the evidence implicating DNA in the process of virus synthesis, and offers no explanation of the restriction of synthesis of RNA and S antigen to the cell nucleus.

The following attempt to analyse the biochemical processes involved may serve to indicate some of the possibilities, demonstrate some of the difficulties, and expose areas in which knowledge is deficient.

## 2. Early Biochemical Events in Influenza Virus Reproduction

The initial union of virus and cell, a result of attachment of haemagglutinin to cell receptor, is followed by penetration as a result of interaction and fusion of the lipoprotein membranes of virus and cell. Some enzymic destruction of phospholipid or lipoprotein may be involved, but any such enzyme must be of host cell origin.

Disintegration of the inner component of the virus with release of the nucleic acid must also be effected by host cell enzyme, and the appearance of aminoacid derived from the virus protein ($H_{206}$) would suggest that the uncoating enzyme is a protease. If, however, the cells normally contain an enzyme capable of digesting S antigen protein it is not easy to understand the considerable stability of S antigen preparations made from infected chorioallantoic membranes by freezing and thawing. It may be that the enzyme is associated with some insoluble cell constituent and does not appear in cell extracts, or the decoating enzyme may not be a protease. Although S antigen in chorioallantoic membrane extracts is serologically stable, the antigen loses its nucleic acid fairly rapidly on storage.

Following the eclipse of the infecting virus an "early protein" appears to be necessary as all production of virus material is prevented by fluorophenylalanine added early in the cycle. It has been assumed by SCHOLTISSEK and ROTT ($S_{67}$)

that this protein is coded by the virus RNA, but this is not necessarily true, it could be a cell protein which is so rapidly turned over that it has to be continually synthesised to produce its effects. At a later stage in the growth cycle fluorophenylalanine is known to prevent escape of S antigen from the cell nucleus, and it is possible that the early protein may be concerned with penetration of the nuclear membrane by S antigen, and that uncoating occurs in the nucleus. Alternatively the protein may be necessary to initiate some step in the synthesis of virus RNA such as the production of a complementary RNA or the uncoupling of a double-stranded DNA molecule.

### 3. Synthesis of Virus RNA

The synthesis of virus RNA has been supposed by ZHDANOV and by SCHOLTISSEK and others ($S_{68}$) to result from the formation of a complementary RNA to give a double-stranded replicating form. Such a replicating form would probably increase logarithmically and any protein produced would probably also increase logarithmically. The production of S antigen in the cell is a linear and not a logarithmic process, equal amounts are produced in equal times, and there is no evidence that virus RNA increases logarithmically. PONS ($XP_1$) has recently detected a ribonuclease-resistant RNA in the infected cell and suggested that it might be a replicating form, but it could be a DNA-RNA hybrid.

There is considerable evidence that DNA is involved in influenza virus synthesis. The fact that virus RNA synthesis only occurs in the nucleus suggests that DNA may be needed. The synthesis is inhibited by actinomycin D which is known to prevent DNA controlled RNA synthesis, and by mitomycin C which is active against DNA. The suppression of virus reproduction by pre-treatment of the cells with UV light cannot possibly be due to action on the virus and must indicate that DNA is needed for virus synthesis. There is however no evidence that synthesis of DNA is necessary ($N_7$), as inhibitors of DNA synthesis such as bromodeoxyuridine or cytosine arabinoside have no effect on virus reproduction. Any DNA required for virus synthesis must be pre-formed in the cell. If therefore virus DNA is synthesised against a DNA template we must make the remarkable assumption that there exists in all cells capable of supporting influenza virus reproduction a DNA molecule which is complementary to the virus RNA. Can this possibly be true? The normal cell contains DNA capable of coding a messenger RNA which causes the cytoplasmic synthesis of a protein which ultimately becomes incorporated in the cell membrane. In the virus-infected cell a protein, V antigen, is produced in the cytoplasm, which although apparently coded by the virus RNA is nevertheless so similar to normal cell membrane protein that it can replace that protein in the cell membrane. It has been suggested by HOYLE and DAVIES ($H_{201}$) on the basis of aminoacid analyses that the virus haemagglutinin is a modification of a protein which is normally incorporated in the cell membrane. ADA and LIND ($A_{20}$) have shown that small amounts of neuraminidase can be detected in normal chorioallantoic membrane, so that the cells must contain a code for the synthesis of neuraminidase although this is not normally very active. The amount of DNA in the cell nucleus is very large and there may well be DNA molecules in reserve which are not normally active. It is possible that such a reserve DNA molecule may be stimulated into activity by the virus RNA. Assuming this to be so, we may develop the following concept for the possible mechanism of virus synthesis.

## 4. Theory of Synthesis Based on the Use of Cell DNA

It is supposed that there is in the cell nucleus a DNA molecule of which one strand is complementary to the virus RNA. Virus RNA entering the nucleus becomes associated with this DNA molecule which uncouples to form a DNA-RNA hybrid molecule and a single-stranded DNA complementary to the DNA in the hybrid. The DNA-RNA hybrid in association with aminoacids uncouples to form virus ribonucleoprotein (S antigen) and free DNA. The free DNA then produces virus RNA and the process is repeated with the continuous production of virus ribonucleoprotein at a constant rate in the cell nucleus.

Meanwhile the complementary DNA strand synthesises a messenger RNA which leaves the nucleus and causes the cytoplasmic synthesis of V antigen in association with cell microsomes. The V antigen complex is coded by cell DNA, but its composition is indirectly coded by viral RNA as the complementary DNA is derived from a double-stranded DNA of which the other strand is complementary to the virus RNA.

While highly speculative the theory does account for the need for pre-formed DNA, the simultaneous synthesis of virus RNA and S antigen in the nucleus, and for the fact that at no stage are S and V antigens associated in a complex except in the final assembly at the cell surface.

There are two objections to this theory. Treatment of virus with hydroxylamine or Bayer A 139 produces damaged RNA and on entry into the cell S antigen is produced but there is no production of haemagglutinin or neuraminidase. The German workers ($S_{68}$) were unable to find evidence of the production of messenger RNA in the infected cell, and by the use of tracer techniques they found that V and S proteins were simultaneously synthesised, but that V antigen only became serologically detectable after it became associated with cell microsomes. These results suggest that all the virus proteins are formed on the virus RNA, but if this is so it is not easy to understand why only the S antigen is found in association with virus RNA.

## 5. Possible Fragmentation and Re-assembly of Virus RNA

Influenza virus particles are only infective if they contain an amount of RNA corresponding to a molecular weight of $2-2\frac{1}{2}$ million. It is perhaps significant that this is the amount which would be needed to line up against a DNA molecule. There is however little reason to suppose that this amount of RNA must enter the cell as a complete molecule, and DAVIES and BARRY ($D_{31}$) have produced evidence to indicate that the nucleic acid in the virus particle is present in several separate units of molecular weight 500,000. The phenomena of genetic recombination and multiplicity reactivation are best explained by supposing that at some stage virus RNA is fragmented and later re-assembled. In this way RNA fragments from separate strains might become mixed and lead to the production of recombinants. It is also possible that RNA fragments from a single virus might assemble against a DNA template in a different order, especially if there are a number of different DNA molecules to select from. Such a process affords many possibilities for variation. It may be that the virus RNA acts as a master key capable of uncoupling a number of different but related double-stranded DNA molecules.

## 6. Assembly of Virus Components at the Cell Surface

The final virus particle is assembled at the cell surface, S and V antigens becoming associated with lipid and polysaccharide of host cell origin. Some selective processes may be involved, especially in the distribution of V antigen components in the particle. Thus if virus is grown in the presence of small amounts of homologous antibody or in partially immune animals strains of altered serological character may be produced. Such effects could only be produced at the cell surface as antibody can not penetrate the cell wall.

## 7. Summary

It is clear from the very speculative nature of the ideas described above that at present our knowledge of the biochemistry of influenza virus synthesis is quite inadequate to formulate a definite theory. The author would stress the point that no theory will be satisfactory that does not account for the great variability of the influenza viruses, and for phenomena such as reactivation, recombination, and production of incomplete virus which do not occur with most other RNA viruses. It may be that more light will be thrown on the functions of RNA in cells by a study of influenza virus reproduction than from the study of simpler viruses which are less closely integrated with the normal processes of cell protein and nucleic acid synthesis.

# XX. Incomplete Growth Cycles

There are a variety of circumstances in which the growth of influenza virus may be incomplete in that although some of the stages of the growth cycle occur normally there is a block at some stage leading to a failure of the cycle to proceed to the formation of infective virus. Such blocks may be produced experimentally by the use of chemical substances interfering with protein synthesis, thus exposure of the infected cells to fluorophenylalanine 2 hours after infection prevents the release of S antigen from the nucleus and prevents synthesis of V antigen in the cytoplasm. Similar effects may be produced by chemical treatments of the virus which produce damage to the virus nucleic acid and lead to a failure to carry out some stage of the growth cycle ($S_{67}$). We are here concerned, however, with incomplete cycles which are not due to any interference with the virus particle or with the cell metabolism.

## 1. Failure to Excrete Infective Virus from the Cell

There are many types of host cell which can be infected by virus and in which synthesis of the virus components proceeds normally but in which no release of infective virus particles occurs. The ability to release infective particles from the cell depends on the ability of the cell to detach fragments of its cytoplasm enclosed in cell membrane into the surrounding fluid. This capacity to release cytoplasmic particles appears to be property of all cells whose function is excretion or secretion, but some types of cell, notably ectodermal cells and fibroblasts appear to lack this capacity. In such cells infective virus is either not released at all or is released only in small amounts probably as a result of cell destruction. The infected cell may, however, show the phenomenon of haemadsorption, and extracts of the cells may contain large amounts of non-infective haemagglutinin and S antigen, usually separable by red cell adsorption.

### a) Incomplete Growth Cycle in Mouse Brain

SCHLESINGER ($S_{47}$) was the first to demonstrate an incomplete growth cycle of this type. He found that the inoculation of PR8 and LEE viruses into mouse brain resulted in the production of S antigen and haemagglutinin but there was no increase in infective virus. The yield of S antigen and haemagglutinin was related to the dose of virus inoculated, and it was supposed that only a single cycle of multiplication occurred. Mouse brain contains a large amount of haemagglutinin inhibitor and it was necessary to destroy this by inoculation of brain extracts with RDE in order to follow the growth cycle. It was found that 1 hour after inoculation there was a considerable fall in the haemagglutinin titre of brain extracts followed by a rise between 4 and 10 hours, and similar rise in S antigen occurred. The results closely resembled those of "one-step" growth curves of PR8 and LEE virus in eggs, the latent period being longer with LEE than with PR8. The morphology of the incomplete virus was studied by adsorption to red cell ghosts followed by shadow casting and electron microscopy ($W_{37}$). Pleomorphic bag-like structures of variable size were seen. Correlation of particle counts with haemagglutinin titre showed that the incomplete and standard virus were equivalent unit for unit.

The neurotropic NWS strain behaved quite differently from the non-neurotropic PR8 and LEE. In this case intracerebral inoculation resulted in an increase of infective virus and the ratio of infectivity titre to haemagglutinin titre was comparable to that of virus grown in eggs.

SCHLESINGER's results were criticised by CAIRNS ($C_2$, $C_4$) who found that 95% of the inoculated virus entered the general circulation and was lost. If allowance was made for this loss a slight increase of infective virus could be demonstrated. However with the exception of the neurotropic strain NWS all the strains died out on passage, and the slight increase of infective virus noted may have been due to release of inoculated virus from union with inhibitor and not to multiplication.

### b) Incomplete Growth Cycle on the Chorionic Surface of the Chick Embryo Chorioallantoic Membrane

ISAACS and FULTON ($I_{24}$, $F_{165}$) inoculated influenza virus to the external surface of the chorioallantoic membrane. There was no detectable increase of infective virus, but with large doses S antigen was produced in amounts depending on the dose of virus inoculated. Relatively little haemagglutinin was produced but complement fixing V antigen was detected. The authors suggested that the chorionic cells would only support a single cycle of multiplication, but as the cells can obviously not distinguish between the inoculated virus and virus produced after one cycle it is clear that the supposed first cycle is in fact an incomplete cycle with no release of infective virus.

### c) Incomplete Cycle in HeLa Cells

HENLE, GIRARDI and HENLE ($H_{65}$) demonstrated that an incomplete growth cycle occurs in HeLa cells (human cervical carcinoma cells). Inoculation with large doses produces a cytopathogenic effect apparently depending on multiple infection of the cells. On serial passage the effects die out. No infective virus was produced in HeLa cells but there was a considerable increase of S antigen, V antigen and haemagglutinin, but these were not released from the cells until cytopathogenic effects became apparent. LÖFFLER, HENLE and HENLE ($L_{111}$) by staining cells with fluorescent antibody showed that S antigen appeared first and was restricted to the nucleus, V antigen appeared in relation to the Golgi apparatus and spread through the cytoplasm to the cell surface.

WHEELOCK and TAMM ($W_{40}$) showed that the neurotropic strain NWS also gave an incomplete cycle in HeLa cells. The presence of virus antigen on the cell surface in infected HeLa cells can be demonstrated by the occurrence of haemadsorption ($R_{68}$).

### d) Other Examples of Incomplete Cycles with Failure to Release Infective Virus

FRANKLIN and BREITENFELD ($F_{125}$) demonstrated an abortive infection of Earle's L cells by fowl plague virus. Non-infective haemagglutinin and S antigen were produced but there was no yield of infective virus. The S antigen was not released from the nucleus. SMITH and MORGAN ($S_{160}$) obtained similar results with PR8 virus. They showed that a lytic cytopathogenic effect occurred with large doses of virus. ISHIDA and OSATO ($I_{39}$) also obtained an incomplete growth cycle in L cells. A similar incomplete cycle occurs in Krebs 2 ascites tumour cells ($L_{117}$). In BHK21 cells the neurotropic strain NWS gives a complete cycle, but MEL and WSE give incomplete cycles with failure to produce infective virus. There is no retention of S antigen in the nucleus ($F_{133}$).

### 2. Production of Extracellular Incomplete Virus — the von Magnus Phenomenon

In 1947 VON MAGNUS observed that if serial passages were made in the allantoic sac of fertile eggs using undiluted infected allantoic fluid inocula there was a reduction in the infectivity titre of the fluids without a comparable reduction in haemagglutinin titre. This reduction did not occur if passages were made with highly diluted allantoic fluid inocula ($M_{19}$). In later studies ($M_{21}$, $M_{22}$) it was shown that serial passage with infected allantoic fluid diluted 1 in 1 million resulted in the regular and uniform production of virus of which the relation of infectivity titre to haemagglutinin titre (I/H ratio) was $10^6$ or more. Such highly infective preparations are usually called "standard virus". On serial passage with undiluted inocula there occurs a reduction in the I/H ratio until by the 4th passage it may be as low as $10^3$. Little reduction in haemagglutinin titre occurs in the first two undiluted passages, but a reduction occurs in the 3rd and 4th passages. At the 5th passage there is usually an increase in the I/H ratio but not to the level of standard virus. The I/H ratio falls again at the 6th passage. VON MAGNUS concluded that in standard virus one haemagglutination unit corresponded to $10^6-10^7$ $EID_{50}$, but in the third undiluted passage only about 1 in 10,000 of the haemagglutinating particles produced was infective.

These results have been repeatedly confirmed and the phenomenon has been extensively studied. By titration of the various passage fluids in mice VON MAGNUS showed ($M_{22}$) that the incomplete virus interfered with the growth of infective virus. In centrifugation experiments GARD and VON MAGNUS ($G_7$) found that a component with a sedimentation constant of 660 S was present in the first 2 serial undiluted passages but was not detected in the 3rd and 4th, while a component of 380 S was present in all the passages. The 660 S component was believed to be infective virus while the 380 S component was non-infective haemagglutinin. In later studies ($G_8$) standard virus was found to contain a major component of 747 S and about 10—20% of a 600 S component. First undiluted passage fluids contained equal amounts of 745 S and 606 S components. At the second passage a small amount of 767 S component was found, but 95% sedimented as a 568 S component. In the third passage all the material was inhomogeneous, S value ranging from 430—679. There appeared to be a range of particles between fully infective virus and fully incomplete material. It was shown that the incomplete virus was not breakdown product of standard virus. By equilibrium sedimentation in caesium chloride, BARRY ($B_9$) found that standard virus formed a sharp band while incomplete virus was slightly less dense and less homogeneous. In de-embryonated eggs BERNKOPF ($B_{58}$) showed that incomplete virus was produced as a result of the introduction of large inocula, that it was released earlier than fully active virus, and that it interfered with the production of active virus. Infective virus was found to have a sedimentation constant of 630 S and incomplete virus a constant of 480 S ($H_{27}$).

FAZEKAS DE ST GROTH and GRAHAM ($F_{43}$) failed to produce incomplete virus by inoculation of eggs with undiluted LEE virus, but VON MAGNUS ($M_{24}$) showed that incomplete virus was produced by serial undiluted passages of LEE virus and that a succession of cycles occurred with maximum production of incomplete virus in passages 6, 18, 26, 31, 38 and 42.

WERNER ($W_{36}$) showed that incomplete virus could be produced in the amniotic sac by inoculation of undiluted seed virus. GINSBERG ($G_{25}$) produced incomplete virus by inoculation of large doses of virus ($10^{6.5}-10^{8.5}$ $EID_{50}$) intranasally in mice, but only with strains which caused extensive lung consolidation. Incomplete virus was found only in the initial cycle, on serial passage only infective virus was formed. The amount of virus antigen produced was the same whether the yield was infectious or incomplete.

### a) Morphology of von Magnus Incomplete Virus

Electron microscope studies of incomplete virus by shadow techniques or by adsorption on red cell ghosts ($P_{17}$, $P_{46}$, $W_{37}$, $V_{33}$, $S_{104}$, $S_{31}$) showed the particles to be pleomorphic, variable in size and shape and often flattened in appearance. Ultrathin sections revealed similar appearances, and the absence of electron-dense centres ($B_{71}$, $P_{46}$). Negative staining methods ($M_{112}$) confirmed the absence of dense internal bodies. MORGAN, HSU and ROSE ($M_{115}$) studied the release of incomplete virus from cells and showed by the use of ferritin-labelled antibody that the surface of the particles contained viral antigen.

### b) Nucleic Acid Content of von Magnus Incomplete Virus

ADA and PERRY ($A_{25}$) found that the nucleic acid content of incomplete virus was less than that of standard virus. LIEF and HENLE ($L_{77}$) found that if the virus produced on serial undiluted passages was disintegrated with ether there was a progressive fall in the amount of S antigen released as compared with haemagglutinin. The author has also found that VON MAGNUS incomplete virus contains less S antigen than standard virus. SCHOLTISSEK, DRZENIEK and ROTT ($S_{61}$) found that when fowl plague virus produced in a series of undiluted passages was inoculated to chick fibroblast cultures there was a sequential loss of capacity to produce infective virus, haemagglutinin and S antigen in that order. They suggested that in incomplete virus there was a loss of RNA starting at one end of the molecule, and not a random loss.

### c) Other Properties of von Magnus Incomplete Virus

UHLER and GARD ($U_1$) showed that incomplete virus contained 54% lipid, nearly double the lipid content of standard virus. SETO ($S_{84}$) found that the sialidase activity of incomplete virus was only 25—50% of that of standard virus. McKEE ($M_{61}$) found that incomplete virus was almost non-toxic, but was as good an antigen as toxic virus. MANIRE ($M_{33}$) also found incomplete virus to be much less toxic than standard virus on intravenous or intracerebral inoculation of mice. SCHÄFER ($S_{30}$) found that vaccines made with incomplete fowl plague virus were less toxic than standard virus vaccines. PAUCKER and HENLE ($P_{20}$) found that third undiluted fluid passage virus had a reduced ability to interfere with the growth of standard virus, and suggested that this was related to the reduced content of S antigen. HAHNEMANN and REINECKE ($H_7$) found that the production of interferon by LEE virus decreased with increasing incompleteness of the inoculum. Interferon was not concerned in the production of incomplete virus, as virus inocula producing only a small quantity of interferon were superior in producing highly incomplete virus to inocula producing more interferon. Inhibition of interferon production by hydrocortisone in serial passages of undiluted LEE virus had no effect on the degree of incompleteness of the final yield ($R_{19}$).

### 3. Analysis of the von Magnus Phenomenon

The mechanism of the VON MAGNUS phenomenon has been the subject of much controversy and is still not fully understood. It seems most probable that the lack of infectivity of VON MAGNUS incomplete virus is a result of its deficiency in nucleic acid content. This deficiency might be a simple quantitative reduction in the amount of RNA in the particles, or it might be a selective deficiency in one particular component of the nucleic acid. In the first case VON MAGNUS incomplete virus should show the phenomenon of multiplicity reactivation, infectivity being restored if cells are inoculated with more than one particle. In the allantoic sac lined by about 100 million cells it would be necessary to introduce between 10,000 and 20,000 particles in order to ensure that one cell became doubly infected, so that the infectivity of a preparation of incomplete virus would appear to be reduced by a factor of $10^3-10^4$ as compared with standard virus if the particles contained only 50% of the minimal infective dose of nucleic acid. BEALE and FINTER ($B_{29}$) compared the infectivity of incomplete virus preparations as tested by the orthodox *in ovo* method and by the 6 hour soluble antigen production test. In this test eggs are inoculated with large doses of virus so that multiple infection of cells can occur. It was found that the preparations appeared more infective by the 6 hour test than by the *in ovo* test and it was concluded that multiplicity reactivation did occur. BARRY ($B_{10}$) studying dose-response curves in single cycle yield experiments in chorioallantoic membrane tissue cultures found that the haemagglutinin response provoked by incomplete virus was Poissonian and concluded that multiplicity reactivation did not occur. However, his results showed that the haemagglutinin yield from incomplete virus was ten times greater than would have been expected from its content of infective virus measured *in ovo*, so that particles which were apparently non-infective were making some contribution to the final yield. ROTT and SCHOLTISSEK ($R_{66}$) studying the formation of incomplete fowl plague virus in chick embryo cell monolayers were unable to detect multiplicity reactivation, but noted that the plaque size decreased with successive undiluted passages. They considered that identical pieces of RNA were lacking in the incomplete virus particles but that the part of the RNA coding for early protein, S antigen and haemagglutinin was still present. They were, however, unable to detect any difference in composition between the RNA of normal and incomplete particles by oligonucleotide analysis. By staining cells with fluorescent antibody they found that during incomplete virus production the S antigen was produced normally in the cell nucleus but that most of it was not released into the cytoplasm.

### a) Role of Multiple Infection in the Production of Incomplete Virus

VON MAGNUS himself ($M_{23}$) considered that multiple infection of the cells was of major importance in the production of incomplete virus, and this has generally been found to be the case. CAIRNS and EDNEY ($C_8$) claimed that multiple infection was not necessary and that incomplete virus might be produced when only 1% of the susceptible cells were infected. This conclusion appears to have been a result of an incorrect assessment of the number of particles in an infecting dose; FINTER, LIU and HENLE ($F_{70}$) and BARRY ($B_{10}$) produced strong evidence that multiple infection of the cells was in fact necessary.

LIU, PAUCKER and HENLE ($L_{106}$) using combined in ovo-de-embryonation methods to study the virus yields at different stages of the growth cycle found that little or no incomplete virus was produced following infection with standard virus even though multiple infection of the cells must have occurred in the later cycles. This result has led some workers to believe that incomplete virus is only produced in the first cycle, but it is more probable that the result means that extracellular virus in the allantoic fluid plays some part in the production of incomplete virus and that the removal of extracellular virus in the *in ovo*-de-embryonation technique prevents its production. LIU, PAUCKER and HENLE showed that if eggs were de-embryonated after infection with saturation inocula of standard virus and then exposed to large doses of infective virus for a period of 2 hours during the early stages of the cycle there was a decrease in the I/H ratio of the yield. Addition of infective virus after 4 hours did not affect the yield.

### b) Role of Heat-inactivated Virus in the Inoculum

HORSFALL found that the yield of infective virus in eggs was greatly reduced if the inoculum contained large amounts of heat-inactivated virus ($H_{165}$, $H_{166}$), and it is probable that the presence of heat-inactivated virus plays a part in the VON MAGNUS phenomenon. PAUCKER and HENLE ($P_{18}$, $P_{19}$) found that serial passage of partially heat-inactivated seed virus resulted in the yield of material with an I/H ratio as low as $10^1$, which is much lower than occurs with serial passage of undiluted fresh seed. They showed that over a wide range of heat inactivation the amount of infective virus produced was always 50—100 times the amount of infective virus in the inoculum and suggested that heat-inactivated virus restricted the infectious process to one cycle. FINTER, LIU and HENLE ($F_{70}$) showed that the non-infective haemagglutinin produced by serial undiluted passage could not be heat-inactivated virus since it was produced simultaneously with infective virus.

### c) Role of Cell Receptors in Production of Incomplete Virus

FAZEKAS DE ST GROTH and GRAHAM ($F_{45}$) showed that strains varied widely in ability to produce incomplete virus and could be placed in a gradient; PR8, WSE, MEL, SW, BON B, BEL, HUT B, CAM, FM1, LEE B, which corresponded closely to the gradient for enzyme action on cell receptors, incomplete virus being more readily produced by the strain with least enzymic activity. It was later shown that if action of virus enzyme on the cell receptors was prevented by treating the cells with periodate, incomplete virus was produced equally well by small or large inocula, more than 90% of the particles produced being incomplete ($F_{44}$).

### d) Possible Role of Premature Release from the Cells

VON MAGNUS ($M_{23}$) considered that incomplete virus interfered with the later stages of virus development. It is possible that the low content of S antigen and nucleic acid in incomplete virus may be due to premature release of particles from the cell before they have incorporated the full amount of S antigen. Infection of cells with large doses of virus leads to such accelerated release.

### 4. Summary

It is clear that the VON MAGNUS effect is a very complicated phenomenon. The lack of infectivity appears to be due to a deficiency in nucleic acid content. The author finds it difficult to believe that this deficiency is a qualitative one with a specific loss of some particular component of the RNA and would suggest that a quantitative reduction in RNA content is enough to explain the results.

The production of incomplete virus may result from a number of different effects produced by large inocula, premature release of particles from the cell, interference by heat-inactivated virus, modification of the cell surface by extracellular virus and interference with release of S antigen from the nucleus may all be involved.

# XXI. The Interference Phenomenon in Influenza Virus Infection

## 1. Interference between Influenza and Unrelated Viruses

ANDREWES ($A_{65}$) using suspended chick embryo cell cultures in Tyrode's solution demonstrated interference between WS and NWS viruses, but found no definite interference between influenza A virus and lymphogranuloma inguinale, between NWS and fowl pox, or between vaccinia and influenza viruses.

HENLE and HENLE ($H_{77}$) found that PR8 virus inactivated by ultraviolet light interfered with the multiplication of epidemic keratoconjunctivitis and Western Equine encephalomyelitis viruses. GROUPÉ and PUGH ($G_{87}$) found that preliminary infection with influenza A virus exerted a sparing effect on eggs subsequently infected with infectious bronchitis virus of chickens, and preliminary infection of eggs with egg-avirulent infectious bronchitis virus reduced the mortality in eggs inoculated with influenza B virus.

TAYLOR ($T_{36}$) found in chorioallantoic membrane tissue cultures that influenza A virus was able to prevent the cytopathogenic effect of Western or Eastern Equine encephalomyelitis viruses, provided the influenza virus was introduced at least 12 hours before the encephalomyelitis viruses.

DEPOUX and ISAACS ($D_{45}$, $D_{46}$) found that if fertile eggs were inoculated to the external chorionic surface of the chorioallantoic membrane and 24 hours later vaccinia virus was inoculated, the pock count was reduced from 60 to 14. The interference was prevented by the inoculation of 1000 units of RDE from 4 hours before to 2 hours after the inoculation of influenza virus. Under these conditions soluble antigen production in the membranes was prevented. The inhibitor of normal ferret serum also prevented interference.

WESSLEN, HERMODSSON and PHILIPSON ($W_{38}$) found that the yield of poliomyelitis virus in monkey kidney tissue cultures was depressed by inoculation of influenza A2 virus 12 hours previously. Virus inactivated by heat or UV light did not interfere.

Other examples of interference between influenza and unrelated viruses are recorded in a review by SCHLESINGER ($S_{50}$). Infection with influenza virus gives protection or interferes with the multiplication of Newcastle disease, mumps, pneumonia virus of mice, St. Louis encephalitis, Bwamba and West Nile viruses, while preliminary infection with Newcastle disease, mumps, pneumonia virus of mice, yellow fever or West Nile virus interferes with the growth of influenza virus.

## 2. Interference between Live Influenza Virus Strains

### a) Interference between Influenza A and Influenza B Viruses

ZIEGLER and HORSFALL ($Z_{38}$) inoculated eggs to the allantoic sac with 10 $EID_{50}$ of PR8 or LEE viruses, and 24 hours later inoculated the same dose of the heterologous virus. After a further 48 hours incubation allantoic fluids were collected and haemagglutination tests carried out, the type of virus present being determined by titration in the presence of ferret antisera to one or other of the two viruses. It was found that preliminary infection with influenza B virus prevented the subsequent growth of influenza A and *vice versa*. By increasing the dose of the second virus inoculated it was found that a preliminary inoculation of PR8 virus prevented the subsequent growth of LEE virus even when the inoculum of LEE virus was very large. In the reverse experiment, however, it was found that preliminary infection with LEE virus only prevented the growth of PR8 if the inoculum of PR8 did not exceed that of LEE by more than 100 times. If large inocula of PR8 were used the allantoic fluids contained both viruses. It was shown that interference did not occur if the second virus was introduced within 4 hours of the first, but did occur if the interval was 12 hours. In general it appeared that the only virus to grow was that inoculated first. However, it was found that the direction of interference could be reversed by the use of very large inocula of virus introduced simultaneously or shortly after a small dose of PR8. In such cases PR8 was suppressed and only LEE virus grew.

SUGG and MAGILL ($S_{320}$, $S_{324}$) made artificial mixtures of PR8 and LEE viruses in equal amounts, and made serial passages of the mixture in eggs, the allantoic fluids being titrated at each passage for haemagglutinin in the presence of antisera to PR8 and LEE. It was found that both viruses multiplied in each passage and could be maintained over a series of 9 passages. With a different strain of virus A, the CC strain, very different results were obtained. In this case serial passage of a mixture of CC and LEE resulted in complete suppression of the LEE virus at the second passage. Serial passage of a mixture of CC and LEE in mice resulted in a considerable reduction on the yield of LEE virus at the 6th passage but it was not completely eliminated. In later work it was shown that mixtures of PR8 and LEE could be maintained through 52 serial passages in eggs.

### b) Interference between Different Strains of Virus A

ANDREWES ($A_{65}$) obtained interference in tissue culture between WS and its neurotropic variant NWS, either strain suppressed the other if given 24 hours previously. NWS was also suppressed by PR8. ZIEGLER and HORSFALL ($Z_{38}$) obtained reciprocal interference between PR8 and SWINE viruses. SINZ ($S_{145}$) found that live PR8 and Asian viruses interfered markedly with each other. KOLCHURINA and BOLOTOVSKII ($K_{82}$) found that inhibitor-resistant strains of A2 virus prevented the growth of inhibitor-sensitive strains in chick kidney or chick embryo tissue cultures. They found that in humans the multiplication of inhibitor sensitive A2 strains in the mucosa of the respiratory tract could sometimes be prevented by inhibitor resistant vaccine strains which possessed high haemagglutinin and neuraminidase activity and were able to multiply at 25°C. FRASER ($F_{128}$) found in recombination experiments with NWS and MEL that recombinants were not produced if the NWS strain was inoculated 2 hours before MEL as a result of interference.

### c) Auto-interference

HENLE and HENLE ($H_{75}$) observed that the yield of virus in eggs inoculated with large doses of virus was less than if small doses were used. The phenomenon was extensively studied by VON MAGNUS and has been described in the previous chapter.

## 3. Interference Produced by Inactivated Virus

### a) Interference by Heat-inactivated Virus

ISAACS and EDNEY ($I_{19-23}$) found that LEE virus inactivated by heating at 56°C interfered with the multiplication of 4 variants of MEL virus and 3 variants of WS virus. In the allantoic sac 400—800 haemagglutinin doses of heated LEE virus completely prevented the growth of all seven strains of active virus in doses of 100—1000 $EID_{50}$. When large doses of active virus were used the strains varied in ability to overcome interference by heated LEE, the enzymically more active strains being more resistant than less active strains. There were also great differences in the amount of heated LEE virus required to produce interference and it was clear that interference required more than one particle of heated LEE virus per cell. It was shown that heated LEE virus underwent eclipse in the cells, and membrane extracts did not combine with specific antibody and were not antigenic in mice. Membranes inoculated with heated LEE still adsorbed active virus one hour later. Interference therefore occurred at a later stage in the growth cycle. The interfering power of heated virus was reduced by treatment with 0.01 to 0.08% formaldehyde and such virus had a reduced affinity for red cells and for chorioallantoic membrane inhibitor. It was suggested that heated virus competes with active virus for intracellular inhibitor and that this inhibitor plays some part in virus reproduction. FAZEKAS DE ST GROTH, ISAACS and EDNEY ($XF_1$) found that when the interfering dose was given 24 hours before 100 $EID_{50}$ of active MEL there was a reduction in both rate of production of virus and final yield. If the interfering dose was given only 45 minutes before challenge the reduction in rate of production of virus did not commence until 14 hours after inoculation, suggesting that interference takes some time to develop.

PAUCKER and HENLE ($P_{18}$, $P_{19}$) found that PR8 virus at 37°C has a half life of $6\frac{1}{2}$ hours. On inoculation of partially heat-inactivated seed there was a difference in the growth curves in de-embryonated eggs as compared with VON MAGNUS incomplete virus of similar content of infective virus and haemagglutinin. In all cases with heat-inactivated virus 50—100 $EID_{50}$ were produced for each $EID_{50}$ inoculated. This suggested that heat-inactivated virus had no effect in the first cycle but interfered in subsequent cycles.

SINZ ($S_{145}$) found that heated Asian virus markedly inhibited active Asian virus. Heated PR8 virus suppressed Asian virus but had little effect on PR8.

### b) Interference by Virus Inactivated by UV Light

ZIEGLER, LAVIN and HORSFALL ($Z_{39}$) found that virus inactivated by UV irradiation interfered with the multiplication of active virus introduced 24 hours later in fertile eggs. Interference was difficult to demonstrate in mice. They suggested that the effect was due to saturation of cell receptors. HENLE and HENLE ($H_{75-78}$) found that UV inactivated virus prevented multiplication of active virus introduced 3—24 hours later. Inactivated PR8, WS, F99, F12, LEE and S15 viruses produced equal interference with active PR8, WS, F99, LEE and S15.

The interfering agent was the inactivated virus itself, it was adsorbed and eluted from red cells and sedimented at the same speed as virus. The effect was neutralized by immune sera. Dialysed infected allantoic fluid irradiated for 3—5 minutes would interfere with active virus even if diluted 81 times, and interference by inactive virus cannot be overcome by excessive amounts of test virus. Henle, Henle and Kirber (H$_{86}$) found that if irradiated virus was introduced into the allantoic sac, the amnion remained fully susceptible to infection, suggesting that interference was a local phenomenon. Cells rendered resistant still adsorbed additional homologous or heterologous virus. Experiments with antiserum showed that interference was induced within 1 minute in the allantoic sac and resistance persisted for 6 days. Henle and Rosenberg (H$_{95}$) showed that by inoculation of large amounts of irradiated heterologous virus after infection the growth of virus could be restricted to a single cycle. Introduction of homologous irradiated virus after infection produced a reduction in the yield of virus, and this effect was type specific but not strain specific. Henle and Paucker (H$_{94}$) showed that the time required to produce solid interference decreased with increasing dosage of inactivated virus. With 850 haemagglutinin units of irradiated virus interference was complete at 9 hours, but with 20—100 doses it was not complete at 24 hours. The minimal interfering dose was about 20 haemagglutinin units, possibly about 1 particle per cell. Barry (B$_{12}$) studying heterologous interference by UV irradiated virus found that interference could be established by a dose of interfering virus corresponding to one particle per cell, but that at this dosage the interference could be overcome by doses of challenge virus sufficiently large to produce a high multiplicity of infection. A curious observation was that in this case the progeny virus was complete in contrast to the usual result of multiple infection of cells. Doses of interfering virus greater than one particle per cell produced solid interference which could not be overcome by multiple infection.

Schulman and Kilbourne (S$_{72}$) found that ultraviolet inactivated virus given to mice as an aerosol interfered with live virus aerosols given 1—3 days later, there being a marked reduction in the extent of lung lesions and a 10-fold reduction of virus content in the lungs.

### c) Interference by Sulphur Mustard Inactivated Virus

Fong (F$_{80}$) found that treatment of influenza virus with $5 \times 10^{-3}$ M sulphur mustard rendered the virus non-infective. Injection of 128 haemagglutinin units of the inactivated virus produced almost complete interference with the growth of 100 EID$_{50}$ of active virus injected later. Larger doses (512 HA units) interfered with $10^3$—$10^5$ EID$_{50}$ but not with $10^7$ EID$_{50}$ of active virus.

### 4. Interference in the Chick Chorion

Isaacs and Fulton (I$_{25}$) were unable to demonstrate interference between influenza viruses when grown on the external surface of the chorioallantoic membrane. Inoculation of mixtures of PR8 and LEE viruses resulted in the production of soluble antigen of both A and B viruses. Virus partially inactivated by heat or ultraviolet light and shown to produce interference in the allantoic sac had no such effect on the chorion, in fact some evidence of multiplicity reactivation was observed.

## 5. Interference in Mouse Brain Tissue

VILCHES and HIRST ($V_{23}$) showed that intracerebral inoculation of large doses of WS virus protected against 2500 lethal doses of Western Equine encephalomyelitis virus. Irradiated virus also protected, and protection occurred even if the viruses were injected on opposite sides of the brain, but not if the interfering virus was given intraperitoneally. Interference occurred if the influenza virus was given up to 15 days before the challenge virus. There was some interfering effect against Eastern Equine encephalomyelitis virus and against St. Louis and Bwamba fever viruses.

BURNET and LIND ($B_{200}$) found that intracerebral inoculation of large doses of WS virus protected against the production of encephalitis by the neurotropic variant NWS. If mixtures of small doses of WS with NWS were inoculated recombinant strains of virus were recovered.

## 6. Mechanism of the Interference Phenomenon

The interference phenomenon has been reviewed and its possible mechanism discussed by HENLE ($H_{69}$), SCHLESINGER ($S_{50}$) and WAGNER ($W_8$). The work of ISAACS and EDNEY ($I_{19-23}$) demonstrated that interference is an intracellular phenomenon and not merely a result of blockade of cell receptors, and there is evidence that interference may result from the entry of a single particle of interfering virus into the cell ($F_{39}$). PAUCKER and HENLE ($P_{20}$) suggested that the actual interfering agent was the S antigen or its nucleic acid component. POWELL and POLLARD ($P_{82}$) studied the inactivation of the interfering property of MEL virus by ionizing radiations, and found that the interfering unit had a radiation-sensitive volume of $2 \times 10^6$ A$^3$ corresponding to a molecular weight of 1,600,000. This would be compatible with action on the RNA. However, the interfering property is more resistant to UV light than the infecting property, and POWELL and SETLOW ($P_{83}$) found that the action spectrum for inactivation of interfering property by UV light had a broad maximum at 2600—2800 Å suggesting light absorption by a protein. The interpretation of the interference phenomenon was much clarified by the discovery of interferon by ISAACS and LINDENMANN in 1957.

## 7. Interferon

ISAACS and LINDENMANN ($I_{28}$) found that if pieces of chorioallantoic membrane were incubated for 2 hours at 37°C with heat-inactivated MEL virus, and were then washed and reincubated for 24 hours, the membranes released an interfering agent into the fluid. The release was maximal at 12 hours and adequate oxygenation of the tissue was necessary, indicating that the agent was newly synthesised. The interfering agent was called interferon. Added to chorioallantoic membrane tissue cultures it was able to prevent haemagglutinin production by active virus. The potency of the agent was low and it failed to produce interference with influenza virus grown in the allantoic sac ($L_{89}$). The properties and characteristics of the action of interferon were elucidated in a series of studies by ISAACS and his collaborators ($I_{28}$, $I_{29}$, $L_{89}$, $I_{12-14}$, $B_{149}$, $B_{150}$, $I_7$, $I_8$, $S_{24}$). Virus inactivated by UV light gave better yields of interferon than heated virus ($L_{89}$, $B_{149}$). VON MAGNUS incomplete virus gave good yields of interferon when grown on the chick chorion ($I_{13}$). No interferon was produced by virus disintegrated by sonic vibration ($B_{149}$). In membranes pre-treated with interferon more interferon was produced by inoculation of live virus ($I_{12}$). The smallest amount of inactivated virus which induced interference also produced interferon ($B_{150}$). Viruses showing different degrees of resistance to the interfering effect of irradiated virus showed similar differences of resistance to interferon ($B_{150}$). Interferon is active not only against the virus causing its production but also against a range of other viruses not necessarily related to the inducing virus, the range of action being the same as that of inactivated virus ($I_8$). Interferon is therefore a non-specific inhibitor of virus multiplication. Some viruses such as Sendai virus are particularly sensitive to interferon and interferon is more readily produced by some viruses, e.g. Newcastle disease virus, than by others. Different types of host cell appear to produce different interferons, and interferon is most active in the homologous host cell and less active or inactive in other types of cell from different animal species. HENLE, HENLE, DEINHARDT and BERGS ($H_{83}$) demonstrated the production of an interferon in MCN cell tissue cultures by inoculation with ultraviolet-irradiated myxoviruses. The substance was not neutralized by antiviral sera, did not agglutinate red cells and was not sedimented by high speed centrifugation. It was inactivated by trypsin. The induction of interference by irradiated viruses in MCN cultures was always associated with the appearance of interferon in the culture media.

WAGNER ($W_8$) found that allantoic fluid of eggs inoculated with WS virus had a powerful inhibiting effect on the multiplication of Eastern Equine encephalomyelitis virus in tissue cultures and on centrifugation at high speed to sediment the virus the interfering effect of the supernatant fluid was equal to that of the original infected allantoic fluid.

Interferon appears to be largely responsible for the recovery of cells from virus infection, thus in infected mouse lungs interferon production occurs at 2—5 days after infection, coinciding with the fall of infectivity of lung extracts ($I_{27}$).

In eggs inoculated to the allantoic sac with influenza virus the multiplication of virus precedes the production of interferon. Interferon production commences about 36 hours after inoculation and the appearance of interferon in large amount in the allantoic fluid is associated with the cessation of virus multiplication ($W_8$).

Interferon may be produced as a result of the introduction of abnormal ribonucleic acid into cells ($I_{14}$). Thus no interferon was produced if tissue cultures of chick embryo fibroblasts, mouse embryo fibroblasts, or rabbit kidney cells were inoculated with homologous RNA, but interferon was produced if the RNA was treated with nitrous acid.

The activity of interferon is greatly reduced by increased oxygenation of the assay system, and the mortality of mice inoculated with WS or LEE virus is increased and the survival period reduced if the animals are held in an atmosphere of 50% oxygen ($S_{24}$).

### a) Properties of Interferon

Interferon does not agglutinate red blood cells and is not sedimented at 100,000 $g$. It is not neutralized by antisera to the inducing strain of virus ($I_{29}$).

It is stable in the pH range 1—10. It is precipitated by saturated ammonium sulphate and is destroyed by trypsin, but is unaffected by ribonuclease, periodate or RDE ($L_{89}$). It is destroyed by peptic digestion, by shaking with amyl-alcohol chloroform, and slowly by UV light ($B_{149}$). It can be concentrated 20-fold by pressure dialysis ($L_{89}$).

Lampson, Tytell, Nimes and Hilleman ($L_{16}$) purified chick interferon from the allantoic fluids of eggs infected with WS virus. Virus and extraneous protein were removed by precipitation with perchloric acid, the fluid was concentrated and the interferon precipitated with zinc acetate. This was followed by two cycles of chromatography on carboxymethyl cellulose and finally by zone electrophoresis on Pevikon. From 50 litres of allantoic fluid a few ml of material with a 47-fold greater activity than the original was obtained. The interferon was a protein stable between pH 2 and 10, destroyed at 76°C and by UV light and proteolytic enzymes. It was powerfully adsorbed to glass, starch, agar and paper but not to polypropylene. The molecular weight was 20—34,000 and the sedimentation coefficient $2.6 \times 10^{-13}$. The isoelectric point was pH 8.0. The material contained 2.3% tyrosine, 2.6% tryptophane, 7.2% arginine and 11.1% lysine. Burke and Ross ($B_{153}$) estimated the molecular weight of chick interferon at 30,000 by sucrose gradient centrifugation and gel filtration through Sephadex G-100.

Merigan, Gregory and Petrally ($M_{78}$) prepared human interferon by infection of human fibroblasts with Newcastle disease virus. It was stable over the pH range 2—10 and could be purified 10-fold by chromatography on XE64 ion exchange resin. By sucrose density gradient centrifugation the molecular weight was estimated at 26,000.

### b) Assay of Interferon

Influenza viruses are relatively resistant to interferon, and interferon preparations only inhibit the growth of influenza virus in tissue cultures when added in large concentrations. The ARBO-group of viruses are much more susceptible, and interferon is usually estimated by determining the dilution which reduces the plaque count of a suitable ARBO-virus on chick embryo cell monolayers to 50% of the count in control cultures ($W_8$).

Finter has described an assay technique using tissue cultures infected with Sendai virus. After growth the cultures are washed with red blood cell suspension and the haemadsorbed red cells are then lysed with water and the haemoglobin estimated. Pre-treatment of the cultures with interferon leads to reduced haemadsorption. Using this method the most potent preparations of interferon were obtained from the brains of mice infected with West Nile virus. These were active against Sendai virus in mouse tissue cultures in a dilution of 1:50,000 ($F_{68}$).

It is to be noted that it is very difficult to demonstrate the effect of interferons in intact animals, only slight reduction in $LD_{50}$ of challenge virus can usually be demonstrated. The author found that in order to produce a significant reduction in the growth of influenza virus in the allantoic sac it was necessary to introduce the whole of the interferon produced in the allantoic fluid of an egg inoculated with the homologous virus. Grossberg, Hook and Wagner ($G_{82}$) found that if eggs were infected with influenza virus by the allantoic route they were protected against the lethal effect of neurotropic virus, NWS, injected intravenously, and attributed the effect to the action of interferon. However, protection against NWS could only be achieved with allantoic fluid interferon by multiple intravenous or intra-allantoic injection.

## c) Inhibitors of Interferon Production

SOLOVYEV and MENTKEVICH ($S_{239}$) found that colchicine reduces the production of interferon by A2 virus in chick embryos and in mouse brain, and also inhibits the action of exogenous interferon in chick embryo cell cultures. SMART and KILBOURNE ($S_{158}$) found that cortisone inhibits interferon synthesis in the chick embryo but does not inhibit the action of interferon. FRIEDMAN and SONNABEND ($F_{150}$) found that interferon action is inhibited by puromycin and suggested that interferon induces a new anti-viral protein in cells.

COTO, GALEOTA and CORAGGIO ($C_{84}$) found that interferon production by UV irradiated influenza virus in chick embryo cells was slightly inhibited by 10 $\gamma$/ml of 5-fluorodeoxyuridine and strongly by 50—100 $\gamma$/ml. It was suggested that interferon synthesis is under the genetic control of cellular DNA and that de novo synthesis of DNA is essential for interferon production. However, BURKE and MORRISON ($B_{152}$) found that small doses of aminopterin and 5-iododeoxyuridine inhibited DNA synthesis in chick embryo cells without any effect on interferon production. Large doses of IUDR inhibited cellular RNA synthesis and also the production of interferon. Irradiation of cells with UV light depressed cellular synthesis of both DNA and RNA and depressed interferon production. Doses of mitomycin C which depressed interferon production also interfered with RNA and protein synthesis. They concluded that DNA synthesis is not necessary for interferon production but RNA synthesis is needed.

## 8. General Discussion of the Interference Phenomenon

There can be little doubt that there is more than one way in which the reproduction of one virus in a cell may be prevented or reduced by previous inoculation of another. External cell receptors may be blocked by the interfering virus, or there may be a destruction of intracellular receptors necessary for the release of virus from the cell. The metabolic processes of the cell may be exhausted by the interfering virus or there may be competition for some essential cell component.

Nevertheless most of the interference phenomena described have features in common which suggest that a similar mechanism is involved. Interference in most cases seems to be an intracellular phenomenon probably induced by the nucleic acid of the interfering virus and taking some time to develop. Virus containing partially damaged nucleic acid appears to be most effective in inducing interference, and interference with live virus only seems to occur after a period of time permitting a complete growth cycle with the possible production of defective nucleic acid at the end of the cycle. All the evidence suggests that, as claimed by ISAACS, interference is mediated by the protein interferon, and that interferon plays an important role in the arrest of reproduction of viruses in cells. Interferon is clearly not a virus component, and it is not detectable in normal cells. Its appearance in the infected cell involves the synthesis of

RNA and protein and is possibly dependent on the presence of DNA, but probably not on the synthesis of new DNA. It appears to be essentially a host-cell product with a wide range of anti-viral activity but showing host cell specificity. It has been suggested that interferon is a defence mechanism of the cells against viral infection. It is to be noted, however, that interferon does not protect cells against infection by viruses, but it does limit the degree of virus reproduction. It is very probable that cells require a mechanism to prevent excessive synthesis of normal RNA and protein constituents, and that the action of interferon in viral infection is incidental to its normal function.

# XXII. Influenza Virus Genetics

Genetic studies of the influenza viruses were initiated in 1951 by the discovery of the recombination phenomenon by BURNET and LIND ($B_{200}$, $B_{201}$) and by the demonstration of the reactivation phenomenon by HENLE and LIU ($H_{91}$).

## A. The Reactivation Phenomenon
### 1. Multiplicity Reactivation

HENLE and LIU ($H_{91}$) found that if influenza virus partially inactivated by UV light was inoculated to the allantoic sac of fertile eggs, the rate of development of both haemagglutinin and infectivity was 10 times greater than was given by untreated virus of the same infectivity titre as measured by standard *in ovo* titration. The phenomenon was seen with both PR8 and LEE virus but only if small doses of UV light were used. If virus was treated with large doses of UV light the capacity to re-activate was lost. The effect was not seen if the irradiated preparation was diluted 1:30 and it was concluded that the reactivation effect only occurred if the cells adsorbed more than one particle — multiplicity reactivation. Partial inactivation by heat at 56°C gave similar results. These results were criticised by CAIRNS ($C_5$) who suggested that the increased rate of virus production was due to receptor destruction by the large inoculum preventing loss of virus by combination with uninfected cells. However, the work of BARRY ($B_{11}$) makes it certain that multiplicity reactivation of UV irradiated influenza virus is a real phenomenon. BARRY studied the yield of virus in single cycle experiments in de-embryonated eggs inoculated with active and UV irradiated virus at various levels of multiplicity (calculated from the pre-irradiation infectivity titre). The results (Fig. 53) showed that with the irradiated virus there was a very steep increase in the yield under conditions of multiple infection of the cells, the rise in yield being much steeper than with active virus. It was concluded that multiplicity reactivation was occurring and was a very efficient phenomenon. Increasing the dose of UV irradiation resulted in an increasing production of incomplete virus until only non-infectious haemagglutinin was produced, beyond which point reactivation ceased.

*Multiplicity Reactivation with von Magnus Incomplete Virus*

BURNET, LIND and STEVENS ($B_{210}$) found that the yield of haemagglutinin in de-embryonated eggs inoculated with VON MAGNUS incomplete virus was greater than would be expected from its *in ovo* infectivity titre, but suggested that the effect was not due to multiplicity reactivation. BARRY ($B_{10}$) found the dose response curve of incomplete virus in de-embryonated eggs had the same slope as active virus and concluded that multiplicity reactivation did not occur. He observed, however, that the yield of haemagglutinin by incomplete virus was greater than expected. BEALE and FINTER ($B_{29}$) found that the yield of S antigen produced in 6 hours by inoculation of eggs with VON MAGNUS incomplete virus was much greater than would be expected from its *in ovo* infectivity titre, and considered the effect to be due to multiplicity reactivation. While the behaviour of VON MAGNUS incomplete virus appears to differ from that of heat or UV inactivated virus, it seems clear that some contribution to the yield of virus material is made by particles which appear to be non-infective by *in ovo* titration.

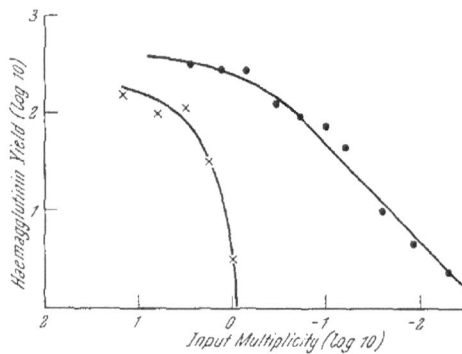

Fig. 53. Multiplication of active and UV irradiated virus in de-embryonated eggs.

The steepness of the response curve with irradiated virus indicates reactivation. Active virus ●——●; UV irradiated virus ×——×. Reproduced from BARRY ($B_{11}$).

## 2. Reactivation by Heterologous Virus

APPLEBY ($A_{92}$) inoculated mixtures of the neurotropic strain NWS and the non-neurotropic strain KUNZ (A) which had been inactivated by UV irradiation into eggs and then inoculated the allantoic fluids intracerebrally to mice. Mouse brain extracts were made after 72 hours and inoculated to eggs. Reactivation was shown by the recovery of several strains of NWS, and one strain was isolated which appeared to be a recombinant. BURNET and LIND ($B_{203}$) inoculated eggs with heat inactivated WS virus in amounts sufficient to infect all the cells. Active MEL virus was then inoculated. Some active WS virus was recovered, but only in the form of recombinants. GOTLIEB and HIRST ($G_{50}$) used a neurotropic and non-neurotropic strain in reactivation experiments. Either strain when inactivated by heat or by UV light could be reactivated by the other, even if the reactivating strain was inoculated as long as 16 hours after the inactive strain. About half the reactivated strains recovered were recombinants. VON MAGNUS incomplete virus could not be reactivated.

## 3. Reactivation of Fowl Plague Virus

PEREIRA and TUMOVA ($P_{34}$) found that UV inactivated fowl plague virus could be reactivated by live influenza A2 virus. Chick embryo fibroblast monolayers were inoculated with 5 million $EID_{50}$ of UV inactivated fowl plague virus and varying amounts of active A2/Singapore/1/57. After 1 hour adsorption the inoculum was recovered and an agar overlay added. Plaques appeared in 3 days in numbers related to the dose of A2 used. The reactivation power could be neutralized by immune sera and it was possible to develop the method to study antigenic relationships between different strains of A2 virus.

## 4. Reactivation and Vaccine Production

FENNER ($F_{63}$) has pointed out the importance of the reactivation phenomenon in vaccine preparation. Apparently inactivated virus may become reactivated if inoculated in large doses, and complications may occur if an inactivated virus preparation is assayed in primary cell cultures which may be
contaminated with some virus capable of reactivating the inactivated virus.

# B. The Recombination Phenomenon

## 1. Marker Characteristics and the Selection of Pure Clones

An essential preliminary to the performance of experiments on genetic interaction between virus strains is the selection of strains with recognisable "marker"
characteristics, and the development of methods of selecting pure clones in which
all the virus particles are identical in properties.

BURNET ($B_{178}$) examined the properties of three sub-strains of the original
WS virus, WSM a strain passaged in the amniotic sac, WSE passaged in the allantoic sac, and the neurotropic strain NWS passaged in mouse brain. These strains
showed a number of common characters, infectivity to eggs and mice, production
of S antigen of A type, production of haemagglutinin and the absence of filamentous forms. In addition each strain possessed "marker" characteristics by
which it could be readily recognized. The strain NWS was neurotropic and
lacked enzymic activity on ovomucin and indicator activity, WSE was firmly
adsorbed to red cells and did not elute, while WSM possessed an unusually heat
stable haemagglutinin.

BURNET suggested that virus strains contain in addition to the predominant
type small numbers of variants from this type. If passages are made in fertile
eggs with dilutions of virus close to the limit of infectivity, $i.e.$ doses in which
only some of the inoculated eggs become infected then the variants will tend
to be eliminated and only the predominant type will grow. If a number of such
cultures are constant in properties it is considered that a "pure clone" of virus
has been obtained. BURNET ($B_{179}$) found that when the NWS strain was passed
in the allantoic sac its original character was maintained if passages were made
at limiting infective dilutions. If, however, passages were made at a low dilution
the strain changed in character, becoming convertible to the indicator state
and developing enzymic activity. By heating the allantoic fluids from low dilution passages at 54°C it was possible to select a variant which was relatively
heat resistant and lacking in intracerebral pathogenicity in mice.

LIU and HENLE ($L_{104}$) studied the limit dilution method of producing pure
clones by inoculating equal mixtures of PR8 and LEE viruses in closely spaced
steps near to the 50% infectivity end point. When less than 2 $ID_{50}$ of mixed
seed was used PR8 and LEE viruses were recovered from the positive eggs in
equal numbers, but with larger inocula PR8 became dominant because of its
greater rate of multiplication. In some eggs haemagglutinin only became detectable after 3—4 days, and in some cases fluids negative at 3 days gave virus on
passage even though no haemagglutinin was produced in some of the eggs on
further incubation. This suggested that infectivity measurements at 3 or even

4 days did not reflect the total amount of virus in the inoculum, so that the limit dilution technique was not an absolutely certain method of producing pure clones.

In a review of the subject BURNET (B$_{186}$) suggested that 2 limit dilution passages normally gave a pure clone, and 3 limit passages always gave a pure clone. Pure clones, while all descendants of one particle, may contain small numbers of mutants or incomplete particles.

## 2. The Recombination Phenomenon

BURNET and LIND (B$_{200}$) inoculated mice intracerebrally with mixtures of the neurotropic strain NWS and a range of dilutions of the non-neurotropic WSM strain. With large doses of WSM interference occurred and no virus was isolated, but with smaller doses the interference was only partial and from the brains of 7 such mice virus was isolated in small amount. Three types of virus were recovered, the original NWS, a strain N$_1$ which differed in being convertible to the indicator state and having greater enzymic activity and heat resistance, and a strain NM in which the haemagglutinin was resistant to heat at 62°C for 30 minutes. The NM strains represented a recombination of the characters of NWS and WSM and differed from both in having a different position in the receptor gradient. By passage in eggs at limiting dilutions it was shown that they were not mixtures. In later work (B$_{201}$) it was found that by inoculation of mixtures of NWS with various serologically different non-neurotropic strains of virus A, recombinants could be isolated in which the neurotropic character of NWS was coupled with the serological type of the other virus. Two such recombinants, neuro-SWINE and neuro-MEL differed from SWINE and MEL in not eluting readily from red cells and having less action on ovomucin. The isolation of these recombinants was greatly facilitated by treatment of the mouse brain suspension with anti-NWS serum before inoculating to eggs.

Inoculation of mixtures of NWS and a filamentous strain Oc1 resulted in the recovery of a neurotropic variant with serological character of Oc1 but which was not filamentous. Recombinants of NWS and MEL showed a considerable range of virulence towards mouse brain (B$_{194}$).

These results were of great importance in that they demonstrated conclusively that on entry into the cell the infective unit became disrupted into smaller units which after replication were later reassembled, recombinants arising by incorporation of units derived from each of the parent strains in conditions of mixed infection.

### Gene-linkage Groups

BURNET, FRASER and LIND (B$_{198}$, B$_{199}$, B$_{202}$, F$_{134}$) found that recombinants of NWS and MEL could be obtained by mixed infection of the chick embryo chorion or in the allantoic sac. An extensive study was made of recombinants produced in the allantoic sac by mixed infection with the strains MEL and WSE. These strains could be differentiated by six marker characteristics, these were designated as ABCDEF in MEL and abcdef in WSE. The character Aa was the serological type; Bb heat resistance (62°C for 30 minutes) of the haemagglutinin, MEL being resistant and WSE sensitive; Cc conversion to indicator state by heat at 56°C, positive with WSE, negative with MEL; Dd reaction of heated virus with salivary gland inhibitor, WSE being inhibited and MEL resistant; Ee production of haemorrhagic lesions in the chick embryo, positive with WSE, negative with MEL; Ff mouse pathogenicity, WSE producing fatal infection and MEL non-fatal lesions. From 37 virus containing fluids from eggs inoculated with mixtures of MEL and WSE at near the limit of infectivity 41 strains of virus were isolated. Of these 22 had the properties ABCDEF and were evidently the original MEL, and 4 had the characters abcdef corresponding to the original WSE. Six strains had the characters ABc DeF and were described as M+, while 9 had the characters abC dEf and were described as WS−. Thus only two characters Cc and Ee appeared to be transferable. The results suggested that the characters were present as two gene linkage groups CE and ABDF in MEL and ce and abdf in WSE. A curious result was that of two "virulence" characters, the ability to produce haemorrhagic lesions in the chick embryo and pathogenicity to mouse lung, only one was transferable; and of two "indicator" characters, reaction with ovomycin and with salivary gland mucoprotein, again only one was transferable.

PERRY and BURNET ($P_{39}$) carried out recombination experiments in mice with two strains of influenza B. Mice were inoculated intranasally with mixtures of LEE and MIL. After 4 days lung extracts were made and eggs inoculated with serial dilutions. Several intermediate types were isolated. If serological type, heat resistance of haemagglutinin, receptor gradient position and mouse virulence were denoted ABCD for LEE and abcd for MIL; the following recombinants were isolated, aBCD, aBcD, abCD, aBcd.

LIND and BURNET ($L_{83}$) were able to recover the original MEL and WSE strains from eggs inoculated with mixtures of the recombinant strains M+ and WS− indicating that the recombination process was reversible — back recombination. In these experiments a number of anomalous results were encountered which were believed to be due to the formation of heterozygotes.

### 3. Production of Doubly Neutralisable Recombinant Strains

HIRST and GOTLIEB ($H_{144}$, $H_{145}$) criticised the work of BURNET and his colleagues in that they used as marker characteristics virus properties which were subject to great variations, and had not conclusively demonstrated that the recombinants isolated were stable over a series of limit dilution passages. They decided to use serological type as the most reliable marker in that it is a very stable virus characteristic but is known to be dependent on the presence of a complex of different antigens. They demonstrated that the haemagglutinin-inhibition test will not detect more than one component of an artificial mixture of serologically different strains, so that the production of a virus strain of which the haemagglutinin could be neutralized by specific antisera to both the original parents would indicate recombination in that the strain must possess antigenic components derived from each of the parents. They found that if eggs were simultaneously inoculated to the allantoic sac with two serologically distinct strains of virus A, some of the eggs yielded virus of which the haemagglutinin was neutralized by antisera to both viruses. In some of the fluids 84% of the total haemagglutinin was doubly neutralisable. The phenomenon was demonstrated with several different virus pairs, and double neutralization was most frequently observed in eggs given large inocula. Using the neurotropic strain WSN and MEL it was found that with large inocula the best yields of doubly neutralisable haemagglutinin (X fluids) were given with equal doses of WSN and MEL, but with small inocula best results were given if the dose of WSN was much greater than that of MEL, this being due to the fact that MEL grows more rapidly in the egg than WSN.

When the doubly neutralisable X fluids were passed serially in eggs only MEL virus was produced, and if passages were made in the presence of anti-MEL serum WSN virus was produced. It appeared that the X type reverted to the parent forms on passage. However, by concentrating the X fluids by centrifugation and inoculating the whole of the virus from six eggs to another six the X form was maintained with great difficulty through several passages. Between the eigth and eleventh passages the number of X fluids obtained increased and could be produced with much smaller inocula. However on passage at limiting dilutions there was reversal to MEL or WSN. This type of fluid was referred to as $X_2$ and its properties were maintained through some 20 passages. After 32 passages a new type of X fluid appeared. This type $X_3$ yielded only $X_3$ fluids on passage, and after several passages at limit dilution was strongly neutralized by both MEL and WSN antisera.

It was concluded that in the course of the experiments three types of combined virus were produced. The first type X, rapidly reverted to the parent types and could only be maintained by the use of large inocula. The second type $X_2$ was somewhat more stable but tended to revert to parent types. The third type $X_3$ appeared to be a true stable recombinant form with its antigenic complex partly derived from WSN and partly from MEL.

## 4. Phenotypic Mixing Heterozygosis and Diploidy

In mixed infections with two different virus strains several types of combined particles may be produced. The genetic complex, genotype, of one strain may be associated with the non-genetic components, phenotype, of the other. Particles may occur which include only one genotype but have both phenotypes — phenotypic mixing. Particles containing some part of the genetic complex of each strain are heterozygous, and particles including the complete genotypes of both strains are diploids.

GOTLIEB and HIRST ($G_{49}$) prepared artificial mixtures of serologically different strains of virus A (MEL and WSN) and artificial mixtures of PR8 (A) and LEE (B) in equal amounts and inoculated limit dilutions of the mixtures to eggs to determine the frequency of occurrence of mixed infection. In each case it was found that when 1 $ID_{50}$ of the mixtures was inoculated about 32% of the infected eggs contained both viruses. By inoculating eggs with large doses of mixtures of PR8 and LEE a number of fluids containing a predominance of doubly neutralisable haemagglutinin (X fluids) were obtained. When such X fluids from mixtures of PR8 and LEE were inoculated at limit dilutions it was found that the number of mixed infections produced was somewhat less than were given by artificial mixtures of PR8 and LEE. It was concluded that the occurrence of doubly neutralisable particles after inoculation of mixtures of PR8 and LEE was due to phenotypic mixing and not to genetic recombination. By contrast X fluids derived from mixtures of MEL and WSN inoculated at limit dilution gave a much larger number of mixed infections than were given by artificial mixtures of MEL and WSN. It was concluded that in mixed infection with MEL and WSN particles were produced which were heterozygous for antigenic type, and as the strains recovered differed from their parents in other properties it was probable that the mixed particles were diploids.

Mixed infection with A and B viruses only gives rise to phenotypic mixing and on subculture only one or other parent type is produced. Mixed infection with two different A strains produces some particles which are heterozygous or diploid and which on subculture yield both parent types.

BURNET and LIND (B$_{205}$) inoculated de-embryonated eggs with mixtures of M+ and WS- in doses sufficient to ensure double infection of the cells, and examined the yields after one cycle. It was found that the recombinants MEL and WSE occurred in the ratio of between 4:1 and 7:1. On purely genetic grounds the ratio should be 1:1. However, it was found that if eggs were inoculated with mixtures of the various strains the MEL tended to outgrow the others while WSE was least dominant. The results could be explained if it was supposed that all influenza particles are diploid. Combinations of MEL, M+, WS-, and WSE could give sixteen possible diploids of which seven will include MEL and will give rise to MEL on limit dilution, while only one will contain only WSE and give rise to WSE on limit dilution. The ratio of MEL to WSE would then be 7:1.

HIRST and GOTLIEB (H$_{146}$) found that in mixed infections of the allantoic sac with large inocula of WSN and MEL the yield on limit dilution produced only the parent types and heterozygotes. On subculture at limit dilution the heterozygotes segregated yielding either MEL+WSN or MEL+non-neurotropic WS, each pair half the time. Heterozygotes from mixed infections with neuro-virulent MEL and non-virulent WS on segregation yielded only neurotropic strains of both MEL and WS. The heterozygotes were believed to be diploids. They give rise to recombinants when they enter a new cell but do not do so in the parent cell indicating that the virus particle reaches maturity as it leaves the cell and not before.

LIND and BURNET (L$_{85}$) carried out single cycle double infections with NWS and MEL. Doubly neutralisable fluids were passed at limit dilution. Of 24 strains isolated 20 were pure clones of parent strains or recombinants and were considered to be homozygous, 4 strains were heterozygous yielding more than one type on passage. It was concluded that the formation of heterozygotes is not an essential preliminary stage in the formation of recombinants.

## 5. Genetics of Virulence in Influenza Viruses

BURNET and LIND payed particular attention to the behaviour of virulence as a marker character in recombination experiments (B$_{206}$). Double infection with MEL and NWS resulted in the production of recombinant NM strains which showed considerable differences in the degree of virulence to mouse brain, some strains being equal in virulence to NWS, some of lower virulence, and some aviru-lent although having other characters suggesting that they were of NM type.

Recombinants of MEL and WSE showed great variations in ability to produce haemorrhagic lesions in the chick embryo, but mouse lung pathogenicity was closely linked to serological type.

With the B viruses LEE and MIL of which only LEE produced mouse lung lesions, there was a wide range of mouse lung virulence in recombinants. It was suggested that virulence is not a function of a single gene and that changes in virulence result from successive gene mutations. HIRST and GOTLIEB ($H_{146}$) also noted that a number of factors appeared to be involved in the virulence factor.

LIND and BURNET ($L_{84}$) found that the neuropathogenicity of several types of recombinant tended to diminish on continued limit dilution passage. Studies of recombinants of the strains NWSE and MEL ($L_{86}$) showed that the neurotropic character was transferred to recombinants in reduced virulence. Mouse lung pathogenicity was absolutely linked to serological type and action on sheep salivary gland mucoid, while egg virulence was linked to action on meconium inhibitor. There were considered to be four gene linkage groups segregating independently.

BURNET and LIND ($B_{207}$) obtained a mutant from MEL by growth in the presence of anti-MEL serum which lacked virulence for mice. It could be maintained by passage at limit dilutions, but inoculation of large doses resulted in recovery of mouse virulence apparently as a single step mutation.

PERRY, VAN DEN ENDE and BURNET ($P_{40}$) in recombination experiments in de-embryonated eggs with the B strains LEE and MIL obtained recombinants which showed much greater variation in mouse lung virulence than the original strains. It was supposed that there were a variable number of virulence genes from 0 to 4, the number present determining the virulence.

A series of papers by FRASER ($F_{129-132}$) describing recombination experiments with MEL and NWS illustrates the complexity of the results. When mouse brain was infected with equal doses and strains isolated by limit dilution it was found that WS types did not appear regularly and were possibly variants, and NM strains originated mostly from particles possessing antigenic properties of both strains; and inoculation of these particles to mouse brain also gave rise to infectious NWS and infectious MEL, although if MEL alone is inoculated intracerebrally an incomplete cycle occurs and no infectious virus is produced. Recombinant NM strains had the same distribution of virulence towards mice and chick embryo, but the intranasal virulence was not related to the intracerebral virulence. When NM recombinants were isolated from successive cycles of double infection with NM and NWS, the NM strains increased in number and virulence at each cycle suggesting a step-like transfer of virulence.

## 6. Recombination Experiments with Plaque-forming Strains

The neurotropic strain WSN gives large clear plaques on chick embryo fibroblast monolayers. SIMPSON and HIRST ($S_{141}$) found that by mixed infection with UV irradiated WSN and an active strain, plaque-forming recombinants could be produced with WS, MEL, PR8, FM1, S1, Jap/305/57, and SWINE. No plaque-forming recombinants were produced with the Asian strain R1/5, NWS, virus N, a strain of equine influenza virus, or with influenza B, C, Sendai, NDV or mumps viruses.

Plaque type recombinants were usually recombinant for a number of other characters and when seven characters were examined a wide variety of combinations were found. Re-activation of the original irradiated WSN occurred in some cases.

The plaque-forming property could thus be transferred to a variety of serological types of virus A and these types could be used in recombination experiments with non-plaque forming types. The results did not support the work of BURNET on the existence of linkage groups, when several marker characters were examined a wide variety of combinations occurred.

The acquisition of plaque-forming power was possibly due to a broadening of the host range of the virus so that it would infect and destroy all the cells in the mixed population of the fibroblast monolayers.

SUGIURA and KILBOURNE ($S_{326}$, $S_{327}$) found that many strains of virus produced plaques in a variant (1-5c-4) of the Chang human heteroploid conjunctival cell line. Sharply defined clear plaques were produced by NWS, while CAM produced red-bordered plaques when neutral red staining was used. Monolayer cultures were inoculated with mixtures of NWS and CAM (4—12 p.f.u. per cell). After 24 hours the cultures were frozen and thawed and plated without antiserum in the overlay and also with overlays of medium containing antisera to NWS or CAM. Reciprocal recombinants of serotype and plaque type were found at about equal frequency with a total recombination rate of about 20%. Back recombination yielded the parents with a frequency of 30%. Some recombinants of mixed antigenic type were found.

STAIGER ($S_{252}$) studied recombination in fowl plague viruses. Wild-type fowl plague virus gives large plaques on chick fibroblast monolayers together with a number of small-plaque mutants. Two small-plaque mutant strains were obtained which differed in rate of production and pathogenicity for chick embryos. Mixed infection of monolayers at high multiplicity with the two small plaque mutants yielded a small proportion (between 0.1 and 0.2% of the total plaque count) of a recombinant with the typical large plaques of the wild-type virus.

TUMOVA and PEREIRA ($T_{70}$) found that the plaque-forming activity of UV inactivated avian viruses (fowl plague and A/Turkey/England/63) could be reactivated by a number of different serotypes of influenza A viruses of human, porcine, equine or avian origin. After repeated cycles of UV inactivation of fowl plague virus followed by rescue with A2 virus, plaque-forming strains were isolated with the heat-labile haemagglutinin of A2, and strain-specific antigens of both viruses. The transmission of the A2 character was gradual and increased in degree with each cycle of cross-reactivation. The A2 haemagglutinin marker was inherited independently of the antigenic marker.

LAVER and KILBOURNE ($L_{41}$) isolated a stable recombinant from NWS and A2/R1/5⁻. This recombinant X-7 contained the specific antigen of NWS but gave a reduced plaque size and contained a minor A2 component demonstrable by complement fixation. It also eluted unusually rapidly from red cells. The various strains were disrupted by dodecyl sulphate, their proteins separated by electrophoresis and peptide mapping experiments carried out. It was found that the recombinant virus possessed the haemagglutinin of NWS and the neuraminidase of the A2 virus.

## 7. Rapid Adaptation of Viruses as a Result of Recombination

KILBOURNE and MURPHY ($K_{49}$) found that passage of freshly isolated filamentous A2 viruses in the presence of PR8 virus inactivated by 17 days exposure to a temperature of 35°C resulted in the rapid emergence of virus with
the serological characters of A2 but with the growth capacity and spherical morphology of PR8. The spherical recombinant produced a greater yield of haemagglutinin and a higher infectivity titre than the original A2 but there was no
evidence of a greater production of total protein.

## 8. Effect of Ribonuclease on Recombination

Ribonuclease is capable of reducing the yield of influenza virus in eggs. If
eggs are inoculated with heat-inactivated inhibitor resistant PR8 virus together
with active inhibitor sensitive A2 virus, and the resulting virus isolated in eggs
in the presence of inhibitor, inhibitor resistant recombinants can be isolated
which are antigenically PR8 but have the filamentous morphology of A2 ($K_{44}$).
The incidence of such recombinants was unaffected by inoculation of the eggs
with ribonuclease before the heated PR8. It was suggested that ribonuclease
does not reduce virus yield by hydrolysis of intracellular viral RNA but by
action on host cell RNA.

## 9. Use of Recombination in Production of Vaccine Strains

Russian workers ($R_{37}$, $F_6$, $S_{186}$) have attempted to produce virus strains with
a greater range of antigens than normal by recombination methods, with a view
to the use of such strains in live vaccines. RITOVA and ZAKSTELSKAYA ($R_{37}$)
passaged a mixture of 5 strains of A1 virus in human embryo tissue culture.
A combined strain was produced which had a complete set of the antigens of
all 5 strains and showed improved survival in the human nasopharynx. Combined
strain vaccines were also produced from 2 different B viruses ($F_6$).

## C. General Discussion of Influenza Virus Genetics

The discovery of the phenomena of reactivation and recombination had a
very great influence on the development of the modern ideas on the nature of
influenza virus reproduction, in particular the phenomena made it quite certain
that some process of disintegration of virus and subsequent reassembly must be
involved. However, it has proved almost impossible to explain the observed facts
in terms of classical genetic theory.

All workers have stressed the extreme complexity of the results. Simple reciprocal crossing of characters is hardly ever observed, in almost all cases many
intermediate forms are isolated in recombination experiments. Much of the complexity may be due to the choice of unsuitable markers such as virulence which
depend on a large number of factors.

The influenza viruses are remarkable in their very great tendency to variation
and the great ease with which recombinants can be produced and the great variety of properties shown by recombinants. If it is accepted that all the properties
of the infectious particle are in the end coded by the virus RNA it seems essential
to assume that on entry into the cell the virus RNA becomes disrupted into a
large number of small pieces which can be reassembled in various ways. But if

this is so it is very difficult to understand why no stable recombinant can be produced between virus A and virus B. These two viruses are so similar that almost any phenomenon shown by one of them can be duplicated with the other. It is very difficult to understand why in a mixed A and B infection involving fragmentation of both types RNA there should be no mixture of the two RNA components and no formation of combined RNA. Possibly there is some agent in the cell, perhaps a DNA molecule, which is involved in the process of synthesis and which is able to accept RNA fragments from one of the two viruses while rejecting the other.

The concepts of classical genetics are dependent on certain assumptions which are possibly not valid when applied to influenza virus reproduction. It is a dangerous assumption that the function of RNA in an influenza virus particle is the same as that of DNA in a bacteriophage particle or in a cell. It is probably a mistake to regard the virus as an independent agent whose reproduction is entirely self-controlled, its properties are almost certainly in part controlled by the host cell. Many of the properties of the virus particle are in fact properties of the cell membrane of the infected cell and only the S antigen component has any real independence. Even this component appears to be reproduced only in close association with the genetic mechanism of the host cell.

The entry of an influenza virus particle into a cell results in the introduction into that cell of certain "information" carried by the virus RNA and possibly also by its protein, which results in a modification of the synthetic processes of the cell which commences in the cell nucleus and which may well be under the control of cell DNA. In most instances the alteration of synthetic processes is temporary and the cell is not killed, in fact there may be some stimulus to cell division. It might be very interesting to try to determine if influenza virus infection results in any permanent alteration in the cell genetic apparatus, but no work on this has been recorded.

The study of genetic interactions between influenza viruses has possibly reached a stage where its further pursuit is likely to add little either to the science of genetics or to the problem of the understanding and control of influenzal infection in man. The author has viewed with some concern the production of recombinants between highly infectious strains of virus such as A2 and highly pathogenic strains such as NWS and fowl plague. The possible production of a highly dangerous recombinant has to be considered.

# XXIII. Adaptation and Variation in Influenza Viruses

Influenza virus strains show great variation in their ability to infect a different animal host, and on introduction into a new host species full pathogenicity is usually only attained after a series of passages.

## A. Adaptation

### 1. Adaptation of Virus to Ferrets

Strains of influenza virus A isolated in 1933—1937 produced fever in ferrets on first inoculation and on serial passage the severity of the disease increased and lung consolidation and death occurred. In later years strains proved more difficult to isolate in ferrets, SUGG and NAGAKI ($S_{325}$) state that a decrease in ferret virulence of influenza A viruses occurred as early as in 1939. They found that on serial passage of the CAM strain (isolated in 1946) by intranasal inoculation of ferrets no evidence of multiplication in the lungs was obtained during 10 passages, a result contrasting with the ready adaptation of the 1933—1937 strains. PAN-THIER, CATEIGNE and HANNOUN ($P_8$) found that a strain of A1 virus which was readily isolated by amniotic inoculation and in mice, failed to infect ferrets. BLASKOVIC and others ($B_{78}$) were able to adapt the Asian strain A2/Bratislava/57 to ferrets after 10 egg passages. Pneumonic changes were produced if the inoculum contained $10^9$ $ID_{50}$ per ml.

Influenza B viruses are more difficult to isolate in ferrets, HIRST ($H_{134}$) found that strains of B virus which failed to infect ferrets on primary isolation did so after isolation in eggs.

## 2. Adaptation of Virus to Mice

If human influenzal washings are inoculated intranasally to mice no lung lesions are produced, but if a series of "blind passages" are made from the lungs of mice killed 2—4 days after inoculation the virus becomes adapted and lung lesions develop ($F_{106}$, $Z_{42}$, $Z_{12}$, $S_{319}$).

Many studies have been made on the adaptation to mice of virus isolated in eggs ($H_{133}$, $W_{16}$, $S_{317-319}$, $L_{124}$, $S_{236}$, $D_{16}$, $M_{40}$, $I_{32}$, $G_{24}$, $D_8$, $L_{52}$, $B_{131}$, $B_{132}$, $L_{50}$, $W_{35}$, $S_{92}$).

If mice are inoculated intranasally with egg adapted virus of high titre, severe lung lesions may be produced at the first passage and the mice may die. On serial passage however the lung lesions disappear. The severe lung lesions produced in the first passage appear to be due to the toxic effects of the inoculum ($S_{317}$, $L_{52}$, $W_{35}$). SUGG ($S_{317}$) showed that the production of lung lesions and death in mice at the first passage was directly related to the egg infective titre of the inoculum, and that the lungs of the dead mice contained relatively small amounts of infectious virus. On serial passage no lung lesions are produced for several passages but between the 4th and 10th passages lesions appear and increase in severity and the mice die.

HIRST ($H_{133}$) observed that although in the early passages of egg-adapted virus in mice no lesions were produced, the virus multiplied and attained maximal titre as measured by *in ovo* titration, and that the appearance of lung lesions and death was not associated with an increase in infectivity titre. WEN and WANG ($W_{35}$) and LEDINKO and PERRY ($L_{52}$) however found that the later passages were associated with an increased yield of virus.

In early passages haemagglutinin production may not be detectable, but haemagglutinin appears in increased amount when lesions begin to develop ($L_{52}$, $S_{236}$). HIRST noted that the haemagglutinin titre for chick red cells was not increased, but that there might be an increase in titre towards guinea-pig cells. LEDINKO and PERRY considered that in the early phases of adaptation a "functionally deficient" form of virus was produced, and HIRST thought that the increase in virulence on passage was due to the replacement of the original egg-adapted virus by an agent with altered pathogenic properties.

Comparisons of the growth rate of egg-adapted and mouse-adapted lines showed that the mouse-adapted strain grows more rapidly, reaching a maximum titre in 24 hours, while the egg-adapted strain takes 48—72 hours to attain maximal titre ($W_{16}$, $B_{131}$). DAVENPORT and FRANCIS ($D_{16}$) found that with unadapted strains there was a lag phase lasting 12 hours in which no multiplication of virus was demonstrable, while the adapted line commenced to multiply in 6 hours. Final titres are usually higher with adapted strains than unadapted, and SUGG ($S_{319}$) found that mice died when the titre of virus reached $10^8$ $ID_{50}$. Occasionally unadapted strains attained this titre and in such cases the mice died. GINSBERG ($G_{24}$) found that mouse-adapted strains were equally infective to mouse lung and chorioallantoic membrane, but with egg-adapted strains the dose of virus required to infect mouse lungs was 100 times as large as the egg infective dose.

DAVENPORT ($D_8$) made a series of decimal dilutions of egg-adapted influenza B virus and initiated a series of passages with each dilution in mice. It was found that the lines became adapted to the mouse in an order corresponding to the dilution of the original virus used to start the line. This suggested that the process of adaptation was one of selection of mutants, there being more mutant particles in the lower dilutions than in the higher. LEDINKO ($L_{50}$) isolated clones by limit dilution at each passage and found that a virus with all the properties of fully adapted virus could be isolated during an early stage of adaptation, suggesting that selection of a pre-existing mutant was occurring.

SOLOVYEV, MARENNIKOVA and GUTMAN ($S_{236}$) found that strains of A1 virus isolated in 1952 adapted more slowly to mice than the 1949 strains.

### 3. Adaptation to Hamsters and Other Animals

FRIEDEWALD and HOOK ($F_{146}$) found that influenza A strains multiplied readily in hamster lung on intranasal inoculation, but no lesions were produced in six passages. There was a sudden increase in virulence at the 7th passage, virus titres rose and haemagglutinin became detectable, at first only against guinea pig cells but at the 11th passage also for fowl cells. The LEE strain of virus B produced lung lesions at the first passage and there was no increase in pathogenicity on passage.

DAVENPORT ($D_6$) adapted the ALLEN strain of virus B to hamsters. Lung lesions were produced at the first passage, then disappeared in later passages, and reappeared as the strain became adapted. Alternate passage of virus in hamsters and eggs did not increase the rate at which the strain became adapted to the hamster. TUMOVA ($T_{66}$) passed PR8, RHODES and A/Prague/51 strains intranasally in hamsters. Virus multiplied to high titre in the lungs after 4 passages, but there were no signs of illness. After 20 passages the strains lost ability to agglutinate chick and guinea pig red cells and did not infect eggs. The hamster strains were more easily adapted to mice than the original egg lines.

CHALKINA ($C_{23}$) found that on passage of mouse-adapted strains in rats a mild illness was produced but the virus died out on serial passage. Virus recovered from rats had reduced virulence for mice. Similar results were obtained by BAYDAKOVA and ZILBER ($B_{26}$) on passing mouse-adapted virus in guinea pigs. SOLOVYEV and MARENNIKOVA noted that mouse-adapted A1 virus was more pathogenic to rats then egg-adapted virus.

## 4. Adaptation to the Chick Embryo Chorion

BURNET ($B_{156}$) obtained only trivial lesions in early passage of virus on the external surface of the chorioallantoic membrane, but virulence increased and by 73 passages all the embryos were killed in 48 hours. There were well-defined local lesions on the membrane and the amount of virus could be estimated by pock-counting. BALENSKI and BRUCKNER ($B_1$) found that after 22—25 passages on the chorioallantoic membrane the pathogenicity of the virus increased for eggs but decreased for mice. BURNET and LUSH ($B_{211}$) adapted WS and MEL to the chorionic surface of the chorioallantoic membrane. After 50 passages both strains killed the embryo in 48 hours, but WS was much more virulent to mice than MEL.

## 5. Adaptation to Growth in Brain Tissue — Neurotropic Variants

STUART-HARRIS ($S_{286}$) obtained a neurotropic variant NWS by serial passage of WS virus in eggs followed by serial passages in the brains of young mice and later of older mice. From the 12th brain passage onwards the mice developed encephalitis, neutralised by WS antiserum. FRANCIS and MOORE ($F_{110}$) also obtained a neurotropic variant of WS virus — WSN, and a similar variant was produced by SHAKH-NAZAROVA and SAKHAROV ($S_{93}$). HENLE and HENLE ($H_{82}$) noted that intracerebral inoculation of influenza viruses produces neurological signs in the form of clonic and tonic convulsions and death usually occurs in 24—72 hours, but the viruses did not multiply and died out on serial passage. APPLEBY ($A_{92}$) obtained a neurotropic variant of the KUNZ strain of virus A by recombination of UV inactivated mixtures of KUNZ and NWS. BRONITSKI and others ($B_{134}$) isolated from the intestine of an infant dead of a fulminating influenza a strain of A1 virus which was readily adapted to mice by intracerebral inoculation.

Several workers have observed that neurotropic strains of virus are pantropic in new born mice, producing a generalized infection with virus multiplication in all the organs on intravenous or subcutaneous inoculation ($W_3$, $W_4$, $O_4$, $T_{46}$). A rapid fall in susceptibility to subcutaneous virus occurred in the first 16 days of life ($T_{76}$). Similar results were obtained in suckling hamsters.

## 6. Adaptation to Growth in the Allantoic Sac — the O-D Variation

In 1943 BURNET and BULL ($B_{192}$) isolated a strain BEL of virus A by amniotic inoculation and found that the amniotic fluid gave a much higher haemagglutinin titre with human or guinea pig red cells than with fowl cells. On passage in eggs the haemagglutinin titre to fowl red cells increased until the fowl/guinea pig haemagglutinin titre ratio approached 1. Virus appeared in the allantoic fluid with an F/G ratio of 1. If passages were made with very small inocula the original state of the virus was preserved.

The change from the original O form to the derived D form was believed to be irreversible, the result of a mutational change.

The O form agglutinated guinea pig red cells to a higher titre than fowl cells, the F/G ratio being 0.05—0.2. It did not grow in the allantoic sac, did not produce lung lesions in mice, caused an increase in the volume of amniotic fluid in eggs inoculated to the amnion, and was not inhibited by human tears. The D form gave an F/G ratio of 0.5—2.0, grew in the allantoic sac, produced lung lesions in mice, caused a reduction in the volume of amniotic fluid, and was inhibited by human tears. The D form multiplied more rapidly in eggs and overgrew the O type.

In later studies BURNET and STONE ($B_{216}$, $B_{217}$) found that large inocula of O phase virus would initiate infection irregularly in mice and large doses would infect the allantoic cavity. Multiplication of O phase virus in the amniotic cavity was more rapid in 13—14 day eggs than in 12 day eggs. During the first 3 days O type virus was recovered from amniotic fluid and lung, later intermediate and D form virus was isolated, and D type only after 5 days incubation. Amniotic fluid and lung extracts both contained virus in the same phase. To maintain the O form on passage the inoculum was prepared from embryo lung emulsion which was adsorbed with fowl red blood cells and dilutions in normal horse serum saline inoculated to the amnion of 13—14 day eggs and lungs harvested after not more than 3 days at 35°C.

ANDERSON and BURNET ($A_{62}$) found that strains MIL and TAN isolated in inter-epidemic periods produced D phase cultures less readily than the epidemic strains BEL or CAM, and their O phase forms grew as readily as D phase in the allantoic sac. STONE ($S_{276}$) using the BEL strain, found that both O and D forms were inhibited by ovarian cyst mucoid, but ovomucin and suspensions of chick embryo lung inhibited D phase but not O phase. Both O and D forms destroyed ovarian cyst mucoid inhibitor on incubation, the other two inhibitors were destroyed by D but not by O phase virus. Both viruses were converted to indicator state by heating at 56°C.

### a) Reversibility of the O-D Change

HIRST ($H_{133}$) found that when egg adapted virus was passed serially in mice the haemagglutinin titre against fowl cells did not increase on passage, but the guinea pig cell titre did increase, as if the original D phase virus was reverting to the O form. GORBUNOVA and MARENNIKOVA ($G_{43}$, $G_{44}$) found that mouse adapted strains which grew well in the allantoic sac gave the same F/G ratios as O variants. ZHDANOV ($Z_{34}$) was unable to find any relation between the O-D state of virus strains and their pathogenicity for mice. MAGILL and SUGG ($M_{18}$) considered that the O-D change was reversible, they found that O variants occasionally appeared at random when D phase strains were grown in the amniotic sac, but not in the allantoic sac. O variants grew as such in tissue cultures of chorioallantoic membrane cells. O phase strains agglutinated fowl cells readily at pH 5.6. BURNET, STONE, ISAACS and EDNEY ($B_{222}$) claimed that MAGILL and SUGG were working with "false O" strains which had already begun to change to D. "True O" virus was said not to agglutinate fowl cells at all, nor does it agglutinate periodate treated fowl cells. "False O" strains agglutinate periodate treated fowl cells, and heated "false O" strains agglutinate fowl cells to increased titre. BRIODY ($B_{127}$) found that some strains of O virus agglutinated fowl cells if heated at 52°C for 10 minutes, or if tested in the presence of oxalate, citrate, arsenate or fluoride ions, or after storage at 4°C for several weeks. Strains modified in this way yielded typical O phase virus on inoculation to eggs and the modification was considered to be a non-inheritable phenotypic variation. BRIODY ($B_{129}$) also found that if D phase strains of WS or MEL were passed several hundred times in tissue cultures they developed an F/G ratio of 1:20. These "pseudo O" strains reverted to the D phase on passage in eggs, or if heated. Borate, arsenate and arsenite inhibited, while phosphate, citrate oxalate and $CaCl_2$ increased the haemagglutinin titre of these strains towards fowl cells.

ISAACS and STONE found that if haemagglutinin inhibition tests were carried out with a mixture of O and D virus, when guinea pig cells were used the serum titre was that expected from the amount of mixed virus used, but when fowl cells were used the titre was that which would have occurred if the O phase virus had not been present. The result suggested that antibody was preferentially adsorbed to the virus which was absorbed on red cells rather than to free virus ($I_{33}$).

KEIL, LIPPELT, MULLER and BRAND ($K_{26}$) found that if D phase virus was mixed with defibrinated human blood and then inoculated to the amnion some fluids with O phase virus were produced and could be maintained in the O phase for one passage. The effect was produced by the human serum and not by red cells.

### b) The O-D Variation with Virus B

BURNET, BEVERIDGE and BULL ($B_{189}$) found that the B virus strain BON did not show the O-D variation, being in the D form when first isolated. HIRST ($H_{134}$) found that B virus did not occur in the O form, though it showed some of the secondary O characteristics. BRIODY ($B_{128}$) found that B viruses did show a change in character on passage. There was an E phase with a fowl/guinea pig haemagglutinin ratio of 1:1 which was maintained by undiluted passages, and an H form with an F/G ratio of 10 or 20 to 1 which appeared if passages were made with dilutions of $10^{-2}$ or $10^{-3}$. The H phase could be maintained by dilute inocula but changed to E on undiluted passage. These results were obtained with the strains BON, MIL and WN, but the LEE strain was in the H phase and did not change to E.

### c) The O-D Variation with Asian A 2 Virus

LÖFFLER ($L_{110}$) isolated 30 strains of Asian virus none of which were in the O phase. The strains differed from previous A strains in agglutinating cow, cat, horse and frog cells. ALBALADEJO and BERENGUER ($A_{41}$) found that newly isolated strains were in the O phase during the first 5 passages, partly in the O and partly D in the next 10 passages, and mostly in the D phase in the next 20 passages.

## B. Variation

### 1. Variations in Haemagglutinin Production during Adaptation

Several workers have observed that, in the course of adaptation of egg adapted virus to mice, in the early stages virus is produced with a high infectivity titre but with little or no haemagglutinating power ($H_{133}$, $W_{16}$, $S_{236}$). In later passages the haemagglutinin titre increases, and this increase is associated with the appearance of lung lesions. TOVARNITSKY ($T_{53}$) noted that in ferrets, nasopharyngeal discharges of high infectivity might fail to agglutinate red cells. FRIEDWALD and HOOK ($F_{146}$) found that in the early stages of adaptation of PR8 virus to hamsters no haemagglutinin was produced in the early passages, but it appeared in later passages along with the appearance of lung lesions. The first haemagglutinin to appear agglutinated human cells but not fowl cells, in later passages both types of cell were agglutinated. There was evidence that in the intermediate stage haemagglutinin to fowl cells was produced but was masked by an inhibitor.

## 2. Variations in Inhibitor Sensitivity and Enzymic Activity

Soon after the discovery of the non-specific inhibitors of haemagglutination it was found that different strains of virus differed in sensitivity to the various inhibitors and in ability to destroy inhibitor. These differences could be used as "marker characters" in recombination studies. Changes in inhibitor resistance and enzymic activity occurred in the course of adaptation of virus to eggs or to animals, and it was considered that these changes were due to the selection of pre-existing mutants. ISAACS and EDNEY ($I_{17}$) by passage of the MEL strain of virus A at high dilution in eggs obtained a variant with reduced enzymic activity, and a similar variant was isolated from PR8 virus. SMITH, WESTWOOD, WESTWOOD and BELYAVIN observed the appearance of an inhibitor-resistant mutant in the course of passage of PR8 virus in the allantoic sac, the new strain appearing at the 80th passage. BRIODY, CASSEL and MEDILL ($B_{132}$) found that on adaptation to mice strains of A1 and B virus developed resistance to the $\beta$ inhibitor of ox serum coincident with the appearance of lung lesions. LEDINKO ($L_{50}$) found that on passage of B virus in mice the strain developed a reduced enzymic action on sheep mucin and mouse lung inhibitor. SIMPSON and MOGABGAB ($S_{142}$) found that the enzymic properties of virus strains grown in human embryonic tissue culture were different from those of the same strains grown in eggs.

MOGABGAB, HOLMES and PELON ($M_{107}$) found that when egg or human embryo tissue lines of A, A1 and B viruses were adapted to monkey kidney cultures they became less sensitive to monkey serum inhibitor while sensitivity to other serum inhibitors increased.

MEDILL-BROWN and BRIODY ($M_{70}$) found that inhibitor sensitive A1 and B strains produced mutants resistant to $\beta$ inhibitor. These mutants could be isolated by growing the virus in the presence of bovine serum in the allantoic fluid. Resistant mutants were stable on passage and were pathogenic to mice on first inoculation. In the adaptation of sensitive strains to mice mixed proportions of resistant and sensitive particles could be detected immediately before the development of extensive lung lesions. It was estimated that in sensitive strains there was 1 resistant mutant in $10^6$ total particles. Resistant virus did not destroy $\beta$ inhibitor. FAUCONNIER, LARTITEGUI and BARUA ($F_{20}$) also obtained resistant variants from A1 virus by growth in the presence of sub-neutralising doses of bovine serum.

Many studies have been made of the variations in inhibitor sensitivity of A2 viruses. LUZYANINA, SALMINEN and SMORODINTSEV ($L_{132}$) found that some Asian strains were strongly inhibited by the heat-stable inhibitors of normal horse, guinea pig, rabbit and rat sera. These were described as "avid" strains. Other "non-avid" strains were not inhibited. CHOPPIN, OSTERHOUT and TAMM ($C_{45}$) found that sensitivity to serum inhibitors was related to reactivity with antibodies. CLEELAND and McKEE ($C_{61}$) isolated 20 strains of Asian virus of which 12 were insensitive to normal serum inhibitors. Haemagglutinin-inhibition tests using the 12 insensitive strains and convalescent sera showed wide variations in reactivity, suggesting the occurrence of antigenic differences. FRIDMAN ($F_{140}$) found that different cultures of the same A2 strain differed in sensitiveness to the heat-stable inhibitor of guinea pig serum. With "avid" strains haemagglutination was almost completely inhibited. While with "non-avid" strains there was no effect. Some cultures appeared to be mixtures of avid and non-avid forms and showed an intermediate degree of inhibitor sensitivity. BLASKOVIC and others ($B_{78}$) found that when inhibitor resistant strains of A2 virus were adapted to ferrets they became more sensitive to both inhibitor and antibody. GOTO ($G_{51}$) found that viruses isolated in 1960 were much more sensitive to normal serum inhibitors than the original A2 strains, and were also highly avid towards antibody.

CHOPPIN and TAMM ($C_{47}$) separated inhibitor sensitive substrains from A2 viruses by passage at limiting dilutions, and isolated resistant substrains by passage in the presence of heated horse serum. The ratio of sensitive to resistant particles was estimated to be 70:1 in one strain and 10,000:1 in another. Changes in the proportions occurred on serial passage in eggs. There were no significant antigenic differences. The inhibitor sensitive particles eluted slowly or not at all from red cells and agglutinated red cells treated with cholera filtrate, while resistant particles eluted rapidly and did not agglutinate cholera filtrate-treated red cells. BOLDASOV ($B_{98}$) found neuraminidase activity to be greater in inhibitor resistant than in inhibitor-sensitive strains, but there was no difference in ability to destroy ovomucin. Strains of A2 which were avid or non-avid towards antibody had the same enzymic activity.

Other workers have observed that inhibitor resistant strains of A2 virus can be produced by passage in eggs in the presence of sub-neutralising doses of horse serum ($J_2$, $L_{72}$). LIEF found that sensitive and resistant particles could be distinguished serologically by the use of the strain-specific complement fixation test ($L_{72}$). COHEN and BIDDLE ($C_{74}$) found that an Asian strain lost its sensitivity to the γ inhibitor of normal horse serum on mouse adaptation. As mouse serum does not contain γ inhibitor this could not be due to selection by growth in the presence of inhibitor.

CHOPPIN and STOEKENIUS ($C_{46}$) found that the characteristics of two variants of A2 virus which differed in reaction with antibody and inhibitors were also shown by ether treated virus and suggested that the differences resided in the "spikes" on the virus surface rather than in the steric arrangement of the spikes.

ISACHENKO, SMIRNOVA and GORBUNOVA ($I_{36}$) obtained haemagglutinins of different inhibitor sensitivity by passage of A2 virus through Sephadex G 75 columns. The more sensitive fraction was non-infective to eggs. COHEN and SMITH ($C_{77}$) separated rapidly eluting and non-eluting fractions from an A1 virus and found that the rapidly eluting fraction was more sensitive to both α and β type inhibitors in spite of its apparently greater enzymic activity. The differences were not stable on passage. PADGETT and WALKER ($P_4$) collected the slowest eluting fractions of LEE virus from red cells and by passage at limiting dilutions isolated a stable variant which had a reduced rate of action on chick mucoproteins. The variant was estimated to represent less that 3% of the original population.

### 3. Variation in Antigenic Structure

*a) Spontaneous Variations in Antigenic Structure of Virus A*

All strains of virus A possess a common S antigen which appears to be invariable, but the specific V antigen complex has shown considerable variations during the period since 1933. The original WS strain isolated in 1933 could be distinguished from the swine virus isolated in 1931. In 1935 strains isolated in Russia were similar to WS, but the strains PR8 and PHILA isolated in America were significantly different though related to WS. Strains isolated in Manchester in 1937 were more closely related to PR8 than to WS. In 1937 BURNET ($B_{157}$) divided virus A strains into three groups, an Old World group represented by WS, BH and Russian strains; a New World group, PR8, PHILA and the Australian MEL strain; and the swine virus S15.

In 1938 MAGILL and FRANCIS ($M_{12}$) studied the antigenic structure of 23 strains using rabbit immune sera. They found antigenic similarities and differences and while a rough grouping was possible the groups tended to merge. Strains from the same epidemic were usually, but not always, similar. They suggested that a mosaic of antigens was present, possibly not all antigens present in every strain. SMITH and ANDREWES ($S_{166}$) studying 28 strains with convalescent ferret sera found that many different antigenic components could be detected but 4 were of major significance. Strains could be divided into three types, a highly specific type with one of the 4 antigens dominant, a less specific type, and a non-specific type with equal amounts of all 4 antigens. ZHDANOV ($Z_{84}$) states that Russian strains isolated before 1936 were of WS type, later the PR8 type became dominant, although occasional strains of WS type were isolated. The PR8 type strains gradually changed between 1941 and 1945.

In 1940 BURNET and LUSH isolated 4 strains in ferrets and adapted them to mice. Two antigenic types were identified but one strain produced a less specific serum in ferrets suggesting that an antigenic change occurred on mouse adaptation. CHAPMAN and HYDE ($C_{33}$) also found two antigenic types in 6 strains isolated in 1940. HIRST ($H_{131}$) isolated 18 strains in 1940—1941, sixteen were indistinguishable, while 2 showed significant difference.

MAGILL and SUGG ($M_{15}$) isolated 5 strains from a localised outbreak in 1943. The strains were more closely related to each other than to PR8, but each showed some degree of specificity, and varied in antigenic potency. MAGILL and SUGG ($M_{16}$) examined the mouse-protective antibodies in 40 human convalescent sera and found quantitative differences in both acute and convalescent sera when tested against 5 strains of virus A. In 1945 LACORTE, MONTEIRO and LOURES ($L_2$, $L_3$) isolated 6 strains of virus in Brazil which were different from PR8. LÖFSTRÖM ($L_{113}$) found that Swedish strains isolated in 1944 and 1947 differed markedly from PR8.

*b) The A Prime or A1 Viruses*

In 1946 and 1947 strains of virus appeared which showed a much greater difference in antigenic structure than had so far been encountered. There appeared to be a new dominant antigen and the new strains were described as the A prime or A1 type. The first A1 outbreak appears to have occurred in 1946 in Japan and Korea, and a strain CAM isolated in 1946 in Australia was of the new type.

In 1947 outbreaks due to the A1 virus occurred in America and the strain FM1 isolated in New Jersey came to be regarded as the prototype strain. FRANCIS, SALK and QUILLIGAN ($F_{119}$) found that vaccination with previously isolated A strains did not produce an antibody rise against A1 and did not protect against infection by it. MORGAN, BARNES and FINLAND ($M_{120}$) found that many convalescents from infection in 1947 developed higher antibody levels against PR8 than against the 1947 strain. MULDER, VAN DER VEEN, ENSERINCK and BRANS ($M_{145}$) noted the great sensitivity of the new strain to inhibitor in ferret serum. In 1947 and the succeeding years the A1 virus completely replaced the previous A strains. Typical A1 strains were widely isolated in 1949, but in this year FINLAND, BARNES, and WELLS ($XF_2$) isolated strains which differed slightly from FM1. In 1950 HILLEMAN, MASON and ROGERS ($H_{119}$) isolated strains which differed from FM1. ISAACS, GLEDHILL and ANDREWES ($I_{26}$) examined 96 strains of A1 type isolated in 1950—1951 and found that they fell into two main serological groups; 47 were closely related to a strain A/England/1/51 isolated in Liverpool and called the Liverpool type, 46 related to A/Sweden/3/50 and called the Scandinavian type, while 3 showed intermediate properties. MULDER, BRANS and MASUREL ($M_{137}$) found that strains isolated in Holland in 1947 and 1949 were related to FM1, while of 17 strains isolated in 1951, 3 were of the Liverpool and 14 of the Scandinavian type. Strains isolated in 1953 differed from both the Liverpool and Scandinavian types and from $FM_1$.

Viruses of A1 type were first isolated in Russia in 1948 and three varieties developed, the KLIM strains in 1949, the PAN strain in 1951—1952, and the RO-STOCK type in 1953—1954 ($Z_{17-20}$, $Z_7$, $G_{92}$). Scandinavian A1 strains were isolated in Italy in 1953 ($B_{119}$). ISAACS, DEPOUX and FISET ($I_{15}$) examining strains received at the World Influenza Centre in 1952—1953 found 175 to be of Scandinavian and 33 of the Liverpool type. By antibody absorption experiments they found that the dominant antigen of A/Sweden/3/50 was contained within the virus particle, while in A/Missouri/303/52 it was a surface constituent. Differences between the 1951 and 1953 strains were noted by FINLAND and BARNES ($F_{67}$).

Extensive studies of the antigenic relations of A and A1 viruses were made by HILLEMAN and his associates ($H_{113}$, $H_{114}$, $H_{117}$, $H_{118}$), MAGILL and JOTZ ($M_{13}$) and JENSEN ($J_5$).

### c) The Asian or A2 Viruses

In 1957 an entirely new type of virus appeared and spread as a pandemic over the whole world. This strain, the Asian or A2 virus had an S antigen of A type but its specific antigen showed no relation to previous A or A1 viruses. The new strain appears to have been first detected in Manchuria, and in the Spring of 1957 strains were isolated in Hong Kong, Japan, Malaya and Formosa ($M_{83}$). Vaccination with pre-1957 strains produced no antibody to the new virus, and the new strain spread rapidly although epidemics of A1 influenza had occurred only 4 months earlier in Japan, U.S.A. and Europe.

Strains of A2 virus were isolated and their properties examined in laboratories all over the world, and some 2000 references to work on the Asian pandemic are listed in a publication by the American Institute of Biological Sciences ($A_{55}$). The A2 virus had a filamentous morphology, was readily isolated in eggs but gave low haemagglutinin titres, and was very sensitive to non-specific inhibitors in human and animal sera. The new virus represented a major antigenic change, though some workers noted a very slight relation to previously isolated A1 strains ($J_7$, $J_5$, $C_{45}$). Strains isolated in the autumn of 1957 in New York showed some antigenic difference from the earlier Asian strains ($C_{45}$), and strains isolated in 1961 in Taiwan differed from the previous A2 strains ($C_{37}$). Antigenic differences occurred in 1963 ($XD_1$, $D_{95}$). PEREIRA, PEREIRA and LAW ($P_{33}$) found that 75% of the strains isolated in Britain in 1963 resembled the prototype A2/Singapore/1/57, while 25% were related to a new variant A2/Netherlands/65/63. A further variant A2/England/12/64 was isolated in 1964.

### 4. Experimental Production of Changes in Antigenic Structure

#### a) Changes Produced by Adaptation to New Hosts

BURNET and LUSH ($B_{212}$) noted that a strain of virus A produced a less specific serum in ferrets after mouse adaptation than before, and suggested that an antigenic change occurred on adaptation to mice. FRANCIS ($F_{95}$) compared two lines of PR8 virus, one passed in mice and one in tissue culture and eggs. The mouse line was 100 times as effective as the TC-egg line as an immunizing agent in mice. Antisera produced in rabbits gave higher protection in neutralization tests with the homologous than with the heterologous line. ISAACS and SAMPAIO ($I_{32}$) also found that mouse-adapted virus was more antigenic in mice than unadapted virus. HIRST ($H_{133}$) noted that the adaptation of egg-adapted strains to mice was accompanied by changes in antigenic pattern as measured by cross haemagglutination-inhibition. SUGG ($S_{318}$) found differences in neutralization tests between egg and mouse lines of the CAM strain of virus using mouse and rabbit antisera, but no antigenic change was detected in 8 mouse passages of egg-adapted CAM although full mouse virulence was attained. MULDER and BRANS ($M_{134}$) using cross haemagglutination-inhibition tests demonstrated antigenic differences between egg and mouse lines of various A1 viruses. Russian workers have also studied the stability of antigenic type on adaptation ($B_{25}$, $B_{26}$, $B_{91}$, $L_{129}$, $F_1$, $F_4$, $M_{40}$). MOGABGAB, SIMPSON and GREEN ($M_{110}$) found that the antigenic composition of recently isolated strains of influenza A and B was not significantly altered by passage in chick embryo or in human tissue culture. HIRST ($H_{134}$) found that antigenic structure was not significantly altered by passage in eggs, and TAMM, GINSBERG and HORSFALL came to similar conclusions ($T_{22}$). JENSEN, MINUSE and FRANCIS ($J_{14}$) adapted virus from a single throat washing to eggs, ferrets, mice, hamsters, and chorioallantoic membrane tissue cultures. The derived lines showed close antigenic relationships, but considerable variations in antigenic potency and in avidity for antibody were noted, especially with the mouse and tissue culture lines.

PAUCKER ($P_{16}$) found that the PR8 virus remained unaltered over 107 egg passages, but an antigenic change occurred at the 130th passage apparently as a result of the appearance of a mutant which became dominant.

#### b) Antigenic Changes Produced by Passage in Partially Immune Animals, or in Presence of Immune Serum

ISAACS and EDNEY ($I_{18}$) isolated a serum resistant mutant of the MEL strain by passage in eggs in the presence of gradually increasing doses of anti-MEL serum. The mutant was stable on passage in the absence of serum. ARCHETTI and HORSFALL ($A_{99}$) produced stable antigenic variants of influenza A strains by serial passage in eggs in the presence of immune serum against different but closely related strains. Antigenic variants were not produced by passage in the absence of serum or in the presence of immune serum against widely different strains.

GAVRILOV ($G_{10}$, $G_{11}$) mixed the Shklauer strain of virus A with dilutions of antiserum and inoculated to the allantoic sac. Serial passages were made in each case using the maximum amount of serum which did not completely neutralize. Even after 50 such passages the antigenic structure was not altered, showing that antigenic variations are not produced by the in vitro action of serum. If, however, virus was passed in immunized mice using large doses to overcome the immunity, stable antigenic variants were produced and the mice developed increasingly severe lung lesions. LUZYANINA ($L_{127}$, $L_{130}$) produced a gradual antigenic change in PR8 virus with increasing pathogenicity by serial passage in actively immunized mice. MAGILL ($M_9$) also produced antigenic variants by serial passage in vaccinated mice. The phenomenon was extensively studied by GERBER, LOOSLI and HAMRE ($G_{17}$, $G_{18}$, $H_{14-16}$, $L_{114}$). They passed PR8 virus serially in mice previously vaccinated with formolised PR8 virus. Antigenic variants were detected in minor amounts up to the 8th passage, and became more prominent on further passage. At the 17th passage it was found that the variant, while sharing some antigenic components with the original PR8, contained a different dominant antigen and could infect vaccinated mice which had a high level of antibody to PR8 in the blood. The variant was then passed serially in mice immunized against it, and this procedure repeated for 4 successive generations, producing a series of variants each derived from the preceding one. There occurred a progressive decrease in reactivity with PR8 antiserum, while the variants retained the ability to elicit antibody to PR8 and to their respective predecessors. Antibody absorption tests showed that the variants possessed specific components increasing in amount with each successive variant, while the amount of antigen related to PR8 progressively decreased. When three further variant generations were produced reaction with PR8 serum almost entirely disappeared, but the strains still elicited antibody to the preceding variants, and the seventh variant produced more antibody to PR8 than did the fifth. A further series of 4 variants was produced by serial passage in mice immunized with both the homologous strain and with PR8. These variants failed to react with PR8 serum and produced less antibody to PR8 than variants of the previous series.

EDNEY ($E_{27}$) passed the CAM strain 7 times in eggs in the presence of antiserum. A serum resistant mutant SP7 was isolated and found to be pathogenic for mice. Analysis of the passage fluids at limiting dilution showed that 2 intermediate mutants had arisen each showing decrease in CAM antigen and an increase in SP7 antigen. The second intermediate was mouse virulent.

MAGILL ($M_{10}$) noted that during passage of virus strains in immunized mice the passage strain developed a capacity to evoke antibodies reacting with virus strains isolated in earlier years, a capacity which was not apparent in the parent strain. The parent strain showed a relatively broad range of surface reactivity which was not apparent in the derived strain. It was suggested that the hereditary change resulted from a spatial rearrangement and quantitative redistribution of antigens in the virus particle, some antigens being present on the surface and some in the inner bulk, and that influenza virus variation results from a rearrangement of existing hereditary elements.

## 5. The P-Q-R Variation

VAN DER VEEN and MULDER ($V_9$) in studies of antigenic structure of viruses by cross haemagglutination-inhibition tests noted that virus strains in the same sub-group behaved differently in the tests. Some (P strains) were inhibited to high titre by homologous antisera and to low titre with heterologous sera, others (Q strains) gave low titres with both homologous and heterologous sera, while a third group (R strains) gave high titres against both homologous and heterologous sera. Q strains gave rise to poorer antisera in the ferret than P and R strains. ISAACS, GLEDHILL and ANDREWES ($I_{26}$) showed that the P-Q change was reversible, P strains were converted to Q by passage in eggs in the presence of homologous serum, and Q strains being converted to P by mouse passage. They found that of the A1 strains isolated in 1947—1949 the Scandinavian types were mostly in the Q phase, while the Liverpool types were all in the P phase. The two sub-types were not P and Q variants of the same virus since they were not interconvertible, and P phase variants of the Scandinavian and Liverpool strains were easily distinguished.

Viruses in the P phase are readily classified by haemagglutination-inhibition tests, while Q phase strains are more difficult to identify. R strains are particularly useful in detection of serum antibody because of their broad reactivity. The PR8 virus has typical R characteristics, reacting with antisera to most A strains.

FISET and DEPOUX ($F_{72}$) and ISAACS, DEPOUX and FISET ($I_{15}$) suggested that the P—Q variation is due to a re-arrangement of antigens, the high avidity for antibody of P strains being due to their dominant antigen being on the surface of the particle, while in the Q strains it lies within the particle. Q variants produced by passage in eggs in the presence of homologous serum react poorly with antibody but can still induce immunity to the parent strain on vaccination. MAGILL ($M_9$) found that the change in antigenic character produced by passage of strains in vaccinated mice was different from the Q—P change which occurs on adaptation of virus to mice since subsequent passage in a homologous immune environment did not induce reversion to the P form.

LEVY and WAGNER ($L_{68}$) found that strains of A2 virus isolated in Baltimore in 1957 gave stronger reactions with homologous and heterologous sera than prototype Asian strains, and concluded that they were in the R phase of the P—Q—R variation. Strains isolated in Taiwan in 1961 ($C_{37}$) were considered to be intermediate between Q and P strains.

Strains of A2 virus vary in sensitivity to normal serum inhibitors, "avid" strains being strongly inhibited while "non avid" strains are not inhibited ($L_{132}$). Avidity to inhibitors is not necessarily related to the avidity towards antibody shown by P strains. BLASKOVIC and others ($B_{78}$) found that passage of A2 virus in ferrets resulted in increased avidity to both antibody and inhibitors, and GOTO ($G_{51}$) found that strains isolated in 1960 were highly avid to antibody and very sensitive to inhibitors, but LOZHKINA ($L_{120}$) found that antibody-avid strains might be sensitive or non-sensitive to inhibitors. ISACHENKO ($I_{34}$), FUKUMI ($F_{160}$) and BOLDASOV ($B_{98}$) also found no relation between P—Q variation and inhibitor sensitivity.

## 6. Antigenic Variation with Virus B

Spontaneous variations in the antigenic structure of B viruses similar to those of A viruses have occurred over the years since the LEE virus was first isolated. In 1944 BURNET, BEVERIDGE and BULL ($B_{189}$) isolated a strain BON which was only distantly related to LEE. TAMM, KILBOURNE and HORSFALL ($T_{24}$) isolated 3 strains in 1950 which showed wide differences from other B strains and from LEE. They considered that B viruses showed as much antigenic variation as A viruses. HILLEMAN ($H_{113}$, $H_{114}$, $H_{118}$) made antigenic analyses of B viruses and found two groups, pre-1943 strains similar to LEE, and later strains resembling a strain WARNER isolated in 1948. MAGILL and JOTZ ($M_{13}$) found that the LEE strain was replaced by strains containing an additional antigen which became dominant in 1949. MULDER, BRANS and DE NOOYER ($M_{136}$) found that strains isolated in 1951—1952 gave low titres in haemagglutination-inhibition tests with LEE but high titres with the BON strain. RoMANENKO and FRIDMAN ($R_{47}$) isolated a strain of B virus distinct from LEE in 1954. TUMOVA and others ($T_{67}$) found that strains isolated in 1959—1961 differed from previous strains and described 2 groups of B viruses, B strains isolated from 1943—1952 and B1 isolated after 1954. ROBINSON and others ($R_{41}$) found that American strains of B virus could be divided into 3 groups, the 1940 LEE strain, strains isolated between 1945 and 1954, and the 1959—1961 strains and suggested that they be named B, B1 and B2. A similar suggestion was made by ZHDANOV and SOLOVYEV ($Z_{34}$). HENNESSY, MINUSE and DAVENPORT ($H_{104}$), while recognizing the existence of three antigenically different prevalent types, considered that the differences were not large enough to warrant classification into separate groups comparable to the A, A1 and A2 viruses.

## 7. Variation in Resistance to Physical and Chemical Agents

### a) Variations in Heat Resistance

BURNET ($B_{179}$) by heating the allantoic fluid of eggs infected with the NWS strain at 54°C and passing serially at low dilutions, was able to select a heat-resistant variant which lacked pathogenicity on intracerebral inoculation in mice. JONES ($J_{19}$) could not adapt PR8 virus to heat by serial passage at a temperature of 41°C, but by heating infected tissue cultures in minced chick embryo and Tyrode's solution for short periods at 50°C followed by subculture at 37°C, after a series of passages with increasing periods of exposure at 50°C the strain was gradually adapted to resist temperature conditions which were normally lethal. SALK ($S_6$) found that there were differences in the heat-resistance of the haemagglutinin of different strains of both A and B viruses, and there might be differences in strains isolated in the same outbreak, and in different lines of the same strain.

BOUDREAULT and PAVILANIS ($B_{114}$) passed virus strains serially at temperatures of 29°C and 41°C. Of 4 strains passed serially 25—50 times at 29°C two developed "cold mutants" which would not grow at 41°C. The same two strains gave "hot mutants" which grew better at 41°C than at 29°C on passage at 41°C. The other 2 strains were unaffected. The hot mutants were more virulent to mice than the cold mutants but the cold mutants were better producers of interferon. SOKOLOV and others (S$_{224}$) obtained similar results, and LOZYNSKAYA (L$_{123}$) described changes in virulence, toxicity, enzymic activity and immunogenic power in cold and hot adapted strains. MAASSAB (M$_4$) found that both A and B viruses could be adapted to growth at 26°C in chick and calf kidney tissue cultures by passage at gradually lowering temperatures, ten passages being required to produce optimal adaptation to 26°C.

### b) Variations in Resistance to UV Light

BJORKMAN and HORSFALL (B$_{73}$) found that by a single exposure to UV light it was possible to produce an alteration in the elution rate of LEE virus which persisted on serial passage. No other property appeared to be affected. SIGEL (S$_{132}$) found that after adaptation to the allantoic sac 5 strains of virus isolated on the same day varied in elution rate and in resistance to UV light. NICOULESCO and others (N$_{19}$) found that the FM1 strain contained two types of particle with different resistance to UV light and different sensitivity to non-specific inhibitors.

### c) Variations in Resistance to Trypsin

CLEELAND and SUGG (C$_{62}$, S$_{322}$) obtained from a single isolate of influenza virus A two lines of different resistance to trypsin. Incubation with trypsin for 1 hour at 37°C destroyed the haemagglutinin and reduced the egg infectivity of the sensitive line by 99%, while the resistant line and several other strains of virus A were unaffected. All preparations of the sensitive lines appeared to contain a small percentage of trypsin resistant particles, but subculture of these particles in eggs yielded primarily trypsin sensitive virus. It was later found that strains of virus A varied in trypsin resistance, and patterns of trypsin sensitivity could be established which showed some relation to the antigenic type.

BORECKY, LACKOVIC and MRENA (B$_{106}$) in the course of adaptation of the strain A1/Moscow/49 to mice obtained a variant which differed from the original egg line in having an extremely high sensitivity to trypsin.

### d) Variations Produced by Chemical Treatment of the Virus

BJORKMAN and HORSFALL (B$_{73}$) treated allantoic fluid infected with PR8 and LEE viruses with 0.01 M Lanthanum acetate. On subculture in eggs altered strains of virus were obtained which eluted from red cells more slowly than the original strains. This property was maintained on serial passage in the absence of Lanthanum.

OBROSOVA-SEROVA (O$_1$) and SOKOLOV and others (S$_{224}$) obtained influenza virus mutants by treatment of infective allantoic fluids with nitrous acid or with 5-methyl-6-oxy-4-diethylaminoethylpyrimidyl sulphide chlorhydrate, followed by limit-dilution passages. The mouse pathogenic strain OM33 gave mutants which were non-pathogenic to mice, had heat-stable haemagglutinins and eluted more slowly than the original strain. Nitrous acid mutants of the WSN strain gave small plaques and had reduced neurovirulence and mouse pathogenicity. The characters were retained on serial passage in eggs but some degree of reversion occurred on mouse passage.

### C. General Consideration of the Phenomena of Adaptation and Variation in Influenza Viruses

The influenza viruses are remarkable in the frequency with which spontaneous variations in their properties occur, and in the ease with which variations can be produced experimentally. This property, described by SMITH ($S_{164}$) as "plasticity" far exceeds in degree the variability shown by other viruses, even by such closely related viruses as the parainfluenza group. It has been supposed by most workers that in the course of reproduction in the cell small numbers of random mutants are produced which differ in character from the predominant form and which under suitable natural or experimental conditions may become selected preferentially with the production of a variant form. There can be little doubt that many of the phenomena of adaptation and variation do involve a mutant-selection process, thus adaptation of virus to a new host tissue is usually only possible if very large inocula are used. But it is difficult to believe that all the findings can be explained by the selection of mutants arising at random, many of the phenomena are more easily understood if the selection process actually causes the formation of mutant forms. Thus in many cases the production of a variant form seems to be a gradual process with a number of intermediate states between the original and variant forms.

It is to be noted that some of the virus properties such as the serological properties of the S antigen appear to be invariable; others such as the antigenic structure of the V component appear to be relatively stable; while other properties such as the enzymic activity, inhibitor sensitivity, and haemagglutinating properties are extremely variable.

The author holds the view that the great variability of the influenza viruses is due to their reproduction involving a degree of integration with the genetic apparatus of the host cell which is not seen in other viruses, with the result that the properties of the progeny virus are partly conditioned by the properties of the infecting virus and partly by the properties of the host cell. Most of the more variable properties of the virus are carried out by the outer component of the virus particle which is derived from the wall of the host cell.

An animal species can only survive if it is able to adapt itself to changes in its environment. The epithelium of the respiratory tract is the major contact of man with his environment and it is very probable that the genetic apparatus of the cells of the respiratory tract is adapted to react to environmental changes and to produce suitable changes in the composition of the surface of the cells. Exposure to influenza virus produces changes in the cell surface which may result in the powers of adaptation of the cells becoming called into action.

# XXIV. The Toxic Properties of Influenza Viruses

The inoculation of large doses of influenza virus into animals may cause severe symptoms, pathological changes, and death in the absence of any apparent virus multiplication, the effects being due to the toxic properties of the inoculum. Intracerebral inoculation in mice produces convulsions and death ($H_{62}$, $H_9$, $H_{63}$). Intravenous or intraperitoneal inoculation in mice causes death with gastro-intestinal haemorrhage and liver necrosis ($H_{79}$, $H_{64}$). A lymphocytopenia is produced by intravenous inoculation in rabbits ($H_{51}$), and a febrile reaction occurs $1\frac{1}{2}$—3 hours after inoculation ($W_9$). In rats the arterial blood pressure is reduced ($K_{31}$). Intranasal inoculation of large doses of unadapted virus in mice produces severe lung lesions ($G_{25}$, $S_{317}$). Intraocular injection in rabbits produces corneal opacities ($E_{41}$).

## 1. The Neurotoxic Property

HENLE and HENLE ($H_{62}$, $H_{63}$) inoculated mice intracerebrally with 0.03 ml of infected allantoic fluid or with virus concentrates produced by adsorption-elution from red cells. The mice developed neurological signs, if suspended by the tail they showed clonic and tonic convulsions. Death occurred in 1—3 days. All strains of virus A and B tested produced these effects if inoculated in sufficiently large doses, and similar reactions were produced in rats, guinea pigs and hamsters. No multiplication of virus in the brain could be detected. The toxic effect could not be separated from the virus particles by differential centrifugation, adsorption-elution from red cells or by protamine precipitation. It was specifically neutralized by homologous antiserum. The toxic effect was related to the amount of virus introduced, preparations with haemagglutinin titres of 2048 or more produced convulsions in 100% of mice inoculated, with preparations of lower titre the percentage of mice developing convulsions became reduced until with titre of 1:16 or less no effects were produced. The toxic effect of the virus was destroyed by heat, exposure to UV light and by formalin. Similar results were obtained by HALE and McKEE ($H_9$).

WAGNER ($W_2$) found that intracerebral inoculation of sub-toxic doses of LEE virus (4—256 haemagglutinin doses) conferred protection against the inoculation 24 hours later of 5 $LD_{50}$ or 3100 haemagglutinin doses. This virus-induced tolerance developed gradually, reaching a maximum at 24 hours and disappearing by the 4th day. Inoculation of WS virus conferred equal protection against both WS and LEE, and inoculation of LEE virus conferred a greater protection against WS than did WS virus. Protection against the neurotoxic effect was also produced by inoculation of crude cholera filtrate and somewhat less effectively by purified RDE. HOOK and WAGNER ($H_{159}$) found that intracerebral inoculation of minute doses of endotoxin from *E. coli*, *S. abortus equi*, or typhoid vaccine protected against the neurotoxic action of influenza virus. Protection developed in 2—4 hours and lasted for 7 days. Cerebral infection by neurotropic strains of influenza virus was not prevented. Intraperitoneal injection of endotoxin had little or no effect. No virus neutralizing factor could be found in the brains of mice rendered refractory by endotoxins.

MIMS ($M_{93}$) examined the brains of mice inoculated with MEL or PR8 virus at intervals by staining frozen sections with fluorescent antibody. Five minutes after inoculation there was marked fluorescence of the cytoplasm of many ependymal cells and of cells lining sub-arachnoid spaces. In 30 minutes the fluorescence was fading and by 3 hours was almost undetectable. After 5—6 hours there was moderate fluorescence in the cytoplasm of most ependymal cells increasing in brightness up to 8 hours. It was suggested that a single cycle of growth of virus occurred, there being no increase in fluorescence after 12 hours. Pre-treatment with RDE greatly reduced the number of fluorescent cells.

SETO ($S_{85}$) found that the neurotoxic effect of A2 virus was neutralized by an anti-sialidase serum, but the serum also contained haemagglutinin inhibiting antibodies.

ADO and ALEKSEYEVA ($A_{35}$) showed that influenza virus exerts a toxic action on the sympathetic nervous system. Perfusion of the superior cervical sympathetic ganglion of ferrets with allantoic fluid from eggs infected with influenza A or B results in a markedly decreased response of the ganglion to acetylcholine. Similar treatment with immunized ferrets results in an increased response to acetylcholine, the effect being type-specific.

## 2. Toxic Effects Produced by Intravenous and Intraperitoneal Inoculation

HENLE and HENLE ($H_{79}$, $H_{64}$) found that intraperitoneal inoculation of 1 ml of infected allantoic fluid in mice caused the death of a varying percentage of mice within 18—96 hours. Different strains of virus varied in toxicity, some strains producing death only if the virus was concentrated. Intravenous inoculation produced more uniform results and the lethal effect was four times greater than by the intraperitoneal route. Some mice died in 8 hours after intravenous inoculation. Similar results were obtained in rats, guinea pigs and rabbits. All strains of virus produced severe necrotic damage to the liver and spleen. Some strains of virus A caused engorgement of the intestinal blood vessels and oedema of the gut. Virus B strains did not produce intestinal reactions but caused the appearance of large pleural effusions. Deaths with A viruses usually occurred within 24 hours, while with the LEE strain most animals died on the 2nd day. Some mice which escaped early death died 6—8 days later with lung consolidation, apparently a result of virus growth. Some mice developed jaundice in 4—8 days.

The toxic effects were associated with the virus particle and were neutralized by homologous serum. Vaccination protected against the toxic effects of the homologous strain but not against heterologous virus. There was no evidence of multiplication of virus outside the respiratory tract. The toxic activity of the virus was slightly more resistant than the infecting property to heat, formalin and UV light.

These results were confirmed by KEMPF and HARKNESS ($K_{31}$), who also noted toxicity to ferrets.

MIMS ($M_{93}$) found that a dose of $10^8$ $ID_{50}$ of virus with an infectivity/haemagglutinin ratio of at least $10^{5.5}$ was required to produce toxic effects in mice, virus with lower I/H ratios were non toxic. By staining liver sections at intervals with fluorescent antibody MIMS found evidence of virus growth in the hepatic cells, but most of the virus inoculated intravenously was taken up by the littoral cells of the liver in which cells it did not grow.

KHOOBYARIAN and WALKER ($K_{36}$) found that inoculation of 0.25 $LD_{50}$ of PR8 virus intravenously in mice protected the mice against the inoculation of 12 $LD_{50}$ 24 hours later. Protection against PR8 virus was also given by small doses of LEE, mumps or NDV. MANIRE ($M_{34}$) obtained similar results and found that after inoculation of sub-toxic doses of PR8 resistance to challenge was maximal at 24—36 hours, while after LEE the resistance to PR8 was maximal at 48 hours and to LEE at 96 hours. Intravenous inoculation of RDE gave partial protection against PR8 but not against LEE.

HENLE and HENLE ($H_{82}$) showed that mice could be protected against the toxic effects of influenza virus by vaccination and suggested that the effect might be used as a test of vaccine potency.

KATO and HARA ($K_{22}$) found that the minimum lethal dose of PR8 virus given intravenously in rabbits was $10^{10.2}$ $ID_{50}$ or 25,000 haemagglutinin units per kilo of body weight. Pathological effects were capillary damage with hyperaemia and oedema of the lungs.

HARRIS and HENLE ($H_{51}$) found that intravenous injection of A or B virus in rabbits produced in 3 hours a lymphocytopenia lasting several hours. The lymphocyte count was reduced by more than 70%. The effect was due to the virus particle and was neutralized by immune serum. It was probably not due to destruction of circulating lymphocytes but to a decrease in delivery of new lymphocytes. A similar lymphocytopenia was produced by injection of mumps virus, rough pneumococci or nucleoproteins from haemolytic streptococci.

KEMPF and CHANG ($K_{31}$, $C_{31}$) produced a precipitous drop in the arterial blood pressure in rats by intravenous inoculation of PR8 virus. The effect was neutralized by antiserum, and it was suggested that it was due to direct action of the virus on the myocardium.

Extensive investigations on the toxic effects produced in animals by intraperitoneal and intravenous inoculation were made in Russia by GOLBERT and PETERSON ($G_{36}$) and ZAKSTELSKAYA ($Z_1$, $Z_2$) with results similar to those described above.

Mice and rats cannot be protected against the toxic effects of intravenous PR8 by previous treatment with adrenocorticotropic hormone ($K_{17}$), in fact cortisone increases the susceptibility of mice to virus toxin ($K_{37}$, $W_{12}$).

## 3. The Pyrogenic Effect of Influenza Viruses

ZELLAT and HENLE ($Z_{14}$) noted that intravenous injection of infected allantoic fluid produced a febrile response in rabbits. WAGNER, BENNETT and LE QUIRE ($W_9$, $W_{10}$) found that injection of 1 ml of PR8 allantoic fluid produced fever commencing $1\frac{1}{2}$—2 hours after inoculation, reaching a peak in 5 hours and returning to normal within 24 hours. Smaller doses of virus produced lower temperature rises after a more prolonged lag period, but even 0.025 ml of allantoic fluid produced some fever. Similar results were produced by LEE virus, and inoculation of lethal doses (40 ml) produced fever in 30 minutes which persisted until death at 18 hours. Adsorption of virus on red cells removed the pyrogenic effect from infected allantoic fluid, while the eluted virus produced fever. The pyrogenic effect resisted heating to a degree which destroyed infectivity, but heat destruction of haemagglutinin also destroyed the pyrogenic effect. The pyrogenic effect was neutralized by specific immune serum. The pyrogenic capacity of virus preparations closely paralleled their ability to induce lymphopenia.

Animals which had suffered a febrile reaction from inoculation of virus gave no response to a second inoculation on the following day. The period of tolerance lasted for about 11 days. Tolerance was produced by heated virus in proportion to the survival of haemagglutinin. Tolerance could be produced by heterologous strains of virus, the results being related to the position of the viruses in the receptor gradient. Thus LEE virus protects against itself, PR8 and NDV; PR8 protects against itself and NDV, while NDV protects only against itself. There was no cross-tolerance between the viruses and bacterial pyrogens. If the initial febrile response was prevented by administration of antipyrine tolerance was still produced.

It was suggested that the fever results from interaction of virus haemagglutinin with receptor substance, that the tolerance effect is produced by destruction of the receptor substance, and that the return of susceptibility is a result of restoration of receptor material. However, FRENCH ($F_{136}$) found that rabbits could not be rendered insensitive by prior inoculation of purified RDE.

BENNET and WAGNER ($B_{43}$) found that if leukopenia was produced in rabbits by inoculation of nitrogen mustard a febrile response was still elicited by intravenous virus, so that the pyrogenic effect is not due to release of pyrogenic material from degenerate leukocytes.

ATKINS and HUANG ($A_{104-106}$) found that after intravenous inoculation of virus a pyrogenic agent appears in the blood which is distinct from the original virus. This pyrogen can be detected by the passive transfer of serum to another animal, in which fever is produced. No pyrogen can be detected in the blood during the latent period after virus inoculation, and the amount is greatly reduced after fever subsides. No pyrogen is produced in the sera of tolerant animals by a second injection of virus, but injections of endogenous pyrogen into tolerant animals produces fever. The endogenous pyrogen is more heat-resistant than virus. Tolerance to virus-induced fever can be produced by prior injection of endogenous pyrogen. Cross-tolerance can be demonstrated between virus and bacterial endotoxin (typhoid vaccine) both of which produce endogenous pyrogens.

SIEGERT and BRAUNE ($S_{130}$, $S_{131}$) confirmed the appearance of an endogenous pyrogen in the blood during virus-induced fever, and that it would induce fever both in normal animals and in animals tolerant to virus-induced fever. Tolerance therefore is not a failure to respond to the pyrogen, but must be due to the blocking of a stage preceding release of pyrogen. They found the pyrogenic effect of virus and its haemagglutinating power equally sensitive to heat, but UV light and formaldehyde destroyed the pyrogenic effect without action on haemagglutinin. The pyrogenic effect was destroyed by ether and it was suggested that it was closely related to the virus lipid.

SOLOV'YEV, ZAKSTELSKAYA and TARUSOV ($S_{244}$) found that the blood was toxic in the acute stage of human influenza.

## 4. Other Manifestations of Virus Toxicity

The intranasal inoculation of large doses of virus in mice may produce severe lung lesions without multiplication of virus. On serial passage lesions disappear ($S_{317}$). Severe lung lesions may even be produced in immunized mice ($G_{26}$).

Corneal opacities were produced by inoculation of virus into the anterior chamber of the eye in rabbits and guinea pigs ($E_{41}$, $W_{52}$). The lesions resolved and recovered corneas were resistant to challenge up to 2 weeks. There was no detectable virus multiplication ($W_{52}$).

The toxic agent of the influenza virus can produce the SHWARTZMAN phenomenon ($K_{24}$). If rabbits are inoculated intradermally with 0.2 ml of PR8, LEE, or A2 infected allantoic fluid and 18—20 hours later a large provocative dose of *E. coli* filtrate is given intravenously the sensitized area of skin becomes haemorrhagic and swollen in 5 hours. The sensitizing potency of PR8 was increased by sonic disintegration or by repeated freezing and thawing, and such preparations could on intravenous inoculation provoke the Shwartzman phenomenon in skin previously sensitized with PR8 or unrelated virus.

### 5. Nature of the Toxic Component of the Virus Particle

The close association of toxicity with haemagglutinating power indicates that the toxic component is part of the outer virus component. However, haemagglutinin fractions produced by ether treatment are not toxic. If virus is disintegrated by sonication ($K_{23}$) toxicity is retained and associated with the haemagglutinating fraction. These findings suggest that the lipid of the outer virus component is involved in the toxic action. Incomplete virus of the VON MAGNUS type is non toxic ($M_{33}$, $M_{61}$), although this virus contains more lipid than standard virus. Non-toxic virus can not be produced by inoculation of eggs with large doses of mouse-adapted virus ($M_{61}$). Intracerebral inoculation of purified brain ganglioside (a macromolecular glycolipid) inhibits the neurotoxic effect of PR8 virus ($B_{96}$) apparently by competing with the cellular ganglioside for the virus. Similar results are produced by intracerebral inoculation of Xerosin ($G_{85}$).

Crude cholera filtrate protects against the neurotoxic effect of virus, but purified RDE is much less effective suggesting that some factor other than neuraminidase is involved, possibly some enzyme acting on lipoprotein.

It appears, therefore, that while the toxic effect of virus is a function of the outer lipomucoprotein component it cannot be with certainty related to the haemagglutinin, neuraminidase or the lipoprotein. The actual toxic symptoms do not appear to be directly produced by the virus but to be a result of its interaction with the host cells resulting in the release of a relatively non-specific endogenous pyrogen.

# XXV. Clinical Features and Pathology of Influenza

No attempt will be made here to review the vast literature on the clinical aspects of influenzal infection, but only to consider how the clinical findings are related to the properties of the virus.

Uncomplicated influenza is a febrile illness of sudden onset occurring in epidemics and with an incubation period of 24—48 hours, accompanied by symptoms indicative of infection of the respiratory tract. The disease commences with a sharp rise of temperature, usually reaching its peak on the first day, and continuing for 2—4 days. The fever is accompanied by headache, sweating, intense lassitude, muscular pains, coryza, cough and congestion of the conjunctivae. Recovery is usually complete within a week, but not infrequently pneumonic complications may occur, the initial fever persisting or being followed by a secondary febrile response associated with the development of pain in the chest, accelerated respiration, cyanosis and production of purulent sputum. These symptoms indicate the occurrence of secondary bacterial infection, and are more common in elderly than in young people.

In severe epidemics such as the 1918 pandemic cases of much greater severity occurred, pneumonic symptoms being present in the very early stages of the disease. Such cases often occurred in young adults and were associated with an intense cyanosis of a peculiar "heliotrope" colour. The mortality in these cases was high and death might occur within 24 hours of onset.

In most epidemics the death rate is low, of the order of 0.5%, most of the deaths resulting from secondary complications in older people, but the attack rate of the disease is so high that the total number of deaths may be large. In the 1918 pandemic the mortality was higher and the disease was estimated to have caused some 20 million deaths throughout the world, many deaths occurring in young people.

In most outbreaks of influenza symptomless cases occur, being recognized by the appearance of increased antibody levels in the blood.

## 1. Relation of the Clinical Disease to the Properties of the Virus

An infected "droplet" from a case of influenza inhaled by a susceptible contact will be deposited at some point in the respiratory tract and will initiate infection of a few cells. After completion of the 6 hour growth cycle these cells will release about 100 times the number of infecting doses which originally entered and a second cycle of growth will be initiated in a larger number of cells. There may, however, be some delay between the first and second cycles owing to the virus being temporarily held up by union with inhibitor substance present in the respiratory secretions. The duration of this delay will vary with different types of virus of different enzymic activity, and this may account for the differences in virulence of the virus in different epidemics. After three or four cycles of growth have occurred the amount of virus released will become large enough to infect   a large proportion of the available cells and at this point symptoms will develop. These symptoms will be of two kinds, those dependent on cellular infection, and those resulting from the toxic properties of the virus.

Cellular infection by the influenza virus does not necessarily result in cell destruction so that the actual damage to the cells of the respiratory epithelium is less than in many other respiratory infections, and the local symptoms in influenza cases are characteristically less noticeable than the toxic effects. The case of common cold, pharyngitis or bronchitis may be able to continue at work, but the influenza patient feels compelled to take to his bed. The toxic symptoms — fever, headache, sweating, lassitude and muscular pains are typically more severe in influenza than in other respiratory diseases. The fulminating cases of influenza associated with rapidly developing pneumonic symptoms, bloodstained sputum, intense cyanosis and early death seen in the 1918 pandemic probably represent an extreme example of the toxic effects of the virus, resembling the similar effects which can be produced in experimental animals by the intranasal inoculation of massive doses of virus.

Although in uncomplicated influenza recovery from infection is rapid, virus being no longer detectable after the third day, and fever being of short duration, the patient is left with a feeling of great prostration out of proportion to the severity of the disease and probably a result of the toxic effects.

In the uncomplicated case defervescence precedes antibody production, and it is probable that the production of interferon is mainly responsible for the arrest of the disease. Persistence of fever after 3—4 days is almost always a result of secondary bacterial infection and it is this secondary infection which is responsible for most of the deaths. Any respiratory pathogen may be involved in secondary infection. In the 1918 pandemic *Haemophilus influenzae* played the major role, but in the epidemics of recent years secondary infection by *Staphylococcus aureus* has been most dangerous.

Any infectious disease which attacks up to half the population may be expected to be associated with numerous types of complication from the exacerbation of previous disease, and many such complications have been recorded, but on the whole it is surprising how little damage is produced by the virus apart from the respiratory effects. There can be no doubt that influenza virus enters the blood stream in the course of infection, but infection of other organs by the virus is probably largely prevented by the adsorption of the virus to red blood cells and to serum inhibitors. Although experimentally the virus can be shown to infect renal cells and to grow in them, apart from a transient albuminuria there is rarely any evidence of renal involvement in human infection. Embryonic tissues are known to be unusually susceptible to virus infection, but damage to the unborn foetus in influenza appears to be a rare phenomenon.

## 2. Pathology of Influenza

Very many studies have been made of the pathological changes in human influenza, especially in fatal cases. The author proposes to consider mainly the experimental pathology of the disease.

### a) The Nasal Mucosa in Influenza

KOLDATITSKAYA ($K_{83}$, $K_{84}$) studied the cytology of the nasal secretion in human influenza by making impression smears from the inferior nasal concha. There was necrosis and desquamation of the columnar epithelium, the cells showing cloudy swelling and pyknotic nuclei. Similar studies were made by other Russian workers ($O_{11}$, $F_{57}$, $U_4$). MALOMUZH ($M_{32}$) found that on the second day of disease the ciliated epithelium underwent degenerative changes and sloughed off. A round cell infiltration occurred in the sub-epithelial layer, and later infiltration by eosinophils and polymorphs occurred. PIGAREVSKY and CHALKINA ($P_{62}$) found inclusion bodies in columnar cells in the nasal mucosa.

*Nasal histology of infection in the ferret.* FRANCIS and STUART-HARRIS ($F_{121}$, $F_{122}$, $S_{292}$) observed necrosis and desquamation of the nasal epithelium in the acute stage, with an inflammatory reaction in the submucosa. Repair commenced on the 4th day. From 6—14 days the denuded area became covered first by transitional, then by stratified squamous, and finally by stratified columnar epithelium. The mucosa returned to normal in 1 month. On the 7th—8th day the mucosa became resistant to re-infection. In second infections in the ferret the local repair process was greatly accelerated.

By staining with fluorescent antibody LIU was able to detect S antigen in the nucleus and V antigen in the cytoplasm of ciliated cells from the nasal turbinates ($L_{98}$, $L_{99}$).

By electron microscopic examination of the nasal mucosa of ferrets 24 hours after inoculation of PR8 virus intranasally HOTZ and BANG ($H_{179}$) found lengthening and swelling of the tips of microvilli, and ballooning of cytoplasm between the cilia, followed by a destructive phase affecting both mucous and ciliated cells up to 4 days. Vesicle formation was prominent in the cells. The endothelial cells of the capillaries showed cytoplasmic processes projecting into the lumen at 6 hours. Regeneration commenced on the 10th day and was complete in 30 days.

CHEVANCE and CATEIGNE ($C_{40}$) found that if the nasal mucosa of mice or ferrets was stripped off and suspended in Hanks' solution the ciliary movement continued for 180 minutes, but if mouse-adapted PR8 virus was added to the fluid the ciliary movement stopped in 25—45 minutes. The effect was prevented by antiserum and did not occur with mucosa from immune animals.

HAFF, SCHRIVER and STEWART ($H_5$) found that nasal washings from infected ferrets showed an increased content of ciliated cells on the 2nd day, and an increased leukocyte content from 2—7 days. Increased nasal capillary permeability was indicated by the finding of an increased recovery of systemically injected dye in the nasal washings, especially marked on the 4th day. Mucin concentration was unaffected.

### b) Tracheal and Bronchial Lesions in Mice

PANTHIER, CATEIGNE and HANNOUN ($P_7$) made saline tracheal washings from mice infected with PR8 virus and stained films by Machiavello's method. They found a large number of distorted vacuolated epithelial cells 48 hours after infection and at 3 days minute red-staining bodies appeared in the vacuolated regions. Polymorphs were also present. PIGAREVSKY ($P_{61}$) considered that in mice the bronchial cells wall off the virus to form eosinophilic inclusions which may be eliminated from the cells. Such cells then regenerate the epithelium. HARFORD and HAMLIN ($H_{46}$) found no damage to the cilia of the bronchial epithelium in mice, but demonstrated basophilic inclusions and suggested that virus was liberated into the lumen of the bronchi by detachment of inclusion-laden globules of cytoplasm. They later made electron microscopic studies ($H_{47}$) and found cytoplasmic inclusions containing many small virus particles. PIGAREVSKY ($P_{63}$) stained epithelial cells from the trachea and bronchi with methyl green-pyronin-orange G followed by basic fuchsin-methylene blue and found acidophil inclusions at 24 hours and sometimes basophil inclusions later. These formations were extruded into the lumen of the bronchi. MAXIMOVITCH and MITCHENKO ($M_{49}$) stained impression smears with acridine orange and found red fluorescent inclusions within 3 hours of infection in the bronchial cells of mice inoculated with PR8 intranasally. The inclusions disappeared on treatment with ribonuclease ($M_{50}$).

FAZEKAS DE ST GROTH ($F_{26}$) found that 48 hours after infection the mucoid inhibitor normally present in the respiratory tract disappeared and did not reappear until the 6th day. From the 8th to 20th day the amount of inhibitor in bronchial washings was ten times as great as in normal mice.

### c) Pathology of the Lung Lesions in Mice

Most of the early workers gave descriptions of the lung lesions produced by intranasal inoculation of virus in mice, the most complete studies were those of STRAUB ($S_{281-283}$), NELSON and OLIPHANT ($N_{14}$) and LAUER and LOZOVA ($L_{31}$). Mice killed on the 3rd or 4th day after intranasal inoculation show purple-red areas of lung consolidation sharply demarcated from the rest of the lung which is emphysematous. The extent of the lesions varies, with the dose of virus and the degree of adaptation of virus to mice, from small areas near the hilum to more extensive lesions involving the whole of one or more lobes of the lung. Microscopical examination shows the basic lesion to be a necrosis of the epithelial lining of the bronchioles and a peribronchial infiltration with mononuclear cells and polymorphs. There may also be a perivascular inflammatory infiltration. At 3—4 days the epithelium of the bronchioles is denuded and the lung tissue shows an intense congestion and oedema, with exudate in the alveoli containing polymorphs and mononuclear cells. With large doses of virus an extreme congestion and oedema of the lungs develops rapidly, with haemorrhage into the alveoli.

With serial passage of virus in mice secondary infection may occur with the development of bronchopneumonia usually due to bacteria of the *Pasteurella* group.

Recovery after mild or moderate degree of lung damage is associated with a characteristic metaplasia of the alveolar and bronchiolar epithelium commencing at about the 6th day. The epithelium proliferates and becomes stratified, filling the terminal air spaces.

The distribution of virus in the respiratory tract of mice inoculated intranasally has been studied by the use of immunofluorescence ($N_6$, $D_{41-43}$). Antigen was detectable in the cytoplasm of bronchial cells at 8 hours and in the alveolar cells at 16 hours. It continued to spread up to the 4th day and then decreased, disappearing by the 10th day ($N_6$). The amount of fluorescence was less with unadapted than with adapted virus ($D_{42}$). In immunized mice fluorescent material was found in the bronchi up to 10 hours followed by a rapid decrease. There was no epithelial destruction ($D_{43}$).

Studies of the relation between virus multiplication and the production of lung lesions have revealed a number of curious facts. TAYLOR ($T_{37}$) showed that gross lesions did not develop until after the maximal production of virus in the lung. Lungs of mice inoculated with sub-lethal doses of virus contained many thousands of lethal doses at the height of infection, and he suggested the existence of some factor appearing at about 48 hours which prevented further spread of infection. This factor is most probably interferon. Mice inoculated with small sub-lethal doses of virus develop fatal pneumonia if they are subsequently inoculated intranasally with fluids such as water, saline, or 10% normal horse serum, but not if immune serum is used. GINSBERG and HORSFALL ($G_{30}$) found that the rate of increase of PR8 virus in mouse lung was independent of the amount introduced and was about 1000-fold per day, but the rate of increase in extent of pulmonary lesions was only 8.5 fold per day. They concluded that the production of lesions is not dependent solely on the multiplication process. RICKARD and FRANCIS ($R_{24}$) inoculated mice intraperitoneally with 50,000 to 1 million intranasal MLD and found that there was a high content of infective virus in the lungs at 48—72 hours, but only mild lesions were produced and there were no deaths. WAGNER ($W_7$) found the virus to be much more infective by the intranasal than by the intravenous route. To produce 50% lung consolidation by intravenous inoculation $10^{9.33}$ $EID_{50}$ was required, while by the intranasal route only $10^{2.5}$ $EID_{50}$ was needed. Intravenous injected virus interfered with virus given intranasally 2 days later.

No lesions are produced by the inoculation of moderate doses of egg-adapted virus intranasally in mice, although virus multiplication occurs, but the inoculation of large doses results in the production of severe lung lesions, and GINSBERG found that large doses of egg-adapted virus would produce severe lesions even in immunized mice ($G_{26}$).

These results would seem to suggest that the production of severe lung damage is not a direct result of virus multiplication in the cells of the respiratory epithelium, but is dependent on the sudden flooding of the respiratory tract with large amounts of virus, and that the effects are produced by the toxic action of the virus.

### d) Epithelial Lesions in the Human Lung

The bronchial lesions in human influenza are very similar to those seen in the mouse. There is a disappearance of the goblet and ciliated columnar cells, but the basal cell layer is preserved and subsequently regenerates to produce a stratified undifferentiated epithelium ($S_{283}$, $M_{146}$). MULDER ($M_{133}$) considers that the influenza viruses have an "intrinsic bronchotropic virulence" and are also able to multiply in the alveolar epithelium.

### e) Combined Virus and Bacterial Infection of the Lung in Mice

The frequent occurrence of secondary bacterial pneumonia in human influenza cases suggested experimental studies of the possible synergism of virus and bacteria. FRANCIS and VICENTE DE TORREGROSA ($F_{123}$) inoculated mice with small doses of virus intranasally and in the next 5 days inoculated strains of *H influenzae* intranasally. If virulent strains of *H influenzae* were given at the same time as the virus most of the animals survived, but if given on each of 5 days after the virus the animals died and *H influenzae* was recovered from the lungs. Similar results were obtained with staphylococci and streptococci.

HARFORD, LEIDLER and HARA ($H_{48}$, $H_{49}$) found that normal mouse lung was able to greatly reduce the numbers of inhaled pneumococci within 3 hours, but in mice with a fully developed influenzal pneumonia the pneumococci were able to grow in the lungs. The effect was believed to be dependent on the presence of pulmonary oedema. GERONE, WARD and CHAPPELL ($G_{21}$) also found that influenza virus infection enhances the pathogenic potential of Type 1 pneumococci for mice, but only if there was an interval of 6 days between exposure to virus and to pneumococci. Mice inoculated by aerosol with small doses of PR8 virus had a death rate of 11%, a pneumococcal aerosol produced no deaths, but an aerosol of PR8 followed 6 days later by pneumococcal aerosol produced 80% deaths.

GALEAZZI and VISIO ($G_2$) found that intranasal inoculation of PR8 virus had no effect on the susceptibility of mice to staphylococcus aureus injected subcutaneously, but increased the susceptibility to air-borne infection. SELLERS and others ($S_{78}$) found that mice inoculated with a non-mouse adapted A2 virus were less able to destroy staphylococci introduced by the respiratory route than normal mice but the persistence of staphylococci in the lung did not alter the pathological reaction to virus. They also found that 40—50% of mice inoculated with influenza virus developed bronchopneumonia due to *Pasteurella* or *Haemophilus* organisms of murine origin.

*Other observations on the enhancing effect of virus on associated infection:* BORDET and QUERSIN-THIRY ($B_{100}$) inoculated guinea pigs intraperitoneally with 5 ml of PR8 infected allantoic fluid, and controls with normal allantoic fluid. After 30 minutes they inoculated 0.5 ml of a 24 hour culture of *H influenzae*. The virus-treated animals died with peritoneal and pleural effusions containing *H influenzae* while the control animals survived. They suggested that the effect was due to an interference with phagocytosis.

DACHY ($D_1$) found that rabbits injected intravenously with staphylococci were 20 times more susceptible if 20,000 haemagglutinin units of influenza virus were given intravenously or intratracheally.

### f) The Blood in Influenza

Uncomplicated cases of influenza usually show a leukopenia with a relative lymphocytosis, but cases developing secondary infection may show a leukocytosis, and leukocytosis may be seen in young children.

*Occurrence of viraemia in influenza:* FAUCONNIER and BEAUCHAMP ($F_{18}$) inoculated mice intranasally with PR8 virus and found that no virus could be detected in the blood up to 8 hours after inoculation but was present in 9 of 16 mice after 12 hours and in 12—37% of mice tested at intervals up to 103 hours. HAMRE, APPEL and LOOSLI found viraemia in 20—50% of mice infected with a PR8 aerosol, the virus being localised in the red cell fraction. MINUSE and others ($M_{99}$) were unable to detect virus in the blood of human cases of Asian influenza by amniotic inoculation, but KAJI and others ($K_2$) detected virus in liver, spleen, kidney and heart in fatal cases. Virus may also be found in the urine ($Z_2$, $Z_{41}$, $Z_{28}$).

ISHIDA, MORIZUKA and HANIENA ($I_{38}$) found that after intranasal inoculation of PR8, infectious virus appeared in the kidneys 72 hours later, reaching $10^4$ $EID_{50}$ per gram in 5 days. Fluorescent antibody demonstrated antigen in the borders of the cells lining the proximal convoluted tubules, but no antigen was ever seen in the nuclei so that the virus was probably not multiplying.

Enormous amounts of virus are present in the blood in fowl plague, mainly associated with the white cells.

*Reaction of virus with white blood cells:* The adsorption and elution of virus from red blood cells has already been described. INGLOT and DAVENPORT ($I_1$) found that leukocytes adsorbed virus more efficiently than red cells, but only a small proportion of the adsorbed virus was eluted. Complexes of virus with fluorescein labelled antibody were phagocyted by leukocytes. SMORODINTSEV and SHISKINA ($S_{206}$) found that the leukocytes of immunized mice were not viricidal and whole blood was no more viricidal than serum. INGLOT and DAVENPORT suggested that leukocytes might be responsible for transport of virus to extrapulmonary sites in infections of man and animals.

GINSBERG and BLACKBURN ($G_{28}$) found that leukocytes adsorbed PR8 and

LEE viruses but only 20% of the adsorbed virus was eluted at 37°C. Adsorption was prevented by RDE and RDE did not remove virus previously adsorbed. Only 1% of the adsorbed virus could be released by disruption of the cells. The anaerobic glycolysis of glucose by leukocytes was markedly inhibited by PR8 and LEE viruses ($F_{74}$, $F_{75}$).

BOAND, KEMPF and HANSON ($B_{93}$, $H_{31}$) found that influenza virus was phagocyted by human and rabbit leukocytes in vitro and showed that the virus penetrated into the leukocyte. Phagocytosis was enhanced by immune serum. By inoculation of purified virus intraperitoneally in mice it was found that the virus was adsorbed by lymphocytes and agglutinated them within a few minutes. Later a neutrophil infiltration occurred, the virus was adsorbed by these cells, agglutinated them in 4 hours, and phagocytosis occurred after 16 hours. In immune mice phagocytosis occurred as early as 8 hours after virus injection and leukocytes from these mice contained no infective virus after 12 hours.

JERUSHALMY, KOHN and DE VRIES ($J_{16}$) found that influenza virus was adsorbed by human blood platelets. Elution was slow and incomplete suggesting that virus became incorporated in the platelets.

### g) Neurotropic and Pantropic Influenza Virus

Neurological symptoms may occur in the course of human influenza, more often in children than in adults, and true encephalitis, though uncommon, does occur. In some cases virus has been isolated from the brain. Neurotropic strains of virus can be produced by serial passage in mouse brain, though only with some difficulty ($S_{286}$, $F_{110}$). These strains multiply in the brain producing a fatal encephalitis. It is possible that some strains of virus possess a neurotropic ability, LACORTE, MONTEIRO and LOURES ($L_7$) isolated a strain of A1 virus, the DL/Rio strain, which showed neurotropic tendency, while other strains isolated at the same time were exclusively pneumotropic.

O'CONNOR and WAGNER ($O_4$) observed that neurotropic strains of virus were much more pathogenic to young mice than to older animals. All of 124 mice aged 1—17 days died after intraperitoneal inoculation of NWS virus, while 100 out of 180 aged 18—35 days survived. In newborn mice NWS produces a generalised infection with persistent viraemia and multiplication of virus in all the tissues ($W_4$). WAGNER studied the growth of influenza viruses in embryonic mouse tissues transplanted into the subcutaneous space of adult hosts and found that while non-neurotropic strains only multiplied in lung tissue, the neurotropic NWS strain grew in lung, intestine, kidney, heart, bladder and spleen tissue but not in skin. The virus therefore showed pantropic properties in embryonic tissues. TYRRELL and CAMERON ($T_{76}$) found that both neurotropic and some non-neurotropic strains produced a fatal infection in suckling mice 1—3 days old on subcutaneous inoculation. There was a rapid fall in susceptibility in the first 10 days of life. Similar results were obtained in suckling hamsters.

### h) Possible Production of Embryonic Malformations by Influenza Viruses

KORNYUSHENKO and MAXIMOVICH ($K_{91}$) isolated influenza A2 virus from the lungs of several infants who died with pneumonia within 2 days of birth. They found that pregnant mice inoculated intranasally with PR8 virus 3—4 days before delivery died in 2—6 days. Virus was isolated from the amniotic fluid and foetal membranes in six cases and was detected by haemagglutination in 19 of 29 embryos and new-born mice. Similar results were found in rats.

COFFEY and JESSOP ($C_{67}$) found that the incidence of congenital malformations in the offspring of 663 women who had influenza in pregnancy was $2\frac{1}{2}$ times greater than in normal women, anencephaly being the most common abnormality. PACHALY and SCHUERMANN ($P_2$) also noted the occurrence of anencephaly which they attributed to influenza in pregnancy, and that in the epidemic month of August 1957 in Chile the perinatal mortality was double that in the previous 16 months.

HAMBURGER and HABEL ($H_{12}$) inoculated PR8 virus over the blastoderm of 48 hour chick embryos. Almost all the embryos died in 3 days with severe abnormalities — small brains, twisted main axis, and impaired growth of amnion. No malformations occurred in 4 day embryos but the embryos were killed. SHEAR and others ($S_{97}$) found that injection of 0.03 ml of dilutions of PR8 allantoic fluid through the vitelline membrane of 48 hours chick embryos caused developmental abnormalities and death in concentrations as low as $10^{-6}$. The susceptibility to teratogenic effect was restricted to the first 48 hours. Neutralizing serum prevented the effects.

WILLIAMSON, SIMONSON and BLATTNER ($W_{54}$) found that inoculation of $10^6$ $EID_{50}$ of PR8 virus to 48 hours chick embryos caused a high incidence of organ defects, retardation of development, microplasia or complete absence of the lens, otocyst or neural tube, twisting of the axis and flattening of the encephalon.

*i) General Accounts of the Clinical and Pathological Features of Influenza*

Numerous references to the clinical and pathological findings in the 1918 pandemic are given by FRENCH ($F_{138}$), OPIE ($O_{10}$), LUKSCH ($L_{125}$), WINTERNITZ, WASON and MACNAMARA ($W_{55}$), and THOMSON and THOMSON ($T_{47}$). More recent work is described by VAN ROOYEN and RHODES ($V_8$), ZHDANOV, SOLOVYEV and EPSHTEIN ($Z_{34}$), SCADDING ($XS_1$), MULDER ($M_{133}$, $M_{141}$), STUART-HARRIS ($S_{288-291}$), and HERS ($H_{107}$).

A complete bibliography of clinical and other studies in the Asian pandemic of 1957 was published by the American Institute of Biological Sciences ($A_{55}$).

# XXVI. Metabolism of the Infected Cell in Influenza

The total amount of virus material produced in an infected cell is very small in relation to the cell volume so that the mere synthesis of virus is unlikely to make any serious demand on the nutritional reserves of the cell, and good growth of virus in de-embryonated eggs or tissue cultures may be obtained with glucose as the only added nutrient.

*a) Protein Synthesis in the Infected Cell*

VAN DEN ENDE ($V_4$) inoculated eggs with virus and at intervals from $1\frac{1}{2}$—18 hours later removed the allantoic fluids and replaced with various amino acid mixtures. The eggs were then further inoculated for 48 hours and the amino acid content of the allantoic fluid determined. If the infected eggs were inoculated for 12—18 hours before addition of the amino acids there was a much more rapid disappearance of amino acid from the allantoic sac than in control un-inoculated eggs, or in eggs inoculated with mumps virus or diphtheria toxin. All the introduced amino acids were reduced equally, there being no evidence of any selective utilisation. If amino acids were added $1\frac{1}{2}$ hours after virus infection there was no marked increase in the rate of disappearance of amino acid in the next 24 hours. This suggested that the effect was not related to virus synthesis but was a non-specific effect possibly related to the inflammatory reaction set up by the virus.

MILLS ($M_{92}$) using de-embryonated eggs found no difference in the uptake of amino acid from casein hydrolysate or from a mixture of 13 amino acids between infected and non-infected eggs. If, however, cysteine, histidine or methionine were omitted from the amino acid mixture there was a greater uptake of the other amino acids in normal than in infected eggs, suggesting an increased need for these three amino acids in the infected cells. Virus production was inhibited by the methionine analogue ethionine, but not by the phenylalanine analogue $\beta$-phenylserine.

ACKERMANN ($A_1$) found that methionine was essential for virus propagation, growth being inhibited by dl-methoxinine and dl-ethionine. Virus synthesis is also inhibited by other amino-acid analogues notably by p-fluorophenyl-alanine.

EATON, SCALA and LOW ($E_{19}$) found that virus production and protein synthesis in Krebs ascites cells was inhibited by excess of lysine, phenylalanine, tryptophane and leucine in the medium. They suggested that amino acid imbalance results in synthesis of abnormal proteins or inefficient enzymes.

### b) Carbohydrate Metabolism in the Infected Cell

DANIELS, EATON and PERRY ($D_3$) found greatly reduced yields of virus in de-embryonated eggs and tissue cultures deprived of glucose. The effect was reversed by addition of glucose. In tissue cultures deprived of glucose oxygen consumption was reduced to 60—70% of that in controls. MILLS ($M_{92}$) found in de-embryonated eggs that glucose uptake was greater in infected than in control membranes, and WIELGOSZ ($W_{50}$) found a similar increase in uptake of pyruvate in infected membranes. LEVINE, BOND and ROUSE ($L_{65}$) found that WS and NWS virus would grow in chorioallantoic membrane tissue in 0.1% glucose, but only WS would grow if the glucose was replaced by pyruvate as carbon source. Infection of cells by NWS occurred in the presence of pyruvate, and if the cells were treated with antiserum after infection growth of NWS occurred if the pyruvate was replaced by glucose. Synthesis of NWS occurred with glucose, xylose or glycerol as sole carbon source, but ribose, dihydroxy-acetone, glycerophosphate, adenosine-5-phosphate, adenosine di- and tri-phosphates, and diphosphopyridine nucleotide were effective only in combination with pyruvate.

KLEMPERER ($K_{58}$) found that aerobic glycolysis accounted for 30% of the glucose taken up by chick embryo cells, less than 10% being oxidized via the pentose cycle. Within $1\frac{1}{2}$ hours of infection when complement fixing antigen first appeared aerobic glycolysis was increased 50—100%. After 3 hours when haemagglutinin first appeared oxidation via the pentose cycle increased 30—50%. The rate of glycolysis was limited by the supply of phosphate and phosphate acceptor. Triphosphopyridine nucleotide was a rate-limiting factor in the direct oxidation of glucose. After infection these materials may be released in the cell as a result of reactions associated with virus synthesis.

VOLUKSKAYA ($V_{27}$) found that in influenzal infection the sugar content of the lungs of mice increased 40—50%, while the glycogen in the liver decreased by 20—30%.

ACKERMANN ($A_3$) found that the growth of influenza virus A in mouse lung was inhibited by intraperitoneal inoculation of doses of sodium fluoracetate which blocked the citric acid cycle as shown by an increased concentration of citrate in the mouse lung, but MOGABGAB and HORSFALL ($M_{108}$) found only slight delay in the multiplication of PR8 virus produced by doses of fluoracetate which caused 10—20% mortality in mice or 100% deaths in chick embryo. With LEE virus there was no effect in eggs, but there was a delay in multiplication in mice, although the final yield was unaffected. Fluoracetate blocks the metabolism of carbohydrate and fat at the tricarboxylic acid stage, and MOGABGAB and HORSFALL considered these processes not absolutely necessary for virus reproduction.

BERRY, MITCHELL and RUBINSTEIN ($B_{60}$) held mice for 3 weeks at reduced atmospheric pressure equivalent to an altitude of 20,000 feet and found their resistance to PR8 virus was increased. The citric acid content of the lungs was reduced to 70% and it was suggested that the resistance was due to an alteration of metabolism affecting the Krebs cycle. KALTER and TEPPERMAN ($K_{12}$) also observed reduced growth of influenza virus in hypoxic mice. SAWICKI, BARON and ISAACS ($S_{24}$) found the mortality of mice inoculated with WS or LEE virus was increased and the survival period reduced if the animals were held in an atmosphere of 50% oxygen. They attributed the effect to a reduced activity of interferon.

KORBECKI ($K_{88}$) found in KB cell cultures infected with an inhibitor-resistant strain of A2 virus that glucose utilization was increased initially, but 48 hours after infection, when the cytopathic effect occurred, it was less than in controls. The virus had no effect on cell aldolase activity, but lactic acid production and dehydrogenase activities were increased. KELLY and GRIEFF ($K_{27}$) found that the allantoic fluid of PR8 infected eggs three days after infection had 18 times the lactic dehydrogenase activity of non-infected fluids probably due to the release of enzyme from injured cells.

ACKERMANN and JOHNSON ($A_{11}$) suggested that the energy required for virus synthesis was derived from the cellular aerobic metabolism and would depend on the presence of high energy phosphate bonds in the host tissue. They found that a concentration of $27 \times 10^{-5}$ M of 2.4 dinitrophenol (DNP) completely inhibits the growth of influenza virus in chorioallantoic membrane tissue suspensions. The substance had no viricidal action and the effect was attributed to its ability to stimulate the enzyme adenosinetriphosphatase with a resulting stimulation of tissue respiration and release of phosphate. VOROB'YEVA ($V_{32}$)

found that addition of adenosine triphosphoric acid increased the yield of virus in de-embryonated eggs and tissue cultures by 1 log. Inhibitors of ATP such as sodium fluoride, malonic acid and dinitrophenol reduced the yield of virus.

### c) Phosphorus and Nucleic Acid Metabolism

Cohn ($C_{78}$) found that infection of the chorioallantoic membrane by influenza virus results in a decrease in the phospholipid content of the membrane and the appearance of large amounts of lipid-bound phosphorus in the allantoic fluid. This would be expected as a result of the excretion of virus enclosed in a lipoprotein membrane from the host cell. Studies with radiophosphorus introduced into the allantoic sac at various times before inoculation of virus indicate that the virus incorporates pre-formed host-cell lipid into its structure, but that the virus nucleic acid is newly synthesized.

Johnson and Ackermann ($J_{17}$) found no significant differences in the content of RNA and DNA phosphorus in normal and infected cells, and it is probable that infection with influenza virus does not significantly modify the normal nucleic acid metabolism in the cells, virus RNA synthesis proceeding as an additional feature in parallel with normal cell RNA synthesis ($C_{34}$, $C_{36}$).

### d) Effects of Endocrines and Vitamins

Kalter, Smolin, McElankey and Tepperman ($K_{11}$) found that treatment of mice with testosterone resulted in increased growth of PR8 virus and an increase in protein anabolism. Castration resulted in a diminished rate of virus growth. Increase of protein catabolism by cortisone or ACTH produced a decrease in virus proliferation. Kilbourne and Horsfall ($K_{47}$), however, found a significant increase in the concentration of PR8 or LEE virus in the allantoic fluid of eggs treated with cortisone acetate. In later studies Kilbourne ($K_{38-42}$) found that cortisone produced an initial inhibition of virus growth in doses of $0.1-1.0$ μg per egg, followed by an eventual increase in the final yield. Cortisone prevents the interfering action of inactivated LEE virus on the growth of a serologically different strain of B virus. Cortisone suppressed the formation of antibody in virus infected mice ($K_{39}$), and mice with an acquired active immunity became infected by challenge virus after administration of cortisone.

Eaton and others ($E_4$, $E_5$, $E_{17}$) found that the oxygen consumption of chorioallantoic membrane tissue was increased by $0.1-10$ μg/ml of thyroxin in the medium in the presence of pyruvate and glucose. Thyroxin stimulated the growth of PR8 virus. Cysteine in doses of $0.5-1.0$ mg/ml interfered with the utilization of pyruvate and glutamate for metabolism and inhibited virus growth. The inhibitory effect of cysteine was reversed by thyroxin. The thyroxin analogue butyl-3-5-diiodo-4-hydroxybenzoate in doses of $2-8$ μg/ml greatly reduced the growth of PR8 virus in tissue culture and in de-embryonated eggs, but the effect was not reversed by thyroxin.

Tani and Hayashi ($T_{27}$) found that vitamin B1 in a dose of 1.5 mg increased the haemagglutinin titre of PR8 virus in infected allantoic fluid by $6.8-14.8$ times, and LEE virus by $2.4-10.5$ times that of controls. Vitamin B2, B12, C and folic acid had no effect.

Underdahl and Young ($U_3$) found that feeding of mice for 4 weeks with vitamin A resulted in a significant decrease in the intensity of lung lesions produced by intranasal inoculation of SWINE virus. Mice fed with vitamin A + vitamine E had more severe infections than mice fed with vitamin A alone or vitamin A + vitamin D.

### e) Other Effects of Host Metabolism on Virus Growth

KALTER ($K_5$) found no difference in the susceptibility to infection by intranasal PR 8 virus in mice in which growth had been retarded by giving a reduced protein diet, nor were there any differences in susceptibility in mice naturally growing at different rates.

BEUTLER and GEZON ($B_{61}$) found that total body irradiation of mice with X rays in a dose of 50—600 r rendered them more susceptible to infection by influenza virus. Irradiation depressed antibody production.

# XXVII. Laboratory Diagnosis of Influenza

The most important methods used in the laboratory diagnosis of influenza are the isolation of virus and the demonstration of an antibody rise during convalescence.

## A. Isolation and Identification of Virus

Methods of isolation of virus have been described in Chapter 4 and need not be further considered. As a diagnostic method virus isolation is comparatively inefficient. Isolation is usually only successful in the first 48 hours of disease, and even then usually in less than 50% of cases. In some outbreaks all attempts to isolate virus may be unsuccessful. Amniotic inoculation of fertile eggs and inoculation of monkey kidney or calf kidney tissue cultures are the most useful methods.

In the author's experience the most rapid and reliable method of identification of an isolate obtained by amniotic inoculation of eggs is by the demonstration of complement fixing S antigen in saline extracts of the amnion and of the embryo lung + trachea. Identification is usually possible in the first egg passage and within 3 days of inoculation of the washings. Differentiation of A, A1 and A2 strains may be possible by the use of complement fixation with specific V antisera ($H_{216}$). The demonstration of haemagglutinin in the amniotic fluid is less reliable as some amniotic fluids show non-specific haemagglutination and in many cases the haemagglutinin titres are low. A second egg passage may be necessary before satisfactory haemagglutination-inhibition tests can be done. FEDOVA and ZELENKOVA ($F_{59}$) found the staining of the epithelial cell sediment from amniotic fluid with specific fluorescent antibody to be very satisfactory, and were able to identify 16 strains of A2 virus at the first egg passage.

In kidney tissue cultures virus is first demonstrable by haemadsorption ($S_{99}$, $R_{68}$) and may be identified by haemadsorption-inhibition with specific antisera ($S_{25}$). At a later stage haemagglutinin appears in the tissue culture fluid. Identification is sometimes possible at the first passage, in other cases a second passage may be necessary.

## B. Serological Diagnosis

The various serological reactions of the influenza viruses have been described in Chapter 17, and any of the methods may be used for serological diagnosis. Two samples of serum are desirable, one taken early in the disease and a second during convalescence, and the demonstration of a four-fold increase in antibody content when the two serum samples are simultaneously tested is usually considered to be diagnostic of infection.

### a) The Group Complement Fixation Reaction

Complement fixation with S antigen preparations from infected chorioallantoic membrane or mouse lung is a most reliable test for the serological diagnosis of influenza and has been very widely used. HOYLE and FAIRBROTHER (H$_{205}$) describe the results of a 10 year experience of the test. With normal human sera examined in inter-epidemic periods using A virus antigen, of 1562 sera tested 54% gave doubtful or negative readings and 46% gave positive readings to low titre. Only three sera gave a titre higher than 1:16. With virus B antigen of 448 sera tested 69% were doubtful or negative, 31% positive but no serum gave a titre higher than 1:16.

Of 138 convalescent sera tested 96% gave a positive reading and 50% gave titres of 1:32—1:256. These results suggest that a titre of 1:32 or more in a convalescent serum is practically diagnostic of recent infection, so that a diagnosis can be made in 50% of cases when only a single serum specimen taken in convalescence is available. However, 50% of convalescent sera have titres lower than 1:32 and in such cases diagnosis can only be made if paired sera are available.

A small number of cases of influenza show no antibody rise in the group complement fixation test. This result is most frequently found in young children (L$_{78}$) and the test is therefore less satisfactory in these cases. It appears that young children behave like experimental animals in that little response to the S antigen occurs in a first infection with virus, though subsequent infections produce a marked rise in antibody.

Very little antibody to the S antigen is produced by vaccination with killed virus, and for this reason the group complement fixation test is particularly useful for the serological diagnosis of influenza occurring in vaccinated individuals (R$_{57}$).

### b) The Specific Complement Fixation Reaction

This test, originally developed because of difficulties experienced with the haemagglutination-inhibition test, has proved very satisfactory in serological diagnosis and in studies of antigenic relations between virus strains. HENLE and LIEF (H$_{89}$) as a result of extensive experience of the method describe the following features.

If homologous V antigen is used for serological diagnosis in an epidemic the proportion of patients showing an antibody rise is greater than if the group S antigen is used. This is due partly to the fact that young children often show little response to S antigen but react well to homologous V antigen, and partly to the fact that the initial antibody level to the V antigen of a new strain of virus may be much lower than the initial antibody level to S antigen. However, in the early stages of an epidemic homologous V antigen may not be available, and if heterologous V antigen is used the results are inferior to the use of the group S antigen, especially in young children.

The antibody response to a new strain of influenza virus occurs more slowly than the response to S antigen or heterologous V antigen to which the patient may have had previous exposure. As a result a rise in antibody titre may be detectable 8—9 days after infection with S or heterologous V antigen but may not be detectable until 17 days or more with homologous V antigen.

Complement fixation with homologous V antigen is very satisfactory for use in assessment of the antibody response to vaccination.

### c) The Haemagglutination-inhibition Test

This test, because of its apparent simplicity, has been the most widely used in the serological diagnosis of influenza. Although it is of great value in the study of differences between individual strains of virus, it is an unsatisfactory test for use in routine diagnosis. Great variation in the titre of both normal and convalescent sera make the use of paired sera essential, and the use of homologous virus is necessary for satisfactory results. But the main difficulty with the test is a result of the presence in normal sera of non-specific inhibitors of haemagglutination, often reacting to very high titre. Different strains of virus show great differences in sensitivity to non-specific inhibitors, and strains isolated in the same epidemic may behave differently. It is therefore necessary to treat sera by some method which will destroy the non-specific inhibitors while leaving specific antibody unaffected, and this removes the main advantage of the test over the use of complement fixation. The considerable number of methods which have been described for the destruction of non-specific inhibitors is some indication of the difficulty which may be experienced, and methods which are satisfactory in one epidemic may be less successful in another.

It has usually been found, in comparisons between the results of haemagglutination-inhibition tests and group complement fixation tests in serological diagnosis, that the complement fixation test yields a larger number of positive results, but the best results are obtained by the use of both tests. Thus KALTER and others ($K_9$) studying paired sera from 422 cases found a rising titre in 89% by complement fixation and in 67% by HI tests. Both tests were positive in 56% of the cases, in 33% the CF test alone was positive, and in 11% only the HI test gave a positive result.

### d) Indirect Serological Diagnosis

Diagnosis may sometimes be made in influenza cases by inoculating ferrets, mice or other animals intranasally with human nasopharyngeal washings, and then demonstrating the appearance of antibody in the animals serum. Positive results may be obtained in cases in which isolation of virus from the washings failed.

### e) Other Methods of Serological Diagnosis

Neutralization and mouse protection tests were used in the early years of influenza research but are too expensive and laborious for general use. Precipitin tests are also unsuitable as they require relatively large amounts of purified virus and false results may be produced by non-specific inhibitors. Virus adsorbed on collodion particles is agglutinated by antibody and a collodion-particle agglutination test was described by BOAND and KEMPF ($B_{92}$). KLEIN, CHAEFSKY and MULLER ($K_{54}$) adsorbed purified virus on a lecithin-cholesterol mixture and resuspended the deposit in a saline solution containing charcoal. Mixtures of this antigen with dilutions of serum were placed on white cards and rocked until dry, when the charcoal showed characteristic agglutination patterns. The results were closely parallel to those of specific complement fixation tests.

## C. Rapid Methods of Laboratory Diagnosis

Russian workers have attempted to increase the speed of influenza diagnosis by methods depending on the direct examination of material from the patient by microscopical or serological methods.

MOROSOV ($M_{124}$) claimed that "elementary bodies" could be detected in smears of nasal mucosa stained by a silver impregnation technique, and MERKULOVA and KAZANSKIY ($M_{80}$, $M_{81}$) reported favourably on the method, but the possibilities of technical error are great and the method has not been accepted by other workers.

### a) Rhinocytoscopy

KOLDATITSKAYA ($K_{83}$, $K_{84}$) studied the cytological characteristics of the nasal secretion in influenza. Impression smears were made with a small glass spatula applied to the inferior nasal concha and stained by Giemsa's method. In normals the smears showed only scanty degenerate cells, but in influenza cases numerous columnar cells were found, often in groups or layers. The cells showed cloudy swelling and pyknotic nuclei. In acute catarrhs columnar cells were rare but leukocytes were numerous. PIGAREVSKY and CHALKINA ($P_{62}$) and UNDRITS and KOLDOBSKIY ($U_4$) modified the method to demonstrate inclusion bodies in the columnar cells.

The potentiality of the rhinocytoscopic method has been much increased in recent years by the introduction of fluorescent antibody staining methods. BLASKOVIC and others ($B_{76}$) took swabs from the nasal concha and soaked them in antibiotic broth for 2 hours at 4°C, centrifuged and stained the deposited cells after formalin fixation with fluorescent antibody. Positive results were obtained in the first three days of disease in 5 of 9 serological positive cases of A2 infection. FEDOVA and ZALENKOVA ($F_{60}$) obtained specific fluorescence in the epithelial cells in 19 of 30 proved cases. Controls were all negative. TATENO, KITAMOTO and KAWAMURA ($T_{28-30}$) obtained positive results in 89% of serologically positive cases of A2 infection. Some cells showed fluorescence only in the nuclei, some only in the cytoplasm, and some in both.

### b) Demonstration of Haemagglutinin in Nasopharyngeal Washings

Diagnosis by the direct demonstration of haemagglutinin in nasopharyngeal washings has been extensively studied by Russian workers ($S_{241}$, $M_{31}$, $R_{20}$, $S_{243}$, $F_{58}$). Although the incidence of positive results was low it was said to accurately reflect the morbidity rate among the population. The test was made more sensitive by the use of large volumes of washing (10 ml of centrifuged washings to 0.5 ml of guinea pig cell suspension), and typing was possible by the use of tubes containing 1:20 dilutions of A or B antisera. Tests were positive in only 25—30% of cases and sensitivity was not increased by attempts to concentrate the virus by adsorption-elution from red cells ($P_{50}$, $L_{118}$). False positive results may occur owing to the presence in human saliva of a haemagglutinin for guinea pig red cells. On the whole the test has proved very unsatisfactory. In 1949 the Moscow laboratories using chick red cells obtained positive results in 25—35% of influenza cases, in 1950—1951 using guinea pig cells they obtained 10—15% positives, while in 1952 using human 0 cells positive reactions occurred in only 0.6—1.2%, a result described by ZHDANOV as "truly pitiful".

### c) Demonstration of Complement Fixing Antigen in Blood, Urine and Nasopharyngeal Washings

SMORODINTSEV ($S_{181}$) found that complement fixing antigen could be demonstrated in the blood of acute cases of influenza. SHISHKINA and YURIKAS ($S_{106}$) found that antigen could be detected in the first few days of disease by using the patients blood serum as antigen, convalescent serum as antibody, and increasing the sensitivity of the test by the use of long fixation at 4°C. ZAKSTELSKAYA ($Z_2$) was able to detect antigen in the urine, especially in children, and found the antigen in the urine to run parallel to the blood level. Antigen was found in the urine of 6 of 17 children but in only 2 of 18 adults tested.

SMORODINTSEV ($S_{184}$) developed a method of detecting complement fixing antigen in nasopharyngeal washings using a long fixation method and unusually large volumes of antigen (5 ml of washings to 0.2 ml each of serum and complement). YURIKAS ($Y_{13}$) and KANTOROVICH ($K_{13}$) state the method to be very sensitive and specific, giving 50% of correct results in influenza diagnosis.

DREYZIN and ZALMANZON ($D_{82}$) using long fixation methods were able to demonstrate complement fixing antigen in frozen and thawed lung suspensions from fatal cases of influenzal pneumonia in 21 out of 30 cases. In only 4 cases could haemagglutinin be detected.

### d) Other Methods of Rapid Diagnosis

FAZEKAS DE ST GROTH ($F_{29}$) attempted diagnosis by measuring the inhibitor titre of nasal secretion against 3 different viruses. The pattern of inhibition was altered in influenzal infection. TOMMILA ($T_{50}$) attempted to demonstrate inhibitor-inactivating activity in throat washings, but found a factor in normal throat washings which inactivated inhibitors.

ZAKSTELSKAYA ($Z_4$) suggested the detection of toxin in the blood of acute cases as a diagnostic method. SEREDA ($S_{81}$) used a method based on the reduction in electrokinetic potential of red cells exposed to nasopharyngeal washings.

# XXVIII. Immunity and Antibody Production in Influenza

## 1. Natural Immunity to Influenzal Infection

Natural infection by viruses of the influenza group has only been recorded in mammals and birds. Virus C appears to be only pathogenic to man under natural conditions. Experimental inoculation results in an antibody rise in ferrets and hamsters without signs of infection, and mice are immune ($T_{40}$). Virus B infection has only been found in man under natural conditions, but the virus will infect ferrets and mice in the laboratory and must be regarded as a potential pathogen of animals. The A group of viruses are more widespread, but great variations occur in the susceptibility of different animals to infection. Man appears to be immune to infection by the avian viruses, and fowls are immune to human viruses although the chick embryo is readily infected. The immunity of the adult fowl to human viruses may be a result of its high body temperature. Amphibia and reptiles, although their cells contain virus receptors, may be immune because of their low body temperature.

Many mammals in which natural infection has not been observed may be infected in the laboratory, and the great adaptability of the influenza viruses makes it difficult to give a list of susceptible and insusceptible animals.

Susceptibility to infection depends in the first instance on the ability of the virus to unite with the cell and is therefore dependent on the presence of a suitable mucoprotein receptor on the cell surface. Secondly, the virus must be able to penetrate the cell membrane and this probably depends on the existence of compatibility between the lipoproteins of the cell membrane and the lipoproteins of the virus membrane permitting a fusion of the two. Finally, susceptibility will depend on the ability of the cell to synthesize virus nucleic acid and protein.

It might be expected that the presence of large amounts of mucoprotein inhibitors of the Francis type in respiratory mucus would afford some protection against infection, but in general this is not so, although infection may be delayed by inhibitors it is not prevented, and the presence of Francis-type inhibitor is an indication of susceptibility to infection and not of resistance. Although serum inhibitors of the $\beta$ or Chu type will neutralize virus *in vitro* there is little evidence that they play much part in resistance to infection.

Resistance to infection may be related to age. KALTER ($K_4$) found that young mice were more susceptible to infection than older mice. Virus multiplied to the same final titre, but the rate of proliferation was greater in the young mice, and more virus was needed to cause death in the older animals. The resistance of the allantoic sac of fertile eggs to influenzal infection also increases with age ($W_{46}$). The final yield of virus was unaffected, but the younger eggs were more susceptible to infection, not because of any failure of virus uptake in the older eggs, but because of some intracellular barrier to infection ($W_{42}$).

## 2. Acquired Immunity

Infection with influenza virus results in the development of immunity which reaches a peak in 2—3 weeks, is maintained at a high level for some weeks or months, and then slowly wanes. The immunity is completely type-specific, infection with virus A does not protect against virus B, and is largely strain-specific, infection with A1 virus conveying little resistance to A2 virus. Immunity to the homologous virus can be produced by vaccination with live or killed virus inoculated subcutaneously.

In the experimental animal immunity is of short duration. SMITH, ANDREWES and LAIDLAW ($S_{168}$) found that convalescent ferrets were immune for 3 months, after 6 months most of the animals were susceptible, and all immunity was lost in 8—12 months. Immunity in mice is also of short duration.

FRANCIS, PEARSON, SALK and BROWN ($F_{111}$) infected human volunteers with B virus by aerosol inhalation and found that most of them could be reinfected 4 months later with the same virus. ZAKSTELSKAYA ($Z_4$) found that after intranasal infection of humans with live virus vaccine immunity lasted for 8—10 months, after which time virus multiplied on a second intranasal instillation.

Under natural conditions influenzal immunity to homologous virus may be of rather longer duration, but does not apparently last more than about 2 years. SIGEL and others ($S_{134}$) observed 2 outbreaks of influenza due to closely related strains of A1 virus in a boarding school in 1947 and 1949. Of 81 boys who were ill in the 1947 outbreak, 28 became infected again in 1949. RITOVA ($R_{31}$) studying the occurrence of influenza in a group of children observed over 6 years found that second infections with A1 virus did not occur in less than 2 years after the first infection, and in the case of B virus the shortest interval between infections was 2 years and 9 months.

Infection produces a stronger immunity than vaccination with killed virus. SOKOLOV and KULIKOVA ($S_{221}$) found that intranasal immunization of mice with $0.1 \, LD_{50}$ of live virus produced resistance to more than $1000 \, LD_{50}$, and a dose of $0.01 \, LD_{50}$ produced immunity to between 100 and $1000 \, LD_{50}$. By contrast it was necessary to inoculate 1 ml of concentrated formolized vaccine subcutaneously to protect against $100 \, LD_{50}$ of live virus intranasally. ZAKSTELSKAYA ($Z_4$) obtained similar results in mice, and also found that in human volunteers intranasal inoculation of live virus produced a much greater immunity than subcutaneous vaccination with killed virus, even though the serum antibody response in the two groups was similar.

It is to be noted that influenza virus is not killed by antibody, active virus can be recovered from neutralized mixtures (Chapter 17). Nor is the virus destroyed by phagocytosis. SMORODINTSEV and SHISHKINA ($S_{206}$) found that whole blood was no more viricidal than serum, and the leukocytes of immunized mice are not viricidal. Although some virus is excreted in the urine, SMORODINTSEV ($S_{183}$, $S_{208}$) considers the main cause of death of influenza virus in the bodies of warm-blooded animals to be a result of thermal denaturation.

### 3. Antibody Production

Great variations occur in the antibody response to influenzal infection. In general the response is greater in severe cases than in mild attacks ($K_{90}$).

#### a) Antibody Production to the S Antigen

HOYLE and FAIRBROTHER ($H_{205}$) using the group complement fixation test found that the average antibody titre of normal human sera examined in interepidemic periods was 2.0 against the S antigen of virus A, while in convalescent sera it was 37.5. The corresponding figures for virus B were 1.25 and 64.8. FRANCIS, MAGILL, RICKARD and BECK ($F_{108}$) found in 42 patients infected with virus A that the average antibody rise between acute and convalescent sera was 10-fold when tested both by complement fixation with S antigen and by neutralization tests. Antibody production to the S antigen is much greater in infection than as a result of vaccination with killed virus. HOYLE and FAIRBROTHER found a 44-fold increase in titre in 18 cases of infection by virus A, while 16 cases vaccinated subcutaneously with a concentrated formolized egg vaccine produced only a 2-fold average increase of titre to S antigen.

In young children and in animals receiving a first infection there is little antibody rise to S antigen, but subsequent infections produce a considerable rise in titre.

### b) Antibody Production to the V Antigen

Infection produces a rise in antibody titre to the homologous V antigen comparable to the rise against S antigen. LIEF, HENLE and SCHRACK ($L_{79}$) found the mean titre of 34 convalescents from A1 infection was 80.5 against S antigen, 85.3 against homologous V antigen, and 7.8 against heterologous A2 antigen.

There is a difference in the rate of development of S and V antibodies. When a patient has had no previous experience of a particular strain of virus the response to V antigen is much slower than to S antigen. HENLE and LIEF ($H_{80}$) studying 98 patients infected for the first time with A2 virus found that in all cases the complement fixation reaction to S antigen became positive in 9 days,

Table 8

| Age group (years) | Percentage of positive sera | | |
| --- | --- | --- | --- |
| | Complement fixation (S antigen) | Neutralization test | |
| | | Human virus | Swine virus |
| $0-\frac{1}{2}$ | 62.5 | 75.0 | 67.0 |
| $\frac{1}{2}-1$ | 0 | 0 | 0 |
| $1-9$ | 17.0 | 27.0 | 13.0 |
| $10-14$ | 30.0 | 45.0 | 53.0 |
| $15-18$ | 42.5 | 50.0 | 57.0 |
| $19-$ | 60.0 | 46.0 | 63.0 |

Age incidence of antibody in normal human sera in 1936 [FAIRBROTHER and HOYLE ($F_9$)].

while the response to homologous V antigen only reached 100% in 19 days. By contrast in 51 patients examined in an A1 outbreak occurring early in 1957 at the end of the period of prevalence of A1 virus, there was little difference in the rate of development of S and V antibodies, all sera becoming positive to both antigens in 9 days. In general the antibody response to heterologous V antigens from viruses prevalent in earlier years is as rapid as that to S antigen.

Vaccination with killed virus produces a greater antibody response to V antigen than to S antigen, the response being greater in individuals who have had previous exposure to the type of virus used. LIEF, HENLE and SCHRACK ($L_{80}$) found that after inoculation of 1 ml of killed A2 vaccine subcutaneously the mean antibody titre against homologous V antigen was 46.1 in individuals with some initial antibody in the serum, while in cases with no demonstrable pre-vaccination antibody it was only 12.4.

Antibody rise to the V antigen may also be demonstrated by neutralization and haemagglutination-inhibition tests. Although it is not certain that the antibodies reacting in the various tests are identical the results are closely comparable to those obtained by complement fixation.

### c) Age Distribution of Antibody in Normal Human Sera

FRANCIS and MAGILL ($F_{105}$) tested 136 human sera for neutralizing antibody and found a high proportion positive at all ages except between 1 and 12 months.

Antibody was present in the sera of new-born infants. FAIRBROTHER and HOYLE ($F_9$) studied the age distribution of antibody in normal human sera by complement fixation with S antigen and by neutralization of WS and SWINE virus (Table 8). Antibody was found in over 60% of the sera of children under 6 months of age. No antibody was found between 6 months and 1 year. In older children the percentage of positive sera increased with age to reach a value of 60% in adults. RAETTIG ($R_1$, $R_2$, $R_3$) studied the age incidence of haemagglutinin-inhibiting antibody in 2230 sera. The average titre of normal sera was 32. Titres were high in new-born children but fell to zero at 8 months. Then there was a rise to reach the average level at 10—15 years. Titres fall off after the age of 40. These results suggested that maternal antibody might be transmitted to new-born children.

### d) Transmission of Maternal Antibody

PILLE ($P_{65}$, $P_{66}$) immunized gravid female rats, rabbits and guinea pigs and found antibodies in the sera of the offspring. It was shown that in rats the antibodies were transmitted partly via the placenta and partly via the milk, but in guinea pigs and rabbits antibody was mainly transmitted through the placenta. MOROZENKO ($M_{125}$, $M_{126}$) obtained similar results and showed that in guinea pigs although antibody is present in both placenta and milk it is only transmissible via the placenta as young guinea pigs do not absorb antibody via the intestinal tract. BIELING and OELRICHS ($B_{70}$) were unable to find evidence of maternal transmission of antibody in mice, but MARINESCU and others ($M_{42}$) found that suckling mice from immunized mothers were resistant to infection with 5 $LD_{50}$ of virus inoculated at 1 week after birth. Antibody was apparently transmitted through the milk as litters from unimmunized mothers became resistant if suckled by immunized animals.

### e) Production of Heterologous Antibody

FRANCIS and MAGILL ($F_{107}$) immunized mice with 2 intraperitoneal injections of live human or SWINE virus and then challenged them by intranasal inoculation of homologous or heterologous virus 1 week later. The immunity was greater to the homologous than to heterologous virus, but if the number of immunizing doses was increased effective immunity to heterologous virus could be achieved.

HARE and RICKEN ($H_{45}$) studied the levels of neutralizing antibody to human and SWINE virus in human sera over the period 1937—1939 during which time two epidemics occurred. Most adults were found to possess neutralizing antibody to both human and SWINE viruses and both increased after infection with human virus. On primary infection in young children only antibody to human virus appeared, but at age 10—11 swine antibody developed after infection with human virus. They considered that antibody to SWINE virus could result from repeated infection with human virus. RICKARD, THIGPEN and ADAMS ($R_{25}$) studied 5 infants aged 9—19 months who became infected with virus A. Acute, early convalescent, and late convalescent sera were tested for haemagglutinating-inhibiting antibody against the infecting strain, PR8, WS and SWINE. Antibody response was greatest to the infecting strain with lower responses to the others. Antibody to SWINE virus was rapidly lost in the late convalescent sera.

MULDER and others (M₁₄₄) found that ferrets inoculated intranasally with influenza virus usually produced highly strain-specific sera, but certain ferrets had some initial antibody possibly as a result of infection and these animals produced on infection antisera of broader specificity. HARBOE (H₃₃) inoculated ferrets intranasally with different strains of virus successively at intervals of several months. Ferrets given successively SWINE, PR8, FM1, and A1/Netherlands/56 did not produce any antibody to A2 virus.

### f) Age Distribution of Antibody to Different Strains of Virus in Man

SHOPE (S₁₁₀) showed that the majority of adult human sera examined in 1936 contained neutralizing antibody to the SWINE virus, but antibody was not found in the sera of children under 12 years. LAIDLAW (L₁₄) and SHOPE suggested that the SWINE virus had been responsible for the pandemic of 1918. Although it was realized that antibody to the SWINE virus could be produced by repeated infection with the human virus this would not explain the presence in certain sera of antibody neutralizing only the SWINE virus.

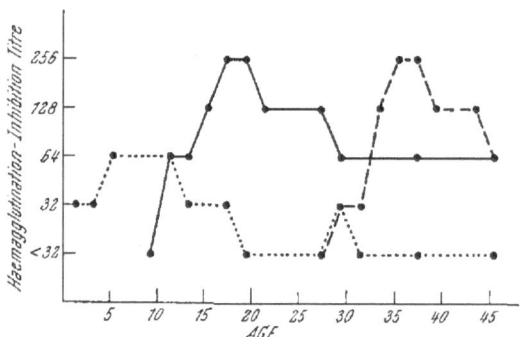

Fig. 54. Age incidence of antibody in normal human sera in 1952.
Age incidence of antibody to A1, A, and Swine viruses in normal human sera in 1952. A1 virus ●·····●; A virus ●——●; Swine virus ○– – –○. Adapted from DAVENPORT, HENNESSY and FRANCIS (D₂₄).

DAVENPORT, HENNESSY and FRANCIS (D₂₄) studied the age incidence of haemagglutinin-inhibiting antibodies in normal human sera collected in 1952, 18 months after the last epidemic of influenza A1. The sera were tested against SWINE, 9 strains of A type and 9 strains of A1 type. The results showed a very striking difference in the age incidence of antibodies to the three groups (Fig. 54). Antibody to the A1 strains, isolated since 1946 was present in young children and maintained the maximal level to the age of 12. After 20 years the antibody was low or undetectable. Antibody to the A strains isolated from 1933—1943 was not detected in children under 11 years, and then there was a sharp rise to a maximum at age 20. Antibody then declined to a constant level after age 28. With the SWINE virus antibody was not demonstrable until age 29. It reached a peak at 33—38 and then declined.

Similar tests with B viruses showed that antibody to the LEE strain isolated in 1940 was not detectable until age 13 after which a high level was maintained till age 50 with a subsequent decline. Antibody to the ALLEN strain isolated in 1945 was first detectable at age 5—6, reached a peak at age 15 and then declined. Antibody to type C virus was present at high levels at all ages.

DAVENPORT, HENNESSY and FRANCIS also measured the antibody in gamma globulin pools collected before and after 1947, the year in which A1 virus first appeared. Both samples contained a high level of antibody to SWINE, WS and PR8. Low levels of antibody to A1 virus were found in the pre-1947 gamma globulin and there was only a 4-fold increase in the post-1947 samples.

As a result of these findings DAVENPORT, HENNESSY and FRANCIS suggested that the antibody response to infection is oriented towards the production of antibodies against the dominant antigens of strains of virus prevalent during the childhood of the patient, and not to the dominant antigen of the infecting strain. Thus in 1952 the predominant antibody was directed against the A1 virus prevalent in 1947—1952 in the case of children, but from the ages of 12—30 the predominant antibody was against the A viruses prevalent from 1933—1943, and from age 30—60 against the SWINE virus which was possibly the prevalent virus in 1918—1920. This concept is sometimes described as the "doctrine of original antigenic sin".

DAVENPORT and HENNESSY ($D_{17}$) studied the antibody response by haemagglutination-inhibition tests in humans of different age groups to three doses of monovalent formalized vaccine of SWINE (1931), PR8 (1933), FM1 (1947), and CUPPETT (1950) viruses. It was found that no matter which vaccine was used the maximum antibody production in children was directed against the A1 viruses, in persons aged 17—26 all the vaccines produced a considerable rise of antibody to PR8 and FM1 in addition to the response to the homologous virus, and in persons over 30 years of age all vaccines produced a response to SWINE virus which was sometimes greater than that to the homologous virus. The production as a result of infection or vaccination of antibody reacting against viruses prevalent in earlier years is known as the "recall phenomenon".

JENSEN, DAVENPORT, HENNESSY and FRANCIS ($J_8$) absorbed human sera of different age groups with SWINE, PR8 or FM1 virus adsorbed on red cells treated with periodate. It was found that the sera of old patients contained three types of antibody all of which reacted with and were adsorbed by SWINE virus. One type reacted only with SWINE, one with SWINE and PR8, and one with SWINE and FM1. Patients aged 17—25 had two kinds of antibody reacting with PR8, one reacting only with PR8 and one which could be absorbed with either FM1 or SWINE in addition to PR8. The FM1 antibody in children's sera reacted only with FM1 and could not be absorbed with PR8 or SWINE. When ferrets were successively infected with WS, WEISS and CAM strains all the antibody produced could be completely removed by adsorption with WS, but not by adsorption with either of the other two strains. It appeared that the antibody was oriented to react with the strain of the first infection. With both human and ferret sera there appeared to be two types of antibody produced, one reacting only with the strain of the first infection, while the other was multivalent, reacting with several strains. The concept that prior exposure predetermines antibody response was further confirmed by the results of vaccination of humans of different age groups ($D_{18}$, $D_{25}$). No matter what type of vaccine was used children always produced high levels of A1 antibody, persons aged 17—25 produced high response to A and A1, while high levels of SWINE and A antibody was produced in the over 30 age group. Children produced little response to SWINE and A, the 17—25 group produced little response to SWINE, and the over 30 group produced little response to A, unless the strains were given in high dosage as a monovalent vaccine.

Although the American studies were confirmed in serological studies in England and Czechoslovakia ($D_{27}$, $B_{79}$) and the "recall phenomenon" has been repeatedly observed ($M_{120}$, $C_{90}$, $P_{36}$, $Z_8$, $W_{26}$, $E_{34}$) the theory that antibody to SWINE virus in human sera is a result of previous infection, and that the SWINE virus was responsible for the 1918 pandemic has not been always accepted. Antibody to SWINE virus has sometimes been found in the sera of people who were too young to have been infected in 1918—1920, ($M_{44}$) and ISAACS ($I_6$) found antibody to WS in the sera of children in Nigeria born after the period of prevalence of A virus. It seems probable that antibody to the SWINE virus may be produced as a result of infection by human viruses. LIEF and HENLE ($L_{80}$) using the specific complement fixation test found that antibodies to a given family of strains were not sharply confined to a particular age group and that the highest levels of such antibodies did not necessarily occur amongst those who were under 10 years of age at the time when the strain was prevalent. Many individuals possessed or developed antibodies to the V antigens of strains to which they could not have been exposed during their lifetimes. They suggested that virus strains contain, in addition to the dominant and minor antigens detectable by in vitro tests, trace quantities of hidden antigens representative of many or all past strains and possibly of future strains. These hidden antigens might be responsible for the recall of antibody in individuals previously exposed to strains in which they were dominant, and also to produce detectable levels of antibody response in repeated infections or in very severe primary infections.

NELSON and LEWIS ($N_{13}$) examining convalescent sera from the 1957 epidemic by neutralization and haemagglutination-inhibition tests found that infection with A2 virus stimulated antibody production to A2 and SWINE but not to A or A1 viruses. They suggested that there was a common antigenic component between A2 and SWINE. SCHULMAN and KILBOURNE ($S_{73}$) found that some resistance to A2 virus was produced by a single intranasal infection of mice with A virus, but this effect was not produced by vaccination with inactivated A virus. EFIMOVA and ZAKSTELSKAYA ($E_{34}$) found that with the exception of the A2 virus, in animals inoculated successively with different A strains the greatest immune response was always to the first used. The A2 virus, however, caused a decrease in production of antibodies to previously inoculated strains, and if used first markedly reduced the response to strains inoculated later.

MULDER and MASUREL ($M_{139}$) found that the sera of a small percentage of persons over the age of 70 examined before the occurrence of the A2 pandemic of 1957 contained antibody to A2 virus, and suggested that the A2 virus might be antigenically related to the virus which caused the 1890 epidemic. DAVOLI and CORSI ($D_{34}$) and DAVENPORT and HENNESSY ($D_{19}$) made similar observations, but MULDER and MASUREL found that persons immunized with A1 vaccine before 1957 sometimes developed antibody to A2 virus. DAVENPORT and others ($D_{23}$) using the DRESCHER haemagglutination-inhibition technique which was claimed to distinguish between homologous and heterologous antibody reactions studied the age distribution of antibodies in normal individuals. Specific antibody to A2 virus was only found in those over the age of 55 in 1957 suggesting that a virus related to A2 was prevalent in 1890, while SWINE antibody was present only in those over 34 suggesting its prevalence in 1918. SCHILD and STUART-HARRIS ($S_{44}$) studied the age incidence of antibody in normal sera collected in 1961 and again obtained results suggesting the relationship of SWINE virus to the 1918 pandemic. They also found that the sera of people over the age of 65 in 1946 contained neutralizing and haemagglutinin-inhibiting antibodies to the equine virus A/Equi/Miami/63 but not to A/Equi/Prague/56 or two duck viruses.

## 4. Relation between Immunity and Antibody Titre

Numerous observations have been made indicating a relationship between serum antibody level and resistance to influenzal infection in both man and experimental animals ($F_{104}$, $H_{202}$, $A_{82}$, $F_{88}$, $F_{122}$, $V_{21}$, $B_{37}$, $S_{206}$, $S_{323}$, $T_{78}$, $Z_4$, $W_{19}$). The serum antibody level after vaccination has been widely used as an index of the protective value of vaccines. Immunity is type specific indicating that the important antibody is that to the V antigen, and in fact S antibody does not neutralize virus.

There are, however, many observations which indicate that the relation between serum antibody and resistance to infection is not a direct one. SMORODINTSEV, GULAMOV and CHALKINA ($S_{203}$) compared the resistance of mice actively immunized with live virus with mice passively immunized with antiserum. Although the blood antibody level was higher in the passively immunized mice they died after intranasal inoculation of 100 MLD, while the actively immunized mice resisted 10,000 MLD. But if antiserum was inoculated intranasally mice became highly resistant although the blood antibody was low. ZAKSTELSKAYA ($Z_4$) considered that a local tissue immunity could be produced by intranasal immunization and SOLOVYEV ($S_{229}$) claimed that antibody could be produced locally in the bronchial tissue.

HERZBERG, MAY and BECK ($H_{109}$) immunized mice with mouse lung preparations of PR8 and FM1, adsorbed on aluminium hydroxide. They were then challenged with heterologous virus and it was found that 35 of 50 mice immunized with FM1 were resistant to PR8, and 25 of 50 mice immunized with PR8 were resistant to FM1. The sera of the mice showed no heterologous neutralizing or haemagglutinin-inhibiting antibody and it was supposed that the cross-resistance was due to some form of tissue immunity.

BURNET, LUSH and JACKSON ($B_{213}$) were the first to demonstrate a virus inactivating agent in human nasal secretion, and FRANCIS ($F_{89}$) found that although the virus inactivation effect of nasal secretion was not closely correlated to serum antibody titre, those with low blood antibody had inactive secretions, while in general those with high blood antibody gave active nasal secretions. FRANCIS and others ($F_{112}$) found that vaccination of humans by subcutaneous inoculation of live virus increased the virus-neutralizing activity of the nasal secretions in 23 of 30 cases. Killed virus was less effective. MULDER, BRANS and HERS ($M_{135}$) obtained similar results and found that the average virus neutralizing titre of nasal secretion was 3% of the average serum titre, but there was no strict correlation between the two. ROLLA ($R_{44}$) found an increase in haemagglutinin-inhibiting activity of the saliva as a result of vaccination, serum titres ranged from 200—750 while saliva titres ranged from 12—24. FAZEKAS DE ST GROTH and DONNELLY (F$_{35}$, F$_{36}$) found that after intranasal inoculation of live virus in mice maximum antibody level in the serum was reached at the 12th day, and bronchial washings contained an average of 5% of the serum anti-haemagglutinin. Subcutaneous inoculation of live virus was much less effective in producing serum antibody and only 0.4% was found in bronchial washings. When serological tests were made on the same day as resistance was challenged it was found that resistance was not closely related to serum antibody level, but was closely related to the anti-haemagglutinin content of bronchial washings. FAZEKAS DE ST GROTH and DONNELLY (F$_{37}$, F$_{38}$, F$_{28}$) described a phenomenon which they called "pathotopic potentiation" in which the specific resistance of mice after subcutaneous or intraperitoneal vaccination could be enhanced by the intranasal inoculation of heterologous virus by a factor of 20 to 100 fold, the increased resistance being associated with a corresponding increase in anti-haemagglutinin content of bronchial washings. Similar results were produced by intranasal inoculation of vaccinia virus, periodate, zinc sulphate and tannin, the effect being attributed to an increase in capillary and tissue permeability. DEPOUX and MUSSET (D$_{47}$) were unable to confirm these results.

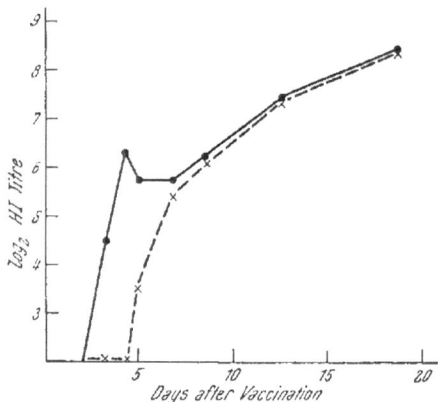

Fig. 55. Early haemagglutination-inhibition responses of mice infected with emulsified influenza virus vaccine.

Untreated serum. ●———● Mercaptoethanol-treated serum. ×———× Reproduced from BERLIN (B$_{52}$).

WEINBERGER, BUESCHER and EDWARDS (W$_{31}$) described 47 cases of infection by A2 virus occurring in individuals with considerable antibody in the blood.

It would appear from these studies that immunity to influenzal infection is only indirectly related to the presence of specific antibody in the blood, but is closely related to the presence of antibody in the respiratory secretions.

## 5. Early and Late Antibodies, Labile Antibodies and the Serum Co-Factor

BERLIN (B$_{52}$) showed that two types of antibody were produced during the primary response of mice to influenza virus vaccine (Fig. 55). One type appeared after an induction period of 2 days and accounted for most of the haemagglutinin-inhibiting activity up to the 7th day. This γM antibody was destroyed by treatment with mercaptoethanol and had a sedimentation constant of 18 S. In passively immunized mice it had a half-life of 12 hours. The second γG antibody appeared after an induction period of 3—4 days, was resistant to mercaptoethanol, had a sedimentation rate of 7 S, and a half-life of 2 days in passively immunized mice. The γM antibody reached a peak in 4 days, while the γG antibody became dominant at 7 days and reached a peak in 35—42 days. Irradiation of mice with X rays before injection of vaccine depressed the production of γM antibody more than γG antibody (B$_{53}$).

## The Serum Co-Factor

STYK, RATHOVA and BLASKOVIC ($S_{315}$) found that part of the specific haem-agglutinin-inhibiting and neutralizing antibody in the serum of mice inoculated with A2 virus was destroyed by heating at 56° C for 30 minutes, but that activity was restored by addition of unheated normal mouse serum which contained no inhibitors of A2 virus. Heating at 56°C reduced the serum titres from 4—8 times, and similar effects were produced by treating the serum with RDE. The heat-labile agent present in normal serum was given the name "serum co-factor" and its properties were extensively studied by the Czechoslovak workers. Anti-bodies requiring co-factor were described as "thermolabile" though it was realised that it is the co-factor which is heat sensitive and not the antibody. Only "non-avid" strains of A2 virus produce antibody requiring co-factor, with "avid" strains the antibody is heat resistant ($B_{82}$). Mouse-adapted A2 virus produced a more heat-resistant antibody than unadapted virus. The heat-labile antibody appeared at an earlier stage of immunization than the heat-resistant type. Co-factor could be titrated by determining the dose of normal serum which would restore the haemagglutinin-inhibiting activity to the heated sera of immunized mice known to contain antibody requiring co-factor ($S_{299}$). The largest amounts of co-factor were found in mouse, rat, bovine, and human sera and bore no relation to their content of Chu or $\beta$ inhibitor. Co-factor was destroyed by M 50 periodate which had no effect on $\beta$ inhibitor, and trypsin destroyed $\alpha$ and $\beta$ inhibitors but had no action on co-factor ($S_{300}$). Treatment with $CO_2$ destroyed both co-factor and $\beta$ inhibitor ($S_{300}$). Co-factor could be separated from $\beta$ inhibitor by chromatography in DEAE cellulose ($H_{23}$). KOKISCOVA, STYK and HANA ($K_{75}$) showed that co-factor was not properdin as it was unaffected by zymosan, and it was not complement as its effect could not be adsorbed out by binding on an antigen-antibody complex ($S_{314}$).

By paper electrophoresis of serum it was found that co-factor was associated with the beta globulins, while antibody was associated with the fast-moving gamma globulins ($S_{305}$). On density gradient centrifugation co-factor was found in the macroglobulin fraction, but both "early" antibody (potentiated by co-factor) and "late" antibodies were found in the 7 S gamma globulins ($S_{310}$). Similar results were obtained by separation on sephadex G 200 ($H_{25}$).

The change between "early" 19 S to "late" 7 S antibody in influenza antisera is more rapid than the change from co-factor requiring to heat resistant antibody, both of which may be of 7 S character ($S_{307}$, $S_{304}$).

ZAKSTELSKAYA and YAKHNO ($Z_9$) found that early in the course of immuniza-tion of guinea pigs, rats, and mice with formalized A2 virus the haemagglutinin-inhibiting antibodies were destroyed by treatment with $KlO_4$ or $CO_2$, but became resistant at the peak of antibody response. Neutralizing antibodies were equally resistant at all stages of immunization. HARBOE and REENAS ($H_{36}$) found in in-fected ferrets that the antibody peak measured with virus of high antibody sen-sitivity was reached several days before that measured with virus of low antibody sensitivity, regardless of the antibody sensitivity of the infecting virus.

SMORODINTSEV and YARBOV ($S_{209}$) described a "non-specific thermolabile stimulator of antibodies" apparently identical with the serum co-factor of STYK. It accelerated the union of virus and antibody and rendered the union firmer.

## 6. Immunological Implications of the Concept of "Original Antigenic" Sin

The finding that immunization with a particular antigen may result in a greater antibody production against a heterologous antigen than against the homologous one, would not be expected on the commonly accepted theories of antibody production. FAZEKAS DE ST GROTH and WEBSTER ($F_{50-52}$) studied the phenomenon with special reference to its bearing on theories of the nature of the immune response. Groups of children who had been infected with A1 virus and had antibody in the serum were vaccinated with SWINE virus. It was found that the number of antibody molecules produced against FM1 and SWINE was the same. Each molecule reacted with both antigens and the antibody could be completely absorbed with either antigen. All the antibody formed was of secondary type. It was suggested that the failure of the vaccinating antigen to produce a primary response was due to its being trapped by cells adapted to produce antibody to the original infecting antigen ($F_{51}$). Further evidence was obtained in animal experiments. Rabbits were inoculated with 5000 HA units of SWINE virus intravenously, and a parallel control group was uninoculated. After $3\frac{1}{2}$ months half the animals were irradiated with X rays in a dose of 350 r, to suppress primary antibody response but not secondary response. Samples of the various groups were then given 31.6 HA units of either SWINE, FM1 or LEE virus intravenously. It was found that with SWINE virus both vaccinated groups produced good antibody response. With FM1 virus the irradiated group produced no antibody (i.e. no primary response) if there had been no previous injection of SWINE virus, but in the group vaccinated with SWINE virus a good response occurred whether the animals were irradiated or not. With LEE virus the irradiated rabbits gave no antibody while the controls gave a good response ($F_{52}$).

The ability of cells previously exposed to one antigen to trap a second antigen and prevent a primary response is not unlimited, it can be overcome by the use of large doses of the second antigen. When rabbits were inoculated with 100 HA units of SWINE virus and then given 3160 HA units of FM1 virus 42 days later it was found that all the antibody produced on the 6th day was cross-absorbable with SWINE, but on the 28th day most of the antibody against FM1 could not be removed by absorption with SWINE.

WEBSTER ($W_{26}$) found that infection of ferrets with virus resulted in the production of large numbers of antibody molecules of low avidity. As time passed the number of molecules decreased but the average avidity increased. A second infection with serologically related virus caused a marked increase in antibody to the first virus. Absorption tests showed that some of the antibody molecules reacted with one virus, some with the other, and some with both.

FAZEKAS DE ST GROTH and WEBSTER ($F_{50}$) suggest that a cell which has once produced antibody is not only predisposed to performing the same task again but does so even when provoked by an incompletely appropriate second antigen, and that this second antigen becomes trapped and is prevented from eliciting a primary specific response, unless given in very large dosage.

## 7. Passive Immunity in Influenza

LAIDLAW, SMITH, ANDREWES and DUNKIN ($L_{15}$) prepared an immune serum in horses by repeated subcutaneous inoculation of emulsion of infected ferret turbinate and lung. The serum neutralized virus in mice and gave some protection when inoculated intraperitoneally 24 hours before intranasal inoculation of virus. SMORODINTSEV and SHISHKINA ($S_{204}$) produced immune sera in horses, pigs and sheep which had some protective properties. HARE ($H_{40}$) found that convalescent ferret serum was superior to immune horse serum.

Several workers used human convalescent serum for the treatment of influenza in the 1918 pandemic ($V_8$) without significant benefit, and STUART-HARRIS and others ($S_{291}$) had no success in the treatment of human cases with immune horse serum.

SMORODINTSEV and SHISHKINA ($S_{204}$) found that the intranasal administration of serum conveyed more protection in mice than inoculation by other routes. TAYLOR ($T_{38}$) found that the intranasal instillation of 0.05 ml of a 1:2 dilution of convalescent ferret serum in mice gave complete protection against 30 $MLD_{50}$ if given 24 hours before virus and had some effect up to 72 hours before virus inoculation. Protection was also given by serum administered 1—6 hours after virus and there was a slight effect at 24 hours.

An American group ($P_{44}$) obtained considerable protection of mice against intranasal infection by previous inhalation of immune serum and the lung lesions were reduced by serum given after infection. No immunity resulted from the inhalation of neutral serum-virus mixtures. SUGG ($S_{321}$) noted that serum did not protect against large doses of virus which caused death by damage to lung cells without virus multiplication. FAZEKAS DE ST GROTH and GRAHAM ($F_{46}$) inoculated mice intraperitoneally with varying doses of mouse, rabbit and ferret antisera to MEL virus, and challenged 2 days later with intranasal MEL. The mouse serum had 30 times the protective effect of the other sera, apparently because it was more effective in reaching the respiratory tract.

KALININA and POSPELOVA ($K_3$) found that mice could be effectively protected against infection by both gamma globulin and beta globulin fractions of influenza immune serum.

The production of passive immunity by intranasal inoculation of immune serum by SMORODINTSEV and his colleagues was confirmed by STOKES and SHAW ($S_{269}$), and the method has been extensively studied as a method of prophylaxis in human influenza by Russian workers, mainly by the use of immune horse serum. SMORODINTSEV, GULAMOV and CHALKINA ($S_{203}$) tested the method in 500 people, 2 ml of atomised serum was inhaled through a gas mask over a period of 15 minutes, the dose being repeated up to 10—15 days. The morbidity in the treated cases was 1.6%, in the controls 8.2%. KOROVIN and DMITROVSKAYA ($K_{93}$) found that inhalation of 1—1.5 ml of immune horse serum every 3—4 days reduced morbidity by 2—2$\frac{1}{2}$ times. KOFMAN, KAL'YANOVA and LYUBINSKAYA ($K_{78}$) found that normal horse serum had some protective effect, morbidity in a group receiving immune serum was 8.4%, in a group receiving normal serum 11.6%, and in controls 15.5%. In general the method was undesirable because of the short duration of the effect and the danger of sensitisation to horse serum.

SERY and others in Czechoslovakia used an immune horse serum to 2 strains of A1 virus in 8000 persons in a dose of 0.5 ml introduced intranasally by an atomizer. The incidence of influenza was reduced from 10.29% in controls to 5.13% in the treated cases ($S_{82}$).

In the treatment of influenza cases by intranasal immune horse serum, SMORODINTSEV recorded a fall in temperature, elimination of toxic symptoms, and prevention of complications in patients receiving serum on the 1st day of disease. Administration after the 2nd day was less effective. SMORODINTSEV later suggested the use of dried serum insufflated as a powder into the nose and combined with penicillin and sulphonamide. Methods of producing immune serum in horses were perfected by SMORODINTSEV, ORLOVA and KALININA, and by DREVO and others ($S_{188}$, $O_{15}$, $D_{75}$). Beneficial results of treatment with intranasal immune horse serum were recorded by several groups of Russian workers ($N_8$, $N_9$, $M_{128}$, $K_{77}$, $G_{93}$).

## 8. Conclusions on the Relation of Antibody to Immunity in Influenza

There can be little doubt that acquired immunity to influenza is mainly a result of antibody production, but antibody is only effective if it reaches the virus before union with the cell, and the important factor is the antibody level in the respiratory secretions. The serum antibody level is of only limited value as an index of immunity. Immunity tends to be strain specific, but repeated infections result in a broadening of the antibody response, and it may be that this is the reason why adults are less susceptible to infection than children. Although the group complement fixing S antibody has no virus neutralizing power it is possibly a better index of resistance than V antibody as it is not strain specific and is reinforced each time an infection occurs.

A variety of virus neutralizing proteins may be found in serum, specific antibodies of two different molecular weights, antibodies directed against one specific virus strain and others of wider activity, antibodies requiring heat-labile cofactor; non-specific heat labile virus inhibitors of the Chu type which behave as if they were natural antibodies, and heat-resistant mucoprotein inhibitors which may neutralize virus strains lacking in effective enzymic activity. The protective value of these agents will probably depend on their ability to penetrate from the blood into the respiratory secretions, and the antibodies of low molecular weight would be most likely to achieve this. Discrepancies in the results of neutralization, haemagglutinin-inhibition, specific complement fixation and protection tests in the measurement of V antibody content of serum are probably due to the various types of virus neutralizing proteins behaving differently in the tests.

The antibody response to infection or vaccination does not depend solely on the antigenic structure of the virus but is largely modified by the patient's previous experience of influenzal infection which may deflect the antibody response along pre-determined channels. Immunity resulting from infection is greater than that produced by vaccination even when the antibody response is of equal extent, and this suggests that some local factor in the respiratory tract may be involved, possibly local antibody production, or the acquisition of some form of cellular resistance.

# XXIX. Vaccination against Influenza

Early studies by SMITH, ANDREWES, and LAIDLAW and their colleagues ($S_{168}$, $A_{81}$, $A_{82}$, $S_{291}$), FRANCIS and MAGILL, ($F_{104}$, $F_{107}$, $XF_8$), FAIRBROTHER and HOYLE ($F_{10}$, $F_{12}$), SMORODINTSEV ($S_{178}$), HORSFALL and LENNETTE ($H_{168}$), and others demonstrated that influenza viruses were good antigens and that protection against infection could be achieved in experimental animals by subcutaneous or intraperitoneal inoculation of comparatively crude preparations of live virus or of virus inactivated by heat or formaldehyde. Immunity was associated with the development of neutralizing antibody in the blood, was greater against homologous than heterologous virus, and was of short duration. These studies suggested the possible use of vaccines in the prevention of human influenza and initiated work on the large scale production of virus and on purification, concentration and methods of inactivation.

## A. Vaccination with Killed Virus

### 1. Preparation of Killed Influenza Virus Vaccines

The cultivation of virus in the allantoic sac of fertile eggs has almost always been used as a method of large scale production of virus. Concentration of the virus was found to be desirable as early studies of vaccines consisting of crude formalized allantoic fluid gave unsatisfactory results because of an inadequate content of virus ($F_{93}$).

HARE, CURL and McCLELLAND ($H_{42}$) concentrated virus from allantoic fluid 10 times by adsorption-elution from red cells and obtained 70—90% recovery of PR8 virus and 63—80% with LEE virus. STANLEY ($S_{258}$, $S_{260}$) used differential centrifugation at 24,000 r.p.m. and 3000 r.p.m. in a Bauer and Pickels air turbine centrifuge and obtained results superior to those of red cell adsorption-elution. Later the Sharples centrifuge was used. Both adsorption-elution from red cells and differential centrifugation have been extensively used in the commercial preparation of influenza virus vaccines. A detailed description of the use of the Sharples centrifuge under industrial conditions is given by TAYLOR and others ($T_{35}$). Eggs were harvested under ultra-violet light and centrifugation was carried out in an air-tight room with sterilised air recirculated through the centrifuge. STANLEY ($S_{260}$) recommended addition of 1:10,000 formalin and 1:100,000 phenyl mercuric nitrate to the allantoic fluid before processing to prevent bacterial contamination.

Viruses may be inactivated by minimal exposure to ultra-violet light ($S_{260}$), but formalin has been almost always used, in a concentration of 1:2000, and it is usual to add a bacteriostatic agent such as phenyl mercuric nitrate 1:100,000 ($F_{93}$), or merthiolate 1:20,000 ($H_{42}$). Other methods of inactivation have been suggested such as the use of 1:1000 sodium ethylmercurithiosalycylate ($N_3$), 0.5% $\beta$-propiolactone at pH 7 ($P_{70}$, $P_{71}$), hydroxylamine ($S_{38}$), nitrous acid ($R_{64}$), or photodynamic inactivation in the presence of toluidine blue ($W_{14}$).

Formolized vaccines can be stored for some months at 4°C ($S_{260}$, $F_{93}$, $P_{29}$, $N_4$), but freezing or drying caused great loss of potency ($S_{260}$, $P_{29}$). Vaccines prepared by adsorption-elution from red cells were more resistant to freezing than vaccines produced by centrifugation methods ($P_{29}$).

### a) Antibody Response to Vaccination with Killed Virus

Very many studies have been made of the antibody response to inactivated influenza virus vaccines. A very complete study of the response of rabbits to various methods of vaccination has been recently made by WEBSTER ($W_{25}$), and studies in mice were made by FAZEKAS DE ST GROTH and DONNELLY ($F_{35}$, $F_{36}$). Antibody production has been described in chapter 28 and it is only necessary to summarize the essential results here.

1. The antibody response increases with increasing doses of virus if the dose of virus is small, but with larger doses a maximal response is elicited which in the case of V antibody is as good as produced in infection ($E_{12}$, $F_{35}$, $H_{151}$, $P_{28}$).

2. Vaccination with killed virus produces only slight increase in S antibody ($M_{129}$, $E_{12}$, $R_{57}$, $F_{35}$, $C_{92}$).

3. In general the antibody response is greater to homologous than to heterologous virus, but in some cases when the individual has been previously infected a recall phenomenon may occur with a greater response to the strain previously encountered ($D_{17}$, $D_{18}$, $D_{50}$, $D_{23}$, $E_{34}$).

4. The degree of antibody increase may not be identical when measured by different serological tests such as neutralization and haemagglutination-inhibition ($B_{219}$, $A_{94}$, $W_{25}$).

5. In humans who have had previous attacks of influenza, if the dose of vaccine is adequate the response to a single dose is as good as to two doses ($B_{148}$, $B_{62}$, $N_{17}$, $G_{32}$, $N_3$), but in young children more than one dose is needed to produce a good response ($Q_5$), and the same result may occur in adults if a new strain of virus such as A2 is used in persons with no previous experience of that type ($H_{155}$, $M_{53}$, $M_{73}$, $M_{77}$), though if an adjuvant type vaccine is used only one dose may be needed ($P_{60}$).

6. Antibody response may vary with different routes of inoculation ($F_{35}$, $W_{25}$).

7. The first antibody produced is of the $\gamma$M (18 S) mercaptoethanol sensitive type. This is soon replaced by the $\gamma$G (7 S) mercaptoethanol resistant type ($B_{49}$, $B_{52}$). With saline vaccines the antibody production reaches a peak in about 2 weeks and commences to decline 6—9 weeks after vaccination ($F_{27}$). With adjuvant vaccines the initial antibody rise may be suppressed and antibody production may occur more slowly reaching a peak in 3 months and declining after 1 year ($H_{151}$, $M_{68}$).

### b) Testing of Potency of Vaccines

The potency of influenza virus vaccines may be tested by measuring their protective effect in mice. Mice are vaccinated with two doses of 0.5 ml of vaccine dilutions intraperitoneally at weekly intervals and 1 week later challenged by inoculation intranasally with 0.05 ml of a range of virus dilutions ($F_{93}$). Alternatively the blood serum of the vaccinated mice may be tested for content of neutralising antibody. EDDY ($E_{23}$) found that with PR8 and WEISS strains of virus

A it required 63—125 times as much vaccine to induce resistance as to produce antibody neutralizing the same amount of virus. With LEE virus 5 times as much vaccine was needed. HARE and MACKENZIE ($H_{43}$), using PR8 and LEE vaccines prepared by adsorption-elution from red cells and with HA titres over 1000, found that two doses of 0.5 ml of a 1:5 dilution of the PR8 vaccine protected against 10,000 $LD_{50}$, and 2 doses of 0.5 ml of 1:500 dilution of LEE vaccine protected against 1000 $LD_{50}$ given intranasally. Protection tests in mice are only satisfactory with mouse adapted strains of virus.

SALK, MENKE and FRANCIS found that the haemagglutinin-inhibition titre produced in humans as a result of vaccination was closely related to resistance to infection by virus strains antigenically similar to the vaccine strain ($S_{13}$), and the production of HI antibody in vaccinated humans has often been used as a test of vaccine potency. But great difficulty may be experienced with non avid strains of virus ($A_{77}$), and HI antibody titres do not always run in parallel to neutralization titres.

HENLE and HENLE ($H_{82}$) tested vaccine potency by immunizing mice by intraperitoneal inoculation of a range of vaccine dilutions and determining the resistance to the toxic effect of allantoic fluid virus inoculated intravenously. Comparing this method with intranasal challenge they found that PR8 vaccine diluted 1:100 protected 50% of the mice against 30,000 $LD_{50}$ intranasally, but vaccine diluted 1:10,000 protected 50% of mice against intravenous inoculation of 2.8 $LD_{50}$, a dose equivalent to 300,000 intranasal $LD_{50}$.

BLASKOVIC and SALK ($B_{81}$) introduced an "antigenic extinction test" which depended on the determination of the smallest dose of vaccine which would produce measurable neutralizing antibody. Mice were inoculated intraperitoneally with one dose of vaccine diluted in two-fold steps. After 2 weeks the serum of the mice was tested for antibody by inoculation to eggs of a mixture of undiluted serum and 1000 $ID_{50}$ of homologous virus.

### c) Toxic Properties of Influenza Vaccines

RATNER and UNTRACHT ($R_{12}$) recorded cases of asthma, urticaria, serum sickness and "shock" in children as a result of inoculation of vaccine. Of 108 known allergic children 11 reacted positively by production of skin erythema on intradermal inoculation of 0.02 ml of an A+B vaccine. All these children also reacted to egg white. One developed a severe local reaction followed by asthma. They suggested that a preliminary intradermal test should precede vaccination. QUILLIGAN, MINUSE and FRANCIS ($Q_5$) also observed reactions in children and suggested that concentrated vaccines should not be used in children. Reactions were related to the dose of virus and consisted of fever up to 105°F, vomiting, prostration and "limb twitching" ($Q_3$). SALK ($S_8$) and NICHOLAS and HENLE ($N_{17}$) also observed febrile reactions in children and considered that they were related to the dose of virus used and not to the presence of egg protein.

Local reactions may be produced by vaccines of virus adsorbed on aluminium phosphate ($M_{68}$), and very extensive and persistent local reactions, sterile abscesses and persistant local granulomas may occur after the use of vaccines with oily adjuvants ($M_{68}$, $H_{70}$).

### d) Adjuvant Vaccines

Many attempts have been made to improve the potency of influenza virus vaccines by adsorption of the virus on various insoluble materials or by emulsification in oil. Such procedures retard the spread of virus from the inoculation site and result in a longer antigenic stimulation, and may reduce the toxic effects of the vaccine. Satisfactory antibody response may result from smaller amounts of virus.

SALK ($S_5$, $S_8$) found that formolized virus adsorbed on calcium phosphate was a more effective antigen than unadsorbed virus. The production of haemagglutinin-inhibiting antibody in mice was doubled, and one dose of adsorbed vaccine protected mice against 1000,000 MID of virus, while unadsorbed virus only protected against 10,000 MID. Similar results were obtained by Russian and Czechoslovak workers ($C_{24}$, $O_{13}$, $P_{27}$).

Vaccines adsorbed on aluminium phosphate, aluminium oxide or hydroxide have also been used ($M_{68}$, $C_{17}$, $D_{69}$, $R_4$, $Z_{31}$). DRESCHER ($D_{69}$) found the antibody response to A2 virus adsorbed on $\gamma$ aluminium oxide was 20 times greater than to unadsorbed virus.

FRIEDEWALD ($F_{143}$, $F_{145}$) emulsified influenza virus in an adjuvant consisting of paraffin oil + killed tubercle bacilli or killed *Myco. butyricum* + Falba (an adsorption base of beeswax, paraffin and oxycholesterin from lanolin). The adjuvant vaccine produced more effective immunity in ferrets and rabbits than vaccine without adjuvant. In mice the controls resisted 100 MLD 4—8 weeks after vaccination but lost immunity in 26 weeks. Mice vaccinated with adjuvant vaccine resisted 1000,000 MLD at 4—8 weeks and 1000 MLD at 26 weeks. Nodular local lesions were produced and not surprisingly Friedewald suggested that this complex witches brew was probably not safe for use in man.

HENLE and HENLE ($H_{80}$) inoculated 2 groups of 80 humans with a saline A+B vaccine and with the same vaccine emulsified in a mixture of Falba and mineral oil. In the saline group HI antibody titres reached a peak in 2 weeks and then declined. In the adjuvant group the titres lagged behind at first but in 3 months were 5—6 times as high as in the saline group and $2\frac{1}{2}$—3 times the 2 week saline maximal titre. The adjuvant vaccine produced subcutaneous nodules and abscesses in two cases. Oily adjuvants were also tried by CHALKINA and ORLOVA ($C_{24}$, $O_{13}$). SALK, LAURENT, BAILEY and others ($S_{9-12}$) made extensive studies of vaccine emulsified with a light mineral oil Bayol F and Arlacel A and inoculated intramuscularly. Local reactions were absent or trivial and high and persistent antibody responses were obtained in monkeys and man. The Salk vaccine was also used by PHILIP and others ($P_{59}$) and by British workers ($M_{68}$) who found it superior to vaccine adsorbed on aluminium phosphate. However, two volunteers developed unusually persistent and extensive local reactions. DAVENPORT and others ($D_{26}$) used a similar vaccine of virus emulsified in 9 parts of a mineral oil Drakeol 6 VR and 1 part of purified Arlacel A. HENNESSY and DAVENPORT ($H_{100}$) recommended two doses of vaccine to produce a broader antibody response. They found that antibody levels might remain elevated even as long as 8—9 years after inoculation ($D_{20}$). BERLIN ($B_{46}$) studied the physical properties of emulsified mineral oil vaccines and found that a high emulsion viscosity reduced the adjuvant effect, while optimal emulsion stability gave maximal adjuvant effect. Exposure of mice to X rays before vaccination reduced the adjuvant effect ($B_{47}$, $B_{48}$).

Local reactions produced by mineral oil adjuvants led to a search for adjuvants which could be metabolised in the human body. WOODHOUR and JENSEN ($W_{60}$) found that A2 virus was adsorbed by cholesterol and cholesterol-adsorbed virus gave an enhanced antibody response. B viruses were not adsorbed. Both A and B strains were adsorbed by stearic acid. Vaccine containing 0.5 mg/ml of hexadecylamine gave a rapid production of antibody exceeding the yield in controls ($W_{61}$). WOODHOUR, JENSEN and WARREN ($W_{62}$) studied the adjuvant effects of various oil-water emulsions. Peanut, sesame and chaulmoogra oils produced a significant elevation of the primary antibody response, while poppyseed and safflower oils had no effect. All the oils tested produced an increased secondary antibody response, but less than with mineral oil. Recently WOODHOUR and others ($W_{63}$, $W_{64}$) have described a metabolisable adjuvant for human use consisting of aqueous influenza vaccine emulsified with an equal volume of a mixture of 86% refined peanut oil, 10% mannide monooleate (Arlacel A) and 4% aluminium monostearate. Results were comparable to mineral oil adjuvant vaccine and reactions were reduced. Peanut oil is non antigenic and probably completely metabolised.

### e) Sub-unit Vaccines

DAVENPORT, ROTT and SCHÄFER ($D_{29}$) found that highly purified preparations of haemagglutinin made by ether treatment of influenza viruses powerfully stimulated the production of haemagglutinin-inhibiting antibody in animals, and DAVENPORT suggested their possible use as vaccines in man ($D_{11}$). DAVENPORT and others ($D_{22}$) found that purified haemagglutinin vaccines produced HI and neutralizing antibodies in man as readily as intact virus and febrile reactions were much reduced. LANGE ($L_{18}$) also found haemagglutinating sub-units to produce antibody and give protection. WEBSTER and LAVER ($W_{27}$) found that while intact viruses were pyrogenic, virus disrupted by ether or deoxycholate were not. Ether treated vaccine produced antibody levels equal to intact virus in rabbits, but deoxycholate treated virus was less antigenic. Antibodies showed the same cross-reactivity as those induced by intact virus. HENNESSY and DAVENPORT ($H_{101}$) immunized human volunteers with haemagglutinin antigens in the form of aqueous or mineral oil adjuvant vaccines and found the results as good or better than with intact virus. Febrile reactions were negligible.

### 2. Results of Vaccination with Killed Virus Vaccines in Man

An extensive study of the use of killed vaccine in influenza prophylaxis was instituted by the Commission on Influenza of the U.S. Army ($F_{93}$).

In 1942 civilian groups were inoculated with a mixed PR8 and LEE vaccine and a good serological response was obtained, average titre rises were 8.9 against the A and 7.7 against the B component. After 3 months the titres had fallen to 65% of the 2 week levels. No influenza epidemic occurred in 1942, but evidence of resistance was obtained by induction of experimental influenza ($F_{118}$, $S_{14}$).

In 1943 a few cases of influenza A infection occurred in the spring. From one of these cases a strain WEISS was isolated which differed somewhat from PR8.

Vaccines were prepared from virus concentrated 10-fold from allantoic fluid by adsorption-elution from red cells. Each dose of vaccine contained 0.5 ml of LEE concentrate and 0.25 ml each of PR8 and WEISS. Vaccination programs were carried out by six different groups of workers, mostly in service personnel between October 19th and December 4th. A severe epidemic of influenza A occurred in November and December 1943 and January 1944. In two of the groups the epidemic commenced before vaccination was complete. All six groups of workers recorded a protective effect of the vaccine, the incidence of influenza in the vaccinated ranging from 1.33—4.26% while in the controls it ranged from 5.97—10.0% ($R_{27}$, $H_8$, $E_{13}$, $H_{149}$, $S_{13}$, $M_{14}$).

A similar degree of protection was recorded in the 1943 epidemic by ANSHELES in Russia ($A_{87}$). In 1945 the Commission on Influenza recorded a high degree of protection conferred by a LEE vaccine in an epidemic of influenza B ($F_{117}$, $H_{152}$), the incidence of influenza being 0.87% in the vaccinated as against 11.21% in controls.

These results, representing a 70% protection in the A epidemic of 1943 and 90—95% in the B epidemic of 1945, were very promising, but in 1947 a pandemic of influenza A occurred due to a new strain of virus, the A1 strain, which was very different serologically from all previous A strains. Vaccination with a mixture of PR8, WEISS and LEE ($F_{119}$, $S_{135}$, $K_{10}$, $T_{39}$) in the autumn of 1946 conveyed no protection whatever against the A1 outbreak in March 1947, and produced no antibody against the new strain. Very poor results were recorded by the use of PR8 vaccine in 1947 by British workers ($M_{75}$).

In succeeding years the A1 strain was included in the vaccine and improved protection rates were observed in A1 epidemics in 1950, 1951, 1953 and 1957 by the Commission on Influenza ($D_{10}$, $D_{21}$). In 1953—1954 Russian workers administered a mixed A and A1 formolized vaccine intranasally to 64,610 people. The incidence of A1 influenza was reduced from 3—4 times in adults as compared with the incidence in controls, and 2.2 times in children. In the same year in Holland ($V_{18}$, $V_{19}$, $F_{85}$) vaccination gave poor results apparently because the infecting strain was considerably different from the A1 strain used in the vaccine.

In 1955—1956 a protection rate of 30—40% against A1 virus was obtained both with a monovalent A1 vaccine and a polyvalent A, A1 and SWINE vaccine by British workers ($M_{69}$).

In 1952 HENNESSY and others used a monovalent LEE vaccine and obtained a protection rate of 2.7 to 1 in an outbreak of influenza B infection. This result was inferior to the results in the 1945 epidemic, but the epidemic strain differed significantly from LEE. The results of the use of an adjuvant type of LEE vaccine in 1955 were even less satisfactory and it was suggested that a new strain of B virus should be used in the vaccine.

In May 1957 cases of influenza occurred in the Far East which were identified by MEYER and others ($M_{83}$) as a new serological variety of the virus, the A2 or Asian strain. Production of vaccine against the new strain was commenced in America and by September 4,600,000 ml were produced, and by the end of October 43,600,000 ml ($S_{157}$). The vaccine was used partly as a monovalent vaccine containing 250 HA units, partly as a monovalent vaccine of 400 HA units, and partly as a polyvalent A2 A1 A B vaccine (C$_{82}$), and four groups of investigators reported a reduction of incidence of infection ranging from 40—77% (S$_{74}$, M$_{73}$, C$_{82}$). A similar reduction in morbidity was noted by a group of Czechoslovak workers using a monovalent A2 vaccine adsorbed on calcium phosphate (P$_{27}$).

Following the pandemic of Asian influenza in 1957 the incidence of influenza has been low and there has been less opportunity to evaluate the results of vaccination.

## B. Vaccination with Live Virus

Early studies showed that subcutaneous inoculation of live virus produced a better immunity in animals than killed virus (C$_{22}$, E$_2$) and in man produced a greater increase in virus-neutralizing power of the nasal secretion (F$_{112}$). Animals could be protected by intranasal inoculation of live virus (E$_6$, S$_{222}$, Z$_4$).

In 1938 CHALKINA (C$_{22}$) inoculated human volunteers intranasally with live virus in the form of dilute suspensions of ferret and mouse lung tissue and obtained a rise in blood antibody. No pathological reactions occurred and it was suggested that the use of live virus intranasally in man might be a promising method for mass immunization. FALKOVICH and others (F$_{13}$) obtained similar results but found that virus could not be recovered from the respiratory tract 24 hours after inoculation.

BURNET and FOLEY (B$_{196}$) inoculated human volunteers with three doses of 0.25 ml of live egg virus at weekly intervals by spraying with an atomiser into the nose and throat. Three strains of virus were used, egg-adapted MEL, partially egg adapted BUR, and unmodified human virus REID AP 1. There were no symptoms after the 1st inoculation, slight nasal symptoms after the second. Three individuals with low initial antibody titres developed definite influenza after the 3rd inoculation and virus was recovered from them. The other cases did not develop any antibody rise. BURNET (B$_{163}$) inoculated 22 volunteers intranasally with 0.2 ml of B virus from the 40th amniotic passage. Only slight symptoms occurred but an antibody rise was obtained. BURNET later inoculated volunteers with a mixture of 2A and 1B viruses. Only 20—30% showed an antibody rise and in a subsequent epidemic there was little protective effect (B$_{164}$). BULL and BURNET (B$_{148}$) inoculated 23 volunteers with low initial antibody titres with a living attenuated B virus. Most developed some symptoms, nasal discharge and headache, which were probably allergic in nature. Virus was demonstrated in the nasal secretions 48 hours after spraying in 14 of 23 cases and an antibody rise occurred in 21 cases. Re-inoculation 3—6 months later produced an antibody rise in only 2 cases and in only 3 could virus be recovered. MAWSON and SWAN (M$_{48}$) obtained similar results to BURNET, and used less attenuated virus but without significant improvement in the results.

The results of the Australian workers were not regarded as encouraging and little subsequent work was done with live virus in the Western laboratories, but the use of live intranasal virus was extensively pursued in Russia.

### 1. Intranasal Immunization with Live Virus in the Form of Mouse Lung Suspension

In 1938 and 1939 SMORODINTSEV and CHALKINA obtained a 50% protection rate as a result of inoculation of live mouse lung virus intranasally, but in 1940 the virus A vaccine gave no protection as the epidemic in that year was due to virus B ($S_{196}$). KHAYT, VINOGRADOVA and NARTSISSOV ($K_{33}$) were unable to recover virus from volunteers inoculated intranasally with live mouse lung virus.

## 2. Intranasal Immunization with Live Virus from Fertile Eggs

KHAYT and others ($K_{32}$) first used egg virus in Russia. No ill effects resulted from intranasal inoculation of two doses of a mouse-virulent A virus grown in 4 day chick embryos given in a dose of 1000 mouse MLD, but no protection resulted in a subsequent epidemic which was due to virus B.

Live egg virus was first used on a large scale in 1948—1949. The following results were obtained in Leningrad by SMORODINTSEV, CHALKINA and AN-SHELES ($S_{185}$, $S_{198}$, $A_{88}$). Three groups of students were studied, one group of 7399 were inoculated intranasally with virus A allantoic fluid, 7324 with virus B and there was an unimmunized control group of 6819. Two months after inoculation an epidemic due to A1 virus occurred and the morbidity rate per 1000 was 27.5 in the group receiving A vaccine and 40.0 in the other two groups. The A vaccine therefore gave some protection against the A1 virus. After 10 months an epidemic due to virus B occurred and protection of the order of 30% was recorded in those receiving the B vaccine.

In Moscow, SOKOLOV, KULIKOVA and SAMVELOVA ($S_{223}$, $S_{219}$) inoculated 7993 persons with a mixed vaccine of three different A strains. In the A1 outbreak which followed 6.2% became ill as against 12.2% of 10,651 controls. A B vaccine (LEE) was given to 1881 people of whom 5.3% became ill in the subsequent B outbreak, while 7.5% of the 4186 controls were attacked. NIKOLAYEV and SAMVELOVA ($XN_1$) immunized 3609 people with a mixed A and A1 vaccine which reduced the morbidity in an A1 epidemic from 11.7 to 7.4%. Immunization with LEE virus was ineffective.

In Kharkov ($S_{79}$, $Z_{16}$) better results were obtained in 674 immunized children. Immunization with B vaccine reduced morbidity rate by 2—$2\frac{1}{2}$ times, and with A virus 1.8 times.

SOLOV'YEV and SOKOLOV ($S_{242}$) found that concentrated formolized egg vaccine produced a 5-fold rise in antibody titre but gave considerable local and some general reactions. The protection rate was about 50%. SOKOLOV ($S_{217}$, $S_{225}$) found that intranasal inoculation produced a greater increase in nasal antibody than subcutaneous vaccination.

These results showed that intranasal immunization with live virus was safe but inconsistent and not very effective. ZHDANOV and SOLOV'YEV ($Z_{16}$, $S_{219}$) suggested that in further work more account should be taken of the antigenic variation of the viruses, that immunogenic properties might be maintained by cultivation in human tissues and that desiccation might be used to maintain the stability of the vaccine. The quality of the vaccine should be tested by determining the duration of its survival in the nasopharynx, the absence of fever, rhinitis or tracheitis, and by measuring the rise in blood and nasal antibody.

## 3. Intranasal Immunization with Human Tissue Culture Vaccines

ZILBER ($Z_{40}$) suggested the use of live virus grown in human tissue. SMORO-DINTSEV ($S_{180}$) found that the immunogenic properties and ability to persist in the human nasopharynx which were lost on adaptation to mice or eggs might be regained by passage in man. FADEYEVA ($F_4$) inoculated human volunteers intranasally with egg lines of the S6 strain of A1 virus, and with egg-human embryo tissue culture lines. Virus survived in the nasopharynx for 2 days in 30% of those receiving the egg line, and in 80% of those receiving the egg-human tissue line. The tissue culture line produced a greater rise in antibody. Contacts became infected by association with those inoculated with the tissue culture line but not with those inoculated with the egg line. Similar results were obtained by RITOVA and ZAKSTELSKAYA ($R_{37}$). Old laboratory strains did not adapt to human tissue culture as easily as freshly isolated strains ($F_6$). Strains fully adapted to human tissue did not change in biological properties in 2 or 3 egg passages. Vaccines were therefore prepared by adaptation of the strains to human embryo tissue cultures followed by one or two egg passages. Three A1 strains and 2 B strains were adapted and survived well in the human nasopharynx. A single dose produced immunity to a subsequent inoculation and there was a rise in blood antibody of 2—32 times, and a larger rise in the neutralizing power of nasal secretions. In the majority of those vaccinated the immunity lasted about 10 months. RITOVA and ZAKSTELSKAYA ($R_{37}$) obtained polyvalent vaccine strains by simultaneous cultivation of several different strains. A polyvalent A1 strain T Ch $E_5$ was obtained by simultaneous cultivation of 5 A1 strains of different antigenic structure in tissue cultures, and a polyvalent B strain T Ch E2 from 2 B strains. Later a strain T Ch E7 was produced from 7 different A1 strains. These strains produced a wider antibody response in animals and man.

It was also shown that in non-immune adults simultaneous immunization with 3 different viruses was satisfactory, all three multiplied and could be isolated from the nasopharynx. A suitable dose was 0.3 ml of allantoic fluid (0.1 ml of each virus). In young children, however, much smaller doses were desirable (0.003 ml). Larger doses produced increased reactions ($R_{36}$).

SOKOLOV ($S_{218-220}$) developed a method of vacuum desiccation at low temperature of the vaccine in a serum-egg-yolk medium. The virus survived for 21 months without loss of virulence at 4°C. It was reconstituted with water or saline. SMORODINTSEV used dry vaccine insufflated as a powder, using starch as an inert base, but SOKOLOVA and GAYLONSKAYA ($S_{227}$) found the liquid form preferable.

SOKOLOV tested reconstituted dried vaccine in the 1952 outbreak. The morbidity rate in the vaccinated was 11.5% and 22.1% in the controls. Similar results were obtained in Stalingrad ($D_{85}$), but in Leningrad the results of the use of the dry vaccine of SMORODINTSEV were very poor ($A_{90}$). In 1952 ZHDANOV and FADEYEVA produced a vaccine consisting of vacuum dried allantoic fluid from eggs infected with virus passed serially in human embryo lung tissue. Virus was reconstituted with water and given in a dose of 1 ml. A general reaction of headache and malaise occurred in 9.3% of those vaccinated and catarrhal symptoms occurred in 8.3%, and 0.2% were off work. Antibody rises occurred in 71% and the vaccine reduced the morbidity rate 2.8 times ($F_8$).

In 1952—1953 a total of 155,626 persons were vaccinated, 67,631 by the Sokolov method, 52,244 with the powdered vaccine of SMORODINTSEV and

CHALKINA, and 21,381 with the vaccine of ZHDANOV and FADEYEVA. All produced some effect, reducing morbidity by 2.2'—5 times, the vaccine of ZHDANOV and FADEYEVA giving the best results ($Z_{27}$, $S_{16}$, $S_{151}$).

In 1953—1954 live dry influenza vaccine was given to over half a million people. In the A1 epidemic which followed morbidity was reduced on the average to about one third of that in control groups. Vaccination given a year before the outbreak was ineffective, and those vaccinated in two successive years were no more resistant than those vaccinated only once ($Z_{34}$).

ZHDANOV and others ($Z_{29}$) showed that A2 viruses isolated in 1957 produced a good antibody response in volunteers receiving intranasal live virus after 4—6 passages in eggs. Febrile reactions occurred in 4 of 30 cases. CHALKINA, BUROV and ILIN ($C_{25}$) inoculated 7683 persons with either live A2 or live B virus and obtained in 1959 a reduction in morbidity of 2.0 fold in the A2 outbreak of January-February and a 3.6 fold reduction in the B outbreak of March 1959. The A2 strains of 1959 were better immunizing agents than the 1957 strains and produced fewer reactions ($L_{59}$).

Live virus vaccines have not been much used outside Russia, but tests in England, America, France and Czechoslovakia ($A_{84}$, $M_{72}$, $P_{15}$, $S_{298}$) have confirmed that antibody rises are produced, though less than with killed vaccine inoculated subcutaneously.

### Live Virus Vaccines in Children

Live vaccine was found to produce severe reactions in young children, necessitating a hundred-fold reduction in dosage. ALEXANDROVA and SMORODINTSEV ($A_{45}$, $A_{46}$, $S_{195}$) attenuated the viruses further by passage in eggs at a temperature of 25—26°C. Cold-adapted viruses were produced in 17 passages and the antigenic properties were unchanged. These vaccines were completely harmless to 3—6 year old children, and caused little reaction in 1—2 year olds. The immunogenic activity of the strains was high and superior to the standard live vaccine used in adults. A 2—3 fold decrease in morbidity was obtained in 20,000 vaccinated children in the A2 epidemics of 1962 and 1965.

### 4. General Principles for Production of Live Virus Vaccines for Use in Man

Many different methods have been used by the Russian workers to produce satisfactory live vaccines, but the general principle would seem to be as follows. Recently isolated strains of virus are adapted to the allantoic sac of fertile eggs by 5—10 passages, when the strains gradually lose human pathogenicity. They are then re-adapted to man either by 5—8 passages in the respiratory tract of susceptible human volunteers ($S_{188}$) or by cultivation in human embryo lung tissue cultures. At each human passage the viruses are inoculated to eggs and the resulting allantoic fluids inoculated to human volunteers to determine the ability of the strain to survive in the nasopharynx and to provoke antibody production and resistance to a second inoculation given 15—20 days later. When a satisfactory level of human pathogenicity has been established the strain is inoculated to eggs for the production of bulk vaccine. It is best to employ a polyvalent vaccine composed of 2 or 3 strains of both A and B virus. The bulk vaccine is freeze-dried and stored at 4°C. It is reconstituted with water before use. Liquid vaccine should not be introduced as a very finely atomised aerosol (particles less than 10 µ) as such aerosols penetrate into the lower respiratory tract and may cause toxic reactions. The vaccine should not be used in children under 10 years. Vaccine should be given twice a year, in September and January.

### C. Summary of the Results of Influenza Vaccination

In spite of the intensive work which has been done on the prophylaxis of influenza the results of vaccination have been unimpressive. Under ideal conditions, when a vaccine containing the correct strain has been used shortly before an epidemic protection rates of the order of 60—80% may be expected with killed vaccine subcutaneously and 50—70% with live vaccine intranasally. But the appearance of strains of virus with new antigenic components may result in failure and it is very difficult to produce a large bulk of vaccine against a new strain in time to use it effectively. The duration of immunity after vaccination is probably less than 1 year so that annual vaccination would be needed to ensure protection, and as the vaccine may produce undesirable reactions vaccination in interepidemic years may result in more absence from work than would occur if a community was left unprotected. However, it is possible that the number of antigenically distinct virus strains may not be unlimited, and the development of non-toxic sub-unit vaccines may make it possible to produce a killed vaccine including all the major antigens, and producing fewer undesirable reactions. The use of polyvalent live virus vaccines is also worthy of further study, it is probable that really extensive coverage of a population can only be achieved by the use of live virus.

The low incidence of influenza in recent years has prevented any extensive study of the newer vaccines under field conditions, but there have been several studies of the antibody response to live vaccines ($XS_5$, $XM_1$, $XR_1$, $XS_6$, $XO_1$, $XB_1$, $XH_2$) and to sub-unit vaccine ($XB_2$).

# XXX. Chemotherapy in Influenza

Large numbers of chemical substances of many varieties have been empirically screened for possible antiviral activity against influenza viruses. COGGESHALL and MAIER ($C_{68}$) tested 67 substances, mostly sulphonamides and sulphones, for activity against A virus given intranasally to mice. ANDREWES, KING and VAN DEN ENDE ($A_{79}$) obtained no prophylactic or therapeutic effect against PR8 virus in mice with a range of 117 substances representative of 15 different classes of compounds. KRUEGER and others ($K_{101}$) tested 26 substances, CUTTING and others ($C_{99}$) tested 150, and WOOLLEY, BOND and PERRINE ($W_{66}$) tested 44 substances against influenza A in mice.

TAKEMOTO, ROBBINS and SMITH ($T_{12}$) tested 182 compounds against PR8 virus in eggs. Twelve compounds had some effect but none were effective in mice. TOYOSHINA, KANO and UEDA ($T_{57}$) tested 38 substances against A and A2 viruses in tissue culture. Many commercial laboratories have carried out screening on a much larger scale ($H_{219}$) with almost entirely negative results. General reviews of viral chemotherapy have been published by HURST and HALL ($H_{219}$) and WHIPPLE and VAN REYON ($W_{41}$).

Although most of the results of empirical screening have been negative, a considerable number of substances have been found to inhibit the growth of influenza viruses to a varying degree, but most of these substances are too toxic to be of value in the treatment of influenza.

## 1. Substances Interfering with Protein Synthesis

Many amino-acid analogues inhibit the reproduction of influenza viruses. Growth of PR8 virus in tissue culture is almost completely inhibited by 0.01 M dl methoxinine and some inhibition is produced by 0.0025 M. Inhibition is also produced by dl ethionine ($A_1$). Both effects are reversed by methionine. The growth of LEE virus in eggs is inhibited by canavanine and the effect is reversed by arginine ($P_{84}$). Canavanine had no effect on PR8 virus in eggs and had no chemotherapeutic effect against LEE virus in mice. A formaldehyde derivative of tyrosine reduces the growth of influenza A virus in eggs if added before or up to 4 hours after inoculation ($D_{54}$).

KUNDIN, ROBBINS and SMITH ($K_{112}$) using tissue cultures of chick embryo lung found that $\beta$ phenylserine inhibited both A and B viruses, inhibition being reversed by phenylalanine. Both viruses were inhibited by norleucine, the inhibition being reversed by methionine. Canavanine was found to have only slight inhibitory activity, but canavanine flavianate inhibited both A and B viruses. Flavianic acid was viricidal but canavanine flavianate was superior as a chemotherapeutic agent to either canavanine or flavianic acid. Allylglycine and S-ethylcysteine inhibited influenza A, the inhibition was reversed by cysteine or cystine. Reversal of the inhibitory effect of amino-acid analogous was accompanied by a parallel reversal of toxicity.

The growth of influenza and fowl plague virus is inhibited by p-fluorophenylalanine, inhibition being reversed by phenylalanine ($A_{16}$, $S_{62}$). BORECKY, RATHOVA and KOCISKOVA ($B_{108}$) found that if mice and hamsters were treated with ethionine and then were inoculated with unadapted virus the virus multiplied more intensively than in controls, but if an adapted line of virus was used ethionine suppressed virus multiplication.

Amino-acid imbalance may result in inhibition of virus growth. EATON and others ($E_{11}$) found that addition of $1-10$ mg/ml of arginine, lysine or ornithine to chorioallantoic membrane tissue cultures retarded the growth of influenza viruses, the effect being reversed by removal of the amino acids. In Krebs ascites cells the production of haemagglutinin and S antigen by PR8 virus is inhibited by excess of lysine, phenylalanine, tryptophane or leucine ($E_{19}$). It was suggested that the effects were due to synthesis of abnormal proteins or inefficient enzymes.

The yield of PR8 virus in fertile eggs is reduced by the inoculation of $0.1-2.0$ mg of synthetic lysine polypeptides, the degree of inhibition depending on the number of amino-acid residues, the shorter molecules being less effective. There was no effect in mice.

## 2. Substances Interfering with Nucleic Acid Synthesis

HANNOUN ($H_{29}$) showed that influenza virus synthesis in intact eggs is inhibited by thiouracil, an inhibitor of RNA synthesis. The inhibition was reversed by uracil. Amos and Vollmayer ($A_{57}$) found that thiouracil in a dose of 100—300 μg/ml suppressed multiplication in chorioallantoic membrane tissue cultures and in de-embryonated eggs. Smaller doses led to the production of incomplete virus.

Some 5-aryl pyrimidines depress the multiplication of influenza virus ($B_{133}$). Daily doses of 1.5 mg of 2 mercapto-4-amino-5(2'chlorophenyl)pyrimidine caused a reduction in mortality and lung disease in mice inoculated with PR8 virus.

Inhibitors of DNA synthesis such as fluoro-, bromo-, or iododeoxyuridine and cytosine arabinoside do not inhibit influenza virus reproduction ($N_7$), but agents such as mitomycin C and actinomycin D which inhibit DNA controlled RNA synthesis inhibit the growth of fowl plague and influenza viruses ($R_{67}$, $B_{14}$, $N_7$). Some inhibition of $A_1$ virus in tissue culture was obtained with 4-aminopterin ($H_3$).

The antiviral activity of some dyes such as proflavine ($S_{66}$) may be due to combination with nucleic acids. The growth of influenza viruses in intact eggs, de-embryonated eggs and tissue cultures is inhibited by ribonuclease ($B_{209}$, $L_{46}$), and Heymann and others ($H_{111}$) studying 66 acidic polymers found that 17 of them inhibited the growth of 100 $ID_{50}$ of PR8 virus in eggs. All 17 were ribonuclease inhibitors.

## 3. Inhibitors of Cell Metabolism

Ackermann ($A_2$) found that malonate which inhibits the citric acid cycle, depresses the oxygen uptake of chorioallantoic membrane cells *in vitro* and inhibits multiplication of PR8 virus. He also found that sub-lethal doses of sodium fluoracetate in mice increased the concentration of citrate in the lungs and inhibited the growth of A virus. Mogabgab and Horsfall ($M_{108}$) found that sodium fluoracetate caused only a slight delay in the growth of PR8 virus in mice or eggs. With LEE virus there was no effect in eggs but there was a significant delay in mice though the final yield of virus was not affected.

The aromatic diamidines pentamidine and stilbamidine which retard oxygen consumption in tissue cultures inhibit the growth of influenza A and B viruses ($E_9$, $E_{15}$).

Eaton, Adler and Perry ($E_5$) showed that the thyroxin analogue butyl-3,5-diiodo-4-hydroxybenzoate in doses of 2—8 μg/ml reduced the growth of PR8 virus in tissue culture and in de-embryonated eggs. The effect was not antagonized by thyroxin. The addition of 0.1—10 μg/ml of thyroxin together with 0.1% pyruvate and 0.02% glucose produced increased oxygen consumption and stimulated the growth of PR8 virus ($E_4$). Kilbourne ($K_{40-42}$) found that cortisone in doses of 0.1—1.0 μg inhibits the growth of influenza viruses in fertile eggs. The effect was temporary and the final yield of virus was unaffected.

The glucose anti metabolic 2-deoxy-D-glucose in doses of 12.5—25 mg in fertile eggs inoculated with a large dose of virus produces an 8—32 fold reduction in haemagglutinin titre at 8 hours, but final yield of virus was not affected ($K_{43}$). The inhibition was reversed by pyruvate but not by glucose.

Vorob'yeva ($V_{32}$) states that substances such as sodium fluoride, malic acid and dinitrophenol which interfere with the synthesis or activity of adenosine triphosphate inhibit virus multiplication. Eaton and Perry ($E_{14}$) found that 2,4 dinitrophenol in a dose of 10 μg/ml in tissue cultures reduced haemagglutinin production by PR8 virus. There was some inhibition in de-embryonated and intact eggs.

PENTTINEN ($P_{30}$) found that polyphloroglucinol phosphate which inhibits alkaline phosphatase and hyaluronidase reduced the yield of influenza virus in eggs, but the effect was probably due to direct union with virus particles with a decrease in the number of free units, as a reduction in haemagglutinin and infectivity titres occurred in vitro.

ROLLY ($R_{45}$) found that 4 doses of 0.2—2.0 mg of isonicotinic acid 3,3-di-(p-chlorophenyl)propylamide reduced virus growth and increased survival time in mice inoculated intranasally with 5000 $LD_{50}$ of PR8, Asian or LEE viruses.

### Benzimidazole Derivatives

Derivatives of benzimidazole have been extensively studied by TAMM and his associates ($T_{17-21}$, $T_{13}$, $T_{25}$). They have no direct effect on the virus and do not prevent union with the cell. Growth of virus is inhibited apparently as a result of some action on the synthetic mechanism of the cell, possibly on the synthesis of nucleic acids. Oxygen uptake by the cells is unaffected and the substances are less toxic than most other viral inhibitory agents. A concentration of 0.0025 M of 2,5-dimethylbenzimidazole inhibits multiplication of PR8 and LEE viruses in chorioallantoic membrane tissue cultures. Added before infection it increased the latent period by 80%, reduced the rate of increase, and reduced the final yield of virus by 99%. Added after inoculation it reduced the yield by an amount inversely proportional to the time of addition.

Substitution of alkyl radicals at various positions in the benzene or imidazole rings produced differences in inhibitory power, and 2,4,5,6,7-pentamethyl-benzimidazole, 5,6-diethylbenzimidazole, and 2-ethyl-5-methylbenzimidazole were highly active producing 75% inhibition of growth at a concentration of 0.0002 M. Chloro derivatives were even more effective, 5,6-dichloro-1-$\beta$-D-ribo-furanosylbenzimidazole (DRB) in a concentration of 0.000055 M prolonged the latent period by 100%, and it inhibited the growth of LEE virus in intact eggs and mice in doses not significantly toxic. Penetration of virus into the cells was not prevented. The amount of virus produced in chorioallantoic membranes eventually reached the same level as in controls, but there was a greatly reduced yield in the medium. The dichloro derivative was 92 times as effective as benzimidazole, and trichloro derivatives had 760 times the effect, inhibiting in a concentration of 1.6 µg/ml.

### 4. Substances Preventing Union of Virus with the Cell

Some mucoprotein inhibitors of virus haemagglutination will prevent or delay the multiplication of influenza viruses in fertile eggs if inoculated before or shortly after the virus. GREEN and WOOLLEY ($G_{75}$) found that inoculation of eggs to the allantoic sac with 25—50 mg/of apple pectin 30 minutes before inoculation of 100 $ID_{50}$ of virus prevented virus multiplication, and some protection was given with a virus dose of 1,000,000 $ID_{50}$. Some protection was given by apple pectin given 1 hour after the virus but not after 2 hours. DREYZIN and KARLINA ($D_{80}$) also obtained protection with apple pectin. Mucin from ox bile retards the growth of PR8 and LEE viruses in the allantoic sac ($T_1$), and a polysaccharide extracted from the edible seaweed Carrageenin protects chick embryos against the lethal action of influenza virus B ($G_{16}$). Kelp extracts have also been studied by KATHAN ($K_{20}$).

Francis-type inhibitors are, however, usually ineffective as they are rapidly destroyed by the virus enzyme ($L_{29}$). HOLLOS ($H_{156}$) coupled Francis inhibitor from human plasma with various aniline derivatives. The products combined irreversibly with virus and inhibited virus multiplication in tissue culture and produced some reduction in virus yield in intact eggs and in mice. The most active coupling materials were p-ethylaniline, diethyl-p-phenylenediamine, and o-methylaniline. SZANTO and others ($S_{345}$, $S_{346}$) prepared an antiserum against chorioallantoic membrane inhibitor and found that it was adsorbed by chorioallantoic membrane cells and would reduce the yield of virus by 99% by preventing adsorption of virus to the cells.

Orozomucoid from urine of nephrotic children has no antiviral action, but a polymer of molecular weight $10^6$ strongly inhibited the growth of a susceptible A2 strain in eggs and mice ($XL_2$).

ANGELA and GUILIANI ($A_{85}$) found that an analogue of sialic acid 2-phenyl-3-$\beta$-chloroethyl-4-keto-2356 tetrahydroxazine completely inhibited the growth of influenza virus in eggs if introduced at the time of inoculation but not after a delay of 6 hours. It had a low toxicity for eggs and mice. The effect was neutralized by previous mixture with RDE.

DREYZIN and SOKOLOVA ($D_{81}$) found that eucalyptus leaf extracts had some inhibiting effect in eggs, the effects apparently being due to tannins which were adsorbed by the cells and prevented entry of virus.

Destruction of cell receptors by V. cholerae filtrate or purified RDE prevents adsorption of virus and inhibits multiplication in eggs and mice ($S_{270-272}$, $C_1$) and in tissue cultures ($M_{52}$). Several N-substituted oxamic acids which inhibit neuraminidase prevent the growth of PR8 virus in chorioallantoic membrane tissue cultures but have no effect in intact eggs ($J_{18}$).

HANAN ($H_{26}$) found that $\alpha$-tocopheryl esters which inhibit the enzymic action of influenza virus on ovomucin if injected into the allantoic sac before virus protect against infection by small doses.

The $\gamma$-inhibitor of normal horse serum administered intranasally to mice protects against the lethal effect of a susceptible A2 strain given later ($C_{69}$, $R_{72}$, $L_{93}$). A 1% preparation of purified inhibitor prevented all deaths in mice inoculated with $10^3$ $LD_{50}$ of a mouse-adapted A2 strain, and in mice inoculated with $10^7$ $ID_{50}$ of an unadapted strain all virus multiplication was prevented. Intravenous inoculation of $\gamma$-inhibitor did not protect.

## 5. Substances Preventing Penetration of Virus into the Cell

ACKERMANN and MAASSAB ($A_4$, $A_{14}$, $A_{15}$) showed that $\alpha$-amino-p-methoxyphenylmethane sulphonic acid prevents the penetration of influenza virus into the cell and inhibits virus synthesis in tissue culture in a concentration of $5 \times 10^{-4}$ M. A dose of 8 mg prevented growth in the intact egg, the toxic dose being 30 mg. Some protection was given in mice.

GREEN and STAHMANN ($G_{74}$) found that $\alpha$-amino-isobutane sulphonic acid and $\alpha$-amino-$\beta$-phenylethane sulphonic acid inhibited the growth of influenza A virus. A peptide derivative of the latter had a decreased inhibiting effect, while an urea derivative was equally effective. They suggested that a requirement for inhibiting activity of α-aminosulphonic acid derivatives was a certain degree of basicity of the α-nitrogen atom.

FAUCONNIER ($F_{16}$) found that addition of 60—100 mg of ammonium citrate or succinate to the allantoic fluid greatly depressed the yield of influenza virus in eggs. The substances were not toxic and had no direct action on the virus. EATON and SCALA ($E_{18}$) observed that glutamine inhibits the growth of influenza virus in Krebs 2 ascites tumour cells. These cells rapidly liberate ammonia from glutamine, and ammonium ion in a dose of 0.5—1.3 mg per cent caused significant suppression of virus growth. Higher concentrations were required to inhibit growth in chorioallantoic membrane tissue. Addition of glutamine or ammonia after the virus had no effect. Ammonium chloride ($J_{10}$) and phosphate ($E_{10}$) produced similar effects in tissue cultures. Several primary and secondary amines especially n-propylamine and tertiary butylamine produced similar effects to ammonia ($J_{13}$), FLETCHER, HIRSCHFIELD and FORBES ($XF_4$) suggested that there was a common mode of action for ammonium ions and various amines. Ammonium chloride blocks the penetration of cells by virus. Virus is adsorbed by the cells but remains neutralizable by serum for as long as 3—75 hours as against less than 45 minutes in controls. Similar results were obtained with propylamine, diethylamine, triethylamine, 1,3 diaminopropane, 1,4 diaminobutane, 1,5 diamino-pentane, and 1-adamantanamine.

## Amantadine

The antiviral activity of 1-adamantanamine (Amantadine) was discovered by DAVIES and others ($D_{32}$, $D_{33}$). Introduced in a dose of 0.5 mg to the allantoic fluid or yolk sac it reduced haemagglutinin production by PR8 and WS viruses and reduced the mortality in eggs inoculated with A2 virus. Given intraperitone-ally in doses up to 40 mg/kilo it increased the survival rate in mice inoculated intranasally. The toxic dose was 233 mg/kilo intravenously or 1080 mg/kilo by mouth. SCHILD and SUTTON ($S_{45}$) tested the activity in monkey kidney and chick fibroblast tissue cultures, adding 2.5 μg/ml 4 hours before infection. It was found that PR8, NWS and FM1 viruses were relatively resistant, but A2 strains were completely inhibited. Strains of animal origin also showed variable resistance but A/Equi/Miami/63 was very sensitive. Intraperitoneal inoculation of 70 mg/kilo daily in mice for 5 days had a slight protective effect against A2 virus.

NEUMAYER, HAFF and HOFFMAN ($N_{15}$) found 25 μg/ml was the highest tole-rated dose in tissue culture, and this concentration inhibited A, A1, A2, and C viruses. Some reduction in haemagglutinin titre was obtained in eggs by in-oculation of 200—500 μg into the yolk sac 30 minutes before allantoic sac in-oculation of virus. Plaque formation by WSN on chick cell monolayers was suppressed by 1.6 μg/ml added before infection, but there was no inhibition by 40 μg/ml added after infection. Amantadine is not viricidal, has no action on neuraminidase and does not prevent adsorption to the host cell. It prevents penetration, virus adsorbed to the cell remaining on the cell surface in an infective state ($H_{154}$). COCHRAN and others ($C_{65}$) demonstrated inhibiting effect against influenza A, B, parainfluenza and rubella viruses, but noted that some strains of influenza B were resistant. In tissue culture influenza viruses grown in the presence of amantadine developed resistance, and strains resistant to amantadine were also resistant to aminophenylmethane sulphonic acid and *vice versa*, indicating a common mode of action of these substances. Other workers failed to demonstrate the development of resistance ($G_{89}$).

Certain derivatives of amantadine, notably α-methyl-1-adamantine methylamine hydrochloride have similar effects ($T_{62}$), and the basic amine 2-ethylaminoethyl-4-methylpiperazine-1-carboxylate is effective in monkey kidney tissue cultures against A, A1, A2, and some strains of B virus. It acts by preventing penetration, ($F_{77}$). The oral toxic dose in mice was 2500 mg/kilo. A dose of 1600 mg/kilo protected 82% of mice against intranasal inoculation but was slightly toxic. A dose of 800 mg/kilo protected 74% and 400 mg/kilo 42% ($L_{90}$). Amantadine was more effective against A2 virus but less effective against PR8. Mixtures of the two agents were more effective than either alone. Aminoadamantane and esculamine (8-dioxydiethylaminomethyl-4-methylesculetin hydrochloride) administered orally in doses of 60 mg/kilo 4 hourly for 2 days increased the survival rate and duration of life of mice infected with $1-5$ $ID_{50}$ of PR8 virus ($C_{38}$). Trials of amantadine in man are described below.

## 6. Effects of Antibiotics

QUILLIGAN and others ($Q_4$) showed that terramycin in doses approaching the toxic level suppresses multiplication of influenza virus in eggs if given before the virus. There was no action in mice. VINSON and WALSH ($V_{24}$) found partial protection against 500 $ID_{50}$ of PR8 virus but not against 1000 $ID_{50}$. The antibiotic had no direct action on the virus.

SHOPE ($S_{120-122}$) obtained an anti-viral substance Helenine from cultures of *Penicillium funiculosum*. It was active against SWINE virus in mice. It was later shown that Helenine acted by stimulating the production of interferon ($S_{123}$, $R_{73}$), which appeared in maximal titre $12-16$ hours after inoculation of Helenine. GINSBERG ($G_{27}$) found a substance Xerosin in culture filtrates of *Achromobacter xerosis* which limited the development of pneumonia in mice resulting from inoculation of large doses, $10^8$ $ID_{50}$, of influenza viruses. It had no effect on the formation of non-infective viruses. It appeared to suppress the secondary reactions leading to oedema, haemorrhage and cellular infiltration.

FURER and others ($F_{166}$) obtained two alcohol stable toxic substances from *Actinomyces violaceus* which inactivated A, A1 and A2 viruses *in vitro*.

LARIN and others ($L_{27}$) found the mould metabolite Gliotoxin, an indole derivative, to delay lung consolidation and increase survival time in mice inoculated with influenza virus, and to reduce the severity of infection in aerosol-infected puppies. Gliotoxin inhibits cytopathic effect of A2 virus in monkey kidney cells ($R_{29}$).

Some streptothricins have a significant inhibiting effect on influenza virus in eggs ($G_{19}$). Virothricin from *streptomyces lavendulae var. virothricinus* in a dose of 5.6 µg/ml caused a 50% reduction in haemagglutinin production in chorioallantoic membrane tissue cultures of PR8 virus, and a dose of 100 µg/ml reduced the haemagglutinin titre from 512 to less than 4 and the infectivity titre from $10^{8.5}-10^{4.6}$. The toxic dose to the cells was 1250—2500 µg/ml. There was little effect on survival time in mice.

The action of actinomycin D and mitomycin C has been described above.

## 7. Antiviral Effect of Dyes

Hoyle ($H_{189}$) examined the effect of a range of dyes used in vital staining on the growth of influenza virus in fertile eggs. Basic dyes of the triphenylmethane group were found to retard the intracellular growth of virus in doses near to the toxic dose. Crystal violet and dahlia violet were the most effective, the green dyes were more toxic and less effective. The dyes probably act by union with nucleoproteins. Fleisher ($F_{76}$) found that Janus green B and a safranin compound diethylsafraninazo-4-(1 phenyl-3-methylpyrazolon-M-sulphonamide) in doses approaching the toxic dose increased the survival rate of mice inoculated with 20 $LD_{50}$ of virus.

Various acridine dyes have some action on influenza virus multiplication. Green, Rasmussen and Smadel ($G_{73}$) found that 0.5 mg of nitroakridin 3582 protects against infection by 1—10 MID of LEE virus in eggs and some reduction in haemagglutinin titre occurred in eggs inoculated with 100 MID. Stabrine and proflavine inhibit the growth of influenza virus in eggs ($Z_{13}$). Atabrine has been used in man by Russian workers (see below).

Fauconnier and Chaigneau ($F_{19}$) obtained marked inhibition of the growth of influenza virus in eggs by thioparatoluidine and four other related dyes.

## 8. Miscellaneous Substances

The aromatic amidines reduce the growth of influenza virus in the allantoic sac ($M_{55}$).

Certain metallic ions, cobalt ($S_{58}$) and copper ($S_{137}$) reduce the growth of virus in eggs.

The thiosemicarbazone cutisone (thiosemicarbazone-p-isopropyl-benzaldehyde) inhibits the growth of influenza A virus in eggs ($Z_{13}$).

A dose of 1/32nd of the lethal dose of N ethylmaleimide inoculated to the yolk sac of eggs before allantoic infection completely inhibits haemagglutinin production ($D_{52}$).

Caprochlorone ($L_{105}$) in a dose of 2 mg inhibits the growth of influenza virus in de-embryonated eggs. Production of S antigen is less inhibited than haemagglutinin or infective virus. Some virus reproduction occurs even with doses close to the toxic dose ($L_{105}$). In mice there was a 10-fold reduction in yield of virus with virus doses of less than 10 $LD_{50}$. Mice infected with 3000 $MLD_{50}$ were saved by combined therapy with caprochlorone and human gamma globulin. Caprochlorone alone protected against only 3 $MLD_{50}$ and gamma globulin against 30 $MLD_{50}$ ($L_{101}$).

Clarke, Isaacs and Walker ($C_{58}$) tested 83 derivatives of 3:4 xylidine in tissue cultures. The compound N-(2 piperidinoethyl)-3:4 xylidine dihydrochloride had a high inhibiting activity with a low toxicity. It was ineffective in intact eggs.

Some derivatives of ω-aminoacetophenone reduce haemagglutinin titres in eggs ($B_{115}$), but have no action if inoculated into the yolk sac or in mice.

Mice infected with virus and treated with an aerosol of dodecyldimethyl-benzyl-ammonium survived longer than controls but all died ultimately ($T_{46}$).

MELANDER ($M_{74}$) described a substance N,N-anhydrobis-($\beta$ hydroxyethyl) biguanide hydrochloride (ABOB) which was said to exert a protective effect in mice inoculated intranasally with PR8 or LEE viruses. Some of the mice, however, were given atropine-like drugs and some were given ammonia inhalations to increase susceptibility! The author found this substance completely ineffective in eggs.

MAGRASSI, CAVALLINI and MASSARINI ($M_{28}$) found that 4 biphenyl-glyoxal bisulphite and $\alpha$-oxo-$\alpha$-(4 biphenylyl)-$\beta$ ethoxy-$\beta$ (4 carboxyphenyl) aminoethane protected mice against intranasal PR8. Given before the virus they protected against 100,000 $LD_{50}$ and were effective up to 24 hours after infection with 5 $LD_{50}$. The second compound was less toxic than the first, and the drugs were effective if given intranasally, subcutaneously or by mouth. A keto-aldehyde derivative of biphenyl, xenalamine, has been used in man ($M_{26}$). Drug-induced virus variants resistant to xenalamine could be produced in mice ($M_{27}$).

LINK, BLASKOVIC and RAUS ($L_{91}$, $L_{92}$) found that the infectivity of A1 virus for tissue cultures was greatly reduced in the presence of 2—4% urethane, and administration of 0.1 ml of 25% urethane to mice at the time of intranasal infection delayed the 50% mortality time from 4 to 6.8 days.

Addition of 0.03 mg/ml of 3,5 diiodo-4-hydroxybenzene sulphonamide reduces the yield of WS virus in chick embryo lung tissue cultures ($K_{113}$). There was no effect in eggs or mice.

Many cyclic acetals of mono and di-$\alpha$-alkoxy substituted derivatives of furan and pyran have anti-viral activity against influenza virus ($P_{41}$).

TOYOSHIMA, KANO and UEDA ($T_{57}$) testing compounds in chorioallantoic membrane tissue cultures against PR8 and Asian virus found a reduction in haemagglutinin production to be produced by 1-($\beta$ decylphenyl)-3-dimethyl-amino-1-propanolhydrochloride and by 6-propyl-3-mercapto-5-hydroxytriazine. A slight effect was observed with caprochlorone, but ABOB and xenalamine were ineffective.

TAKANO and others ($T_2$) found that intranasal instillation of yeast nucleic acid solution containing 20—60 mg/ml increased the resistance of mice to influenza virus B. Protection persisted for several days, but inoculation of the material by other routes than the intranasal was ineffective. The authors consider the effects to be partly due to stimulation of interferon production and partly to a non-specific anti-viral response of the pulmonary epithelium. Nucleic acid from herring sperm also gives some protection. Russian workers have used a material ccmoline derived from fish milt and obtained a protective effect in mice and eggs ($Y_{10}$, $D_{98}$).

## 9. Chemotherapy of Influenza in Man

Apart from the use of anti-bacterial agents to control secondary infection, a number of substances with anti-viral action have been tried in human influenza.

## a) Ecmoline

The material ecmoline obtained by YERMOLEVA from fish milt has been much used by Russian workers ($Y_{10}$, $R_{31}$, $K_{34}$, $L_{107}$, $S_{101}$). Instilled intranasally or given as an aerosol it shortens the duration of fever in influenza cases and reduces the incidence of complications. The preparation was unstandardized and gave variable results in different epidemics, being most successful in 1949 ($Z_{34}$). Some workers used it in combination with penicillin.

## b) Atabrine

Atabrine in a dosage of 0.3 g per day was found by Russian workers to reduce the febrile period and duration of influenza if given early in the disease. It was ineffective after the 3rd day ($S_{79}$, $E_{39}$, $B_{54}$). AHSBEL, STOLPETSKAYA and SOKOLOVA ($A_{39}$) used atabrine as an aerosol, in a dose of 0.1 g for inhalation. Atabrine by mouth or as an aerosol appears to have some value as a prophylactic during an epidemic ($A_{39}$, $R_{71}$, $G_3$). AHSBEL administered it in a dose of 0.1 g by aerosol at 7 to 9 day intervals in an industrial group. Only 1 of 108 workers became infected, while 14 cases occurred in a control group of 90 workers.

## c) Cutisone

The thiosemicarbazone cutisone introduced by ZEYTLENOK and VAGZDANOVA ($Z_{13}$) was used in the treatment of influenza by FILIPPOVA-NUTRIKINA ($F_{65}$) in a dose of 0.015 g per day. It was said to reduce the duration of fever and toxaemia but was inferior to the use of antiserum.

## d) Xenalamine

MAGRASSI and others ($M_{26}$) treated 228 patients with A2 or B influenza with xenalamine in a dosage of 2—2.5 g per day for 4—7 days. There were 206 controls. Treatment produced some reduction in the febrile period and duration of illness.

## e) ABOB, Flumidin, Virugon

The biguanide ABOB introduced by MELANDER ($M_{74}$) was claimed by Swedish workers to produce a considerable reduction in the severity of cases of influenza A2 and B infections, reducing the fever and duration of disease ($M_{74}$, $S_{147}$, $S_{148}$, $S_{262}$). The commercially manufactured products also included small amounts of atropine or scopolamine and it is possible that these drugs were responsible for the beneficial effects.

## f) Amantadine and Related Substances

STANLEY, MULDOON, AKERS and JACKSON ($S_{257}$) tested the value of amantadine in experimental Asian influenza in human volunteers. Various schedules of treatment were used. Subjects with an initial low antibody titre and given treatment starting 20 hours before infection showed beneficial results, the rate of infection being reduced to the level found in controls with an initial high antibody. Administration of the drug after infection was of no value. WOOD ($W_{59}$) obtained beneficial results both in human volunteers and in a natural influenza epidemic. About 90% of the orally administered amantadine was excreted in the urine in man and there were no toxic effects, but in mice, rats, dogs, and monkeys the rate of excretion was less and toxic effects occurred. TYRRELL, BYNOE and HOORN ($T_{75}$) carried out a double-blind trial on 33 volunteers infected with A2 virus. Amantadine in a dose of 400 mg per day showed no anti-viral effect. HALONEN ($H_{11}$) studied the effect of amantadine in 47 military trainees during an A2 epidemic. One of the treated cases developed antibody rise to A2 while 5 such cases occurred in the control group. Influenza B infections also occurred and were the same in both groups. Amantadine is of value in prophylaxis of equine influenza ($XB_3$).

### g) Interferon and Interferon Stimulators

The species specificity of interferon, and the relative resistance of influenza viruses to its action, make the direct use of interferon impracticable for chemotherapy of influenza in man. It might require the whole of the interferon produced in a human amnion to protect a single individual against influenza. Stimulation of endogenous interferon production might be more promising. ZALMANSON and others ($Z_{10}$) have recently used aerosol inhalations of UV inactivated SWINE virus as a stimulator of interferon production during an outbreak of influenza and suggest that the method might be of value in the urgent prophylaxis of influenza during an epidemic.

## 10. General Considerations of Influenza Chemotherapy

In spite of an extensive search for chemotherapeutic agents active against influenza viruses no substance has been found which has a prophylactic or therapeutic value equal to that of immune serum.

Many substances interfering with cell metabolism or with synthesis of protein or nucleic acid exert an inhibiting effect on the growth of virus under experimental conditions, but in all cases the effective dose is too close to the toxic dose to permit their use in man. The intracellular synthesis of influenza virus components is so closely integrated with normal synthetic processes that there seems no immediate prospect of the discovery of any virus-specific inhibitor of intracellular synthesis.

Non-specific inhibitors of haemagglutination may delay or prevent infection of cells but require to be introduced into the respiratory tract in large amounts in order to compete with virus.

Substances preventing or delaying penetration of virus into the cell appear more promising, and the basic amines, some of which can be administered orally or parenterally and are relatively non-toxic, are worthy of further study. BAUER ($B_{23}$) suggested a possible correlation between virus particle size and the minimal effective dose of potentially active chemotherapeutic agents, and according to his concept it should be possible to find substances more active against influenza virus than amantadine.

The examination of substances for chemotherapeutic activity presents considerable difficulty. Screening by tissue culture methods readily reveals antiviral activity, but cells in tissue culture may be relatively resistant to toxic effects which become apparent when intact animals are used. Screening in fertile eggs is simple and convenient and is especially useful in determining the stage of the growth cycle at which a substance is active. Screening in mice is a procedure which is more closely similar to the conditions of natural infection in man. But the penetration of virus into the cells is affected both by the properties of the virus and of the host cell, so that in the end it will be necessary to test the efficacy of substances acting on the penetration stage in human volunteers and with a range of virus strains. It is quite possible that substances without effect in non-human systems might be effective in man.

Influenza is a disease with a high infectivity but a low mortality and it has been usual to test the value of chemotherapeutic agents by determining their ability to prevent infection. What is really required is an agent which prevents death, and it might be more valuable to search for substances neutralizing the toxic properties of the virus. This could be easily done in experimental animals but it would be very difficult to demonstrate effects in man because of the low death rate in human influenza.

# XXXI. Influenza C

In 1947 TAYLOR ($T_{39}$) isolated a strain of influenza virus, 1233, which appeared to be antigenically unique. In 1950 FRANCIS, QUILLIGAN and MINUSE ($F_{115}$) isolated a strain JJ which proved to be identical to 1233, and they suggested that these strains should be designated influenza virus C. The properties of the C virus differ from those of the A and B viruses to a greater extent than the A and B viruses differ from each other, so that it is necessary to consider the C virus separately.

## 1. Distribution of Influenza C

The C virus appears to be a common pathogen of man. FRANCIS, QUILLIGAN and MINUSE ($F_{115}$, $M_{97}$) found that antibody to virus C was widely distributed in adults, and a rising titre was observed in a few cases occurring during the 1950 A1 epidemic and also in paired sera collected in 1947. Antibody was also detected in sera collected as far back as 1936. TAYLOR ($T_{40}$) found haemagglutinin-inhibiting antibody present in a titre of 1:64 or more in 62% of 204 adults in 1951. STYK ($S_{290}$) isolated 4 strains in an epidemic in a childrens home in Bratislava in 1952. DE MEIO and others ($D_{39}$) isolated 31 strains in Navy recruits in 1953—1954. SOLOV'YEV ($S_{231}$) examining 900 paired sera in an A1 epidemic in 1956 found 6.6% to show an antibody rise to influenza C. SETH and KALRA ($S_{83}$) isolated the virus in an outbreak in Delhi in 1963.

Infection with influenza virus C has often been encountered at the same time as epidemics of infection by A1 virus, and MINUSE and DAVENPORT ($M_{95}$) record the simultaneous recovery of both A1 and C viruses from the same patient.

## 2. Isolation, Cultivation and Pathogenicity

Virus C is readily isolated by inoculation of eggs to the amniotic sac and on passage attains high titres in the amniotic fluid, but it grows poorly or not at all in the allantoic sac ($T_{40}$, $M_{97}$). No obvious pathological effects are produced in eggs and the embryos develop normally ($B_{141}$). MOGABGAB ($M_{104}$) isolated a strain from man in primary rhesus monkey kidney cultures, but found that most strains of virus C could not be adapted to mammalian tissue culture. BUD-DINGH ($B_{141}$) found that combined infection by virus C and *Haemophilus influenzae* type B in the amniotic cavity increased the severity of the lesions produced by the haemophilic bacillus. KORYCH and FRANKOVA ($K_{94}$) found that large doses of influenza C produced a cytopathic effect in primary pig kidney cells, but the virus did not multiply and could not be maintained on passage. Intranasal inoculation of ferrets, mice, hamsters and monkeys produced no signs of illness but antibody was produced ($T_{40}$, $M_{97}$).

MINUSE, QUILLIGAN and FRANCIS carried the virus through 34 passages in hamsters but the virus could not be adapted to mice. TAYLOR found that mice could not be infected either intranasally or intracerebrally.

In man infection with C virus demonstrable by antibody rise may occur without obvious signs of illness ($D_{39}$), and the disease in general is very mild. In children in addition to the usual symptoms of influenza STYK ($S_{296}$) noted the occurrence of conjunctivitis. QUILLIGAN, MINUSE and FRANCIS ($Q_6$) inoculated 6 volunteers intranasally with $10^{2.5}$ $ID_{50}$ of JJ amniotic fluid. No illness developed but virus was detected in garglings up to 4 days and 4 cases developed an antibody rise. The operator who administered the atomized spray developed a mild illness in 7 days, JJ virus was isolated on the 2nd day and an antibody rise occurred.

### 3. Serology of Virus C

All strains of virus C appear to be serologically identical. The complement fixing antigen is distinct from those of virus A and B. No relation is found by neutralization or haemagglutination-inhibition tests between virus C and A, A1, A2, B or other viruses ($F_{115}$, $T_{40}$, $M_{97}$).

Vaccination of humans with A, A1, or B vaccines produces no antibody rise to virus C and *vice versa* ($T_{40}$, $M_{97}$). Inoculation of hamsters previously infected with A, A1 or B virus with virus C does not produce a secondary antibody rise ($T_{40}$).

Human convalescents from C virus infection show antibody titres ranging from 1:8 to 1:256 by complement fixation, and from 1:64 to 1:4096 by haemagglutination-inhibition tests.

### 4. Haemagglutination by Virus C

The interaction of virus C with red blood cells differentiates the virus sharply from virus A and B. Virus C does not agglutinate guinea pig red cells, and elutes so rapidly from fowl and human cells that haemagglutination tests have to be done in the cold. At 6°C the virus agglutinates fowl, hamster, human and mouse cells, but does not agglutinate guinea pig, monkey, ferret or sheep cells ($M_{97}$). Only mouse cells are strongly agglutinated at 24°C, but some agglutination of hamster cells occurs.

Adsorption of virus by red cells is very poor, even heated virus is not readily adsorbed. Adsorption with 10% red cells in the cold may produce no measurable reduction in haemagglutination titre, though the titre falls off on serial adsorption. HIRST ($H_{140}$) showed that treatment of red cells with virus C does not destroy the receptors for other influenza viruses or for mumps and Newcastle disease viruses, and treatment of cells with influenza A, B, mumps or NDV virus does not remove receptors for virus C. Egg white contains an inhibitor of haemagglutination by virus C and some destruction of this inhibitor is produced by Newcastle disease virus. WHITE ($W_{47}$) states that strain 1233 falls between WSE and LEE in the human red cell receptor gradient.

Virus C is unaffected by most normal serum inhibitors, but STYK ($S_{296}$) found that normal rat sera contain a powerful inhibitor of haemagglutination

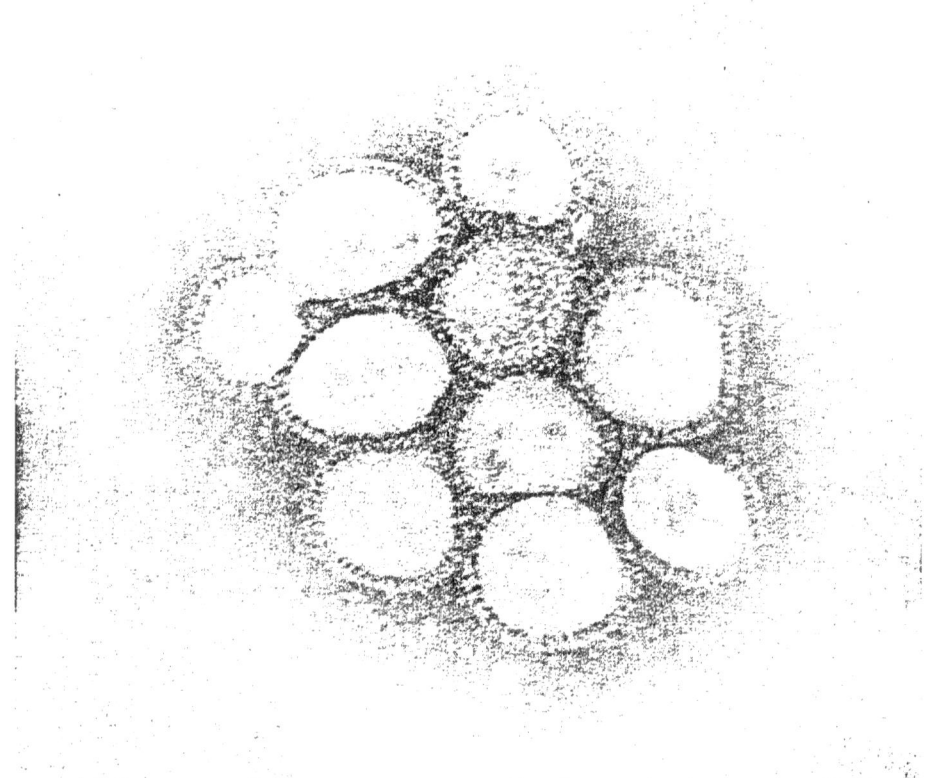

Figs. 56 and 57. Morphology of influenza virus C.
Fig. 56. Spherical particles of influenza virus C. Photograph by Dr. I. ARCHETTI.

by virus C, only active with fowl cells. The inhibitor had no action on A, A1, B or SWINE viruses but slightly inhibited mumps and NDV viruses ($S_{297}$). The C inhibitor is a thermostable glycoprotein which is not destroyed by cholera filtrate, RDE, periodate or $CO_2$, but is destroyed by trypsin ($H_{19}$, $S_{303}$).

## 5. Morphology of Influenza Virus C

Electron microscopic examination of amniotic fluids infected with virus C reveals the presence of spherical and filamentous particles of size and general appearance similar to those of virus A and B. Negative staining methods, however, show a striking difference in the surface structure. The surface of virus C shows projections which appear to arise from the corners of a hexagonal lattice structure resembling wire netting (Figs. 56, 57). In the filamentous particles the lattice is uniformly hexagonal, but in the spherical particles a combination of hexagonal and pentagonal lattice is seen ($W_{22}$, $A_{100}$). The appearance is quite different from that seen with virus A in which the surface projections are more randomly distributed. It is, however, to be noted that the virus C preparations were made with virus grown in the amniotic sac, while electron microscope pictures of virus A have been made with allantoic fluid virus. It is therefore possible that the

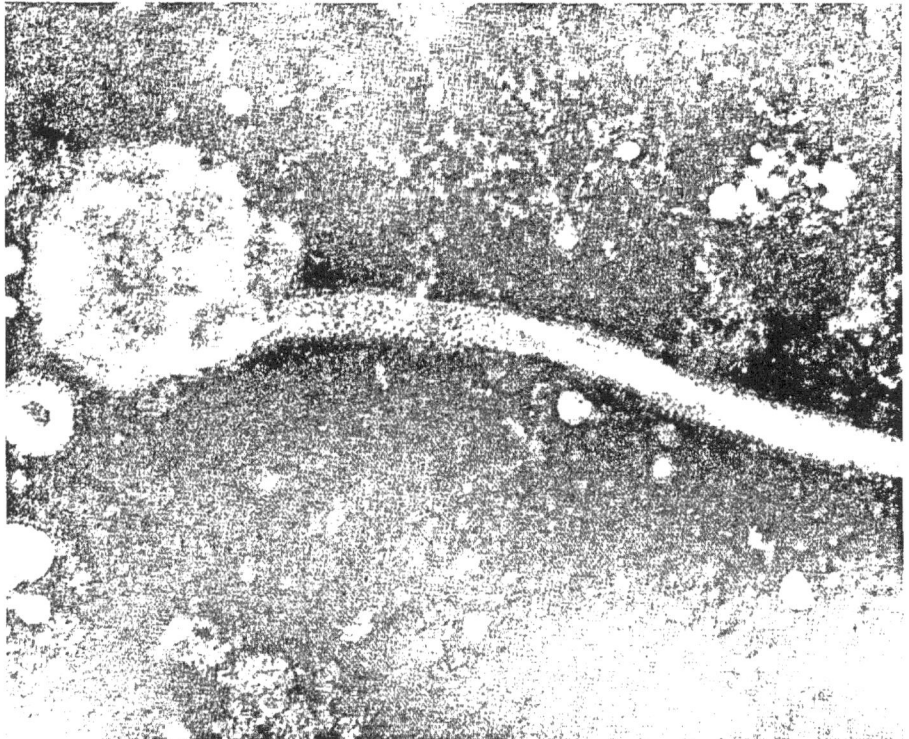

Fig. 57. Filamentous particles of virus C. Photograph by Dr. A. P. WATERSON.

differences are produced by growth in a different host cell and are not peculiar to virus C. Further work is necessary to determine this point.

At present nothing is known of the internal component of virus C.

## XXXII. Animal Influenza

Only the A viruses appear to produce natural infection in animals. Experimentally some animals can be infected by B viruses, and with more difficulty by C virus.

### A. Experimental Influenza in Animals

Many experiments have been made on the transmission of influenza to animals. Results have varied with different workers and with different strains of virus. Young animals tend to be more susceptible than older ones.

In ferrets a typical influenza-like illness is readily produced by intranasal inoculation and is transmissible from animal to animal by contact. Contact transmission has also been noted in hedgehogs ($S_{284}$), which are highly susceptible.

Mice are very susceptible to infection and serial passage results in rapid adaptation, but contact transmission in mice is infrequent ($K_{100}$), probably because mice do not sneeze.

Hamsters ($T_{42}$, $L_{122}$, $F_{116}$, $S_{133}$) are somewhat less susceptible than mice.

Results with rats have been very variable ($A_{38}$, $S_{285}$, $C_{23}$, $H_{50}$, $S_{237}$). Contact transmission has been recorded ($C_{26}$) but in general the virus is less pathogenic to rats than mice.

Various Tunisian rodents especially gerbilles are susceptible to infection ($D_{97}$), as are various squirrels ($T_{64}$, $T_{26}$), Chinese mink ($T_{26}$), lemmings and field voles ($S_{127}$). Susliks ($V_{14}$) are less susceptible and infection is not readily transmitted in series.

Monkeys have been inadequately studied. Macacus Rhesus ($S_{22}$), Macacus mulatta ($S_{23}$) and cercopithecus monkeys ($P_8$) are susceptible, but cynomolgus monkeys are relatively resistant ($B_{161}$). Guinea pigs do not develop any illness but an antibody rise may occur after infection ($H_{230}$, $B_{26}$, $S_{285}$). Rabbits appear to be totally resistant ($F_{104}$, $H_{230}$). Chipmunks and Chekian squirrels ($T_{26}$) are resistant. Adult fowls ($T_{26}$, $E_{37}$) and other birds are resistant to human viruses.

Among domestic animals dogs are highly susceptible ($T_{48}$, $A_{37}$). Sheep respond with an antibody rise without signs of infection ($M_{65}$). Influenza infection of cats has been recorded by Lozovaya ($L_{122}$). There have been unpublished reports of infection of calves.

## B. Natural Infection with Influenza Viruses in Animals

A search for influenza viruses in animals was initiated by the World Health Organization in view of the possible existence of an animal reservoir of influenza ($K_{15}$). Hungarian workers ($R_{50}$) isolated an A 2 virus in sheep affected with lamb pneumonia, and there have been unpublished reports of the isolation of a PR 8-like virus from sheep and an S 15-like virus from cattle. In view of the frequency with which laboratory contamination by influenza viruses has been encountered great caution is necessary in the interpretation of these results. The demonstration of antibody in animal sera presents considerable difficulties, the presence of non-specific haemagglutinin inhibitors may lead to error, and non-specific results may also occur in complement fixation tests. Apart from swine, horses, and birds, there is at present no certain evidence of the natural occurrence of influenza virus infection in animals.

## 1. Swine Influenza

Swine influenza first appeared, apparently as a new disease, at the time of the 1918 pandemic of human influenza. It was first recognized by Koen ($K_{76}$) in Iowa, who considered that the disease arose from man. Millions of swine were affected in late 1918, and since that time the disease has occurred annually in the autumn and early winter in the U.S.A., outbreaks being usually preceded by bad weather or a change in husbandry. The onset occurs simultaneously in most of the animals, arising at the same time in different foci over a wide area. The animals develop fever up to 106.5°F, with cough and anorexia. The infection rate approaches 100% in animals under 1 year old. The death rate is 4% or less. Swine influenza outbreaks are common in Europe, but the disease is rare in England or Japan, and in recent years has been milder in America.

The aetiology of the disease was elucidated by SHOPE in 1931 ($S_{107}$, $L_{69}$, $S_{108}$). SHOPE found that the disease could be transmitted by filtrates of infective material which produced a mild form of the disease. If the filtrate was combined with *Haemophilus influenzae suis* isolated from the respiratory tract of swine the resulting disease was more severe, resembling the natural illness. The presence of *Haemophilus influenzae suis* undoubtedly increases the severity of swine influenza in swine, and a synergistic effect may also be seen in eggs, combinations of virus and organisms having a more lethal effect on the chick embryo than either alone ($B_2$). Human strains of *Haemophilus influenzae* do not increase the severity of swine influenza and are unable to establish themselves in the respiratory tract of swine ($M_{132}$). However, there is no doubt that swine influenza is primarily a virus disease, and with sufficiently large doses of virus a severe disease may result in the absence of *Haemophilus influenzae suis*.

Following the transmission of human influenza to ferrets and mice it was found that the swine virus would infect these animals and was more readily adapted to mice than the human virus ($S_{109}$). Many workers demonstrated the presence of neutralizing antibody to the swine virus in many normal human sera, and HOYLE and FAIRBROTHER ($H_{203}$) showed that the human and swine viruses possessed a common S antigen. The swine virus therefore is a strain of virus A, and studies of the age incidence of antibody to the swine virus in human sera have indicated that it may have been responsible for the 1918 pandemic of human influenza (see Chapter 28).

### a) The Swine Influenza Cycle

Certain features of the epidemiology of swine influenza, the regular annual occurrence and simultaneous appearance of the disease in widely separated herds, were difficult to explain on the basis of contagion from animal to animal.

SHOPE in a long series of investigations described a remarkable sequence of events in the development of swine influenza ($S_{114}$, $S_{107}$, $L_{69}$, $S_{108}$, $S_{115-119}$, $S_{124}$). The virus appears to survive in the nine month interepidemic period in an intermediate host, the swine lungworm. Lungworms present in swine suffering from swine influenza become infected with virus and a latent or masked infection occurs in the lungworm ova which escape from the animal and are ingested by earthworms. The eggs pass through a developmental cycle in the earthworm and re-enter swine when the earthworms are eaten. Migrating lungworm larvae set up an occult infection in the lungs of the swine. Swine influenza then develops as a result of some kind of non-specific stimulation, usually the occurrence of bad weather. The virus can apparently persist for 32 months in lungworm larvae in the earthworm. SHOPE fed embryonated lungworm larvae from pigs infected with swine influenza to earthworms. Numerous third-stage lungworm larvae were found in the hearts and glands 2—12 months later. The earthworms were then minced and fed to swine. After 1 month the swine were exposed to "raw" weather. Of 25 such swine 4 developed clinical swine influenza and 6 developed serological evidence of infection ($S_{124}$). Clinical disease could also be stimulated in swine fed with lungworm infected earthworms by intramuscular injection of *Haemophilus influenzae suis* ($S_{114}$). Exposure of such swine to a mixture of PR 8 virus and *Haemophilus influenzae suis* resulted in clinical disease from which only the swine virus was isolated, though an antibody response to both human and swine virus occurred ($S_{119}$).

SHOPE's results were confirmed by SEN and others ($S_{80}$) who found that virus in the lungworms is present in a masked non-infective form and a provocative stimulus is needed to initiate infection in swine. Multiple injection of ascaris extract or migration of ascaris larvae furnished the needed provocation, and in pathogen-free colostrum-deprived pigs the virus could be elicited in the spring and summer as well as in the autumn. PETERSON, DAVENPORT and FRANCIS ($P_{54}$) found that lungworms contain receptor substances capable of adsorbing influenza virus. They were unable to demonstrate any multiplication of the virus in lungworms or in earthworms, and no virus receptor substance was found in earthworms.

### b) Infection of Swine by Human Influenza Viruses

SHOPE and FRANCIS ($S_{125}$, $S_{111}$) showed that swine could be infected experimentally with PR 8 virus. SHOPE found that of 8 swine inoculated intranasally with human virus + *Haemophilus influenzae suis* 6 developed mild swine influenza. Human virus alone produced a mild "filtrate disease" in 9 swine. Swine previously infected with swine virus were immune to the human virus, but previous infection with the human virus did not protect against swine virus. KONO and others ($K_{86}$) infected swine intranasally with FM 1 virus and were able to pass the infection through 5 generations. The growth of young swine was reduced and 4 died after 5—10 weeks. Virus was isolated from the lungs of the fatal cases.

GULRAJANI ($G_{90}$) found that a recently isolated strain of human influenza A 1 produced a mild disease in pigs. Virus was recovered from a pig killed at 7 days, but not from one killed on the 28th day. Mouse-adapted swine virus could be detected in nasal washings and lung suspension of pigs only up to 1 week after inoculation. WALLACE and KISSLING ($W_{13}$) infected 6—8 week old pigs with A 2 virus. No clinical disease developed but antibodies were produced.

Hungarian workers ($R_{50}$) reported the isolation of PR 8-type viruses from 5 herds of swine in outbreaks of swine influenza in 1959, and an A 2 strain in a sixth outbreak. In one herd from which a PR 8-like virus was isolated 60% of the animals showed high levels of anti-PR 8 antibody. No PR 8 virus had been isolated from man in Hungary since 1941 and it was suggested that influenza virus strains prevalent in previous epidemics of human influenza might persist in swine. There have also been reports of the isolation of A 2 virus from pigs in China and Spain ($K_{14}$), and of a WS-like virus in Russia. Some of these results may have been examples of laboratory contamination.

Swine in contact with human cases of influenza in 1936—1937 developed antibodies neutralizing the human virus but not the swine virus ($S_{112}$, $S_{96}$).

Japanese workers ($A_{40}$) collected sera from pigs before and during the human

A2 epidemic of 1957—1958. The sera contained non-specific inhibitors of A2 haemagglutinin removed by periodate treatment. No antibody was found in swine sera collected before the epidemic or in 427 sera collected during the epidemic.

Antibody to swine virus was found in 9% of the sera. It was concluded that swine played no part in the epidemiology of Asian influenza in Japan.

### c) Infection of Humans by Swine Virus

Swine-type antibodies are said to be more frequent in the sera of humans associated with swine than in the general population ($B_{64}$). A woman in Czechoslovakia is said to have developed infection with swine virus as a result of laboratory exposure, and the disease spread to other members of the household ($K_{62}$).

### d) Antigenic Relations of the Swine Influenza Virus

The S antigen of swine virus is similar to that of the human A viruses. SHOPE found that strains of swine virus isolated in different epizootics appeared serologically identical ($S_{113}$). SHAW, KENNEY and STOKES ($S_{96}$) found that convalescent ferret sera from animals infected with human 1937 viruses protected mice against the homologous strain, PR8 and WS but not against SWINE. GOMPELS ($G_{37}$) found 4 American swine strains to be closely related and not related to the human strains WS, PR8 and FM1. As swine are not usually allowed to live long enough to develop an immune population there is little opportunity for the development of variants as a result of exposure to antibody. However, variants do occur.

Recently isolated American strains differ from the original S15, minor antigens present in the original strain having become dominant. A strain "Oti" isolated in Korea in 1939 resembled human A and A1 strains more closely than the original swine strains ($F_{114}$). Sera of children infected with human influenza contained antibody neutralizing Oti virus in mice and inhibiting its haemagglutination. Ferrets inoculated with Oti developed neutralizing antibody to Oti, A and A1 viruses but not to the swine strain 1976. Sera of ferrets inoculated with A or A1 virus did not neutralize Oti.

Strains of virus isolated from swine in Britain in 1938 and 1941 showed little relation to S15 and appeared to be intermediate between later American swine strains and human A strains ($H_{214}$, $G_{37}$).

NELSON and LEWIS ($N_{13}$) have postulated a relation between SWINE and A2 virus. They found that in human sera SWINE antibodies are detected more frequently in sera containing A2 antibody than in sera not containing it.

### e) Diagnosis of Swine Influenza by the Use of Fluorescent Antibody

BLASKOVIC and others ($B_{84}$) inoculated 8 week old pigs with a large dose of the swine strain Iowa 15. Clinical symptoms developed 2 days after inoculation and from the 3rd day viral antigen could be demonstrated by fluorescent antibody in sections of the turbinates and in the lungs. Antigen appeared to be mainly concentrated in bronchial and alveolar cells. All the animals were suffering from the endemic pneumonia of pigs and the influenzal antigen was particularly evident in areas of pre-existing lung disease.

## 2. Equine Influenza

Outbreaks of influenza-like disease in horses have been described frequently in past centuries, but the first scientific evidence of infection of horses by influenza viruses was obtained in Sweden in 1955 when HELLER, ESPMARK and VIRIDEN ($H_{60}$) found that the sera of 21 convalescent horses gave complement fixation titres of $1:8—1:64$ against the S antigen of virus A. There was no reaction with virus B. A rising titre was found in 4 of 5 cases where paired sera were available. No V antibody could be detected by haemagglutination-inhibition tests against A, A1, or SWINE virus. No virus was isolated in Sweden.

### a) The Equi 1 Virus

In 1956 there was a large outbreak in Czechoslovakia and eastern Europe, and SOVINOVA and others ($S_{247}$) isolated a new strain of virus A designated A/Equi/1/Prague/56 which was demonstrated by serological tests to be the cause of the disease. Antibodies to the equine virus were subsequently found in the blood of horses in many other countries. No equine influenza occurred in the U.S.A. in 1955—1957, but serological studies showed that the Equi 1 strain had been present in horse populations for some time ($D_{58}$), antibody was present in over 90% of 4 year old horses which had raced for 2 or more years ($D_{63}$).

A strain of Equi 1 virus was isolated in Canada in 1960 ($D_{57}$) and in May 1963 a strain A/Equi/1/Lexington/1/63 was isolated from a horse epizootic in Kentucky ($XB_3$). The disease only occurred in young horses, the older ones had antibody and were resistant. Virus was also isolated in Pennsylvania ($L_{73}$).

In 1962 antibody to Equi 1 virus was present in a small percentage of horses in Northern Ireland ($M_{71}$). In the autumn of 1953 a widespread outbreak occurred in horses in England, "Newmarket cough". Similar outbreaks had occurred in 1935, 1947, 1955 and 1959. BEVERIDGE, MAHAFFEY and ROSE ($B_{66}$) isolated virus with some difficulty, several passages in the amniotic cavity of fertile eggs was necessary before satisfactory haemagglutination occurred. The virus, A/Equi/1/Cambridge/63 was closely similar to A/Equi/1/Prague/56. Antibodies against the virus were found in horse sera collected as far back as 1948, and were present also in the sera of two zebras collected in 1960 at the Whipsnade zoo. The disease in horses was very infectious, spreading to most of the horses in an affected stable. Coughing was the main and sometimes the only symptom of the disease, and usually recovery occurred within 2 weeks. Some horses developed fever and a watery nasal discharge. Pulmonary involvement and high fever only occurred if the horses were kept in training or were raced.

### b) The Equi 2 Virus

Early in 1963 cases of respiratory disease occurred in Florida and later spread to cause a severe epizootic all over the U.S.A. and Canada. The disease first appeared in Miami in January 1963 and a new strain of virus, A/Equi/2/Miami/63 was isolated by WADDELL, TEIGLAND and SIGEL ($W_1$). Many other isolations were made in the next six months ($K_{70}$, $S_{59}$, $M_{67}$). The virus contained the S antigen of virus A but was not related by haemagglutination-inhibition tests to other human and animal strains of virus A.

The disease was extremely infectious and attacked horses of all ages, in one outbreak 82% of 1200 horses were attacked. There was nasal catarrh, coughing and fever of $102-106°$F. The illness lasted $2-7$ days and there were no deaths ($M_{66}$).

It was a first thought that the disease had been imported from Argentina as epidemics of laryngotracheitis had occurred in horses near Buenos Aires in 1962 and 1963. Blood samples of Argentine horses, however, showed no antibody to either Equi 1 or Equi 2 viruses. Equi 2 virus was first isolated in South America six months after its appearance in the U.S.A., and in September 1963 an outbreak occurred in Uruguay from which a strain of A/Equi/2 was isolated ($S_{245}$). The disease then spread northward to Brazil.

In 1965 severe outbreaks of Equi 2 influenza occurred in England ($R_{55}$) and Switzerland ($P_1$) and later in France and Germany. No antibody to Equi 2 virus was found in horses before 1965 but infection with Equi 2 virus in 1965 stimulated a rise in titre of antibodies to Equi 1 ($R_{56}$).

### c) Serological Relations of the Equine Viruses

In 1942 JONES and MAURER ($J_{20}$) found that the sera of horses collected in a severe outbreak of equine influenza in $1939-1940$ contained no neutralizing antibody for S15, MEL, WS, PR8 or a strain of B virus.

In recent studies no haemagglutinating-inhibiting antibody active against SWINE or human A and A1 viruses has been found in horse sera ($R_{51}$). Serological evidence of the presence of A2 antibody in the sera of horses is uncertain because of the presence of non-specific inhibitor of A2 virus in horse serum. Dutch workers ($V_{20}$) found low titres of antibody to A2 in 21 of 79 horse sera collected at the time of the 1957 epidemic of Asian influenza. Using the specific complement fixation test LIEF and COHEN ($L_{73}$) found no evidence of antibody to human viruses in the sera of normal or convalescent horses.

The Equi 1 and Equi 2 viruses appear to share some minor antigens. These are not detectable by ordinary cross HI tests, but animals infected with Equi 2 virus develop some rise in titre to Equi 1 virus if they have any residual antibody to that virus ($M_{96}$, $R_{56}$). LIEF and COHEN detected a minor V antigen component shared by Equi 1 viruses and fowl plague virus ($L_{73}$).

CAMERON and others ($XC_1$) produced experimental infections of Chincoteague ponies with A/Equi/2 virus, of 24 animals inoculated 15 developed fever, virus was isolated from 22, and all 24 developed antibody rise to A Equi 2. There was no antibody rise to A Equi 1 of which virus the animals had had no previous experience.

VOTH and FELDMAN ($V_{34}$) and MINUSE and others ($M_{96}$) found that antibody to the Equi 2 virus is present in the sera of $30-50\%$ of humans over the age of 70 in 1958. Low titre antibody was also present in a few human sera of age $55-70$. Antibody to the Asian virus had been detected in the sera of old people collected before the occurrence of Asian influenza and it had been suggested that this was a result of infection in the pandemic of 1890. No relation was found between the occurrence of antibody to Equi 2 and to $A_2$ viruses in the sera of old people. The possibility was suggested that a virus related to Equi 2 was prevalent in man before the $1889-1890$ outbreak ($M_{96}$).

### d) Infection of Man by Equine Virus and of Horses by Human Virus

VERSTEEG, MOUTON and VERLINDE ($V_{20}$) infected a horse intratracheally with a mouse-adapted Asian virus and found that haemagglutinin-inhibition and neutralizing antibody was produced in 2 weeks. Although there is a tradition among veterinarians that horses are attacked in the course of human epidemics there is little serological evidence of infection of horses by human viruses.

DOMRACHEVA ($D_{59}$) states that an A2 epidemic in Moscow in 1959 attacked both horses and man and virus was isolated from both and antibody rises occurred.

Humans exposed to equine influenza do not contract the disease ($M_{66}$, $B_{66}$) but human volunteers have been infected with Equi 2 virus by intranasal inoculation of large doses ($K_{16}$). Virus was recovered from 5 volunteers 2—4 days after inoculation and one developed fever. Four developed antibody rises to Equi 2 and three also developed antibody rise to human A2 virus.

### e) Other Properties of the Equine Viruses

TUMOVA and FISEROVA-SOVINOVA ($T_{68}$, $T_{69}$) observed some similarities between A2 human strains and A/Equi/1/Prague in that they were able to agglutinate the red cells of swine, calves and horses which are not agglutinated by A, A1, B and C viruses.

SORINOVA and LUDVIK ($S_{246}$) made electron microscope pictures of A/Equi/Prague/56 absorbed on lysed red cells. Spherical particles of 50—150 mμ diameter and filaments $80 \times 200$—3500 mμ were seen.

## 3. Avian Influenza

The work of SCHÄFER and his colleagues in Tübingen revealed a close resemblance in morphology, structure and growth cycle between fowl plague virus and influenza viruses, and the discovery that fowl plague and influenza virus A had a common S antigen indicated a relationship between the viruses that had not formerly been suspected.

In recent years several different strains of avian influenza viruses have been isolated from birds. All these viruses have contained S antigen of A type.

### a) Fowl Plague

Fowl plague was first described by PERRONCITO in Italy in 1878 and was shown to be a virus disease by CENTANNI and SAVUNOZZI in 1900 ($C_{20}$). It caused in Italy an extremely severe and fatal disease in fowls but did not spread to other countries until about 1890. It later spread over most of the world, but tended to become less severe and has progressively become less prevalent. At the present time the disease is thought to be endemic in Egypt and to be spread from time to time by wild birds. A severe outbreak occurred in the U.S.A. in 1924 but the disease has not occurred in America since 1929.

The disease only attacks birds, chiefly fowls, turkeys and geese, but many other birds are susceptible. Waterfowl are relatively resistant. The illness runs a very rapid course, fowls become semi-comatose and show dyspnoea, catarrhal discharges from the respiratory tract and cyanosis. Nervous symptoms are prominent — spasmodic twitching movements of head and neck, staggering and inability to stand. There is fever at the height of the disease, but the temperature becomes subnormal before death. Mortality rate varies from 40—100%. At autopsy haemorrhagic lesions are found in almost any part of the body.

The disease is essentially a septicaemia and the blood may be infective when diluted 1 to 1 million or more. All the body tissues are highly infective. Infection may be transmitted by subcutaneous, intramuscular, or intravenous injection or by instillation into the respiratory tract, conjunctivae, or cloaca. Fowls which recover are immune and their blood serum contains neutralizing antibody.

A description of the early studies of fowl plague is to be found in the article by TODD and RICE in Vol. 7 of the System of Bacteriology, Medical Research Council, 1930 ($T_{49}$).

An intensive study of the structure and multiplication of fowl plague virus was made by SCHÄFER and his colleagues ($S_{32-37}$) and has been described in previous chapters. The virus is readily cultivated in the fertile egg and its growth is similar to that of influenza virus. Its structure resembles that of the influenza virus and it possesses the same S antigen in its inner ribonucleoprotein component. The fowl plague virus differs from human influenza virus A in its different host range and in its pronounced pantropic properties. Although fowl plague virus only naturally infects birds, it can be grown in monkey kidney tissue cultures. On prolonged passage in human tissue cultures it loses pathogenicity for the fowl ($H_{10}$).

*Avian strains related to fowl plague:* A strain "virus N" was isolated by DIXTER from chickens in 1949 ($D_{56}$) which is serologically related to but distinct from classical fowl plague virus. In 1963 a strain "Langham" was isolated from turkeys in England by WELLS ($W_{33}$). It is designated A/Turkey/England/63 and is related serologically to fowl plague but not to virus N ($P_{35}$).

### b) The Duck Viruses

In 1956 ducklings on a farm in Slovakia were attacked by a respiratory disease with inflammation of the infra-orbital sinuses. It was usually fatal in 3—4 days. A virus A/Duck/1/Czech./56 was isolated which had the S antigen of virus A but by haemagglutination test was distinct from fowl plague or other myxoviruses ($K_{87}$, $B_{80}$). In 1956 a strain A/Duck/2/Eng/56 unrelated to the Czech strain was isolated from a mild disease on an English farm ($A_{83}$). In 1962 a second outbreak occurred on the same farm due to a strain A/Duck/England/62 related to the Czechoslovak strain but not identical to it.

### c) The Chick-Tern Viruses

In 1959 WILSON isolated a strain A/Chick/Scotland/59 from an illness in chickens clinically similar to classical fowl plague. The virus was serologically unrelated to fowl plague virus. In 1961 an explosive epizootic occurred in the Common Tern in South Africa, and a virus was isolated A/Tern/S. Africa/61 by BECKER ($B_{33}$) which was serologically identical with A/Chick/Scotland/59. It produced a very severe disease in chickens, the virus having the pantropic properties of fowl plague virus ($U_6$). The disease was also produced experimentally in Terns ($B_{34}$). BECKER made a special study of the morphology of Tern virus

($B_{32}$, $B_{35}$) and showed that it was identical to that of influenza virus A, and that it could be disintegrated by ether treatment with the production of haemagglutinating sub-units and an S antigen component showing the same double-helix construction as in influenza virus.

### d) Other Avian Viruses

In 1963 an A-type virus "Wilmot virus" was isolated from turkeys in Canada ($L_{17}$). The disease was mild with a mortality less than 1%. The virus was highly lethal to chick embryos and allantoic fluids agglutinated chick and turkey red cells. It was serologically quite distinct from other avian viruses by haemagglutination-inhibition tests, but specific CF antigen tests revealed a relationship to Duck/England/62 and A/Equi/Miami/63 viruses. It resembles A2 and A/Equi 2/63 viruses in agglutinating horse and cow red cells. Other isolations of Wilmot type virus have been reported in 1963 and 1965.

A haemagglutinating virus was isolated from turkeys in California in 1965 ($B_4$) which is more pathogenic than Wilmot virus and has a greater heat resistance.

### 4. Serological and Genetic Relations of the Avian Viruses

The serological relations of the avian viruses were studied by PEREIRA, TUMOVA and LAW ($P_{35}$). They describe four antigenic groupings, (1) Fowl plague, virus N, and Turkey/England/63, (2) two duck strains, Duck/Czech./56 and Duck/England/62, (3) the duck strain Duck/Eng/56, (4) the chick-tern group, Chick/Scotland/59 and Tern/South Africa/61. The Wilmot turkey virus probably respresents a fifth group.

There appears to be a slight antigenic relationship between fowl plague virus and A/Equi 1/56 ($L_{73}$) and between the Wilmot virus and A/Equi 2/63.

TUMOVA and PEREIRA ($T_{70}$) demonstrated genetic interactions between avian and other influenza viruses. The plaque-forming activity of UV inactivated fowl plague virus could be reactivated by several different serotypes of human, porcine and equine A viruses.

## C. Relations between the Human and Animal Influenza Viruses

Two major problems are presented by the existence of influenza virus infection in animals, the possible existence of an animal reservoir maintaining the human viruses between epidemics, and the problem presented by the great differences in virulence of the various infections.

### 1. Possible Existence of an Animal Reservoir of Human Influenza Virus

Many workers have suspected that the influenza viruses which survive outside the body for only brief periods and which apparently disappear from man between epidemics, might be maintained by passage in some animal host. Suspicion has fallen mainly on swine as a possible reservoir. There is no doubt that swine can be infected by the human strains of virus A, and if we accept the theory that the SWINE virus was the virus responsible for the human pandemic of 1918—1919, there is no doubt that this virus persisted in swine much longer than it did in man. The peculiar SWINE lungworm-earthworm cycle affords an explanation of this persistence of the virus in swine. There is, however, great difficulty in accepting swine as a source of human influenzal infection. There is little evidence outside the laboratory for the transmission of SWINE virus to man, and there is no evidence of the occurrence in swine of viruses of the A, A1, or A2 type before the appearance of these viruses in man. All the evidence suggests that man is the source of the SWINE virus and not the reverse.

There is even less reason to suspect horses or birds as a source of human infection, man appears to be immune under natural conditions to the equine and avian viruses.

The search for a possible reservoir in wild animals presents great difficulties. The occurrence of influenzal infection in wild animals would very probably be unrecognized in time for successful isolation of virus, and the possibility of laboratory contamination would make it difficult to be certain that a virus isolated, perhaps only from a single animal, was really of animal origin. The demonstration of the existence of antibody in the sera of wild animals offers a more promising line of attack, but the existence of non-specific inhibitors of many types in animal sera makes the use of haemagglutination-inhibition tests unsatisfactory. Complement fixation tests also present difficulties because of the occurrence of heterophile antibodies in many animal sera which may give non-specific reactions with insufficiently purified viral antigens or may react with the host cell component of purified virus preparations. Probably the most reliable test would be the use of complement fixation with purified S antigen obtained by ether fractionation of purified virus.

The problem of the persistence of influenza virus B between epidemics is as great as that of virus A, and there is no evidence at present to implicate the B virus as a cause of disease in animals.

At the present time therefore there is practically no evidence of the existence of an animal reservoir of human influenza, all the evidence suggests that man is the primary host.

## 2. The Problem of the Difference in Virulence between Mammalian and Avian Influenza Viruses

Natural infection by influenza viruses in mammals is usually a comparatively mild disorder, the disease process being restricted to the cells of the respiratory tract and the mortality being low and mostly a result of secondary bacterial infection. By contrast most of the avian viruses produce a severe disease, the virus shows pantropic properties, most of the organs are attacked and there is a very high mortality. It is possible that the avian viruses possess some special virulence factor which is responsible for their pantropic properties. Alternatively there may be some defence mechanism in the mammal which is not present in birds. Pantropism is not peculiar to avian viruses, neurotropic strains of human viruses can be produced in the laboratory and such viruses show pantropic properties in new born mice or hamsters ($W_4$, $T_{76}$). In general the influenza virus produces a much more severe disease with a greater mortality in very young animals than in older ones.

There is a fundamental difference between mammals and birds which may possibly account for the development of pantropic properties in the avian viruses.

A large proportion of the sera of new-born children contain antibody to the influenza viruses which is transmitted from the mother via the placenta and milk. As a result in naturally occurring influenza in man the majority of new-born children have some degree of immunity and it is therefore unlikely that serial transmission of virus in the new-born can occur, and conditions favourable to the development of pantropic properties will be unusual. Also new-born children are not usually found in large numbers in close association so that there are few opportunities for direct spread of infection between them.

By contrast new-born chicks are commonly herded together in large numbers so that case to case spread of infection is easy. While it is possible that some maternal antibody may be present in the egg yolk, transmission of such antibody to the chick is probably less efficient than the dual transmission of antibody via placenta and milk in man. It is possible that in avian influenza pantropic properties have developed as a result of serial passage in very young birds. BIELING and OELRICHS ($B_{70}$) state that eggs from fowls immunized against influenza or Newcastle disease virus can be infected with these viruses, but HALLAUER ($XH_1$) found that in fowl plague yolk sac antibody is transferred to the embryo, and chicks hatched from the eggs of immunized fowls were immune for 3 to 4 weeks. Other workers obtained similar results ($XS_2$).

# XXXIII. Epidemiology of Influenza

The author does not intend to review the whole of the large volume of literature dealing with the epidemiological aspects of influenza virus infection, but rather to describe the main facts and discuss the problems they present. Influenza epidemics have been frequently recorded in past centuries, and lists of the major outbreaks are given by HIRSCH ($H_{124}$), CREIGHTON ($C_{87}$) and MOTE ($M_{131}$). The disease occurs in a pandemic form affecting a very large proportion of the population and spreading rapidly over the entire world, in an epidemic form of more limited spread, and in the form of small localized outbreaks.

A remarkable change in the incidence of influenza occurred in 1889 in Britain and probably also in other countries. In the previous 30 years influenza had become rare and in 1880—1889 had almost disappeared. A severe pandemic occurred in 1889—1890 and since that time the disease has remained much more prevalent though fluctuating in incidence from year to year (Fig. 58). This change of incidence in 1889 is attributed by ZHDANOV ($Z_{34}$) to the growth of large industrial centres and the increasing development of rapid methods of communication by rail and sea which occurred in the late nineteenth century, leading to much greater possibility of transmission from case to case and from country to country.

## 1. The Pandemic of 1889—1890

According to BURNET and CLARK ($B_{193}$) epidemics of influenza occurred in Russia in 1886 and 1887. In 1889 the disease appeared in Bokhara, Greenland and Athabasca in Canada. They suggest that the pandemic originated in Siberia or the circumpolar region in May—June 1889. It then spread via Russia reaching Tomsk in early October and St. Petersburg in late October. It spread all over western Europe in November and reached northern Africa in December. Eastern

U.S.A. was attacked in late December and in January 1890 San Francisco, Mexico and central America were involved. South America was attacked in February— April. The disease reached Egypt and Persia in January, India in February and March, and Australia and New Zealand in March—May 1890. The disease spread at a speed equivalent to that of human travel. The incidence was high, 15—70% of the population being attacked in different areas, and there were many deaths. Three waves occurred in England, in January 1890, May 1891 and January 1892. The incidence of the disease was highest in the first wave in which there was a high percentage of deaths in young adults. The mortality was highest in the last wave, deaths occurring mainly in old people. MULDER and MASUREL (M$_{139}$)

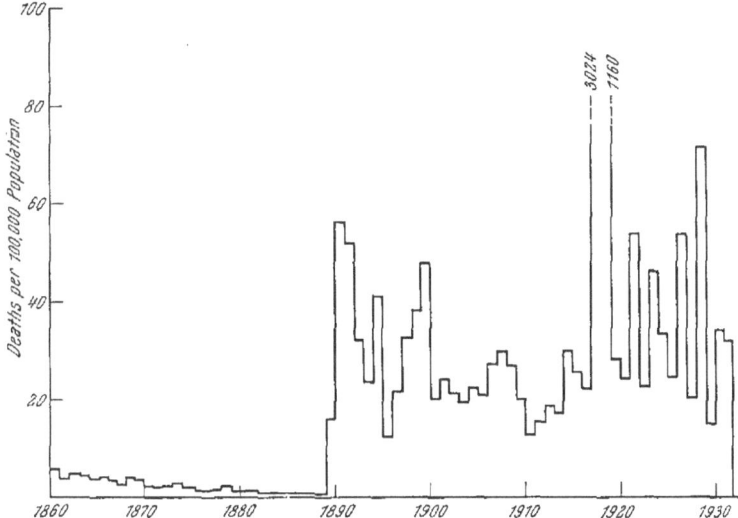

Fig. 58. Death rate from influenza in England and Wales 1860—1932.

suggested that the pandemic of 1889—1890 might have been due to a virus strain related to the A2 virus, as they found that a small proportion of sera of persons over the age of 70 contained antibody to the A2 virus before the occurrence of the A2 pandemic of 1957.

After the 1889—1890 pandemic the incidence of influenza remained much higher than in the pre-pandemic period and between 1892 and the outbreak of the great pandemic of 1918—1919 several epidemics of influenza occurred, notably in 1893, 1898—1899, 1908 and 1915.

## 2. The Pandemic of 1918—1919

This was the worst epidemic in human history and is believed to have caused some 20 million deaths. The exact origin of the pandemic is uncertain. Outbreaks of influenza of a localized nature had occurred in the Eastern cities of the U.S.A. in January of 1918 and in central U.S.A. in March, but were possibly not related to the pandemic disease. In April 1918 numerous cases of influenza occurred amongst American and French troops in France. Cases appeared in the British army in June and the troops then spread the disease to Britain and America, a severe outbreak occurring in Boston in September. The disease attacked France and Italy in April; Spain, Portugal and Greece in May; England and Switzerland in June; Germany, Denmark and Norway in July, and Holland and Sweden in August. Africa and India were involved, the disease first appearing in the seaports. Outbreaks of influenza occurred in China in May and June and there was a severe epidemic in Chungking in July. BURNET and CLARK suggest that there may have been an Asian development independent of the European, and that the disease might have been brought to France by Chinese labourers.

The first wave of the pandemic had been comparatively mild and the morbidity decreased in August, but in the autumn a second wave of much greater severity occurred, reaching a peak in October with attack rate up to 50% of the population and death rates up to 6—8% of those attacked. Almost the whole world was involved with the exception of Australia where a strict quarantine system applied to arriving ships delayed the disease until January 1919. The incidence of the disease declined in December and January, but a third wave occurred in February and March of 1919. The disease finally ended in May 1919.

Over the whole period probably 50% of the population of the world was attacked, with a mean death rate of the order of 3%. Only New Guinea, St. Helena and some of the South Pacific Islands escaped the disease.

The disease was characterized by the occurrence of high death rates in young adults, by secondary infection by *Haemophilus Influenzae* in the second and third waves which probably played a large part in the increased mortality, and by the occurrence in many of the more severe cases of an intense cyanosis.

It is very probable that the 1918—1919 pandemic was due to a strain of virus A related to or possibly identical with the SWINE virus isolated by SHOPE in 1931. This suggestion, first made by LAIDLAW ($L_{14}$) and SHOPE ($S_{110}$) has been strongly supported by studies of the age incidence of antibody to the SWINE virus in human sera by DAVENPORT, HENNESSY, FRANCIS and others ($D_{24}$, $H_{102}$, $H_{99}$, $D_{23}$).

BURNET and CLARK ($B_{193}$) suggest that the 1918 pandemic may have been caused by a strain of virus of unusual virulence, and ANDREWES ($A_{71}$) considers that a number of different mutants may have arisen during the course of the pandemic. ZHDANOV ($Z_{34}$) however suggests that the results of attempted transmission of the disease to volunteers during the pandemic did not support the idea of an unusual virulence of the 1918 strain.

An extensive review of over 4,500 papers dealing with the 1918—1919 pandemic was made by THOMSON and THOMSON ($T_{47}$).

### 3. The Era of Prevalence of the A Type Viruses, 1933—1946

After the pandemic of 1918—1919 epidemics of influenza occurred usually at intervals of 2—3 years in various parts of the world, the severity of the disease tending to decline and the deaths to occur mainly in old people. This decline in severity occurred more rapidly in America than in Britain, where some of the characters of the pandemic disease were maintained for about 10 years. In 1927 the author, then a medical student, saw 6 cases of influenzal pneumonia which showed the "heliotrope" cyanosis described by observers of the 1918 pandemic. The cyanosis did not respond to intensive oxygen therapy and all died.

If we accept the theory that the pandemic of 1918—1919 was due to a virus of the SWINE type, then at some time between 1918 and 1933 a change of serological type occurred with the replacement of the SWINE strain by the A type virus first isolated in 1933. The appearance of the new type was not associated with a pandemic spread of the disease, possibly because there was some degree of serological relation between the SWINE and A type viruses resulting in some degree of protection being conferred by previous exposure to the SWINE form.

Between 1933 and 1946 several epidemics occurred which failed to attain a world-wide spread. These outbreaks were all due to strains related to the WS strain isolated in 1933, but minor changes in antigenic structure occurred. Strains isolated in 1934—1939 were mostly related to the PR 8 virus which showed a broader antigenic structure than the WS strain. In 1943 strains were isolated which differed from the PR 8 virus though clearly related to it, and in that year the disease almost attained a pandemic spread, though the disease was clinically not severe. It was in this period that protective vaccination against the A virus gave its best results.

Although no full-scale pandemic occurred in this period the A type virus became distributed all over the world as a result of a series of epidemics in different areas at different times. In any one area outbreaks tended to occur at 2 year intervals, almost always in the winter months.

## 4. The Period of Prevalence of A Prime or A1 Viruses, 1946—1956

Between February and April 1947 the U.S.A. experienced an epidemic of influenza due to an A-type virus which was serologically markedly different from all previous A strains. Vaccination with the older A strains gave no protection and produced no antibody rise to the new strain, which became designated as the A-prime or A1 virus. The origin of the A1 virus is very uncertain. In May 1946 a strain CAM isolated in Australia proved to be an A1 virus and is the first certain isolation of the new type. But in 1945 LACORTE, MONTEIRO and LOURES ($L_2$, $L_3$) reported the isolation of six strains in Brazil which did not correspond to PR 8 or LEE viruses, and LOFSTROM ($L_{113}$) reported that a strain isolated in Sweden in 1944 differed markedly from PR 8 and was similar to a strain later isolated in 1947. In January of 1947 an epidemic occurred in a prisoner of war camp in Britain and also several localised outbreaks in schools ($D_{90}$). Virus strains isolated proved to be of the A1 type as did a strain isolated in Holland. The A1 viruses isolated in 1947 were not all precisely identical and it is possible that in the period between the last major outbreak of A virus infection in 1943—1944 and the extensive establishment of the A1 type in 1948—1949 the virus was in a transitional state in which a number of different variants were present. Although the new A1 virus ultimately completely replaced the older A strains there may have been a period in which both types were present. In 1949 both A and A1 viruses were isolated in Canada ($V_7$, $N_2$). In 1948 French workers ($L_{60}$) isolated 10 strains of virus of which 7 were type A, 2 type A1 and one mixed. WARD and EDDY ($W_{18}$) isolated 5 strains in Baltimore in 1946—1948 which were related to the HYDE strain isolated in 1935 and were pathogenic to mice on first isolation. PAYZIN and OKKEN ($P_{25}$) isolated 3 strains in Turkey in 1949 by serial passage in mice which were distinct both from A1 and from all laboratory strains of virus

A. Isaacs and Andrewes ($I_9$) consider that isolations of A type viruses since 1947 may have been due to laboratory contamination, especially when strains proved pathogenic to mice on first isolation. French workers ($C_{18}$, $L_{61}$) noted that patients infected with A1 virus in 1948—1949 developed antibody rises to both A and A1 viruses, and in some cases a rise only occurred to A virus. This example of the recall phenomenon indicates some slight serological relation between the A1 virus and virus A.

In 1948—1949 the A1 virus caused a widespread epidemic in western Europe. The epidemic commenced in September and October 1948 in Sardinia, where Magrassi ($M_{25}$) stressed the simultaneous onset of the disease in a number of isolated areas of the island. The outbreak had been preceded by a small outbreak in southern Italy, Sicily and Sardinia in the spring of 1948. The epidemic reached Italy and Sicily in November with a peak incidence in January 1949. Switzerland, Austria and France were attacked in December and January, Belgium and Holland in January, and West Germany in January and February. Apart from Iceland where an unusually prolonged outbreak occurred, the Scandinavian countries and Britain were not seriously attacked and there was little spread to eastern Europe. A sharp outbreak occurred in Turkey. Strains of virus isolated in 1948—1949 formed a homogeneous group differing slightly from the 1947 FM1 Strain ($C_{55}$, $M_{143}$).

In the winter of 1949—1950 practically no influenza occurred in Europe, but in May and June 1950 small outbreaks occurred in Sweden and Norway ($S_{340}$). In October 1950 a high incidence of influenza occurred in Denmark and Norway and Sweden were attacked in November and December. The Scandinavian disease had a high incidence but a low mortality. The disease spread to Holland, England, Southern Ireland and Iceland in late December and January. In these areas the disease was mild, resembling the Scandinavian illness, but in addition a much more severe form of influenza with a high death rate appeared in Liverpool and spread over northern England and North Ireland. Strains isolated in Liverpool differed from the Scandinavian strains. Both types were related to FM1, but the Liverpool strains were all very similar and their haemagglutinin was inhibited to high titre by homologous serum, while the Scandinavian strains were more heterologous and were poorly neutralized even by homologous serum. The Liverpool strains resembled P-phase virus while the Scandinavian strains were apparently in the Q phase ($I_9$). Virus strains isolated in France, Spain, Greece, Turkey and Palestine proved to be of the Liverpool type, and viruses of this type had been isolated in Australia and South Africa six months earlier. Liverpool type virus was later isolated in the U.S.A. and Canada. Some Scandinavian-type strains were isolated in Australia and South Africa.

It appeared that there were two prevalent types of A1 virus in 1950—1951, one originating in Scandinavia and one which originated in the southern hemisphere and spread to southern Europe and to Liverpool. Isaacs, Gledhill and Andrewes ($I_{26}$) examined 96 strains of A1 virus isolated in 1950—1951, and found 46 to be of the Scandinavian, 47 of the Liverpool, and 3 of intermediate type. All the Liverpool strains were in the P phase and most of the Scandinavian strains were in the Q phase. The Liverpool and Scandinavian strains were not P-Q phases of the same virus, with both strains the P-Q transformation could be

produced by passage in eggs in the presence of homologous antiserum, and the
reverse change by passage in mouse lung. The Scandinavian type was closely
related to the previous A1 strains of 1947 and 1949, while the Liverpool strains
contained a new antigen not present in previous A1 strains.

In 1952—1953 influenza due to A1 viruses again became very prevalent.
Outbreaks occurred in Finland in the autumn of 1952, in Japan and U.S.A.
in December, and in early 1953 the disease became widespread in Europe, North
Africa, Central America and the Pacific Islands. Of 208 strains examined by
Isaacs, Depoux and Fiset ($I_{15}$) 175 were of the Scandinavian and 33 of the
Liverpool type. An interesting finding was that all 19 strains from Portugal were
of the Liverpool type, while all 17 from Italy were Scandinavian. Liverpool strains
were replaced by Scandinavian in America and Japan in 1952—1953.

In 1955 no extensive epidemic occurred but in the spring a number of strains
were isolated which were distinct from the previous A1 strains though related
to the Liverpool type. This type, the Eire/55 type, was isolated in Eire, England,
India and New York, and in the following year in England, India and South
Africa, but the strain never caused a serious epidemic.

In February and March of 1956 a further serologically distinct variant of the
A1 virus was isolated in Holland, the Dutch/56 strain. This strain also appeared
in India, Canada, Tanganyika and Berlin. Of 101 virus strains isolated by the
Public Health Laboratory Service in England and Wales in 1955—1956, 97 were
of the 1953 Scandinavian type, 2 of the Eire/55 and 2 of the Dutch/56 type
($H_{52}$).

Russian outbreaks of A1 infection appear to have been out of step with most
of the rest of the world. No A1 strains were detected in Russia in 1946—1948
($Z_{34}$), but the strain appeared in 1949 and from then on only A1 type viruses were
isolated ($R_{46}$, $R_{30}$, $R_{32}$, $M_{30}$, $M_{127}$). Outbreaks occurred in the spring of 1949,
1952, 1954 and 1956. No outbreak occurred in Russia in the winter months of
1950—1951 or 1952—1953 when the A1 virus reached its most widespread distribution in other parts of the world.

The evolution of the A1 virus has been discussed by Isaacs and Archetti
($I_{10}$, $I_5$). They suggest that two separate lines developed from the original 1947
virus, one evolving into the Liverpool and Eire 55 types, and the other to the
Scandinavian and Dutch/56 type. In some countries only the Scandinavian type
developed, in others only the Liverpool type, some showed both types simultaneously, and in others the Liverpool type became replaced by the Scandinavian.
The Eire and Dutch types failed to completely establish themselves, possibly
because of the appearance of the pandemic Asian strain.

## 5. The 1957 Pandemic of Asian or A2 Influenza

In 1957 a new strain of influenza virus appeared, the Asian or A2 virus.
This strain possessed the S antigen of A type, but its specific antigen differed
in a major degree from all previous A type viruses, and the strain immediately
attained a pandemic spread. Very extensive investigations were made by workers
in the influenza reference laboratories which had been established under the guid-
ance of the World Health Organization, and a bibliography compiled by the

American Institute of Biological Sciences (WHO) includes over 2000 references to work on Asian influenza and some 660 references to its epidemiology.

The pandemic began in February 1957 in the Kweichow province of China where virus of the new type was first isolated ($C_{54}$, $C_{57}$). The disease spread all over China in March, reached Hongkong and Singapore in April and then spread to Japan, India and the rest of Asia. The Middle East was attacked in July. In August pilgrims returning from Mecca introduced the disease into Sudan and West Africa, and in this month the virus reached tropical America, appearing in the Caribbean and in Ecuador. Australia, South Africa and Chile were also attacked in August and Brazil in September and October. Although the virus was certainly introduced into Europe and North America at an early stage the pandemic did not break out in these areas until September and October. In Russia the peak incidence was reached in October in Leningrad and in December in Moscow. After the initial outbreak in April Japan was attacked again in September and October, and other countries also experienced a second wave.

Although the morbidity rate was high, the disease attacking 30—80% of the population, the illness was mild and the death rate was very low, usually less than 0.1%.

The unique characters of the Asian virus were recognized very early ($C_{57}$) and were fully described by American workers before the pandemic occurred in the U.S.A. By haemagglutination-inhibition tests the strain showed no relation to previous A type viruses and personnel vaccinated before 1957 had no antibody to the Asian strain ($M_{83}$, $J_7$, $J_9$). Several countries were attacked by the Asian virus within a few months of previous outbreaks of A1 infection ($M_{83}$). In Russia both A1 and A2 viruses were isolated in an outbreak in April 1957, and in the subsequent A2 outbreak in the summer it was found that the incidence of disease in those who had had A1 infections in the spring was only half that in those who had not been infected in the spring ($S_{152-154}$). In Hungary, where an A2 outbreak in the autumn followed an A1 epidemic in the spring, the A2 virus isolated differed from the Singapore strain in having a minor antigenic component shared with A1, and the possibility of recombination between A1 and A2 viruses was considered ($T_5$). Strains of A1 virus isolated in the spring of 1957 in Vladivostock are stated to have had some antigenic components of the A2 virus ($G_{20}$). Strains with both A1 and A2 antigenic components were also isolated in Rumania ($D_{49}$, $D_{50}$). In an examination of 196 Asian strains sent to the International Influenza Centre for the Americas a few cross reactions with A1 and A strains were observed ($J_9$).

There may, therefore, have been some overlap in incidence of A1 and A2 infection, and possibly some intermediate types of virus in early 1957, but in general the A2 viruses isolated were a very homogeneous group and the new strain replaced the previously prevalent A1 virus. All strains of A virus isolated since 1957 have been of the A2 type.

DAVENPORT and HENNESSY ($D_{19}$) noted that the attack rate of Asian influenza was highest in children and declined with advancing age, suggesting that some degree of resistance might have resulted from previous infections with other types of A virus. They also confirmed the finding of MULDER and MASUREL ($M_{139}$) that antibody to the A2 virus was present before the Asian pandemic in the sera of persons over the age of 70. The Asian pandemic was apparently not the first time that the specific A2 antigen had been encountered by man. Similar observations were made by other workers ($D_{34}$, $H_{116}$), and the possibility was suggested that the pandemic of 1889—1890 might have been due to a strain related to the A2 virus.

In 1959 Asian influenza again became very prevalent, especially in Russia and eastern Europe, the outbreak being complicated by the simultaneous occurrence of an epidemic of influenza B. The 1959 strains were identical with the 1957 strains, but the illness produced was more severe than in 1957 ($A_{44}$, $B_7$, $E_{33}$, $N_{12}$, $R_7$, $S_{191}$).

Since 1959 only small epidemics of influenza have occurred, the community having acquired a considerable resistance to the A2 virus. Antigenic changes have occurred in recent years. In an outbreak in Canada in 1963 strains were isolated which differed antigenically from earlier A2 strains ($D_{95}$). Some patients developed antibodies to the epidemic strain alone, some also to the earlier A2 virus. Antisera to A2 neutralized the epidemic strains, but antibody to the epidemic strain did not neutralize A2 virus of previous years.

PEREIRA, PEREIRA and LAW ($P_{33}$) consider that a gradual antigenic shift occurred in 1962—1964. Of A2 strains isolated in Britain in 1963, 75% resembled A2/Singapore/1/57, while 25% were closer to a new variant A2/Netherlands/65/63. A further variant A2/England/12/64 was isolated in 1964 and of 63 strains examined in that year 59 resembled the new variant.

## 6. The Epidemiology of Influenza B Infection

Many of the features of the epidemiology of influenza virus B are similar to those of virus A, the occurrence of small outbreaks in between the major epidemics, the absence of the disease in the summer months and variations in serological type with existing types being replaced by others. Differences between the two diseases may possibly be a result of the fact that the B virus appears to be a rather better antigen in man than the A virus. In any one area epidemics of B infection tend to occur at greater intervals of time than with A infections. Changes of antigenic structure have been less dramatic than with A virus and a new type has taken longer to establish itself, so that although B virus infection has often become very widespread explosive pandemics have not occurred.

The B virus was first isolated by FRANCIS in 1940 ($F_{90}$) but serological studies by several workers showed that it had been responsible for epidemics of influenza in America in 1936 and 1939 ($F_{92}$, $N_{21}$) and in Australia in 1938 and 1939 ($B_{162}$). The B virus was isolated in various localized outbreaks in 1940—1943. Most of the strains were closely related to the original LEE strain but minor differences were noted ($E_7$, $G_{46}$, $H_{121}$).

In 1944, BURNET, BEVERIDGE and BULL ($B_{189}$) isolated a strain BON in Australia which was only distantly related to LEE and in 1945—1946 infection by strains of this type became very widespread. The major outbreak was preceded by small localized outbreaks in the spring of 1945 ($C_{81}$) in the U.S.A., and in the Pacific Islands in June. Epidemics then occurred in Australia in November, in America in November and December, in Holland and Belgium in December and in Britain in January 1946 ($D_{91}$). BURNET and STONE noted that in children the antibody response against the current strain was higher than against LEE, but in adults the response to LEE was equal to or greater than that to the current strain ($B_{219}$). Although the 1945 strain was distinct from LEE, vaccination with LEE virus gave good protection ($F_{117}$, $H_{152}$). In the winter of 1946—1947 an epidemic due to the new type of B virus occurred in Russia.

In the period 1949—1952 influenza B was again very prevalent. Outbreaks occurred in eastern Europe and Russia in 1949 ($F_{14}$, $L_{70}$, $B_{55}$, $K_{89}$, $N_{11}$, $R_5$, $R_6$, $S_{243}$, $S_{102}$). Strains isolated such as the Russian KR1 strain and the Czechoslovak strain B/Czechoslovakia/1/49 were related to the BON strain and slightly to LEE.

Strains were isolated in the U.S.A. in 1950 ($T_{24}$) and in 1951—1952 influenza B epidemics occurred in western Europe, Britain, America and other countries. Strains isolated in 1951—1952 were all very similar and related to the BON strain ($M_{136}$, $H_{97}$, $B_{118}$). MAGILL and JOTZ ($M_{13}$) suggested that over the period 1940—1952 the LEE strain became replaced by strains containing an additional antigen which became dominant in 1949 as the LEE antigen disappeared. Similar conclusions were reached by KOSYAKOV and ROVNOVA ($K_{95}$).

Strains isolated in 1953—1954 were closely related to the 1950 strains ($W_{68}$), but in 1955 an epidemic in Germany yielded strains of B virus which were not related to the BON strain ($H_{110}$).

Between 1959 and 1962 the B virus again caused widespread disease and strains isolated in South Africa in 1958, England, Russia, Czechoslovakia in 1959, Bulgaria in 1959—1960, and Czechoslovakia and Norway in 1961, were different from both LEE and BON ($T_{67}$). American strains isolated in 1961—1962 also differed from previous B strains, and both American and Russian workers suggested that the B viruses could be divided into three groups, B, B1, and B2 ($R_{41}$, $Z_{34}$), though HENNESSY, MINUSE and DAVENPORT ($H_{104}$) pointed out that the differences were less than occurred with the A viruses. The B type represented by the LEE strains was prevalent from 1936—1944, the B1 represented by the BON and KRI strains from 1945—1954, and the B2 type in 1959—1961.

In 1964 a strain B/India/363/64 was isolated which differed considerably from all previous B strains ($K_{61}$).

## 7. Problems Presented by the Epidemiology of Influenzal Infection

The epidemiology of influenza virus infection presents several problems which have been the subject of much speculation. The solution of these problems might be of great importance both in regard to the possible control of the disease and in our understanding of the nature of the virus and its relation to man.

### a) Survival of the Virus between the Epidemics

Influenza cases are infectious to others for very short periods, and outside the human body the virus has very poor powers of survival. Influenza outbreaks do not occur in the summer months, and in any one area there may be long periods in which no evidence of the presence of virus can be found. Searches for carriers of influenza virus in interepidemic periods have usually given negative results. HORIKAWA, SUZUKI and FUZITA ($H_{161}$) failed to isolate virus in an inter-epidemic period from 433 healthy boys, 115 patients with acute febrile respiratory disease, and 33 samples of tonsils and adenoids removed at operation. McKee and Hale examining nasopharyngeal washings from 50 normal individuals and tracheal mucosa from 100 autopsy cases isolated only one strain of virus A which was probably a laboratory contaminant ($M_{63}$) as it was not of the prevalent virus type. Three methods of survival between epidemics have been suggested.

### α) Survival by Case to Case Spread

According to this theory, which has been widely supported, influenza does not really disappear between the epidemics but is endemic and is maintained by a low degree of case to case transfer possibly producing mainly subclinical infections. Survival over the summer months is explained by the virus becoming transferred to the southern hemisphere and returning to the northern hemisphere in the following winter -- the "trans-equatorial swing". A considerable body of evidence can be adduced to support this view. There is no doubt that case to case spread is the major factor in the spread of epidemics, which often travel from place to place at a rate comparable to that of human travel. Isolated communities may escape infection even in pandemics, and even large communities may be protected by a quarantine system such as was used in Australia in 1918. In recent years the iron curtain seems to have offered a considerable barrier to the spread of epidemics between eastern and western Europe. New serological types have more than once appeared in the southern hemisphere during the months of June—September and appeared in the northern hemisphere in the following November—February. Serological studies have always indicated a greater incidence of influenzal infection in epidemics than was evident on clinical grounds so that subclinical attacks are common. There is no doubt also that occasional sporadic cases of influenza occur between epidemics. The concept of survival by case to case spread has been supported by such distinguished authorities as the American Commission on Acute Respiratory Diseases ($C_{80}$), Burnet ($B_{193}$), Zhdanov ($Z_{34}$) and others. Andrewes ($A_{72}$), however, finds difficulty in accepting the existence of any widespread subclinical interepidemic case to case transmission and there are in fact great difficulties inherent in such a concept.

Influenza cases appear to be infectious only for about 48 hours and the incubation period of the disease is also about 48 hours. Case to case spread therefore necessitates the occurrence of a new infection every four days, and each case would have to successfully transmit the virus to at least one other. There will be a delicate balance between the infectivity of the virus and the degree of herd immunity. If the herd immunity is low a case of influenza may be expected to infect more than one other individual and in such a situation the number of cases will rapidly build up an epidemic. Thus, if each case infects on the average 2 others there will be 1000 cases, and if 4 are infected there will be over 500,000 cases in six weeks. If on the other hand the herd immunity is high then sooner or later transmission will fail and the disease will rapidly die out.

It is very difficult to believe in the development of an endemic distribution in such a knife-edge situation, and especially to believe that the virus might spread widely without producing an obvious epidemic. Nor does the idea of trans-equatorial spread offer much help, influenza epidemics do not occur in the southern hemisphere and northern hemisphere at 6 monthly intervals. Between the last widespread epidemic of A virus infection in 1943—1944 and the A1 epidemic of 1947 the supposed endemic virus would have to cross the equator four times and undergo a considerable antigenic modification on the way.

BLASKOVIC, BORECKY and RATHOVA ($B_{77}$) attempted to obtain direct evidence of the circulation of influenza virus in a pre-epidemic period. They exposed ferrets for 5—7 days in public rooms in Bratislava (post-office, station and hospital waiting rooms). One month later 2 animals were found to have a 2-fold and 1 a 4-fold rise in HI antibody to virus A, and 5 showed a 2—4-fold rise to virus B. The significance of such borderline titre changes is very doubtful.

### β) Survival in a Non-infectious "Masked" Form

Most of the difficulties disappear if we suppose that while in the epidemic form influenza virus spreads from case to case, in the inter-epidemic periods it survives in an intracellular non-infective form. While there is no direct evidence of survival in such a form our knowledge of the growth cycle of the virus makes it not only possible but probable that such a persistence might occur. Incomplete growth cycles of various types occur naturally and can be produced experimentally. Some types of cell never produce infective virus although the virus components are all synthesized. How long virus antigens might persist in a cell not producing infective virus is not known, but we know that infection of cells does not necessarily result in cell death and does not necessarily prevent cell division. In the intra-cellular phase virus might remain for long periods protected from the action of antibody. The survival of swine virus in a masked form in pig lungworms illustrates the possibility. Infection of pigs with infected lungworms does not result in an immediate attack of influenza, the disease only develops as a result of some non-specific stimulus such as exposure to bad weather. Human influenza presents numerous examples of the simultaneous onset of disease in remotely separated areas as if a latent virus was being stimulated into activity by some climatic factor. There are also many examples of the introduction of a new serological type of virus into a community in the spring without the immediate occurrence of an epidemic, only to be followed by an explosive outbreak in the autumn. We might suppose that there is some factor present in the summer months which favours the development of some form of incomplete growth cycle, while a "winter factor" provokes the appearance of the infective form. If the "winter factor" is a climatic one it is difficult to understand its nature, neither temperature nor humidity changes can be the sole factor. Influenza can occur in any area from the Arctic to the Tropics, and in all conditions of humidity. There is much evidence that overcrowding is a major factor in accelerating the spread and increasing the incidence of influenza during epidemics, and may be involved in the winter prevalence of the disease.

### γ) Survival in an Animal Reservoir

The difficulty of detecting influenza in non-epidemic periods led many observers to believe that the disease was always imported from somewhere else, and the ultimate extension of this idea led to the suggestion of the possible existence of an animal reservoir of the disease. Attempts to find such a reservoir were instituted by the World Health Organization ($K_{15}$, $P_{23}$, $P_{24}$). A reservoir in Central Asian swine has been suggested for two reasons. Firstly, the existence of the lungworm—earthworm cycle in swine influenza affords a possible mechanism for the persistence of virus from year to year, and secondly it is claimed that all the major pandemics of influenza can be traced back to a Central Asian origin. The idea of the influenza virus lurking in some remote Mongolian pigsty and bursting out from time to time to sweep over Asia and Europe in the manner of the Golden Horde is an entertaining one, but will scarcely bear critical examination. The 1889—1890 pandemic began in Bokhara, but appeared at almost the same time in Greenland and Canada. Are we to believe that the Eskimos spread it over the North Pole? The 1918 pandemic began in France in April, and while, according to BURNET and CLARK, outbreaks of influenza occurred in China in May and June it is difficult to believe that the European outbreak was secondary to the Chinese. The origin of the A1 virus is unknown, but if it originated in China or Mongolia it is surprising that the first isolation of the A1 virus should have been made in Australia in 1946 while it was not detected in the USSR until 1949. The A2 virus certainly arose in China, but Japanese workers ($A_{40}$) were unable to find any evidence of infection of swine by A2 virus either before or during the A2 epidemic. If the A2 virus arose in Chinese swine it is surprising that it should have been unable to infect Japanese swine 2 months later.

### b) The Origin of the Serological Variations in the Influenza Viruses

Throughout their history the influenza viruses have shown a remarkable sequence of variations in antigenic structure. These variations have been of two distinct types, the sudden appearance of new types at long intervals of time, and the slower "antigenic drift" which occurs from year to year.

### α) The Phenomenon of "Antigenic Drift"

All the main types of A virus, A, A1 and A2, have shown this phenomenon and it also occurs with virus B. A good example is afforded by the A1 viruses. The original FM1 type of 1947 underwent in succeeding years a series of changes of antigenic structure in small steps, with the appearance of new antigens which slowly became dominant. The changes developed in at least two different directions leading to the emergence of the Scandinavian and Liverpool types, both of which underwent further changes. A similar type of antigenic modification can be produced in the laboratory by serial passage of virus in partially immunized animals, and there can be little doubt that "immunological pressure" explains the natural occurrence of the "antigenic drift" phenomenon in man, antigenic change resulting from serial infection of humans possessing some degree of residual immunity from past infection.

The P-Q variation is possibly a stage in the process of antigenic drift. A virus originally in the P phase with a dominant antigen reacting strongly with antibody may as a result of passage in a partially immune host become converted into the Q phase in which no dominant antigen is present. Further serial passage may then result in the appearance of a strain with a new dominant antigen.

Most workers have supposed that the serological variations are produced as a result of a process of random mutation. It is supposed that during the process of virus reproduction a small number of random mutants are produced which differ in antigenic structure from the parent type. The growth of such a mutant will then be favoured in an immune environment, so that it will come to replace the original strain. The theory does not explain how the mutants arise, and has been strongly criticized by ZHDANOV ($Z_{34}$) who maintains that the immune environment produces the antigenic change. The author has considerable sympathy with the Russian view. If the effect of the immune environment is to select a pre-formed mutant then the mutant selected should be the one most remote from the parent type, but in fact the serological changes appear to occur in short steps with many intermediate stages before a significantly different antigenic type is evolved. The serological changes occurring in the course of the antigenic drift appear to be forced on the virus, and the only objection to the obvious conclusion that the environment produces the antigenic change is the reluctance of classical geneticists to accept the inheritance of an acquired character.

### β) The Appearance of the Pandemic Types

While the antigenic drift phenomenon is essentially a slow change, the appearance of the pandemic types such as the A 2 represents a sudden and extensive change of antigenic structure and has much more of the characteristics of a mutational change. The Asian virus appeared to arise spontaneously without any apparent relation to the previously prevalent virus. While it may possibly be serologically related to the virus of the 1889 pandemic as suggested by the finding of antibody to it in the sera of elderly people, it is impossible to believe that the 1889 virus remained dormant for over 60 years to re-emerge in 1957. The appearance of the Equi 2 strain in horses in Miami in 1963 is another example of a new strain appearing apparently from nowhere.

Although the change from A to A 1 in 1946—1947 is less dramatic than the appearance of the pandemic virus in 1918 or the A 2 virus in 1957 it also would appear to represent a greater and more sudden change than would be expected from the antigenic drift process.

If these serological changes are in fact mutations we may ask how and where the mutation occurs. A virus particle in the extracellular phase might be modified by some external mutagenic agent such as radiation or chemical mutagen, and there have been suggestions that the appearance of new types of virus might be related to such factors as cosmic radiation, fall-out from atomic bomb testing, use of poison gases in warfare etc. However, it is very unlikely that any mutation arises in the influenza virus outside the human body. All the evidence from studies of the effects of ionizing radiation on virus particles indicates that a single quantum hit is lethal. Chemical mutagens are similarly essentially destructive in action, they might modify a virus by producing loss of some component or property, but would be very unlikely to produce an extra property such as a new antigen. If mutations occur there can be little doubt that they occur during the active intracellular phase of virus reproduction.

### γ) Number of Possible Antigenic Variants

One of the major problems in the epidemiology of the influenza virus is that of the possible line of future development of antigenic modifications. Can the virus go on producing new serological types indefinitely, or is there a limit to the number of possible types? Many workers have thought that the virus is potentially able to produce only a limited number of specific antigen components and that the only type changes possible are produced by variations in the amount and spatial distribution of these components in the particle. It is supposed that after a number of antigens have become dominant in turn the available antigens for future type changes will become exhausted and the older types will then reappear in dominant form. There will thus be a cyclic change. The idea is supported by the results indicating a possible serological relationship between the A2 virus and the virus of the 1889—1890 pandemic. In 1953 JENSEN and FRANCIS ($J_{12}$) estimated the probable number of different antigenic components to be about 18. This was before the appearance of the A2 virus and the subsequent changes of that type, so that at the present time a larger number than 18 would be needed. We may find ourselves in a difficulty in that the number of different antigens which can be produced is greater than can be coded by the amount of RNA in the virus particle, especially when we remember that the virus RNA must also code for the S antigen, neuraminidase, and possibly other proteins as well.

### δ) Disappearance of the Old Types on the Appearance of a New Form

It is extremely difficult to understand why when a new serological type of virus appears and establishes itself widely the previously prevalent virus is completely suppressed and never appears again. It is not that the old virus gradually loses virulence and slowly dies out. Thus the A virus produced a very extensive epidemic in 1943—1944 and was in that year at least as active as in any previous year, yet on the appearance of the A1 virus in 1947 the A virus was immediately suppressed except in areas such as Russia which the A1 virus did not reach until 1949. After 1949 the A viruses disappeared. Similarly the A1 viruses were active up to the spring of 1956 only to be completely suppressed by the A2 virus six months later. It appears that infection with A2 virus produced a complete resistance to A1 virus, while not producing complete resistance to a second infection by A2 in later years, and this effect was produced not only in adults but also in children with no previous exposure to A1 virus. It is impossible to explain this phenomenon without reference to the host, in some way the human race suddenly becomes resistant to a virus which has been producing regular outbreaks over a period of several years.

### c) Role of the Host Cell in Influenza Epidemiology

Almost all the serological and other variations to which the influenza virus is subject are variations in the properties of the outer component of the virus. The virus ribonucleoprotein component appears to be stable. The outer component is derived from the wall of the infected cell and it is quite possible that some of its characters arise under the genetic control of the host cell and not of the virus RNA. It may be that the evolutionary phenomena which we have been considering represent an evolution of the host cell and not of the virus, if, in fact, such a separate consideration is possible.

In the period 1933—1945 the human respiratory mucosa was potentially capable of synthesizing influenza virus of type A. In 1946—1947 a change occurs and from 1949—1956 the mucosa is no longer able to synthesize virus A but has acquired the ability to produce a new type — the A1 virus. After 1957 the A1 virus is no longer produced but the mucosa now synthesises A2 virus. A process of evolution of the human respiratory tract appears to be occurring, and it may be that the changes in properties of influenza virus over the years reflect the evolution of the host cell.

As the mucosa of the respiratory tract represents the major contact of man with his environment it may be necessary for changes in the properties of the cell surface to occur in order to adapt to changes in the atmospheric environment. If we regard the influenza virus as a cell product we see that it possesses the remarkable property of being able to transmit "information" from one cell to another and from one individual to another in a very efficient manner, so that an adaptation of the cell surface occurring in one human is rapidly spread to the rest of the race.

It is possible therefore to regard the influenza virus not as a parasite but as a mechanism evolved by the host cell as an emergency method of transmitting information from cell to cell. It may seem absurd to treat an infectious agent which on occasion can produce millions of deaths as anything other than a parasite, but we have many examples in Nature where the survival of a species is ensured only at an enormous cost in the lives of individual members of the species.

Further speculation on these lines is probably undesirable, but it might be rewarding to compare the properties of the host cell surface before and after influenza virus infection.

## References

A₁   ACKERMANN, W. W.: The role of l-methionine in virus propagation. J. exp. Med. 93, 337—343 (1951).

A₂   ACKERMANN, W. W.: Concerning the relation of the Krebs cycle to virus synthesis. J. biol. Chem. 189, 421—428 (1951).

A₃   ACKERMANN, W. W.: The relation of the Krebs cycle to virus synthesis. 2. The effect of sodium fluoroacetate on the propagation of influenza virus in mice. J. exp. Med. 93, 635—642 (1951).

A₄   ACKERMANN, W. W.: α-aminosulphonic acids and viral propagation. Proc. Soc. exp. Biol. (N.Y.) 80, 362—367 (1952).

A₅   ACKERMANN, W. W.: Mechanisms of persistent and masked infections in tissue cultures. Ann. N.Y. Acad. Sci. 67, 392—402 (1957).

A₆   ACKERMANN, W. W.: Cellular aspects of the cell-virus relationship. Bact. Rev. 22, 223—239 (1958).

A₇   ACKERMANN, W. W.: Biochemical studies of virus-infected cells. Ann. N.Y. Acad. Sci. 81, 188—192 (1959).

A₈   ACKERMANN, W. W.: Comparative study of the reactive surfaces of polyoma and influenza viruses. Ext. Acta Un. int. Cancer 19 (1963).

A₉   ACKERMANN, W. W., and S. DINKA: Concerning the nature of the viral neutralising activity of bovine serum. Acta virol. 9, 526—533 (1965).

A₁₀   ACKERMANN, W. W., N. ISHIDA, and H. F. MAASSAB: Growth characteristics of influenza virus concerning the binding of virus by host cells. J. exp. Med. 102, 545—554 (1955).

A₁₁   ACKERMANN, W. W., and R. B. JOHNSON: Some energy relations in a host-virus system. J. exp. Med. 97, 315—322 (1953).

A$_{12}$ ACKERMANN, W. W., and H. KURTZ: A new host-virus system. Proc. Soc. exp. Biol. (N.Y.) **81**, 421—423 (1952).

A$_{13}$ ACKERMANN, W. W., and P. C. LOH: Chemistry of viral infected cells. Ann. N.Y. Acad. Sci. **88**, 1298—1307 (1960).

A$_{14}$ ACKERMANN, W. W., and H. F. MAASSAB: Growth characteristics of influenza virus: The influence of a sulphonic acid. J. exp. Med. **99**, 105—117 (1954).

A$_{15}$ ACKERMANN, W. W., and H. F. MAASSAB: Growth characteristics of influenza virus. Biochemical differentiation of stages of development. J. exp. Med. **100**, 329—339 (1954).

A$_{16}$ ACKERMANN, W. W., and H. F. MAASSAB: Growth characteristics of influenza virus. Biochemical differentiation of stages of development. 2. J. exp. Med. **102**, 393—402 (1955).

A$_{17}$ ADA, G. L.: Ribonucleic acid in influenza virus. Ciba Found. Symp. "The nature of viruses" 104—122 (1957).

A$_{18}$ ADA, G. L., M. DONNELLEY, and J. PYE: Studies on the complement fixing antigen of influenza virus. 1. Purification of antigen. Aust. J. exp. Biol. med. Sci. **30**, 301—311 (1952).

A$_{19}$ ADA, G. L., and A. GOTTSCHALK: The component sugars of the influenza virus particle. Biochem. J. **62**, 686—689 (1956).

A$_{20}$ ADA, G. L., and P. E. LIND: Neuraminidase in the chorioallantois of the chick embryo. Nature (Lond.) **190**, 1169—1171 (1961).

A$_{21}$ ADA, G. L., P. E. LIND, L. LARKIN, and F. M. BURNET: Failure to recover infective ribonucleic acid from myxovirus preparations. Nature (Lond.) **184**, 360—361 (1959).

A$_{22}$ ADA, G. L., P. E. LIND, and W. G. LAVER: An immunological study of avian, viral, and bacterial neuraminidase based on specific inhibition of enzyme by antibody. J. gen. Microbiol. **32**, 225—232 (1963).

A$_{23}$ ADA, G. L., and B. T. PERRY: Studies on the soluble complement fixing antigens of influenza virus. 3. The nature of the antigens. Aust. J. exp. Biol. med. Sci. **32**, 177—185 (1954).

A$_{24}$ ADA, G. L., and B. T. PERRY: The nucleic acid content of influenza virus. Aust. J. exp. Biol. med. Sci. **32**, 453 (1954).

A$_{25}$ ADA, G. L., and B. T. PERRY: Infectivity and nucleic acid content of influenza virus. Nature (Lond.) **175**, 209—210 (1955).

A$_{26}$ ADA, G. L., and B. T. PERRY: Specific differences in the nucleic acids from A and B strains of influenza virus. Nature (Lond.) **175**, 854 (1955).

A$_{27}$ ADA, G. L., and B. T. PERRY: Influenza virus nucleic acid: Relationship between biological characters of the virus particle and properties of the nucleic acid. J. gen. Microbiol. **14**, 623—633 (1956).

A$_{28}$ ADA, G. L., and B. T. PERRY: Properties of the nucleic acid of the Ryan strain of filamentous influenza virus. J. gen. Microbiol. **19**, 40—54 (1958).

A$_{29}$ ADA, G. L., B. T. PERRY, and A. ABBOT: Biological and physical properties of the Ryan strain of filamentous influenza virus. J. gen. Microbiol. **19**, 23—39 (1958).

A$_{30}$ ADA, G. L., B. T. PERRY, and M. EDNEY: Infectivity of influenza virus filaments. Nature (Lond.) **180**, 1134 (1957).

A$_{31}$ ADA, G. L., B. T. PERRY, and J. PYE: Studies on the soluble complement fixing antigen of influenza viruses. 2. Serological behaviour of the antigen. Aust. J. exp. Biol. med. Sci. **31**, 391—403 (1953).

A$_{32}$ ADA, G. L., and J. D. STONE: Effect of haemagglutinating viruses on the electro-phoretic mobility of human erythrocytes. Nature (Lond.) **165**, 189—190 (1950).

A$_{33}$ ADA, G. L., and J. D. STONE: Electrophoretic studies of virus-red cell interaction: Mobility gradient of cells treated with influenza viruses and the receptor destroying enzyme of V. cholerae. Brit. J. exp. Path. **31**, 263—274 (1950).

A$_{34}$ ADDAMIANO, L., B. BABUDIERI, and C. MOSCOVICI: Morphological forms of the influenza virus (Italian). R.C. Ist. sup. Sanità **19**, 161—167 (1956).

A$_{35}$ ADO, A. D., and T. M. ALEKSEYEVA: On the toxic effect of influenza virus on the sympathetic nervous system. Acta virol. 1, 161—166 (1957).

A$_{36}$ ADO, A. D., T. A. ALEKSEYEVA, A. KH. KANCHURIN, and S. M. TITOVA: Problems of the pathogenesis of influenza in the light of pathophysiological investigations (Russian). "Problemy grippa i ostrykh respiratornykh zabolevanii." Tez. dokl. nauch. Sess. Inst. AMN SSSR, Moscow (1959).

A$_{37}$ ADO, A. D., and S. M. TITOVA: A study of influenza in dogs. Probl. Virol. (N.Y.) 4, 36—41 (1959).

A$_{38}$ AGAPOV, S. I.: The pathogenic properties of swine and human influenza viruses (Russian). Zh. Mikrobiol. (Mosk.) 17, 543 (1935).

A$_{39}$ AHSBEL, S. I., N. G. STOLPETSKAYA, and V. G. SOKOLOVA: Experience in the treatment of virus influenza and the chemoprophylaxis of this disease under industrial conditions (Russian). Sovetsk. Med. 3, 22—23 (1954).

A$_{40}$ AKAO, Y., T. MATSUI, R. SANO, H. SHIMOJO, A. SUGIURA, and C. ENOMOTO: Bull. Inst. publ. Hlth Tokyo 9, 27—34 (1960).

A$_{41}$ ALBALADEJO, J., and G. BERENGUER: The pH reaction and its value for the diagnosis of the phases of influenza virus (Spanish). Microbiol. esp. 11, 119—133 (1958).

A$_{42}$ ALBRECHT, P., O. KRIZANOVA, and J. SZANTO: Location of influenza virus haemagglutination inhibitor demonstrated in chick embryo cells by fluorescent antibodies. Acta virol. 5, 232—235 (1961).

A$_{43}$ ALEKPEROV, M. A.: Functional state of adrenal cortex in influenza (Russian). Sovetsk. Med. 23, 70—73 (1959).

A$_{44}$ ALEKSANDROVA, G. I.: Comparative characteristics of A2 influenza virus strains isolated in the epidemics of 1957 and 1959 (Russian). "Problemy grippa i ostrykh respiratornykh zabolevanii". Tez. Dokl. nauch Sess. Inst. AMN SSSR Moscow 20—21 (1959).

A$_{45}$ ALEKSANDROVA, G. I.: Reactogenic and immunogenic properties and epidemiological effectivity of additionally attenuated vaccine strains of influenza virus (Russian). Vop. Virus 1, 67—73 (1965).

A$_{46}$ ALEKSANDROVA, G. I., and A. A. SMORODINTSEV: Obtaining of an additionally attenuated vaccinating cryophil influenza strain. Rev. Roumaine Inframicrobiol 2, 179—186 (1965).

A$_{47}$ ALEKSEEVA, T. E., A. I. MEL'NIKOVA, M. A. NOGICHEVA, and A. S. SHADRIN: A study of the anti-influenza antibody content in different epidemic periods in the sera of blood donors, placental sera, and in gamma globulins (Russian). "Problemy grippa i ostrykh respiratornykh zabolevanii." Tez. dokl. nauch. Sess. Inst. AMN SSSR, Moscow 159—160 (1959).

A$_{48}$ ALFORD, R. H., J. A. KASEL, P. J. GERONE, and V. KNIGHT: Human influenza resulting from aerosol inhalation. Proc. Soc. exp. Biol. (N.Y.) 122, 800—804 (1966).

A$_{49}$ ALLEN, R., R. A. FINKELSTEIN, and S. E. SULKIN: Viral inhibitors in normal animal sera. Tex. Rep. Biol. Med. 16, 391—421 (1958).

A$_{50}$ ALLISON, A. C.: Observations on the inactivation of viruses by sulphydryl reagents. Virology 17, 176—183 (1962).

A$_{51}$ ALLISON, A. C., F. E. BUCKLAND, and C. H. ANDREWES: Effects of sulphydryl reagents on infectivity of some viruses. Virology 17, 171—175 (1962).

A$_{52}$ ALMEIDA, J. D., and A. P. WATERSON: Some observations on the envelope of an influenza virus. J. gen. Microbiol. 46, 107—110 (1967).

A$_{53}$ ALTUCCI, P., F. CORAGGIO, G. MARGHERITA, and C. GALEOTA: The multiplication of influenza virus A (PR8) in chorioallantoic tissue culture (Italian). Riv. Ist. sieroter. ital. 36, 90—110 (1961).

A$_{54}$ AMELUNXEN, R. E., and A. A. WERDER: Quantitative considerations of influenza virus neutralisation. 1. Number of antibody molecules per virus particle. J. Bact. 80, 271—277 (1960).

A$_{55}$ American Institute of Biological Sciences. "An annotated bibliography of influenza". Waverly Press, Baltimore 1957—1960.

A₅₆ AMINOFF, D.: Methods for the quantitative estimation of N-acetyl-neuraminic acid and their application to hydrolysates of sialomucoids. Biochem. J. 81, 384–392 (1961).

A₅₇ AMOS, H., and E. VOLLMAYER: Inhibition of influenza virus synthesis by 2-thiouracil. Virology 6, 337–347 (1955).

A₅₈ ANANTHANARAYAN, R.: The fabric of virus elementary bodies. Brit. J. exp. Path. 35, 381–388 (1954).

A₅₉ ANANTHANARAYAN, R., and C. K. J. PANIKER: Non-specific inhibitors of influenza viruses in normal sera. Bull. Wld Hlth Org. 22, 409–419 (1960).

A₆₀ ANDERSON, S. G.: The breadth of antibody response following influenza A infection. Aust. J. exp. Biol. med. Sci. 25, 243–246 (1947).

A₆₁ ANDERSON, S. G.: Mucins and mucoids in relation to influenza virus action. 1. Inactivation by R.D.E. and by viruses of the influenza group of the serum inhibitor of haemagglutination. Aust. J. exp. Biol. med. Sci. 26, 347–354 (1948).

A₆₂ ANDERSON, S. G., and F. M. BURNET: Sporadic and minor epidemic incidence of influenza A in Victoria 1945–46. 1. Phase behaviour of influenza A strains in relation to epidemic characteristics. Aust. J. exp. Biol. med. Sci. 25, 235–242 (1947).

A₆₃ ANDERSON, S. G., F. M. BURNET, S. FAZEKAS DE ST GROTH, J. F. Mc CREA, and J. D. STONE: Mucins and mucoids in relation to influenza virus action. 6. General discussion. Aust. J. exp. Biol. med. Sci. 26, 403–411 (1948).

A₆₄ ANDERSON, S. G., F. M. BURNET, and J. D. STONE: A modified Salk test for in vitro titration of influenza antibodies. Aust. J. exp. Biol. med. Sci. 24, 269–275 (1946).

A₆₅ ANDREWES, C. H.: Interference by one virus with the growth of another in tissue culture. Brit. J. exp. Path. 23, 214–220 (1942).

A₆₆ ANDREWES, C. H.: Adventures among viruses. 1. Some properties of viruses.
A₆₇ 2. Epidemic influenza. New Engl. J. Med. 242, 161–166 and 197–203 (1950).

A₆₈ ANDREWES, C. H.: Epidemiology of influenza in the light of the 1951 outbreak. Proc. roy. Soc. Med. 44, 803–804 (1951).

A₆₉ ANDREWES, C. H.: The place of viruses in Nature. Proc. roy. Soc. Med. 139, 313–326 (1952).

A₇₀ ANDREWES, C. H.: Prospects for prevention of influenza. Trans. Coll. Phycns. Philad. June, 20 (1952).

A₇₁ ANDREWES, C. H.: Epidemiology of influenza. Bull. Wld Hlth Org. 8, 595–612 (1953).

A₇₂ ANDREWES, C. H.: Influenza: Theme and variations. Calif. Med. 84, 375–380 (1956).

A₇₃ ANDREWES, C. H., F. B. BANG, and F. M. BURNET: A short description of the myxovirus group (influenza and related viruses). Virology 1, 176–184 (1955).

A₇₄ ANDREWES, C. H., and R. E. GLOVER: Spread of infection from the respiratory tract of the ferret. 1. Transmission of influenza A virus. Brit. J. exp. Path. 22, 91–97 (1941).

A₇₅ ANDREWES, C. H., R. E. GLOVER, F. HIMMELWEIT, and W. SMITH: Influenza virus as a laboratory contaminant. Brit. J. exp. Path. 25, 130–134 (1944).

A₇₆ ANDREWES, C. H., and D. M. HORSTMANN: The susceptibility of viruses to ethyl ether. J. gen. Microbiol. 3, 290–297 (1949).

A₇₇ ANDREWES, C. H., and A. ISAACS: Antigenic variation among influenza viruses in relation to the production use and standardisation of vaccines. Proc. 3rd int. Meet. Biol. Stand. Opatija 73–75 (1957).

A₇₈ ANDREWES, C. H., A. ISAACS, and B. P. MARMION: Neutralising action of human nasal secretions on neurotropic influenza virus. Brit. J. exp. Path. 35, 264–269 (1954).

$A_{79}$ ANDREWES, C. H., H. KING, and M. VAN DEN ENDE: Chemotherapeutic experiments with the viruses of influenza A, lymphogranuloma venereum and vaccinia. J. Path. Bact. 55, 173—181 (1943).

$A_{80}$ ANDREWES, C. H., P. P. LAIDLAW, and W. SMITH: The susceptibility of mice to the viruses of human and swine influenza. Lancet 859—862 (1934).

$A_{81}$ ANDREWES, C. H., and W. SMITH: Influenza: Further experiments on the active immunisation of mice. Brit. J. exp. Path. 18, 43 (1937).

$A_{82}$ ANDREWES, C. H., and W. SMITH: The effect of foreign tissue extracts on the efficacy of influenza virus vaccines. Brit. J. exp. Path. 20, 305—315 (1939).

$A_{83}$ ANDREWES, C. H., and G. WORTHINGTON: Some new or little known respiratory viruses. Bull. Wld Hlth Org. 20, 435—443 (1959).

$A_{84}$ ANDREWS, B. E., A. S. BEARE, J. C. MCDONALD, A. J. ZUCKERMANN, and D. A. J. TYRRELL: Further trials of live influenza vaccine. Brit. med. J. 637—640 (1966).

$A_{85}$ ANGELA, G. C., and G. GIULIANI: The antiviral activity of an analogue of sialic acid (Simo). Action on the influenza virus. Panminerva med. 4, 187—189 (1962).

$A_{86}$ ANGULO, J. J.: On the identity of the so-called filamentous forms of influenza and fowl pest viruses. Arch. ges. Virusforsch. 4, 199—206 (1951).

$A_{87}$ ANSHELES, I. M.: Material for the epidemiological characterisation of the winter outbreak of influenza in Leningrad 1943/44 (Russian). Rabot. Leningrad vrach. v. god. Otech. voyny Leningrad 6, 36 (1945).

$A_{88}$ ANSHELES, I. M.: Material on the epidemiology and specific prophylaxis of influenza Type A (Russian). "Virusnyye infektsii." Tr. Inst. epidemiol. mikrobiol. i gig imeni Pasteria i Inst. eksper. med. AMN SSSR Leningrad 142—160 (1953).

$A_{89}$ ANSHELES, I. M., E. A. FRIDMAN, T. A. KLUSHINA, E. S. STENINA, L. B. KHAZENSON, and E. F. TARASOVA: The influenza pandemic in 1957. Some epidemiological and virological features of influenza in Leningrad (Russian). Vop. Virus 4, 38—43 (1959).

$A_{90}$ ANSHELES, I. M., N. N. ROMANENKO, N. S. KLYUCHEREVA, and Z. N. DANILOVA: Experience in the study of immunisation against influenza in Leningrad in 1952—53 (Russian). Zh. Mikrobiol. (Mosk.) 10, 63 (1953).

$A_{91}$ APOSTOLOV, K., and T. H. FLEWETT: Internal structure of influenza virus. Virology 26, 506—508 (1965).

$A_{92}$ APPLEBY, J. C.: The isolation and properties of a modified strain of neurotropic influenza A virus. Brit. J. exp. Path. 33, 280—287 (1952).

$A_{93}$ APPLEBY, J. C., and C. H. STUART-HARRIS: The use of filtrates of vibrio Cholerae in the classification of influenza virus strains. Brit. J. exp. Path. 31, 797—805 (1950).

$A_{94}$ ARCHETTI, I.: Laboratory trial of vaccination against influenza (Italian). Ext. R.C. Ist. sup. Sanità 17, 991—998 (1954).

$A_{95}$ ARCHETTI, I.: Appearances associated with filamentous forms of influenza viruses. Arch. ges. Virusforsch. 6, 29—35 (1955).

$A_{96}$ ARCHETTI, I.: Further studies on antigenic variation of influenza A virus. Proc. Soc. exp. Biol. (N.Y.) 80, 212—214 (1952).

$A_{97}$ ARCHETTI, I., and D. S. BOCCIARELLI: On the structure of filamentous forms of influenza virus. Arch. ges. Virusforsch. 11, 599—606 (1961).

$A_{98}$ ARCHETTI, I., and D. S. BOCCIARELLI: Contribution to the study of the structure of influenza virus (Italian). Ext. R.C. Ist. sup. Sanità 24, 546—552 (1961).

$A_{99}$ ARCHETTI, I., and F. L. HORSFALL: Persistent antigenic variation of influenza A viruses after incomplete neutralisation in ovo with heterologous immune serum. J. exp. Med. 92, 441—462 (1950).

$A_{100}$ ARCHETTI, I., A. JEMOLO, and D. S. BOCCIARELLI: On the fine structure of influenza viruses. Arch. ges. Virusforsch. 20, 133—136 (1967).

$A_{101}$ ARI, A. B.: Attempt at rapid isolation of influenza virus from throat washings collected from patients during the 1952—53 influenza epidemic by inoculation of embryonated eggs with concentrated viruses on red blood cells. Türk Ij. tecr. Biyol. Derg. **15**, 226—230 (1955).

$A_{102}$ ARKHANGELSKY, D. C., U. I. SAVCHENKO, and KH. DJ. ZHUMATOV: Cytomorphology and cytochemistry of experimental RNA influenza infection. Abstr. 9th int. Congr. Microbiol. Moscow **391** (1966).

$A_{103}$ ARYA, S. C.: Haemadsorption followed by fluorochrome staining. Brit. med. J. 98—100 (1963).

$A_{104}$ ATKINS, E., and W. C. HUANG: Studies on the pathogenesis of fever with influenzal viruses. 1. The appearance of an endogenous pyrogen in the blood following intravenous injection of virus.

$A_{105}$ 2. The effect of endogenous pyrogen in normal and virus-tolerant recipients.

$A_{106}$ 3. The relation of tolerance to the production of endogenous pyrogen. J. exp. Med. **107**, 383—401, 403—414, 415—435 (1958).

$A_{107}$ AVETISIAN, V. A.: Data on the etiology of influenza in Erivan and on some properties of strains isolated in 1958 and 1959 (Russian). "Problemy grippa i ostrykh respiratornykh zabolevanii." Tez. dokl. nauch. Sess. Inst. AMN SSSR Moscow 33—34 (1959).

$A_{108}$ AZADOVA, N. B.: Synthesis of structural proteins of myxoviruses. Abstr. 9th int. Congr. Microbiol. Moscow 498—499 (1966).

$B_1$ BADINSKY, G., and I. BRUCKNER: Contribution to the study of filtrable viruses. Attempts to cultivate the viruses of human influenza, climatic bubo, and Aujeszky's disease on the chorioallantoic membrane of chick embryos (French). Arch. roum. Path. exp. **10**, 341—368 (1937).

$B_2$ BANG, F. B.: Synergistic action of Haemophilus Influenzae Suis and the swine influenza virus on the chick embryo. J. exp. Med. **77**, 7—20 (1943).

$B_3$ BANG, F. B., and A. ISAACS: Morphological aspects of virus cell relationships in influenza, mumps and Newcastle (myxovirus). Ciba Found. Symp. "The nature of viruses" 249—262 (1957).

$B_4$ BANKOWSKI, R. A., and T. MIKAMI: An apparently new respiratory disease of turkeys. Proc. 68th Ann. Meet U.S. Livestock Sanit. Ass. 495—517 (1965).

$B_5$ BARB, K., G. J. KOTELER, F. ANTONI, and G. TAKATSI: Studies on the nucleic acid metabolism of chorioallantoic membrane cells after influenza virus infection. Acta microbiol. Acad. Sci. hung. **11**, 185—192 (1964).

$B_6$ BARON, S., and K. E. JENSEN: Evidence for genetic interaction between non-infectious and infectious influenza A viruses. J. exp. Med. **102**, 677—697 (1955).

$B_7$ BAROYAN, O. V.: Basic epidemiological characteristics of influenza after the 1957 pandemic (Russian). "Problemy grippa i ostrykh respiratornykh zabolevanii." Tez. Dokl. nauch Sess. AMN SSSR·Moscow 74—76 (1959).

$B_8$ BAROYAN, O. V., and V. G. SLEPUSHKIN: Evaluation of a live monovalent anti-influenza vaccine (Russian). "Problemy grippa i ostrykh respiratornykh zabolevanii." Tez. Dokl. nauch Sess. AMN SSSR Moscow 119—121 (1959).

$B_9$ BARRY, R. D.: Equilibrium sedimentation of influenza virus in caesium chloride density gradients. Aust. J. exp. Biol. med. Sci. **38**, 499—507 (1960).

$B_{10}$ BARRY, R. D.: The multiplication of influenza virus. 1. The formation of incomplete virus. Virology **14**, 389—397 (1961).

$B_{11}$ BARRY, R. D.: The multiplication of influenza virus. 2. Multiplicity reactivation of ultraviolet irradiated virus. Virology **14**, 398—405 (1961).

$B_{12}$ BARRY, R. D.: The multiplication of influenza virus. 3. Heterologous interference. Virology **14**, 406—409 (1961).

$B_{13}$ BARRY, R. D.: Effects of inhibitors of nucleic acid synthesis on the production of myxoviruses. Ciba Found. Symp. "Cellular Biology of Myxovirus infections" 51—75 (1964).

B₁₄ BARRY, R. D., D. R. IVES, and J. G. CRUIKSHANK: Participation of deoxy-ribonucleic acid in the multiplication of influenza virus. Nature (Lond.) 194, 1139—1140 (1962).

B₁₅ BARTOLOMEI-CORSI, O.: Relations between A 2 influenza virus and non-specific inhibitors (Italian). Sperimentale 109, 549—572 (1959).

B₁₆ BARUA, D., and B. FAUCONNIER: Non enzymatic elution of influenza virus (French). Ann. Inst. Pasteur 97, 479—496 (1959).

B₁₇ BARYKIN, V., Z. SHAKLMALIYEVA, and YE. BAUER: Cultivation of the influenza virus (Russian). Zh. Mikrobiol. (Mosk.) 17, 548 (1936).

B₁₈ BATEMAN, J. B., M. S. DAVIS, and P. A. McCAFFREY: Influence of influenza haemagglutinin on haemolysis rates. Proc. Soc. exp. Biol. (N.Y.) 80, 720—722 (1952).

B₁₉ BATEMAN, J. B., M. S. DAVIS, and P. A. McCAFFREY: Rate of reaction of red blood cells with influenza virus haemagglutinin. Amer. J. Hyg. 62, 342—348 (1955).

B₂₀ BATEMAN, J. B., M. S. DAVIS, and P. A. McCAFFREY: A note on the reaction of heated influenza virus protein with red blood cells. Amer. J. Hyg. 62, 349—354 (1955).

B₂₁ BATEMAN, J. B., A. ZELLNER, M. S. DAVIS, and P. A. McCAFFREY: The electrophoretic properties of red blood cells after reaction with influenza virus haemagglutinin. Arch. Biochem. 60, 384—391 (1956).

B₂₂ BAUER, D. J.: Multiplication of the animal viruses. Nature (Lond.) 164, 767 (1949).

B₂₃ BAUER, D. J.: Possible correlation between virus particle size and activity of antiviral agents. Nature (Lond.) 209, 639—640 (1966).

B₂₄ BAYANDINA, S. A., A. K. SVETLOVA, and V. V. RITOVA: Clinical peculiarities of the course of influenza A 2 in infants (Russian). Pediatriya 36, 38—43 (1958).

B₂₅ BAYDAKOVA, Z. L., and L. A. ZIL'BER: Adaptive variability of influenza virus (Russian). Zh. Mikrobiol. (Mosk.) 12, 86 (1948).

B₂₆ BAYDAKOVA, Z. L., and L. A. ZIL'BER: Influenza virus variability (Russian). Vop. Virus. 2, 204 (1949).

B₂₇ BAYER, M. E., and E. MANNWEILER: Antigen antibody reactions of influenza virus as seen in the electron microscope. Arch. ges. Virusforsch. 13, 541—547 (1963).

B₂₈ BEALE, A. J.: A comparison of the soluble antigen production by tissues infected with preparations of extracellular and intracellular influenza virus. J. Hyg. (Lond.) 52, 225—229 (1954).

B₂₉ BEALE, A. J., and N. B. FINTER: The infectivity of chorioallantoic membrane influenza virus and incomplete influenza virus by the six hour soluble antigen production test. J. Hyg. (Lond.) 54, 68—78 (1956).

B₃₀ BEARD, J. W., D. G. SHARP, A. R. TAYLOR, I. W. McLEAN, D. BEARD, E. FELLER, and J. H. DINGLE: Ultracentrifugal, chemical, and electron microscopic identification of the influenza virus. Sth. med. J. 37, 313—320 (1944).

B₃₁ BECKER, W. B.: Electron microscopy of Tern virus. S. Afr. J. Lab. clin. Med. 8, 163—164 (1962).

B₃₂ BECKER, W. B.: The morphology of Tern virus. Virology 20, 318—327 (1963).

B₃₃ BECKER, W. B.: The isolation and classification of Tern virus. Influenza virus A/Tern/South Africa/1961. J. Hyg. (Lond.) 64, 309—320 (1966).

B₃₄ BECKER, W. B.: Experimental infection of common terns with Tern virus. J. Hygiene (Lond.) 65, 61—65 (1967).

B₃₅ BECKER, W. B., C. DAUGUET, and F. DELAHAYE: Electron morphology of Tern virus. Enantiomorphic nucleocapsid (French). Ann. Inst. Pasteur 106, 575—580 (1964).

B₃₆ BEDSON, S. P.: The mode of virus multiplication and the susceptibility of these agents to the antibiotics. J. clin. Path. 9, 83—93 (1956).

$B_{37}$ BELIKOV, P. F.: Complement fixation reaction as an index of immunity in influenza (Russian). Arkh. biol. nauk. **59**, 44 (1940).

$B_{38}$ BELYAVIN, G.: Influenza complement fixation. A simple quantitative micro method. J. Hyg. (Lond.) **51**, 492—501 (1953).

$B_{39}$ BELYAVIN, G.: The direct flocculation of influenza virus. Lancet, 698—701 (1955).

$B_{40}$ BELYAVIN, G.: Normal serum inhibitors of influenza virus flocculation. Brit. J. exp. Path. **37**, 75—80 (1956).

$B_{41}$ BELYAVIN, G.: The influenza virus flocculation reaction as a method of antigenic typing. J. Hyg. (Lond.) **55**, 281—289 (1957).

$B_{42}$ BELYAVIN, G., and A. COHEN: Chromatography of influenza virus haemagglutination inhibitors. Virology **18**, 144—145 (1962).

$B_{43}$ BENNET, I. L., and R. R. WAGNER: Production of fever by influenza viruses. 5. Effect of nitrogen mustard $HN_2$. Bull. Johns Hopk. Hosp. **97**, 43—46 (1955).

$B_{44}$ BENNET, I. L., R. R. WAGNER, and V. S. LeQUIRE: The production of fever by influenza viruses. 2. Tolerance in rabbits to the pyrogenic effect of influenza viruses. J. exp. Med. **90**, 335—347 (1949).

$B_{45}$ BERGANINI, L., and O. BARTOLOMEI-CORSI: Activity of a bacterial mucinase on some inhibitors of influenza virus haemagglutination (Italian). Sperimentale **106**, 76—87 (1956).

$B_{46}$ BERLIN, B. S.: Gross physical properties of emulsified influenza virus vaccines and the adjuvant response. J. Immunol. **85**, 81—89 (1960).

$B_{47}$ BERLIN, B. S.: Antibody responses induced by aqueous and emulsified influenza virus vaccines in mice exposed daily to X rays. J. Immunol. **89**, 906—909 (1962).

$B_{48}$ BERLIN, B. S.: Antibody responses after injection of aqueous and emulsified influenza virus vaccines in mice chronically exposed to gamma rays. Radiation Res. **18**, 223—230 (1963).

$B_{49}$ BERLIN, B. S.: The nature of antibodies in mice injected with influenza virus vaccine. Proc. Soc. exp. Biol. (N.Y.) **113**, 1013—1016 (1963).

$B_{50}$ BERLIN, B. S.: Interruption of macroglobulin synthesis in mice by exposure to X rays. J. Immunol. **93**, 315—318 (1964).

$B_{51}$ BERLIN, B. S.: Sparing effect of X rays for mice inoculated intranasally with egg-adapted influenza virus, CAM strain. Proc. Soc. exp. Biol. (N.Y.) **117**, 864—869 (1964).

$B_{52}$ BERLIN, B. S.: Radiosensitivity of YM antibody response in mice injected with killed influenza virus. Radiation Res. **26**, 554—566 (1965).

$B_{53}$ BERLIN, B. S.: Early antibody responses of mice given a single injection of emulsified influenza virus vaccine. J. Immunol. **96**, 424—429 (1966).

$B_{54}$ BERLYANT, M. L., and F. G. EPSHTEYN: Some material on influenza therapy (Russian). Vo.-med. Zh. **12**, 74—75 (1953).

$B_{55}$ BERLYANT, M. L., and N. N. ORLOVA: Epidemiological observations in influenza (Russian). Vo.-med. Zh. **4**, 79 (1951).

$B_{56}$ BERNKOPF, H.: Differences in haemagglutination by strains of influenza virus in the presence of egg white and normal serum. Proc. Soc. exp. Biol. (N.Y.) **71**, 160—163 (1949).

$B_{57}$ BERNKOPF, H.: Cultivation of influenza virus in the chorioallantoic membrane of de-embryonated eggs. Proc. Soc. exp. Biol. (N.Y.) **72**, 680—682 (1949).

$B_{58}$ BERNKOPF, H.: Study of infectivity and haemagglutination of influenza virus in de-embryonated eggs. J. Immunol. **65**, 571—583 (1950).

$B_{59}$ BERNKOPF, H.: Experiences with the laboratory diagnosis of Asian influenza in Israel. Israel med. J. **17**, 209—212 (1958).

$B_{60}$ BERRY, L. J., R. B. MITCHELL, and D. RUBENSTEIN: Effect of acclimatisation to altitude on susceptibility of mice to influenza A virus infection. Proc. Soc. exp. Biol. (N.Y.) **88**, 543—548 (1955).

$B_{61}$ BEUTLER, E., and H. M. GEZON: The effect of total body X irradiation on the susceptibility of mice to influenza A virus infection. J. Immunol. **68**, 227—242 (1952).

B₆₂　BEVERIDGE, W. I. B.: Lack of increase of antibody after second injection of influenza virus in man. Aust. J. exp. Biol. med. Sci. **22**, 301—305 (1944).

B₆₃　BEVERIDGE, W. I. B.: Some topical comments on influenza in horses. Vet. Rec. **77**, 427—428 (1965).

B₆₄　BEVERIDGE, W. I. B.: Public health significance of animal influenza. Int. Congr. comp. Path. Beirut (1966).

B₆₅　BEVERIDGE, W. I. B., and F. M. BURNET: The cultivation of viruses and Rickettsia in the chick embryo. Med. Res. Counc. Rep. 256 (1946).

B₆₆　BEVERIDGE, W. I. B., L. W. MAHAFFEY, and M. A. ROSE: Influenza in horses. Vet. Rec. **77**, 57—59 (1965).

B₆₇　BIDDLE, F., and G. BELYAVIN: The haemagglutination inhibitor in edible bird nest: Its biological and physical properties. J. gen. Microbiol. **31**, 31—44 (1963).

B₆₈　BIDDLE, F., and A. COHEN: Erythrocyte receptors for type A 2 influenza viruses. Nature (Lond.) **196**, 297—298 (1962).

B₆₉　BIDDLE, F., and J. P. STEVENSON: Flocculation of influenza viruses by horse serum inhibitor. Nature (Lond.) **209**, 1223—1225 (1966).

B₇₀　BIELING, R., and L. OELRICHS: On the influenza virus inhibitor of hens eggs and on the immunity of the offspring of parents immunised against that virus. Z. Hyg. Infekt.-Kr. **127**, 592—603 (1948).

B₇₁　BIRCH-ANDERSEN, A., and K. PAUCKER: Studies on the structure of influenza virus. 2. Ultrathin sections of infectious and non-infectious particles. Virology **8**, 21—40 (1959).

B₇₂　BIRKOVSKAYA, E. A.: Oxygen treatment of patients with influenza and pneumonia during the influenza pandemic in 1957 (Russian). Aziatskii gripp (Sborn. nauch. rabot) Kiev, 237—240 (1958).

B₇₃　BJORKMANN, S. E., and F. L. HORSFALL: The production of a persistent alteration in influenza virus by Lanthanum or ultraviolet irradiation. J. exp. Med. **88**, 445—461 (1948).

B₇₄　BLASKOVIC, D.: Influenza A and B antibodies in mothers and their infants up to six months after birth. (Czech). Čas. Lék. čes. (Prague) **89**, 1195—1199 (1950).

B₇₅　BLASKOVIC, D.: Some properties of S and G antigens and haemagglutinins of myxoviruses (German). Zbl. Bakt. I. Abt. Ref. **175**, 396 (1960).

B₇₆　BLASKOVIC, D., P. ALBRECHT, V. LACKOVIC, J. LESSO, V. RATHOVA, and B. STYK: The use of fluorescent antibody method for rapid diagnosis during an epidemic of influenza (Czech). Čs. Epidem. **12**, 129—139 (1963).

B₇₇　BLASKOVIC, D., L. BORECKY, and V. RATHOVA: An attempt to prove the circulation of influenza virus in a pre-epidemic period by means of susceptible animals. Čs. Epidem. **8**, 376—387 (1959).

B₇₈　BLASKOVIC, D., N. A. MAKSIMOVIC, B. STYK, and P. ALBRECHT: Course of adaptation of the inhibitor resistant influenza virus A 2 to ferrets. Čs. Epidem. **10**, 158—165 (1961).

B₇₉　BLASKOVIC, D., and V. RATHOVA: Influenza virus A, A prime, B, C, and Shope Iowa 15 antibody titres in population of Czechoslovakia. Čs. Epidem. **5**, 113 (1956).

B₈₀　BLASKOVIC, D., V. RATHOVA, and L. BORECKY: Antigenic relationships to myxoviruses of a virus isolated from a respiratory infection in ducks. Acta virol. **3**, 17—24 (1959).

B₈₁　BLASKOVIC, D., and J. E. SALK: A method applicable to the standardisation of influenza virus vaccines. Proc. Soc. exp. Biol. (N.Y.) **65**, 352—359 (1947).

B₈₂　BLASKOVIC, D., A. A. SMORODINTSEV, and B. STYK: Problems of pathogenesis and immunology of virus infections. Acta virol. **3**, 97—106 (1959).

B₈₃　BLASKOVIC, D., and F. SOKOL: Some properties of the G antigen and haemagglutinin of influenza virus A/FE/Singapore/1957 (Czech). Biol. Prace **4**, 75—77 (1958).

$B_{84}$ BLASKOVIC, D., J. SZANTO, P. ALBRECHT, E. SADECKY, and V. LACKOVIC: Demonstration of swine influenza virus in pigs by the fluorescent antibody method. Acta virol. 8, 401—409 (1964).

$B_{85}$ BLOUGH, H. A.: The role of the surface state in the morphogenesis of influenza virus filaments. Virology 19, 112—114 (1963).

$B_{86}$ BLOUGH, H. A.: The effect of vitamin A on influenza virus in ovo: Morphological aspects. Biochem. J. 88, 11 (1963).

$B_{87}$ BLOUGH, H. A.: The effect of vitamin A alcohol on the morphology of myxoviruses. 1. The production and comparison of artificially produced filamentous virus. Virology 19, 349—358 (1963).

$B_{88}$ BLOUGH, H. A.: A molecular approach to teratogenesis: Effect of vitamin A on influenza virus in ovo. Nature (Lond.) 199, 33—35 (1963).

$B_{89}$ BLOUGH, H. A.: Role of the surface state in the development of myxoviruses. Ciba Found. Symp. "Cellular biology of myxovirus infections" 120—151 (1964).

$B_{90}$ BLOUGH, H. A.: Selective inactivation of biological activity of myxovirus by glutaraldehyde. J. Bact. 92, 266—268 (1966).

$B_{91}$ BLYUMBERG, N. A.: Adaption variability of the influenza virus (Russian). Dissertation, Moscow (1949).

$B_{92}$ BOAND, A. V., and J. E. KEMPF: The identification of influenza virus antibodies by means of the collodion particle agglutination reaction. J. Lab. clin. Med. 37, 914—923 (1951).

$B_{93}$ BOAND, A. V., J. E. KEMPF, and R. J. HANSON: Phagocytosis of influenza virus. 1. In vitro observations. J. Immunol. 79, 416—421 (1957).

$B_{94}$ BODILY, H. L., M. COREY, and M. D. EATON: Precipitation and concentration of influenza virus with alum. Proc. Soc. exp. Biol. (N.Y.) 52, 165—168 (1943).

$B_{95}$ BODILY, H. L., and M. D. EATON: Specificity of the antibody response of human beings to strains of influenza virus. J. Immunol. 45, 193—204 (1942).

$B_{96}$ BOGOCH, S., P. LYNCH, and A. S. LEVINE: Influence of brain ganglioside upon the neurotoxic effect of influenza virus in mouse brain. Virology 7, 161—169 (1959).

$B_{97}$ BOHM, P., J. ROSS, and L. BAUMEISTER: The cleavage of N-acetyl neuraminic acid from serum by influenza B virus and by Clostridium Welchii A toxin (German). Hoppe-Seylers Z. physiol. Chem. 308, 181—182 (1957).

$B_{98}$ BOLDASOV, V. K.: Studies on the enzymic properties of inhibitor sensitive and inhibitor resistant strains of A2 influenza viruses. Acta virol. 7, 361—367 (1963).

$B_{99}$ BOLDYREVA, A. S., and I. A. KISELEVA: Comparative evaluation of methods used for the removal of non-specific inhibitors in the HI reaction (Russian). "Problemy grippa i ostrykh respiratornykh zabolevanii." Tez. dokl. nauch. Sess. Inst. AMN SSSR. Moscow, 50—51 (1959).

$B_{100}$ BORDET, P., and L. QUERSIN-THIRY: Enhancing effect of influenza virus on associated infections (French). Ann. Inst. Pasteur 81, 394—406 (1951); 84, 695—702 (1953).

$B_{101}$ BORECKY, L.: On the question of virus receptors. The selective agglutinability of erythrocytes of the ground squirrel (Citellus Citellus). Folia biol. (Prague) 2, 344—350 (1956).

$B_{102}$ BORECKY, L.: The sensitivity of A-PR influenza viruses to inhibitors in normal animal sera (Czech). Biol. Prace 4, 32—38 (1958).

$B_{103}$ BORECKY, L.: A factor destroying virus receptors in pneumococcal cultures. 2. The effect of pneumococcal factor 103 on erythrocyte receptors and inhibitors. Acta virol. 2, 201—207 (1958).

$B_{104}$ BORECKY, L.: A factor destroying virus receptors in pneumococcal cultures. 3. The effect of pneumococcal factor 103 in experiments in vivo. Acta virol. 3, 9—16 (1959).

$B_{105}$ BORECKY, L., D. KOCISKOVA, and L. HANA: An attempt to affect the level of serum inhibitors of myxoviruses *in vivo*. 1. Effect of ethionine administration on the level of inhibitors in guinea-pig sera. Acta virol. 5, 236—244 (1961).

$B_{106}$ BORECKY, L., V. LACKOVIC, and E. MRENA: The effect of trypsin on influenza virus. Acta virol. 8, 555 (1964).

$B_{107}$ BORECKY, L., and V. RATHOVA: Investigations of possible cross-reactions between preparations of salmonellae containing Vi antigen and inhibitors or antibodies against myxoviruses. Acta virol. 7, 339—349 (1963).

$B_{108}$ BORECKY, L., V. RATHOVA, and D. KOCISKOVA: Attempt to affect the level of serum inhibitors of myxoviruses *in vivo*. 2. Effect of ethionine administration on virus multiplication. Acta virol. 6, 97—104 (1962).

$B_{109}$ BORECKY, L., V. RATHOVA, D. KOCISKOVA, and L. HANA: Effect of dl ethionine on some factors of non-specific resistance to myxoviruses. J. Hyg. Epidem. (Praha) 6, 65—70 (1962).

$B_{110}$ BORECKY, L., and J. ZAVADA: Effect of anti-cellular sera on infection with myxoviruses. 1. The effect of antichorioallantoic serum. Acta virol. 4, 110—123 (1960).

$B_{111}$ BORETTI, G., A. DI MARCO, and P. JULITA: Preparation and characterisation of an influenza virus inhibitor from horse serum. G. Microbiol. (Naples) 12, 55—69 (1964).

$B_{112}$ BORGHANS, J. G. A., O. MAKSTEINIEKS, J. VERSTEEG, and J. D. VERLINDE: A mixed epidemiological outbreak of acute respiratory disease caused by influenza A virus and APC virus in a group previously vaccinated against influenza A (Dutch). Ned. T. Geneesk. 100, 2110—2116 (1956).

$B_{113}$ BOUDREAULT, A., and V. PAVILANIS: Relationship between cytopathogenic, haemadsorption, and haemagglutination properties of influenza virus (French). Canad. J. Microbiol. 7, 347—353 (1961).

$B_{114}$ BOUDREAULT, A., and V. PAVILANIS: The development of some biological properties of 4 influenza virus strains after successive passage at 29°C and 41°C. Research on cold mutants. Rev. Canad. Biol. 23, 253—260 (1964).

$B_{115}$ BOUG, J., C. A. DE BOCK, H. D. MOED, and A. J. KLEIN: Antiviral action of derivatives of w-aminoacetophenone. Brit. J. Pharmacol. 13, 404—410 (1958).

$B_{116}$ BOVARNICK, M., and P. M. DE BURGH: Virus haemagglutination. Science, 105, 550—552 (1947).

$B_{117}$ BOWERS, R. H., O. L. DAVIES, and W. HURST: A comparison between intranasal instillation and natural inhalation as a means of infecting mice with influenza virus. Brit. J. exp. Path. 33, 601—609 (1952).

$B_{118}$ BOZZO, A.: Studies of the antigenic composition of influenza B viruses. Bull. Wld Hlth Org. 5, 149—163 (1952).

$B_{119}$ BOZZO, A., A. JACONO, S. RICCIARDI, and G. STARACE: Serological study of four strains of influenza virus isolated in various parts of south Italy in 1953 (Italian). Rev. Inst. Sieroter. ital. 28, 424—435 (1953).

$B_{120}$ BRANDAO, H.: Use of the Do unit (zero mortality) in virus titrations (Portuguese). Rev. Inst. Med. trop. S. Paulo 4, 343—350 (1962).

$B_{121}$ BRANS, L. M.: Studies on the antigenic composition of influenza virus B strains. Onderzoek. Mededel. Inst. praevent. Geneesk. (Leiden) No 9 (1952).

$B_{122}$ BRANS, L. M., E. HERTZBERGER, and J. L. BINKHORST: Studies on the elimination of non-specific inhibitors in sera against influenza viruses with the aid of filtrates of Vibrio Cholerae. Antonie v. Leeuwenhoek 19, 309—323 (1953).

$B_{123}$ BREITENFELD, P. M., and W. SCHAFER: The formation of fowl plague virus antigens in infected cells as studied with fluorescent antibodies. Virology 4, 328—345 (1957).

$B_{124}$ BRENNER, S., and R. W. HORNE: A negative staining method for high resolution electron microscopy of viruses. Biochim. biophys. Acta (Amst.) 34, 103—110 (1959).

B₁₂₅    BRIGHTMAN, I. J., and J. D. TRASK: Recovery of a filtrable virus from children sick with influenza. 1. Epidemiologic and clinical observations.

B₁₂₆    2. The experimental disease in ferrets. Amer. J. Dis. Child. 52, 67—77 and 78—91 (1936).

B₁₂₇    BRIODY, B. A.: Haemagglutination by influenza virus. 1. Modification of the 0 phase of influenza A virus. J. infect. Dis. 83, 283—287 (1948).

B₁₂₈    2. The existence of two phases of influenza B virus.

B₁₂₉    3. The XO phase of influenza A virus. J. exp. Med. 83, 288—292 and 293—298 (1948).

B₁₃₀    BRIODY, B. A.: Variation in influenza viruses. Bact. Rev. 14, 65—95 (1950).

B₁₃₁    BRIODY, B. A., and W. A. CASSEL: Adaptation of influenza virus to mice. 2. Changes in the growth curve of an A prime strain of influenza virus. J. Immunol. 74, 37—40 (1955).

B₁₃₂    BRIODY, B. A., W. A. CASSEL, and M. A. MEDILL: Adaptation of influenza virus to mice. 3. Development of resistance to B inhibitor. J. Immunol. 74, 41—45 (1955).

B₁₃₃    BRONCOVA, O., and V. BYDZOVSKY: Effect of some 5-arylpyrimidines on influenza, vaccinia and Newcastle disease viruses (Czech). Čs. Epidem 11, 179—188 (1962).

B₁₃₄    BRONITKI, A., A. PETRESCU, P. PETRESCU, L. ROTHSCHILD, and P. ATHANASIU-STORESCU: The isolation of a strain of influenza virus of predominantly neurotropic character (Rumanian). Stud. Cercet. Inframicrobiol. 8, 561—569 (1957).

B₁₃₅    BRUDNAYA, E. I., L. YA. ZAKSTEL'SKAYA, and K. K. LUNEVA: Serological reactions in persons vaccinated a live anti-influenza A2 vaccine (Russian). Vop. Virus 5, 133—135 (1958).

B₁₃₆    BRUNNER, K. T.: The formation of anti endotoxin in rabbits after injection of influenza virus (German). Path. et Microbiol. (Basle) 25, 91—100 (1962).

B₁₃₇    BRUNO-LOBO, M.: "Gripe Asiatica" Inst. Microbiol. Universidade do Brasil (Portuguese). Rio de Janeiro (1958).

B₁₃₈    BRUNO-LOBO, G. G., M. BRUNO-LOBO, J. V. VASCONCELLOS, and J. TRAVASSOS: Studies on Asian influenza in Rio de Janeiro. 3. Sensitivity of the virus to non-specific inhibitors (Portuguese). An. Microbiol. 5, 69—93 (1957).

B₁₃₉    BRUNO-LOBO, M., and J. TRAVASSOS: Studies on Asian influenza in Rio de Janeiro. 6. Post-vaccinal immunity (Portuguese). An. Microbiol. 6, 151—176 (1958).

B₁₄₀    BRYANS, J. T.: Viral respiratory disease of horses. Scientific Proceed. 101st Ann. Meet. Amer. vet. Med. Ass. July 1964.

B₁₄₁    BUDDINGH, G. J.: Experimental combined viral and bacterial infection (influenza C and Haemophilus influenzae type B) in embryonated eggs. J. exp. Med. 104, 947—958 (1956).

B₁₄₂    BUKRINSKAYA, A. G.: Direct detection of influenza virus in human embryo tissue cultures (Russian). Vop. Virus. 3, 162—165 (1958).

B₁₄₃    BUKRINSKAYA, A. G.: Behaviour of influenza virus type A2 in tissue cultures of man and animals (Russian). Vop. Virus. 5, 61—68 (1958).

B₁₄₄    BUKRINSKAYA, A. G.: Formation of incomplete influenza A2 virus and multiplication of virus in tissue cultures of variable sensitivity. Probl. Virol. (N.Y.) 5, 481—485 (1960).

B₁₄₅    BUKRINSKAYA, A. G.: Resistance of synthesis of influenza virus ribonucleic acid to the action of actinomycin D. Nature (Lond.) 204, 205—206 (1964).

B₁₄₆    BUKRINSKAYA, A. G., E. A. VLADIMERCEVA, A. K. GITELMAN, and Y. A. SMIRNOV: Myxovirus induced early RNA and early proteins. Abstr. 9th int. Congr. Microbiol. Moscow, 494 (1966).

B₁₄₇    BUKRINSKAYA, A. G., and G. K. VORBUNOVA: Reproduction of influenza virus RNA in the presence of low concentrations of actinomycin D (Russian). Vop. Virus. 6, 657 (1964).

B₁₄₈    BULL, D. R., and F. M. BURNET: Experimental immunisation of volunteers against influenza virus B. Med. J. Aust. 1, 389—394 (1943).

B₁₄₉ BURKE, D. C., and A. ISAACS: Further studies on interferon. Brit. J. exp. Path. 39, 78—84 (1958).

B₁₅₀ BURKE, D. C., and A. ISAACS: Interferon: Relation to heterologous interference and lack of antigenicity. Acta virol. 4, 215—219 (1960).

B₁₅₁ BURKE, D. C., A. ISAACS, and J. WALKER: The nucleic acid content of influenza virus. Biochim. et biophys. Acta (Amst.) 26, 576—584 (1957).

B₁₅₂ BURKE, D. C., and J. M. MORRISON: Interferon production in chick embryo cells. 2. The role of DNA. Virology 28, 108—114 (1966).

B₁₅₃ BURKE, D. C., and J. ROSS: Molecular weight of chick interferon. Nature (Lond.) 208, 1297—1299 (1965).

B₁₅₄ BURNET, F. M.: Influenza virus isolated from an Australian epidemic. Med. J. Aust. 2, 651—653 (1935).

B₁₅₅ BURNET, F. M.: Propagation of the virus of epidemic influenza on the developing egg. Med. J. Aust. 2, 687—689 (1935).

B₁₅₆ BURNET, F. M.: Influenza virus on the developing egg. 1. Changes associated with the development of an egg-passage strain of virus. Brit. J. exp. Path. 17, 282—293 (1936).

B₁₅₇ BURNET, F. M.: Influenza virus on the developing egg. 5. Differentiation of two antigenic types of human influenza virus. Aust. J. exp. Biol. med. Sci. 15, 369—374 (1937).

B₁₅₈ BURNET, F. M.: Influenza virus infections of the chick embryo lung. Brit. J. exp. Path. 21, 147—153 (1940).

B₁₅₉ BURNET, F. M.: Influenza virus infection of the chick embryo by the amniotic route. 1. General character of the infections.

B₁₆₀ 2. Titration and serum neutralisation tests. Aust. J. exp. Biol. med. Sci. 18, 353—360, and 19, 39—44 (1940).

B₁₆₁ BURNET, F. M.: Influenza virus A infections of Cynomolgous monkeys. Aust. J. exp. Biol. med. Sci. 19, 281—290 (1941).

B₁₆₂ BURNET, F. M.: Influenza virus B. 1. Observations on the growth in chick embryos and on the occurrence of antibodies in Australian serum.

B₁₆₃ 2. Immunisation of human volunteers with living attenuated virus. Med. J. Aust. 1, 671—673 and 673—674 (1942).

B₁₆₄ BURNET, F. M.: Immunisation against epidemic influenza with living and attenuated virus. Med. J. Aust. 1, 385—389 (1943).

B₁₆₅ BURNET, F. M.: Characteristics of the influenza virus-antibody reaction as tested by the method of allantoic inoculation. Aust. J. exp. Biol. med. Sci. 21, 231—238 (1943).

B₁₆₆ BURNET, F. M.: Modification of human red cells by virus action. 3. A sensitive test for mumps antibody in human serum by the agglutination of human red cells coated with a virus antigen. Brit. J. exp. Path. 27, 244—247 (1946).

B₁₆₇ BURNET, F. M.: Mucins and mucoids in relation to influenza virus action. 3. Inhibition of virus haemagglutination by glandular mucins.

B₁₆₈ 4. Inhibition by purified mucoid of infection and haemagglutination with the virus strain WSE.

B₁₆₉ 5. The destruction of "Francis inhibitor" activity in a purified mucoid by virus action. Aust. J. exp. Biol. med. Sci. 26, 371—379, 381—387, and 389—402 (1948).

B₁₇₀ BURNET, F. M.: The initiation of cellular infection by influenza and related viruses. Lancet 7—11 (1948).

B₁₇₁ BURNET, F. M.: Inactivation of the receptor destroying enzyme of V. Cholerae by immune sera. Aust. J. exp. Biol. med. Sci. 27, 217—222 (1949).

B₁₇₂ BURNET, F. M.: The effect of periodate on the virus-inhibiting qualities of mucoids in solution. Aust. J. exp. Biol. med. Sci. 27, 361—374 (1949).

B₁₇₃ BURNET, F. M.: Inhibition of influenza virus haemagglutination by mucoids. 3. Specific relation between active virus and the corresponding indicator strains in studies of soluble inhibitors. Aust. J. exp. Biol. med. Sci. 27, 575—579 (1949).

B₁₇₄ BURNET, F. M.: The interaction of virus and cell surface. Proc. roy. Soc. Ser. B 138, 47—44 (1951).

B₁₇₅ BURNET, F. M.: Some biological implications of studies on influenza viruses. 1. The process of infection by the virus.

B₁₇₆ 2. Reproduction and variation in influenza viruses. Bull. Johns Hopk. Hosp. 88, 119—137 and 137—157 (1951).

B₁₇₇ BURNET, F. M.: Mucoproteins in relation to virus action. Physiol. Rev. 31, 131—150 (1951).

B₁₇₈ BURNET, F. M.: A genetic approach to variation in influenza viruses. 1. The characters of three sub-strains of influenza A virus (WS).

B₁₇₉ 2. Variation in the strain NWS on allantoic passage. J. gen. Microbiol. 5, 46—53 and 54—58 (1951).

B₁₈₀ BURNET, F. M.: Irreversible binding of influenza virus to the red cell surface. Aust. J. exp. Biol. med. Sci. 30, 119—128 (1952).

B₁₈₁ BURNET, F. M.: Influenza virus group. Ann. N.Y. Acad. Sci. 56, 567 (1953).

B₁₈₂ BURNET, F. M.: Virulence in animal viruses. Lancet, 559—563 (1954).

B₁₈₃ BURNET, F. M.: The riddle of influenza virus. Endeavour 14, 5—11 (1955).

B₁₈₄ BURNET, F. M.: Filamentous forms of influenza. Nature (Lond.) 177, 130 (1956).

B₁₈₅ BURNET, F. M.: Structure of influenza virus. Science 123, 1101—1104 (1956).

B₁₈₆ BURNET, F. M.: Genetics of animal viruses. Handbuch Virusforsch. 4, 3, Erg.-Bd. (1958). Springer-Verlag, Wien.

B₁₈₇ BURNET, F. M., and S. G. ANDERSON: Modification of human red cells by virus action. 2. Agglutination of modified human red cells by sera from cases of infectious mononucleosis. Brit. J. exp. Path. 27, 236—244 (1946).

B₁₈₈ BURNET, F. M., and W. I. B. BEVERIDGE: Titration of antibody against influenza by allantoic inoculation of the developing chick embryo. Aust. J. exp. Biol. med. Sci. 21, 71—77 (1943).

B₁₈₉ BURNET, F. M., W. I. B. BEVERIDGE, and D. R. BULL: Study of a strain of influenza B virus isolated by chick embryo inoculation. Aust. J. exp. Biol. med. Sci. 22, 9—16 (1944).

B₁₉₀ BURNET, F. M., W. I. B. BEVERIDGE, D. R. BULL, and E. CLARKE: Investigation of an influenza epidemic in military camps in Victoria, May 1942. Med. J. Aust. 2, 371—376 (1942).

B₁₉₁ BURNET, F. M., W. I. B. BEVERIDGE, J. McEWIN, and W. C. BRAKE: Studies on the Hirst haemagglutination reaction with influenza and Newcastle disease viruses. Aust. J. exp. Biol. med. Sci. 23, 177—192 (1945).

B₁₉₂ BURNET, F. M., and D. R. BULL: Changes in influenza virus associated with adaptation to passage in chick embryos. Aust. J. exp. Biol. med. Sci. 21, 55—69 (1943).

B₁₉₃ BURNET, F. M., and E. CLARKE: Influenza; A survey of the last 50 years in the light of modern work on the virus of epidemic influenza. Monograph No. 4, Walter and Eliza Hall Inst. Melbourne, 1942.

B₁₉₄ BURNET, F. M., and M. EDNEY: Recombinant viruses obtained from double infections with the influenza A viruses. Aust. J. exp. Biol. med. Sci. 29, 353—362 (1951).

B₁₉₅ BURNET, F. M., and M. EDNEY: Influence of ions on the interaction of influenza virus and cellular receptors or soluble inhibitors of haemagglutination. Aust. J. exp. Biol. med. Sci. 30, 105  118 (1952).

B₁₉₆ BURNET, F. M., and M. FOLEY: The results of intranasal inoculation of modified and unmodified influenza virus strains in human volunteers. Med. J. Aust. 2, 655—659 (1940).

B₁₉₇ BURNET, F. M., and M. FOLEY: Two methods for the detection of influenza virus in human throat washings without the use of ferrets. Med. J. Aust. 3, 68—72 (1941).

B₁₉₈ BURNET, F. M., and K. B. FRASER: Studies on recombination with influenza viruses in the chick embryo. 1. Invasion of the chick embryo by influenza viruses. Aust. J. exp. Biol. med. Sci. 30, 447—558 (1952).

B₁₉₉ BURNET, F. M., K. B. FRASER, and P. E. LIND: Genetic interaction between influenza viruses. Nature (Lond.) 171, 163 (1953).

B₂₀₀ BURNET, F. M., and P. E. LIND: A genetic approach to variation in influenza viruses.

B₂₀₁ 3. Recombination of characters in influenza virus strains used in mixed infections. 4. Recombination of characters between influenza virus A strain NWS and strains of different serological sub-types. J. gen. Microbiol. 5, 59—66 and 67—82 (1951).

B₂₀₂ BURNET, F. M., and P. E. LIND: Studies on recombination with influenza viruses in the chick embryo. 3. Reciprocal genetic interaction between two influenza virus strains. Aust. J. exp. Biol. med. Sci. 30, 469—477 (1952).

B₂₀₃ BURNET, F. M., and P. E. LIND: Reactivation of heat-inactivated influenza virus by recombination. Aust. J. exp. Biol. med. Sci. 32, 133—143 (1954).

B₂₀₄ BURNET, F. M., and P. E. LIND: Recombination of influenza viruses in the de-embryonated egg. 1. The use of periodate-treated sera for in vitro characterisation of influenza virus strains. Aust. J. exp. Biol. med. Sci. 32, 145—151 (1954).

B₂₀₅ BURNET, F. M., and P. E. LIND: Recombination of influenza viruses in the de-embryonated egg. 2. The conditions for recombination and the evidence for the possible existence of diploid influenza virus. Aust. J. exp. Biol. med. Sci. 32, 153—163 (1954).

B₂₀₆ BURNET, F. M., and P. E. LIND: Genetics of virulence in influenza viruses. Nature (Lond.) 173, 627—630 (1954).

B₂₀₇ BURNET, F. M., and P. E. LIND: Reversion to virulence in an influenza virus mutant. Aust. J. exp. Biol. med. Sci. 35, 225—239 (1957).

B₂₀₈ BURNET, F. M., and P. E. LIND: Studies on the filamentous forms of influenza virus with special reference to the use of dark-ground microscopy. Arch. ges. Virusforsch. 7, 413—428 (1957).

B₂₀₉ BURNET, F. M., P. E. LIND, and B. PERRY: The action of ribonuclease on the multiplication of influenza viruses in the de-embryonated egg. Aust. J. exp. Biol. med. Sci. 35, 517—529 (1957).

B₂₁₀ BURNET, F. M., P. E. LIND, and K. M. STEVENS: Production of incomplete influenza virus in the de-embryonated egg. Aust. J. exp. Biol. med. Sci. 33, 127—141 (1955).

B₂₁₁ BURNET, F. M., and D. LUSH: Influenza virus on the developing egg. 8. A comparison of two antigenically dissimilar strains of human influenza virus after full adaptation to the egg membrane. Aust. J. exp. Biol. med. Sci. 16, 261—274 (1938).

B₂₁₂ BURNET, F. M., and D. LUSH: Influenza virus strains isolated from the Melbourne 1939 epidemic. Aust. J. exp. Biol. med. Sci. 18, 49—55 (1940).

B₂₁₃ BURNET, F. M., D. LUSH, and A. V. JACKSON: A virus inactivating agent from human nasal secretion. Brit. J. exp. Path. 20, 377—385 (1939).

B₂₁₄ BURNET, F. M., and J. F. McCREA: Inhibitory and inactivating action of normal ferret sera against an influenza virus strain. Aust. J. exp. Biol. med. Sci. 24, 277—282 (1946).

B₂₁₅ BURNET, F. M., J. F. McCREA, and J. D. STONE: Modification of human red cells by virus action. 1. The receptor gradient for virus action in human red cells. Brit. J. exp. Path. 27, 228—236 (1946).

B₂₁₆ BURNET, F. M., and J. D. STONE: The significance of primary isolation of influenza virus by inoculation of mice or of the allantoic cavity of chick embryos. Aust. J. exp. Biol. med. Sci. 23, 147—150 (1945).

B₂₁₇ BURNET, F. M., and J. D. STONE: Further studies on the O-D change in influenza virus A. Aust. J. exp. Biol. med. Sci. 23, 151—160 (1945).

B₂₁₈ BURNET, F. M., and J. D. STONE: A method for the isolation of influenza virus from throat washings without filtration. Aust. J. exp. Biol. med. Sci. 23, 161—163 (1945).

$B_{219}$ BURNET, F. M., and J. D. STONE: Serological response to influenza B infection in human beings. Differentiation of specific and non-specific type reactions. Aust. J. exp. Biol. med. Sci. 24, 207—212 (1946).

$B_{220}$ BURNET, F. M., and J. D. STONE: The receptor destroying enzyme of V. cholerae. Aust. J. exp. Biol. med. Sci. 25, 227—233 (1947).

$B_{221}$ BURNET, F. M., J. D. STONE, and S. G. ANDERSON: An epidemic of influenza B in Australia. Lancet 807—811 (1946).

$B_{222}$ BURNET, F. M., J. D. STONE, A. ISAACS, and M. EDNEY: The genetic character of the O-D change in influenza A. Brit. J. exp. Path. 30, 419—425 (1949).

$B_{223}$ BUSLENKO, A. I.: Epidemiologic and clinical characteristics of the course of influenza in T.B. patients (Russian). "Problemy grippa i ostrykh respiratornykh zabolevanii." Tez. dokl. nauch, Sess. Inst. AMN SSSR Moscow, 201—202 (1959).

$B_{224}$ BUZZELL, A., F. B. BRANDON, and M. A. LAUFFER: Effects of X radiation on influenza virus. Arch. Biochem. 58, 318—331 (1955).

$C_1$ CAIRNS, H. J. F.: Protection by receptor destroying enzyme against infection with a neurotropic variant of influenza virus. Nature (Lond.) 167, 335 (1951).

$C_2$ CAIRNS, H. J. F.: The growth of influenza viruses and Newcastle disease virus in mouse brain. Brit. J. exp. Path. 32, 110—117 (1951).

$C_3$ CAIRNS, H. J. F.: Quantitative aspects of influenza virus multiplication. 3. Average liberation time. J. Immunol. 69, 168—173 (1952).

$C_4$ CAIRNS, H. J. F.: The limited growth of influenza viruses in mouse brain. Aust. J. exp. Biol. med. Sci. 32, 123—128 (1954).

$C_5$ CAIRNS, H. J. F.: Multiplicity reactivation of influenza virus. J. Immunol. 75, 326—329 (1955).

$C_6$ CAIRNS, H. J. F.: Site of the chorioallantoic inhibitor of influenza virus haemagglutinin. Nature (Lond.) 178, 744—745 (1956).

$C_7$ CAIRNS, H. J. F.: The asynchrony of infection by influenza virus. Virology 3, 1—14 (1957).

$C_8$ CAIRNS, H. J. F., and M. EDNEY: Quantitative aspects of influenza virus multiplication. 1. Production of incomplete virus. J. Immunol. 69, 155—160 (1952).

$C_9$ CAIRNS, H. J. F., and S. FAZEKAS DE ST GROTH: The number of allantoic cells in the chick embryo. J. Immunol. 78, 191—200 (1957).

$C_{10}$ CAIRNS, H. J. F., and P. J. MASON: Production of influenza A virus in the cell of the allantois. J. Immunol. 71, 38—40 (1953).

$C_{11}$ CAJAL, N., O. BURDUCEA, S. MATEASCU, G. MARINESCU, M. CEPLEANU, and Y. CEPELOVICI: The development of experimental virus infections (influenza, poliomyelitis, coxsackie, rabies, and herpes) under the influence of radioactive phosphorus P 32 and radioactive iodine I 131 (Rumanian). Stud. Cercet. Inframicrobiol. 12, 29—37 (1961).

$C_{12}$ CANCHOLA, J. G., R. M. CHANOCK, B. C. JEFFRIES, E. E. CHRISTMAS, H. W. KIM, A. J. VARGOSCO, and R. H. PARROT: Recovery and identification of human myxoviruses. Bact. Rev. 29, 496—503 (1965).

$C_{13}$ CARLINFANTI, E.: Studies of the receptor destroying factor from different viruses (Italian). Riv. Ist. sieroter. ital. 23, 141—150 (1948).

$C_{14}$ CARLINFANTI, E.: Rapid liberation of influenza virus from red cells by Newcastle disease virus. Riv. Ist. sieroter. ital. 23, 151—155 (1948).

$C_{15}$ CARSON, S., and A. W. FRISCH: The inactivation of influenza viruses by tannic acid and related compounds. J. Bact. 66, 572—575 (1953).

$C_{16}$ CARUBELLI, R., V. P. BHAVANANDAN, and A. GOTTSCHALK: Studies on glycoproteins. 11. The O-glycosidic linkage of N-acetylgalactosamine to seryl and threonyl residues in ovine submaxillary gland glycoprotein. Biochim. biophys. Acta (Amst.) 101, 67—82 (1965).

$C_{17}$ CATEIGNE, G.: Anti-influenza vaccine (French). Rev. Prat. (Paris) 8, 1446—1451 (1958).

$C_{18}$ CATEIGNE, G., C. HANNOUN, and R. PANTHIER: Antigenic characters of influenza virus in the 1948—49 epidemic (French). C. R. Soc. Biol. (Paris) **143**, 812—814 (1949).

$C_{19}$ CATEIGNE, G., and P. RECULARD: Work on the adaptation of two strains of influenza virus A to chick embryo cells *in vitro* (French). Ann. Inst. Pasteur **11**, 229—234 (1966).

$C_{20}$ CENTANNI, E., and O. SAVUNOZZI: (1900) Cited by Stubbs, E. L. in "Diseases of poultry" 4th edit. Biester and Schwartz, Iowa State Univ. Press.

$C_{21}$ CHAGNON, A., J. C. GILKER, and A. BOUDREAULT: Infectivity and haemagglutinating properties of influenza virus at varying pH (French). Canad. J. Microbiol. **11**, 845—851 (1965).

$C_{22}$ CHALKINA, O. M.: Immunological changes in the blood of people vaccinated with the virus of epidemic influenza (Russian). Arkh. biol. nauk. **52**, 125 (1938).

$C_{23}$ CHALKINA, O. M.: Susceptibility of white rats to the influenza virus (Russian). Zh. Mikrobiol. (Mosk.) **10—11**, 23 (1943).

$C_{24}$ CHALKINA, O. M.: Increase in the immunogenic power of influenza vaccine with a biologic stimulator (Russian). Vop. Virus **2**, 235 (1949).

$C_{25}$ CHALKINA, O. M., S. A. BUROV, and N. A. ILIN: The evaluation of the epidemiologic effectiveness of live anti-influenza vaccine during the 1959 outbreaks of influenza A2 and B (Russian). "Problemy grippa i ostrykh respiratornykh zabolevanii." Tez. dokl. nauch. Sess. Inst. AMN SSSR, Moscow, 80—82 (1959).

$C_{26}$ CHALKINA, O. M., and A. G. GULAMOV: The use of white rats for the laboratory diagnosis of influenza (Russian). Zh. Mikrobiol. (Mosk.) **10—11**, 26 (1943).

$C_{27}$ CHALKINA, O. M., and O. S. ZAK: Virological characterisation of a Type A influenza outbreak (Russian). "Bibliografiya i annotatsii rabot otdela virusologii." Inst. Eksper. med. AMN SSSR Za, 1947—1949, Leningrad (1950).

$C_{28}$ CHAMBERS, L. A., and W. HENLE: Precipitation of active influenza A virus from extra-embryonic fluids by protamine. Proc. Soc. exp. Biol. (N.Y.) **48**, 481—483 (1941).

$C_{29}$ CHAMBERS, L. A., and W. HENLE: Studies on the nature of the virus of influenza. 1. The dispersion of the virus of influenza A in tissue emulsions and in extra-embryonic fluids of the chick.

$C_{30}$ 2. The size of the infectious unit in influenza A. J. exp. Med. **77**, 251—264, and 265—276 (1943).

$C_{31}$ CHANG, H. T., and J. E. KEMPF: The hypotensive action of influenza virus on rats. J. Immunol. **65**, 75—84 (1950).

$C_{32}$ CHANG, S. S., and C. Y. CHEN: Inhibitor spectra and effects of various treatments on the inhibitors of influenza virus haemagglutination. J. Formosan Med. Ass. **59**, 1110—1122 (1960).

$C_{33}$ CHAPMAN, J., and R. R. HYDE: Antigenic differences in viruses from cases of influenza and colds. Amer. J. Hyg. **31**, 46—68 (1940).

$C_{34}$ CHEBURKINA, N. V.: Study of cell metabolism in virus infections (Russian). Annot. rabot. Inst. virusol. AMN SSSR za 1949, Moscow, 8 (1950).

$C_{35}$ CHEBURKINA, N. V., and L. YA. ZAKSTELSKAYA: Metabolism in the chick embryo in influenzal infection (Russian). Zh. Mikrobiol. (Mosk.) **9**, 15 (1950).

$C_{36}$ CHEBURKINA, N. V., and L. YA. ZAKSTELSKAYA: Metabolism in the developing chick embryo in influenzal infection (Russian). Probl. obshch. virusol. Tez. dokl. 6-y Inst. virusol. AMN SSSR Moscow 58 (1953).

$C_{37}$ CHEN, C. Y., W. F. KUO, and S. S. CHANG: Studies on the strains of A2 (asian) influenza virus isolated in Taiwan. Mem. Coll. Med. Taiwan Univ. **11**, 1—10 (1965).

$C_{38}$ CHERNUKH, A. M., I. S. RUDAKOVA, N. S. TOLMACHEVA, and A. M. PROCHOROVA: A comparative study of the effect produced by esculamine and aminoadamantane on the experimental influenza. Abstr. 9th int. Congr. Microbiol., Moscow 545 (1966).

$C_{39}$ CHESBRO, W. R., and L. R. HADRICK: Comparison of the tendency to stick to glass and haemagglutination produced in erythrocyte suspensions by ferric chloride and by influenza virus. Nature (Lond.) 182, 1164 (1958).

$C_{40}$ CHEVANCE, L. G., and G. CATEIGNE: A new test for the study of influenza virus (French). Ann. Inst. Pasteur 104, 264—273 (1963).

$C_{41}$ CHOPPIN, P. W.: Plaque formation by influenza A2 virus in monkey kidney cells. Virology 18, 332—334 (1962).

$C_{42}$ CHOPPIN, P. W.: On the emergence of influenza virus filaments from host cells. Virology 21, 278—281 (1963).

$C_{43}$ CHOPPIN, P. W.: Multiplication of two kinds of influenza A2 virus particles in monkey kidney cells. Virology 21, 342—352 (1963).

$C_{44}$ CHOPPIN, P. W., J. S. MURPHY, and W. STOEKENIUS: The surface structure of influenza virus filaments. Virology 13, 548—550 (1961).

$C_{45}$ CHOPPIN, P. W., S. OSTERHOUT, and I. TAMM: Immunological characteristics of N.Y. strains of influenza virus from the 1957 pandemic. Proc. Soc. exp. Biol. (N.Y.). 98, 513—520 (1958).

$C_{46}$ CHOPPIN, P. W., and W. STOEKENIUS: Interactions of ether-disrupted influenza A2 virus with erythrocytes, inhibitors, and antibodies. Virology 22, 482—492 (1964).

$C_{47}$ CHOPPIN, P. W., and I. TAMM: Studies of two kinds of virus particles which comprise influenza A2 virus strains. 1. Characterisation of stable homogeneous substrains in reactions with specific antibody mucoprotein inhibitors, and erythrocytes.

$C_{48}$ 2. Reactivity with virus inhibitors in normal sera.

$C_{49}$ 3. Morphological characteristics: Independence of morphological and functional traits. J. exp. Med. 112, 895—920, 921—944, and 945—952 (1960).

$C_{50}$ CHU, C. M.: Agglutination of red blood cells of different animal species by influenza and Newcastle disease viruses. J. Hyg. (Lond.) 46, 239—246 (1948).

$C_{51}$ CHU, C. M.: Enzyme action of viruses and bacterial products on human red cells. Nature (Lond.) 162, 606—607 (1948).

$C_{52}$ CHU, C. M.: Inactivation of haemagglutinin and infectivity of influenza and Newcastle disease viruses by heat and by formalin. J. Hyg. (Lond.) 46, 247—251 (1948).

$C_{53}$ CHU, C. M.: The action of normal mouse serum on influenza virus. J. gen. Microbiol. 5, 739—757 (1951).

$C_{54}$ CHU, C. M.: The etiology and epidemiology of influenza: An analysis of the 1957 epidemic. J. Hyg. Epidem. (Praha) 2, 1—8 (1958).

$C_{55}$ CHU, C. M., C. H. ANDREWES, and A. W. GLEDHILL: Influenza in 1948—49. Bull. Wld Hlth Org. 3, 197—214 (1950).

$C_{56}$ CHU, C. M., I. M. DAWSON, and W. J. ELFORD: Filamentous forms associated with newly isolated influenza virus. Lancet 602 (1949).

$C_{57}$ CHU, C. M., C. SHAO, and C. C. HOU: Study of strains of influenza viruses isolated during the 1957 epidemic in Changchun (Russian). Vop. Virus 2, 278—281 (1957).

$C_{58}$ CLARK, R. J., A. ISAACS, and J. WALKER: Derivatives of 3:4 xylidine and related compounds as inhibitors of influenza virus: Relationships between chemical structure and biological activity. Brit. J. Pharmacol. 13, 424—435 (1958).

$C_{59}$ CLARKE, E., and F. P. O. NAGLER: Haemagglutination by viruses. The range of susceptible cells with special reference to agglutination by vaccinia virus. Aust. J. exp. Biol. med. Sci. 21, 103—106 (1943).

$C_{60}$ CLARKE, S. K. R., and D. A. J. TYRRELL: The neutralisation of influenza and other viruses by homologous immune serum. Studies in roller tube tissue cultures. Arch. ges. Virusforsch. 8, 453—468 (1958).

$C_{61}$ CLEELAND, R., and A. P. McKEE: Antigenic variation of Asian influenza virus. Proc. Soc. exp. Biol. (N.Y.) 99, 371—374 (1958).

$C_{62}$ CLEELAND, R., and J. Y. SUGG: Effect of trypsin on a susceptible strain of influenza virus. J. Immunol. 85, 539—545 (1960).

$C_{63}$ CLEELAND, R., and J. Y. SUGG: Inactivation of haemagglutinating capacity of type A influenza virus by trypsin treatment. Proc. Soc. exp. Biol. (N.Y.) **112**, 913—917 (1963).

$C_{64}$ CLEELAND, R., and J. Y. SUGG: Serologic and antigenic properties of type A influenza virus after trypsin treatment. J. Immunol. **93**, 414—419 (1964).

$C_{65}$ COCHRAN, K. W., H. F. MAASSAB, A. TSUNODA, and B. S. BERLIN: Studies on the antiviral activity of amantadine hydrochloride. Ann. N. Y. Acad. Sci. **130**, 432—439 (1965).

$C_{66}$ COCITO, C., P. LADURON, and P. DE SOMER: Nucleotide composition of RNA viruses. Proc. Soc. gen. Microbiol. April 1963.

$C_{67}$ COFFEY, J. P., and W. J. JESSOP: Maternal influenza and congenital deformities. Lancet 935—938 (1959).

$C_{68}$ COGGESHALL, L. T., and J. MAIER: Effect of various sulphonamides, sulphones, and other compounds against experimental influenza and poliomyelitis infection in white mice. J. Pharmacol. **76**, 161—166 (1942).

$C_{69}$ COHEN, A.: Protection of mice against Asian influenza virus infection by a normal horse serum inhibitor. Lancet 791—794 (1960).

$C_{70}$ COHEN, A., and G. BELYAVIN: Haemagglutination-inhibition of Asian influenza viruses: A new pattern of response. Virology **7**, 59—74 (1959).

$C_{71}$ COHEN, A., and G. BELYAVIN: The influenza virus haemagglutination inhibitors of normal rabbit serum. 1. Separation of the inhibitory components. Virology **13**, 58—67 (1961).

$C_{72}$ COHEN, A., and G. BELYAVIN: An automatic titration method for influenza haemagglutination inhibitors and its application to large scale antibody surveys. Abstr. 9th int. Congr. Microbiol. Moscow 379 (1966).

$C_{73}$ COHEN, A., G. BELYAVIN, and S. E. NEWLAND: Analysis of the horse serum inhibitor of A 2 influenza virus haemagglutination and infectivity. Abstr. D 30.4. 8th int. Congr. Microbiol. (1962).

$C_{74}$ COHEN, A., and F. BIDDLE: The effect of passage in different hosts on the inhibitor sensitivity of an Asian influenza virus strain. Virology **11**, 458—473 (1960).

$C_{75}$ COHEN, A., F. BIDDLE, and S. E. NEWLAND: Analysis of horse serum inhibitors of A 2 influenza virus haemagglutination. Brit. J. exp. Path. **46**, 497—513 (1965).

$C_{76}$ COHEN, A., and J. DORMAN: Non-specific inhibitors of A 2 influenza virus in human sera. Acta virol. **9**, 519—525 (1965).

$C_{77}$ COHEN, A., and W. SMITH: The heterogeneity of influenza virus particles within a single influenza virus strain. Brit. J. exp. Path. **38**, 385—395 (1957).

$C_{78}$ COHN, Z. A.: Quantitative distribution of phosphorus in chorioallantoic membrane as affected by infection with influenza virus. Proc. Soc. exp. Biol. (N.Y.) **79**, 566—568 (1952).

$C_{79}$ COLVILLE, J. M., J. M. DUNBAR, and H. R. MORGAN: Temperature and host cell factors in the growth of influenza virus (PR 8 strain) in avian tissues *in vitro*. J. Immunol. **76**, 264—269 (1956).

$C_{80}$ Commission on acute respiratory diseases. Endemic influenza. Amer. J. Hyg. **47**, 290—296 (1948).

$C_{81}$ Commission on acute respiratory diseases. Influenza B. Study of a localised outbreak preceding the 1945 epidemic. Amer. J. Hyg. **47**, 297—303 (1948).

$C_{82}$ Commission on influenza. Vaccination against Asian influenza. J. Amer. med. Ass. **165**, 2054—2058 (1957).

$C_{83}$ Committee on standard serological procedures in influenza diagnostic studies. An agglutination-inhibition test proposed as a standard of reference in influenza diagnostic studies. J. Immunol. **65**, 347—353 (1950).

$C_{84}$ COTO, V., C. A. GALEOTA, and F. CORAGGIO: Inhibition of interferon production by a halogenated pyrimidine derivative 5-fluoro-2-deoxyuridine. Tex. Rep. Biol. Med. **24**, 158—163 (1966).

C₈₅   Cox, H. R., J. Van der Scheer, S. Aiston, and E. Bohnel: The purification and concentration of influenza virus by means of alcohol precipitation. J. Immunol. **56**, 149—166 (1947).

C₈₆   Cox, C. D., A. Zerschling, and R. Melroy: Haemolysis of PR8 virus sensitised erythrocytes. Proc. Soc. exp. Biol. (N.Y.) **84**, 693—696 (1953).

C₈₇   Creighton, C.: A history of epidemics in Britain. Cambridge University Press 2 vols. 1891 and 1894.

C₈₈   Csaky, T. Z., F. Lanni, and J. W. Beard: Purification of the egg white inhibitor of influenza virus haemagglutination by filtration. Proc. Soc. exp. Biol. (N.Y.) **75**, 105—108 (1950).

C₈₉   Csonka, A., and A. Koch: Effects of formaldehyde on influenza virus. 3. Effects on the virus as antigen. Acta microbiol. Acad. Sci. hung. **9**, 89—96 (1962).

C₉₀   Culver, J. D., E. H. Lennette, G. Navarre, and G. A. Donahue: The recall phenomenon in the antibody response to influenza virus vaccines. J. Immunol. **81**, 452—459 (1958).

C₉₁   Culver, J. O., E. H. Lennette, G. Navarre, and C. H. Kempe: The response to influenza vaccines as determined by levels of complement fixing antibody in the acute phase of patients with respiratory disease. J. Immunol. **84**, 98—105 (1960).

C₉₂   Culver, O., E. H. Lennette, T. E. Stevens, and R. E. Nitz: Antibody response to influenza vaccines containing the Asian strain. J. Immunol. **85**, 197—202 (1960).

C₉₃   Curtain, C. C.: The action of urea on the urinary inhibitor of influenza virus haemagglutination. Aust. J. exp. Biol. med. Sci. **31**, 615—622 (1953).

C₉₄   Curtain, C. C.: The nature of the bond split by the influenza virus enzyme. Aust. J. exp. Biol. med. Sci. **31**, 623—630 (1953).

C₉₅   Curtain, C. C.: The adsorption of influenza virus by haemagglutination inhibitors coupled to powdered cellulose. Brit. J. exp. Path. **35**, 255—263 (1954).

C₉₆   Curtain, C. C., E. L. French, and J. Pye: The preparation and properties of an inhibitor of influenza virus haemagglutination from human meconium. Aust. J. exp. Biol. med. Sci. **31**, 349—360 (1953).

C₉₇   Curtain, C. C., B. P. Marmion, and J. Pye: Separation of an inhibitor of influenza virus haemagglutination from human sputum. Nature (Lond.) **171**, 33—34 (1953).

C₉₈   Curtain, C. C., and J. Pye: A mucoprotein from bovine submaxillary glands with restricted inhibitory action against influenza virus haemagglutination. Aust. J. exp. Biol. med. Sci. **33**, 315—322 (1955).

C₉₉   Cutting, W. C., R. H. Dreisbach, R. M. Halpern, E. A. Irwin, D. W. Jenkins, F. Proescher, and H. B. Tripi: Chemotherapy in virus infections. J. Immunol. **57**, 379—390 (1947).

D₁   Dachy, H. A.: Influenza virus and staphylococcal infection (French). 17th Congr. Ass. pediat. lang. franç. Montpellier, 1958.

D₂   Dales, S., and P. W. Choppin: Attachment and penetration of influenza virus. Virology **18**, 489—493 (1962).

D₃   Daniels, J. B., M. D. Eaton, and M. E. Perry: Effect of glucose on the growth of influenza virus in de-embryonated eggs and in tissue cultures. J. Immunol. **69**, 321—329 (1952).

D₄   Danon, D., Z. Jerushalmy, A. Kohn, and A. De Vries: *In vitro* phagocytosis by leucocytes of influenza virus adsorbed onto erythrocytes. Electron microscopic and cinematographic observations. Virology **15**, 258—262 (1961).

D₅   Danon, D., Z. Jerushalmy, and A. De Vries: Incorporation of influenza virus in human blood platelets *in vitro*. Electron microscopical observations. Virology **9**, 719—722 (1959).

D₆   Davenport, F. M.: Alternate passage of influenza virus. A comparison of the effects of serial passage and alternate passage of type B influenza virus in hamsters. J. Immunol. **67**, 83—92 (1951).

$D_7$ DAVENPORT, F. M.: Reliability of 50% egg infectivity titrations with suspensions of influenza viruses prepared from mouse lungs. Proc. soc. exp. Biol. (N.Y.) **82**, 1—5 (1953).

$D_8$ DAVENPORT, F. M.: The inequality of potential in influenza virus for adaptation to mice. J. Immunol. **72**, 485—488 (1954).

$D_9$ DAVENPORT, F. M.: Vaccination against respiratory diseases caused by viruses. J. Mich. med. Soc. **57**, 702—705 (1958).

$D_{10}$ DAVENPORT, F. M.: Control of influenza by vaccination. Proc. 6th int. Congr. trop. Med. Malaria **5**, 454—466 (1959).

$D_{11}$ DAVENPORT, F. M.: Inactivated influenza virus vaccines. Amer. Rev. resp. Dis. **83**, 146—150 (1961).

$D_{12}$ DAVENPORT, F. M.: Pathogenesis of influenza. Bact. Rev. **25**, 294—300 (1961).

$D_{13}$ DAVENPORT, F. M.: Influenza. Industry and Tropical Hlth 4 (1961).

$D_{14}$ DAVENPORT, F. M.: Current knowledge of influenza vaccine. J. Amer. med. Ass. **182**, 11—13 (1962).

$D_{15}$ DAVENPORT, F. M.: Factors of importance in the control of influenza. Medical clinics of North America 47, 1185—1190 (1963).

$D_{16}$ DAVENPORT, F. M., and T. FRANCIS: A comparison of the growth curves of adapted and unadapted lines of influenza virus. J. exp. Med. **93**, 129—137 (1951).

$D_{17}$ DAVENPORT, F. M., and A. V. HENNESSY: A serologic recapitulation of past experiences with influenza A; Antibody response to monovalent vaccine. J. exp. Med. **104**, 85—97 (1956).

$D_{18}$ DAVENPORT, F. M., and A. V. HENNESSY: Predetermination by infection and by vaccination of antibody response to influenza virus vaccines. J. exp. Med. **106**, 835—850 (1957).

$D_{19}$ DAVENPORT, F. M., and A. V. HENNESSY: The clinical epidemiology of Asian influenza. Ann. intern. Med. 49, 493—501 (1958).

$D_{20}$ DAVENPORT, F. M., A. V. HENNESSY, and J. A. BELL: Immunologic advantages of emulsified influenza virus vaccines. Milit. Med. **127**, 95—100 (1962).

$D_{21}$ DAVENPORT, F. M., A. V. HENNESSY, S. H. BERNSTEIN, O. F. HARPER, and W. H. KLINGENSMITH: Comparative incidence of influenza A-prime in 1953 in completely vaccinated and unvaccinated military groups. Amer. J. publ. Hlth **45**, 1138—1146 (1955).

$D_{22}$ DAVENPORT, F. M., A. V. HENNESSY, F. M. BRANDON, R. G. WEBSTER, C. D. BARRETT, and G. O. LEASE: Comparisons of serologic and febrile responses in humans to vaccination with influenza A viruses or their haemagglutinins. J. Lab. clin. Med. **63**, 5—13 (1964).

$D_{23}$ DAVENPORT, F. M., A. V. HENNESSY, J. DRESCHER, J. MULDER, and T. FRANCIS: Further observations on the relevance of serologic recapitulation of human infection with influenza viruses. J. exp. Med. **120**, 1087—1097 (1964).

$D_{24}$ DAVENPORT, F. M., A. V. HENNESSY, and T. FRANCIS: Epidemiologic and immunologic significance of age distribution of antibody to antigenic variants of influenza virus. J. exp. Med. **98**, 641—656 (1953).

$D_{25}$ DAVENPORT, F. M., A. V. HENNESSY, and T. FRANCIS: Influence of primary antigenic experience upon the development of a broad immunity to influenza. Trans. Ass. Amer. Phycns 70, 81—90 (1957).

$D_{26}$ DAVENPORT, F. M., A. V. HENNESSY, H. B. HOUSER, and W. F. CRYNS: An evaluation of an adjuvant influenza virus vaccine tested against influenza B in 1954—55. Amer. J. Hyg. **64**, 304—313 (1956).

$D_{27}$ DAVENPORT, F. M., A. V. HENNESSY, C. H. STUART-HARRIS, and T. FRANCIS: Epidemiology of influenza: Comparative serological observations in England and the USA. Lancet 469 (1955).

$D_{28}$ DAVENPORT, F. M., and F. L. HORSFALL: The associative reactions of pneumonia virus of mice (PVM) and influenza viruses: The effects of pH and electrolytes upon virus-host cell combinations. J. exp. Med. **88**, 621—644 (1948).

$D_{29}$  DAVENPORT, F. M., R. ROTT, and W. SCHAFER: Physical and biological properties of influenza virus components obtained after ether treatment. Fed.
Proc. 18, 563 (1959).

$D_{30}$  also: J. exp. Med. 112, 765—782 (1960).

$D_{31}$  DAVIES, P., and R. D. BARRY: Nucleic acid of influenza virus. Nature (Lond.)
211, 384—387 (1966).

$D_{32}$  DAVIES, W. L., R. R. GRUNERT, R. F. HAFF, J. W. McGAHEN, E. M. NEU-
MAYER, M. PAULSTOCK, J. C. WATTS, T. R. WOOD, E. C. HERMANN, and
C. E. HOFFMANN: Antiviral activity of 1-adamantanamine (amantadine).
Science 144, 862—863 (1964).

$D_{33}$  DAVIES, W. L., R. R. GRUNERT, and C. E. HOFFMANN: Influenza virus growth
and antibody response in amantadine treated mice. J. Immunol. 95, 1090—
1094 (1965).

$D_{34}$  DAVOLI, R., and O. BARTOLOMEI-CORSI: Antibodies against the Asian influenza
virus in the normal population (Italian). Sperimentale 107, 358—362 (1957).

$D_{35}$  DAWSON, C. R., M. A. EPSTEIN, and K. HUMMELER: Cytochemical and electron
microscopical observations on the presence and origin of adenosine triphos-
phatase-like activity at the surface of two myxoviruses. J. Bact. 89, 1526—
1532 (1965).

$D_{36}$  DAWSON, I. M., and W. J. ELFORD: Electron microscope studies on the interaction of certain viruses with fowl red cell membranes. Nature (Lond.) 163,
63—64 (1949).

$D_{37}$  DAWSON, I. M., and W. J. ELFORD: The investigation of influenza related viruses in the electron microscope by a new technique. J. gen. Microbiol. 3,
298—311 (1949).

$D_{38}$  DE BURGH, P. M., P. C. YU, C. HOWE, and M. BOVARNICK: Preparation from
human red cells of a substance inhibiting virus haemagglutination. J. exp.
Med. 87, 1—9 (1948).

$D_{39}$  DEMEIO, J. L., R. L. WOOLDRIDGE, J. E. WHITESIDE, and J. R. SEAL: Epidemic
influenza B and C in Navy recruits 1953—54. 2. Antigenic studies on influenza virus type C. Proc. Soc. exp. Biol. (N.Y.) 88, 436—438 (1955).

$D_{40}$  DEMPSTER, G., and B. BUCHNER: Serological pattern of influenza virus strains
isolated in Canada from 1948—1951. Canad. med. Ass. J. 69, 119—124
(1953).

$D_{41}$  DENK, H., and W. KOVAC: Immunofluorescence studies of the growth of influenza virus in mouse lung. 1. Studies of infection with mouse-adapted virus
(German). Arch. ges. Virusforsch. 17, 641 (1965).

$D_{42}$  DENK, H., and W. KOVAC: Immunofluorescence studies of the growth of influenza virus in mouse lung. 2. Studies of infection with non mouse-adapted
virus. Arch. ges. Virusforsch. 18, 244—252 (1966).

$D_{43}$  DENK, H., and W. KOVAC: Immunofluorescence studies on intranasal influenza
virus infection in immunised mice (German). Arch. ges. Virusforsch. 18,
294—303 (1966).

$D_{44}$  DEPOUX, R., and P. FISET: Serological behaviour of influenza viruses. 1. The
effect of different fowl erythrocytes on the results of serological tests. Bull.
Wld Hlth Org. 11, 981—986 (1954).

$D_{45}$  DEPOUX, R., and A. ISAACS: Interference between influenza and vaccinia
viruses. Brit. J. exp. Path. 35, 415—418 (1954).

$D_{46}$  DEPOUX, R., and A. ISAACS: Effect of inhibitors on interference between influenza and vaccinia viruses. Brit. J. exp. Path. 35, 419—425 (1954).

$D_{47}$  DEPOUX, R., and M. V. MUSSETT: Attempts to potentiate immunity to influenza
in mice. J. Hyg. (Lond.) 52, 469—474 (1954).

$D_{48}$  DEREVICI, A.: Action of certain metals on haemagglutination (Rumanian).
Comunicarile Acad. R.P.R. 2, 375—376 (1952).

$D_{49}$  DEREVICI, A., A. PETRESCU, L. ROTHSCHILD, and A. BRONITKI: Epidemiologic
and ecologic aspects of summer influenza in 1957 in the Romanian Peoples
Republic (Rumanian). Stud. Cercet. Inframicrobiol. 9, 21—30 (1958).

D₅₀ DEREVICI, A., D. SARATEANU, A. BRONITKI, A. PETRESCU, L. ROTHSCHILD, N. DRAGANESCU, C. SATMARI, J. PETRUSCA, A. STANGU, A. TIMERMAN, and M. PIRONCOF: The dynamic of serial influenzal antibodies in infants and adults vaccinated with "autochtone" vaccine (Rumanian). Stud. Cercet. Inframicrobiol. 6, 429—441 (1955).

D₅₁ DEREVICI, A., D. SARATEANU, L. ROTHSCHILD, A. PETRESCU, A. BRONITKI, J. SELEVARTZ, and O. MOSCOVIVI: Laboratory investigations during the 1957 influenza epidemic in the Romanian Peoples Republic (Rumanian). Stud. Cercet. Inframicrobiol. 8, 507—516 (1957).

D₅₂ DE RITIS, F., G. GIUSTI, and V. VILLARI: The effect of N-ethyl maleimide on influenza virus multiplication (Italian). Arch. ges. Virusforsch. 5, 432—440 (1954).

D₅₃ DICK, G. W. A., and R. M. TAYLOR: Bovine plasma albumin in buffered saline as a diluent for viruses. J. Immunol. 62, 311—317 (1949).

D₅₄ DICKINSON, L.: The formaldehyde derivatives of aminoacids and the mode of action of a formaldehyde-tyrosine derivative against influenza A virus in ovo. Brit. J. Pharmacol. 10, 56—60 (1955).

D₅₅ DIEBEL, R., and J. E. HOTCHIN: Quantitative application of fluorescent antibody technique to influenza virus infected cell cultures. Virology 8, 367—380 (1959).

D₅₆ DINTER, Z.: A variant of the fowl plague virus in Bayern (German). Tierärztl. Umsch. 4, 185—186 (1949).

D₅₇ DITCHFIELD, J. N. W. MACPHERSON, and A. ZBITNEW: Upper respiratory disease in thoroughbred horses: Studies of its viral etiology in the Toronto area 1960 to 1963. Ann. J. comp. Med. Vet. Sci. 29, 18—22 (1965).

D₅₈ DOLL, E. R.: Influenza of horses. Amer. Rev. resp. Dis. 83, 48—50 (1961).

D₅₉ DOMRACHEVA, Z. V.: An influenza A2 outbreak in men and horses (Russian). "Problemy grippa i ostrykh respiratornykh zabolevanii." Tez. dokl. nauch. Sess. Inst. AMN SSSR Moscow 47 (1959).

D₆₀ DONALD, H. B., and A. ISAACS: Some properties of influenza virus filaments shown by electron microscopic particle counts. J. gen. Microbiol. 11, 325—331 (1954).

D₆₁ DONALD, H. B., and A. ISAACS: Counts of influenza virus particles. J. gen. Microbiol. 10, 457—464 (1954).

D₆₂ DORMAN, D. C.: Effect of urea on A2 influenza virus. Nature (Lond.) 203, 789—791 (1964).

D₆₃ DOWDLE, W. R., W. B. YARBROUGH, and R. Q. ROBINSON: U.S. epizootic of equine influenza 1963. Publ. Hlth Rep. (Wash.) 79, 398—402 (1964).

D₆₄ DRAGANOV, K. I.: Hypothesis on the nature of elementary bodies and "NV" bodies in cultures of influenza virus (Russian). J. gen. Biol. Moscow 17, 13—22 (1956).

D₆₅ DRAGANOV, K. I.: New electron microscopic data on the study of the influenza virus (Russian). Byull. Eksper. biol. i med. 10, 53 (1953).

D₆₆ DRAGANOV, K. I., and A. N. SLEPUSHKIN: Comparative electron microscopic study of the influenza virus cultivated in the chorioallantoic cavity and in tissue culture (Russian). Vop. Virus. 4, 64 (1954).

D₆₇ DRESCHER, J.: Determination of haemagglutinin and antibody content. 1. Description of photometric procedure for determining virus content.

D₆₈ 2. Investigations on inhibitory antibody and description of a photometric method for determining antibody content (German). Zbl. Bakt. 1. Abt. Orig. 169, 314—348 and 461—470 (1957).

D₆₉ DRESCHER, J.: The influenza pandemic of 1957. Reports and discussions at a meeting of specialists Nov. 1957 in Berlin. Concerning A/Asia/57 Y-aluminium oxide vaccines. Abhandl. Bundesgesundheitsamt 1, 42—49 (1958).

D₇₀ DRESCHER, J.: Adsorption phenomena as a means of differentiating viral species. Zbl. Bakt. I. Abt. Orig. 176, 295—301 (1959).

D₇₁ DRESCHER, J.: Comparison of the adsorption of influenza virus strain B/Berlin/ 2/55 on aluminium oxide and on aluminium hydroxide. Amer. J. Hyg. 74, 104—118 (1961).

D₇₂ DRESCHER, J., F. M. DAVENPORT, and A. V. HENNESSY: Photometric methods for the measurement of haemagglutinating viruses and antibody. 2. Further experience with antibody determinations and the description of a technique for analysis of virus mixtures. J. Immunol. 89, 805—814 (1962).

D₇₃ DRESCHER, J., A. V. HENNESSY, and F. M. DAVENPORT: Photometric methods for the measurement of haemagglutinating viruses and antibody. 1. Further experience with a novel photometric method for measuring haemagglutinins. J. Immunol. 89, 794—804 (1962).

D₇₄ DRESCHER, J., and F. C. LANGE: The determination of haemagglutinin and antibody. 4. Description of a photometric procedure for the determination of specific and unspecific inhibiting antibodies (German). Zbl. Bakt. I. Abt. Orig. 173, 185—195 (1958).

D₇₅ DREVO, M., A. SPOUSTA, B. BLASCO, and D. SLONIM: Preparation of specific equine serum against influenza (Czech.). Čs. Epidem. 7, 175—181 (1958).

D₇₆ DREYZIN, R. S.: Method of preparing concentrated and purified influenza virus by the method of elution with formolised erythrocytes (Russian). Gripp. Tr. AMN SSSR Moscow 28, 82 (1953).

D₇₇ DREYZIN, R. S.: Conditions of reaction of the influenza virus with susceptible erythrocytes in the presence of specific antibodies (Russian). Gripp. Tr. AMN SSSR Moscow 28, 198 (1953).

D₇₈ DREYZIN, R. S.: The methods of studying the effect of chemotherapeutic preparations on an experimental influenza model (Russian). Vop. Virus. 4, 344 (1954).

D₇₉ DREYZIN, R. A., A. A. DAVYDOVA, E. E. ZOLOTARSKAYA, and A. G. BUKRINSKAYA: Differential diagnostic methods for adenoviruses and the viruses of herpes, influenza and poliomyelitis (Russian). Vop. Virus. 5, 178—183 (1958).

D₈₀ DREYZIN, R. S., and M. I. KARLINA: The effect of an iodine preparation and apple pectin on experimental influenzal infection (Russian). Vop. Virus 4, 331 (1954).

D₈₁ DREYZIN, R. S., and N. N. SOKOLOVA: The effect of eucalyptus leaf extracts on the multiplication of influenza virus in the developing chick embryo (Russian). Vop. Virus 4, 323 (1954).

D₈₂ DREYZIN, R. S., and YE. S. ZALMANZAN: Serological diagnosis of fatal cases of influenzal pneumonias. Vop. Virus 4, 290 (1954).

D₈₃ DRZENIEK, R.: Inhibition of neuraminidase by polyanions. Nature (Lond.) 211, 1205—1206 (1966).

D₈₄ DRZENIEK, R., J. C. SETO, and R. ROTT: Characterisation of neuraminidases from myxoviruses. Biochim. biophys. Acta (Amst.) 128, 547—558 (1966).

D₈₅ DUBAKINA, A. V., I. A. SUTINA, and B. S. VUL'VOVICH: Efficacy of immunisation against influenza with dry live vaccine in Stalingrad in 1952 (Russian). Zh. Mikrobiol. (Mosk.) 10, 63 (1953).

D₈₆ DUBAKINA, A. V., and E. E. KUSHNAREVA: Some features of the 1957 influenza outbreak in the Stalingrad district (Russian). Vop. Virus 4, 162—164 (1959).

D₈₇ DUBINAKOVA, A. M.: Data on haematology and rhinocytology of influenza in 1957 (Russian). Vop. Virus 5, 83—86 (1959).

D₈₈ DUBROVINA, T. YA.: Conditions of isolation of infectious ribonucleic acid from influenza virus. Acta virol. 10, 117—123 (1966).

D₈₉ DUC-NGUYEN, H., H. M. ROSE, and C. MORGAN: An electron microscopic study of changes at the surface of influenza-infected cells as revealed by ferritin conjugated antibodies. Virology 28, 404—412 (1966).

D₉₀ DUDGEON, J. A., H. MELLANBY, R. E. GLOVER, and C. H. ANDREWES: Influenza in Great Britain in 1946—47. Brit. J. exp. Path. 29, 132—141 (1948).

D₉₁ DUDGEON, J. A., C. H. STUART-HARRIS, R. E. GLOVER, C. H. ANDREWES, and W. H. BRADLEY: Influenza B in 1945—46. Lancet 627—631 (1946).

D$_{92}$ DUJARRIC DE LA RIVIERE, R., and J. CHEVE: Influenza virus in ferrets. Study of different French strains of influenza virus (French). Ann. Inst. Pasteur 59, 445—456 (1937).

D$_{93}$ DULBECCO, R., and M. VOGT: Plaque formation and isolation of pure lines with poliomyelitis viruses. J. exp. Med. 99, 167—182 (1954).

D$_{94}$ DULBECCO, R., and M. VOGT: One step growth curves of Western Equine encephalomyelitis virus on chicken embryo cells grown *in vitro* and analysis of virus yields from single cells. J. exp. Med. 99, 183—199 (1954).

D$_{95}$ DUNCAN, I. B. R., M. O. CREIGHTON, and B. JEANS: Epidemic Type A influenza in Southwestern Ontario in 1963. Canad. med. Ass. J. 91, 541 (1964).

D$_{96}$ DUNHAM, W. B., and W. J. MacNEAL: Inactivation of influenza virus by mild antiseptics. J. Immunol. 49, 123—128 (1944).

D$_{97}$ DURAND, P.: The susceptibility of certain Tunisian rodents to influenza virus (French). C.R. Soc. Biol. (Paris) 139, 718—719 (1945).

D$_{98}$ DZAGUROV, S. G.: Experimental study of the effect of antibiotics on microbes encountered in influenza and on the influenza A virus (Russian). Tr. ob'yed. nauch. Sess. AMN SSSR, Moscow, 162—171 (1953).

E$_{1}$ EATON, M. D.: Transmission of epidemic influenza virus by contact in mice. J. Bact. 39, 220—241 (1940).

E$_{2}$ EATON, M. D.: Experimental immunisation of mice with the virus of epidemic influenza. 1. Quantitative studies on the antigenicity of active and inactive virus. J. Immunol. 39, 43—45 (1940).

E$_{3}$ EATON, M. D.: Strain specificity of complement fixation with sera of mice immune to the virus of influenza A and swine influenza. J. Immunol. 41, 383—389 (1941).

E$_{4}$ EATON, M. D., L. T. ADLER, P. BOND, and A. R. SCALA: The effect of thyroxin on respiration and virus growth in chick embryo tissue cultures. J. infect. Dis. 98, 239—248 (1956).

E$_{5}$ EATON, M. D., L. T. ADLER, and M. E. PERRY: Virus growth and cellular energy production. Effect of substances chemically related to thyroxin on influenza viruses. Proc. Soc. exp. Biol. (N.Y.) 84, 57—60 (1953).

E$_{6}$ EATON, M. D., and M. D. BECK: Experimental immunisation of mice with the virus of epidemic influenza. 2. Immunity after intranasal inoculation of mouse passage, tissue culture, and ferret passage strains. J. Immunol. 39, 57—64 (1940).

E$_{7}$ EATON, M. D., and M. D. BECK: A new strain of virus of influenza B isolated during an epidemic in California. Proc. Soc. exp. Biol. (N.Y.) 48, 177—180 (1941).

E$_{8}$ EATON, M. D., M. COREY, W. VAN HERICK, and G. MEIKLEJOHN: A comparison of various methods of demonstrating influenza virus in throat washings. Proc. Soc. exp. Biol. (N.Y.) 58, 6—9 (1945).

E$_{9}$ EATON, M. D., C. G. LEVENSON, and M. PERRY: Inhibition of the growth of viruses in the chick embryo by aromatic diamidines. J. Immunol. 68, 1—9 (1952).

E$_{10}$ EATON, M. D., I. E. LOW, A. R. SCALA, and S. URETSKY: Inhibition by ammonium ion of the growth of influenza virus in chorioallantoic tissue. Virology 18, 102—108 (1962).

E$_{11}$ EATON, M. D., B. MAGASANIK, M. E. PERRY, and D. KARIBIAN: Inhibition of influenza and mumps viruses in tissue culture by basic aminoacids. Proc. Soc. exp. Biol. (N.Y.) 77, 505—508 (1951).

E$_{12}$ EATON, M. D., and W. P. MARTIN: Personnel of Naval Lab. Res. unit No. 1. Immunisation with inactive virus of influenza B. Comparison of antibody response with that produced by infection. Publ. Hlth. Rep. (Wash.) 57, 445—451 (1942).

E$_{13}$ EATON, M. D., and G. MEIKLEJOHN: Vaccination against influenza: A study in California during the epidemic of 1943—44. Amer. J. Hyg. 42, 28—44 (1945).

$E_{14}$ EATON, M. D., and M. E. PERRY: Further observations on the effect of 2,4-dinitrophenol on the growth of influenza virus. J. infect. Dis. **93**, 269—277 (1953).

$E_{15}$ EATON, M. D., M. E. PERRY, C. G. LEVENSON, and I. M. GOCKE: Studies on the mode of action of aromatic diamidines on influenza and mumps virus in tissue culture. J. Immunol. **68**, 321—334 (1952).

$E_{16}$ EATON, M. D., and A. R. SCALA: Reversible effect of hypotonic solutions on the growth of influenza virus in tissue cultures. Proc. Soc. exp. Biol. (N.Y.) **92**, 289—297 (1956).

$E_{17}$ EATON, M. D., and A. R. SCALA: Interrelated effects of cysteine, thyroxin, and substrate on the growth of influenza virus in tissue cultures. Arch. ges. Virusforsch. **7**, 274—283 (1957).

$E_{18}$ EATON, M. D., and A. R. SCALA: Inhibitory effect of glutamine and ammonia on replication of influenza virus in ascites tumour cells. Virology **13**, 300—307 (1961).

$E_{19}$ EATON, M. D., A. R. SCALA, and I. E. LOW: Aminoacid imbalance and incomplete virus replication. Arch. ges. Virusforsch. **14**, 583—598 (1964).

$E_{20}$ ECKERT, E. A.: Envelope proteins derived from influenza virus. J. Bact. **91**, 1907—1910 (1966).

$E_{21}$ ECKERT, E. A.: Characterisation of a low molecular weight antigenic protein from the envelope of influenza virus. J. Bact. **92**, 1430—1434 (1966).

$E_{22}$ ECKERT, E. A., F. LANNI, D. BEARD, and J. W. BEARD: Effect of swine influenza virus on the viscosity of the egg-white inhibitor of haemagglutination. Science **109**, 463—464 (1949).

$E_{23}$ EDDY, B. E.: A study of influenza virus vaccines by a serum-virus neutralisation test and by active immunisation. J. Immunol. **57**, 195—202 (1949).

$E_{24}$ EDDY, B. E., and R. W. G. WYCKOFF: Influenza virus in sectioned tissues. Proc. Soc. exp. Biol. (N.Y.) **75**, 290—293 (1950).

$E_{25}$ EDMUNDS, P. N.: Immunity to old and new strains of influenza A virus. Brit. med. J. 155—156 (1951).

$E_{26}$ EDNEY, M.: The influence of calcium ion on the reaction of the V. Cholerae enzyme RDE and influenza virus with specific mucopolysaccharides. Aust. J. exp. Biol. med. Sci. **27**, 253—264 (1949).

$E_{27}$ EDNEY, M.: Genetic analysis of the development of serum resistance in an influenza virus strain. J. gen. Microbiol. **17**, 25—38 (1957).

$E_{28}$ EDNEY, M., and F. M. BURNET: Influence of ions on thermal inactivation and modification of an influenza virus. Aust. J. exp. Biol. med. Sci. **30**, 129—138 (1952).

$E_{29}$ EDNEY, M., and A. ISAACS: Interference between active and inactive influenza viruses in the chick embryo. 3. Inhibition of virus haemagglutination in the chorioallantoic membrane. Aust. J. exp. Biol. med. Sci. **28**, 603—612 (1951).

$E_{30}$ EDWARD, D. G. ff.: Resistance of influenza viruses to drying and its demonstration on dust. Lancet 664—666 (1941).

$E_{31}$ EDWARD, D. G. ff., W. J. ELFORD, and P. P. LAIDLAW: Studies on air-borne infection. 1. Experimental technique and preliminary observations on influenza and infectious ectromelia. J. Hyg. (Lond.) **43**, 1—10 (1943).

$E_{32}$ EDWARD, D. G. ff., D. LUSH, and R. B. BOURDILLON: Studies on air-borne virus infections. 2. The killing of virus aerosols by ultraviolet radiation. J. Hyg. (Lond.) **43**, 11—15 (1943).

$E_{33}$ EFIMOVA, K. V., L. V. ZAKHAROVA, and E. N. MILYAEVA: Epidemiologic characteristics of influenza in Kuibyshev from 1957—1959 (Russian). "Problemy Gripp. i ostrykh respiratornykh zabolevanii." Tez. dokl. nauch. Sess. Inst. AMN SSSR, Moscow, 90—92 (1959).

$E_{34}$ EFIMOVA, V. A., and L. YA. ZAKSTEL'SKAYA: The mechanism of anamnestic reactions to A influenza virus strains (Russian). "Problemy grippa i ostrykh respiratornykh zabolevanii." Tez. dokl. nauch. Sess. Inst. AMN SSSR, Moscow, 167—168 (1959).

$E_{35}$ ELFORD, W. J., and C. H. ANDREWES: Centrifugation studies: Viruses of vaccinia, influenza, and Rous sarcoma. Brit. J. exp. Path. **17**, 422—430 (1936).

$E_{36}$ ELFORD, W. J., C. H. ANDREWES, and F. F. TANG: The sizes of the viruses of human and swine influenza as determined by ultrafiltration. Brit. J. exp. Path. **17**, 51—53 (1936).

$E_{37}$ ENDERS, J. F., and H. E. PEARSON: Resistance of chicks to infection with influenza virus. Proc. Soc. exp. Biol. (N.Y.) **48**, 143—146 (1941).

$E_{38}$ EPSHTEYN, F. G., A. S. GORBUNOVA, and P. M. KARAKUYUMCHAN: Skin test in influenza (Russian). Vop. Virus. **4**, 298 (1954).

$E_{39}$ EPSHTEYN, F. G., and V. S. PANASYUK: Experience in the treatment of influenza with atabrine (Russian). Vo.-med. Zh. **4**, 32—34 (1951).

$E_{40}$ EPSHTEYN, F. G., N. V. SERGEEV, and E. J. SOROKINA: The clinical course of Asian influenza in adults (Russian). Klin. med. Moskva. **36**, 35—42 (1958).

$E_{41}$ EVANS, C. A., and E. R. RICKARD: The toxic effect of influenca virus in the rabbit eye. Proc. Soc. exp. Biol. (N.Y.) **58**, 74—75 (1945).

$F_1$ FADEYEVA, L. L.: Study of the properties of the influenza virus cultivated in tissues outside the body (Russian). Problemy grippa i OKVDP. Tez. dokl. ob'yed. Sess. Inst. AMN SSSR, Moscow 12 (1952).

$F_2$ FADEYEVA, L. L.: Isolation of the influenza virus by means of the tissue culture method (Russian). Problemy grippa i OKVDP. Tez. dokl. ob'yed. Sess. Inst. AMN SSSR, Moscow 22 (1952).

$F_3$ FADEYEVA, L. L.: Biological properties of the influenza virus cultured in animal and human tissues (Russian). Dissertation, Moscow (1952).

$F_4$ FADEYEVA, L. L.: Tissue cultures of the influenza virus (Russian). Zh. Mikrobiol. (Mosk.) **2**, 47 (1953).

$F_5$ FADEYEVA, L., and A. I. DAVIENTZO: Epidemiological testing of influenza tissue vaccine (Russian). Zh. Mikrobiol. (Mosk.) **10**, 25 (1953).

$F_6$ FADEYEVA, L. L., V. V. RITOVA, and L. YA. ZAKSTEL'SKAYA: Perfection of an influenza tissue vaccine (Russian). Vop. Virus **4**, 393 (1954).

$F_7$ FAILLARD, H.: Isolation of N-acetylneuraminic acid from mucin by means of receptor destroying enzyme of V. Cholerae (German). Hoppe-Seylers Z. physiol. Chem. **305**, 145—148 (1956).

$F_8$ FAILLARD, H.: Liberation of N-acetylneuraminic acid from mucin by the receptor destroying enzyme of V. Cholerae (German). Hoppe-Seylers Z. physiol. Chem. **307**, 62—86 (1957).

$F_9$ FAIRBROTHER, R. W., and L. HOYLE: Observations on the aetiology of influenza. J. Path. Bact. **44**, 213—223 (1937).

$F_{10}$ FAIRBROTHER, R. W., and L. HOYLE: Active immunisation against influenza: The use of heat-killed elementary body suspensions. Brit. J. exp. Path. **18**, 430—436 (1937).

$F_{11}$ FAIRBROTHER, R. W., and A. E. MARTIN: Serologic studies in epidemic influenza with particular reference to the persistence of antibodies after infection. Lancet 718 (1938).

$F_{12}$ FAIRBROTHER, R. W., and A. E. MARTIN: Further observations on the value of heated elementary body suspensions in immunisation against experimental influenza. Arch. ges. Virusforsch. **1**, 114—119 (1939).

$F_{13}$ FALKOVICH, L. I., M. I. IGNETOVA, G. A. PISKUNOVA, and YE. V. ARKHINA: Study of the virus of epidemic influenza. Experiment in vaccination of people with the influenza virus (Russian). Tr. Mosk. oblet. Inst. Infekt. bolezn. imeni. Mechnikova, Moscow, 239 (1939).

$F_{14}$ FARKAS, E., and G. TAKATSY: Isolation of strains of influenza virus (Hungarian). Nepegeszsegugy Budapest **30**, 402—409 (1949).

$F_{15}$ FAUCONNIER, B.: Changes in pH of allantoic fluid inoculated with influenza virus, growth of virus in slightly acid allantoic fluid (French). Ann. Inst. Pasteur **85**, 222—229 (1953).

$F_{16}$ FAUCONNIER, B.: Development of influenza virus and pH of the allantoic fluid of embryonated egg. 5. Acidifying and inhibitory effect of ammonium salts (French). Ann. Inst. Pasteur 90, 171—181 (1955).

$F_{17}$ FAUCONNIER, B.: Spontaneous variation of type A influenza virus towards serum inhibitors of normal animals (French). Ann. Inst. Pasteur 97, 377—384 (1959).

$F_{18}$ FAUCONNIER, B., and H. BEAUCHAMP: A study of the duration of viraemia in mice inoculated with influenza virus (French). Ann. Inst. Pasteur 86, 660—663 (1954).

$F_{19}$ FAUCONNIER, B., and M. CHAIGNEAU: Inhibition of influenza virus multiplication by thioparatoluidine and certain of its derivatives (French). Ann. Inst. Pasteur 94, 514—516 (1958).

$F_{20}$ FAUCONNIER, B., R. M. LARTITEGUI, and D. BARUA: Experimental adaptation of influenza virus to inhibitors of normal sera. 1. Isolation of a non-sensitive variant by cultivation of the sensitive strain in the presence of its inhibitor (French). Ann. Inst. Pasteur 97, 473—478 (1959).

$F_{21}$ FAZEKAS DE ST GROTH, S.: Destruction of influenza receptors in the mouse lung by an enzyme from V. Cholerae. Aust. J. exp. Biol. med. Sci. 26, 29—36 (1948).

$F_{22}$ FAZEKAS DE ST GROTH, S.: Regeneration of virus receptors in mouse lungs after artificial destruction. Aust. J. exp. Biol. med. Sci. 26, 271—285 (1948).

$F_{23}$ FAZEKAS DE ST GROTH, S.: Viropexis: The mechanism of influenza virus infection. Nature (Lond.) 162, 294—295 (1948).

$F_{24}$ FAZEKAS DE ST GROTH, S.: Modification of virus receptors by metaperiodate. 1. The properties of $IO_4$ treated red cells. Aust. J. exp. Biol. med. Sci. 27, 65—81 (1949).

$F_{25}$ FAZEKAS DE ST GROTH, S.: Influenza: A study in mice. Lancet 1101—1105 (1950).

$F_{26}$ FAZEKAS DE ST GROTH, S.: Studies in experimental immunology of influenza. 1. The state of virus receptors and inhibitors in the respiratory tract. Aust. J. exp. Biol. med. Sci. 28, 15—29 (1950).

$F_{27}$ 6. The duration of induced immunity. Aust. J. exp. Biol. med. Sci. 28, 559—568 (1950).

$F_{28}$ 9. The mode of action of pathotopic adjuvants. Aust. J. exp. Biol. med. Sci. 29, 339—352 (1951),

$F_{29}$ FAZEKAS DE ST GROTH, S.: Quick test for the early diagnosis of influenza. Nature (Lond.) 167, 43 (1951).

$F_{30}$ FAZEKAS DE ST GROTH, S.: Production of non-infective particles among influenza viruses: Do changes in virulence accompany the Von Magnus phenomenon? J. Hyg. (Lond.) 53, 276—291 (1955).

$F_{31}$ FAZEKAS DE ST GROTH, S.: Evaluation of quantal neutralisation tests. Nature (Lond.) 191, 891—893 (1961).

$F_{32}$ FAZEKAS DE ST GROTH, S.: Steric inhibition: Neutralisation of a virus-borne enzyme. Ann. N.Y. Acad. Sci. 103, 674—687 (1963).

$F_{33}$ FAZEKAS DE ST GROTH, S., and H. J. F. CAIRNS: Quantitative aspects of influenza multiplication. 4. Definition of constants and general discussion. J. Immunol. 69, 173—181 (1952).

$F_{34}$ FAZEKAS DE ST GROTH, S., and M. DONNELLEY: Studies in experimental immunology of influenza. 2. Production of viral antigens. Aust. J. exp. Biol. med. Sci. 28, 31—44 (1950).

$F_{35}$ 3. The antibody response. Aust. J. exp. Biol. med. Sci. 28, 45—60 (1950).

$F_{36}$ 4. The protective value of active immunisation. Aust. J. exp. Biol. med. Sci. 28, 61—75 (1950).

$F_{37}$ 5. Enhancement of immunity by pathotopic vaccination. Aust. J. exp. Biol. med. Sci. 28, 77—85 (1950).

$F_{38}$ 8. Pathotopic adjuvants. Aust. J. exp. Biol. med. Sci. 29, 323—337 (1951).

$F_{39}$ FAZEKAS DE ST GROTH, S., and M. EDNEY: Quantitative aspects of influenza virus multiplication. 2. Heterologous interference. J. Immunol. 69, 160—168 (1952).

$F_{40}$ FAZEKAS DE ST GROTH, S., and A. GOTTSCHALK: Ovomucin: A substrate for the enzyme of influenza virus. 2. Demonstration of its inhomogeneity. Brit. J. exp. Path. 32, 21—24 (1951).

$F_{41}$ FAZEKAS DE ST GROTH, S., and A. GOTTSCHALK: Studies on glycoproteins. 10. Equilibrium measurements on influenza virus-glycoprotein systems. Biochim. biophys. Acta (Amst.) 78, 248—257 (1963).

$F_{42}$ FAZEKAS DE ST GROTH, S., and D. M. GRAHAM: Modification of virus receptors by metaperiodate. 2. Infection through modified receptors. Aust. J. exp. Biol. med. Sci. 27, 83—98 (1949).

$F_{43}$ FAZEKAS DE ST GROTH, S., and D. M. GRAHAM: Absence of incomplete virus production in certain influenza strains. Nature (Lond.) 172, 1193 (1953).

$F_{44}$ FAZEKAS DE ST GROTH, S., and D. M. GRAHAM: Artificial production of incomplete influenza virus. Nature (Lond.) 173, 637 (1954).

$F_{45}$ FAZEKAS DE ST GROTH, S., and D. M. GRAHAM: The production of incomplete virus particles among influenza strains: Experiments in eggs. Brit. J. exp. Path. 35, 60—74 (1954).

$F_{46}$ FAZEKAS DE ST GROTH, S., and D. M. GRAHAM: Studies on experimental immunology of influenza. 10. Passive immunity and its enhancement. Aust. J. exp. Biol. med. Sci. 32, 369—386 (1954).

$F_{47}$ FAZEKAS DE ST GROTH, S., and D. M. GRAHAM: The production of incomplete virus particles among influenza strains: Chemical induction of the Von Magnus phenomenon. Brit. J. exp. Path. 36, 205—213 (1955).

$F_{48}$ FAZEKAS DE ST GROTH, S., and R. G. WEBSTER: The neutralisation of animal viruses. 3. Equilibrium conditions in the influenza virus antibody system. J. Immunol. 90, 140—150 (1963).

$F_{49}$ 4. Parameters of the influenza virus-antibody system. J. Immunol. 90, 151—164 (1963).

$F_{50}$ FAZEKAS DE ST GROTH, S., and R. G. WEBSTER: The antibody response. Ciba Found. Symp. "Cellular biology of myxovirus infections" 246—271 (1964).

$F_{51}$ FAZEKAS DE ST GROTH, S., and R. G. WEBSTER: Disquisition on original antigenic sin. 1. Evidence in man.

$F_{52}$ 2. Proof in lower creatures. J. exp. Med. 124, 331—345, and 347—361 (1966).

$F_{53}$ FAZEKAS DE ST GROTH, S., and D. O. WHITE: An improved assay for the infectivity of influenza viruses. J. Hyg. (Lond.) 56, 151—162 (1958).

$F_{54}$ FAZEKAS DE ST GROTH, S., and D. O. WHITE: The dose-response relationship between influenza viruses and the surviving allantois. J. Hyg. (Lond.) 56, 523—534 (1958).

$F_{55}$ FAZEKAS DE ST GROTH, S., and D. O. WHITE: Comparison of the infectivity of influenza viruses in two host systems: The allantois of intact eggs and surviving allantois-on-shell. J. Hyg. (Lond.) 56, 535—546 (1958).

$F_{56}$ FAZEKAS DE ST GROTH, S., J. WITHELL, and K. J. LAFFERTY: An improved assay method for neutralising antibodies against influenza viruses. J. Hyg. (Lond.) 56, 415—426 (1958).

$F_{57}$ FEDOROV, YE. A.: Cytology of impressions from the nasal mucosa in influenza and in catarrhs of the respiratory tract (Russian). Zh. Mikrobiol. (Mosk.) 6, 25 (1951).

$F_{58}$ FEDOROV, YE. A.: Clinical-laboratory diagnosis of influenza (Russian). Zh. Mikrobiol. (Mosk.) 1, 56 (1952).

$F_{59}$ FEDOVA, D., and L. ZELENKOVA: The use of the fluorescent antibody method for the rapid identification of influenza virus in the epithelial cell sediment of allantoic or amniotic fluid of infected chick embryos.

$F_{60}$ 2. The identification of influenza virus in nasal smears by the fluorescent antibody technique. J. Hyg. Epidem. (Praha) 9, 127—134, and 135—146 (1965).

$F_{61}$ FELDMAN, H. A., and S. S. WANG: Sensitivity of various viruses to chloroform. Proc. Soc. exp. Biol. (N.Y.) 106, 736—738 (1961).

$F_{62}$ FELTYNOWSKI, A.: An aluminium replica method for the study on the red cell surface, especially of influenza virus adsorption on intact chicken red cells (Polish). Medycyna Doswiadczalna 4, 423—438 (1953).

$F_{63}$ FENNER, F.: The reactivation of animal viruses. Brit. med. J. 135—142 (1962).

$F_{64}$ FENNER, F., and J. F. SAMBROOK: The genetics of animal viruses. Ann. Rev. Microbiol. 18, 47—94 (1964).

$F_{65}$ FILIPPOVA-NUTRIKRINA, Z. L.: Therapy with cortisone during virus influenza (Russian). "Kliniya i Meditsina." Moscow, 149—153 (1954).

$F_{66}$ FINLAND, M., and M. W. BARNES: Laboratory experiences with influenza in Boston during the winter 1949—50. J. infect. Dis. 97, 48—56 (1955).

$F_{67}$ FINLAND, M., and M. W. BARNES: Influenza A in Boston in 1951 and 1953. Isolation of virus and serological studies in patients, including a study of antigenic relationships among strains of influenza A virus isolated in Boston from 1943 through 1953. Amer. J. Hyg. 62, 1—20 (1955).

$F_{68}$ FINTER, N. B.: Studies on interferon. Abstr. Proc. Soc. gen. Microbiol. Jan. 1965.

$F_{69}$ FINTER, N. B., and A. J. BEALE: The 6-hour soluble antigen production test for comparing the infectivity of influenza virus preparations. J. Hyg. (Lond.) 54, 58—67 (1956).

$F_{70}$ FINTER, N. B., O. C. LIU, and W. HENLE: Studies on host-virus interaction in the chick embryo influenza virus system. 10. An experimental analysis of the Von Magnus phenomenon. J. exp. Med. 101, 461—478 (1955).

$F_{71}$ FINTER, N. B., O. C. LIU, M. LIEBERMANN, and W. HENLE: Studies on host-virus interaction in the chick embryo influenza virus system. 8. An experimental analysis of various de-embryonation techniques. J. exp. Med. 100, 33—52 (1954).

$F_{72}$ FISET, P., and R. DEPOUX: Serological behaviour of influenza viruses. 2. Patterns of antigenic relationship. Bull. Wld Hlth Org. 11, 987—993 (1954).

$F_{73}$ FISET, P., and H. B. DONALD: A method for influenza antibody absorption. Brit. J. exp. Path. 34, 616—620 (1953).

$F_{74}$ FISHER, T. N., and H. S. GINSBERG: Reactions of influenza viruses with guinea-pig polymorphonuclear leucocytes. 2. The reduction of white blood cell glycosis by influenza viruses and receptor destroying enzyme (RDE).

$F_{75}$ 3. Studies on the mechanism by which influenza viruses inhibit phagocytosis. Virology 2, 637—655 and 656—664 (1956).

$F_{76}$ FLEISCHER, M. S.: The effect of various dyes in vitro and in vivo upon influenza virus infection in mice. J. Immunol. 62, 245—255 (1949).

$F_{77}$ FLETCHER, R. D., J. E. HIRSCHFIELD, and M. FORBES: Effect of 2-diethyl-aminoethyl-4-methylpiperazine-1-carboxylate on influenza viruses in tissue culture. Proc. Soc. exp. Biol. (N.Y.) 121, 68—72 (1966).

$F_{78}$ FLICK, J. A., B. SANFORD, and S. MUDD: The effect of salt concentration on the interaction of influenza A virus and erythrocytes. J. Immunol. 61, 65—77 (1949).

$F_{79}$ FLORMAN, A. L., A. B. WEISS, and F. E. COUNCIL: Effect of large doses of streptomycin and influenza viruses on chick embryos. Proc. Soc. exp. Biol. (N.Y.) 61, 16—18 (1946).

$F_{80}$ FONG, J.: Sulphur mustard-inactivated influenza virus as interfering agent. J. Immunol. 71, 241—245 (1953).

$F_{81}$ FONG, J., and E. BERNAL: In vitro action of mustards on the infective and toxic components of influenza A (PR 8) virus. J. Immunol. 70, 89—96 (1953).

$F_{82}$ FONG, J., and E. BERNAL: In vitro action of sulphur mustard and chloroethyl-amine derivatives upon antigenicity of influenza A (PR 8) virus. J. Immunol. 70, 97—102 (1953).

$F_{83}$ FONG, J., and B. BLUM: Action of mustards on interfering, haemagglutinating and eluting capacities of influenza A (PR 8) virus. J. Immunol. 71, 26—29 (1953).

$F_{84}$ FONG, J., R. LOUIS, and C. CHING: Effect of interfering virus and RDE upon development of toxic reactions. Proc. Soc. exp. Biol. (N.Y.) 84, 246—248 (1953).

$F_{85}$ FORTUIN, G. J., and J. D. VERLINDE: Vaccination against influenza (Dutch). Ned. T. Geneesk. 98, No. 31 (1954).

$F_{86}$ FRANCIS, T.: Transmission of influenza by filtrable virus. Science 80, 457—459 (1934).

$F_{87}$ FRANCIS, T.: Immunological relationships of strains of filtrable viruses recovered from cases of human influenza. Proc. Soc. exp. Biol. (N.Y.) 32, 1172—1175 (1935).

$F_{88}$ FRANCIS, T.: Quantitative relationship between the immunising dose of epidemic influenza virus and the resulting immunity. J. exp. Med. 69, 283—300 (1939).

$F_{89}$ FRANCIS, T.: The inactivation of epidemic influenza virus by nasal secretions of human individuals. Science 91, 198—199 (1940).

$F_{90}$ FRANCIS, T.: A new type of virus from epidemic influenza. Science 91, 405—408 (1940).

$F_{91}$ FRANCIS, T.: Differentiation of influenza A and influenza B by the complement fixation reaction. Proc. Soc. exp. Biol. (N.Y.) 45, 861—863 (1940).

$F_{92}$ FRANCIS, T.: The problem of epidemic influenza. Trans. Coll. Phycns. Philad. 8, 218—227 (1941).

$F_{93}$ FRANCIS, T.: The development of the 1943 vaccination study of the Commission on influenza. Amer. J. Hyg. 42, 1—11 (1945).

$F_{94}$ FRANCIS, T.: Dissociation of haemagglutinating and antibody measuring capacities of influenza virus. J. exp. Med. 85, 1—7 (1947).

$F_{95}$ FRANCIS, T.: Apparent serological variation within a strain of influenza virus. Proc. Soc. exp. Biol. (N.Y.) 65, 143—147 (1947).

$F_{96}$ FRANCIS, T.: The significance of multiple immunological types of influenza virus. Cincinn. J. Med. 31, 97—101 (1950).

$F_{97}$ FRANCIS, T.: Plan for the evaluation of vaccination against influenza. Amer. J. publ. Hlth 41, No. 8 (1951).

$F_{98}$ FRANCIS, T.: Significance of antigenic variation of influenza viruses in relation to vaccination in man. Fed. Proc. 11, 808—812 (1952).

$F_{99}$ FRANCIS, T.: Influenza: The newe acquayntance. Ann. intern. Med. 39, 203—221 (1953).

$F_{100}$ FRANCIS, T.: The current status of the control of influenza. Ann. intern. Med. 43, 534—538 (1955).

$F_{101}$ FRANCIS, T.: Mobilisation against influenza. Science 126, 1—2 (1957).

$F_{102}$ FRANCIS, T.: Influenza in perspective. Amer. Rev. resp. Dis. 83, 98—104 (1961).

$F_{103}$ FRANCIS, T.: Epidemic influenza. Amer. Rev. resp. Dis. 88, 148—151 (1963).

$F_{104}$ FRANCIS, T., and T. P. MAGILL: Immunological studies with the virus of influenza. J. exp. Med. 62, 505—516 (1935).

$F_{105}$ FRANCIS, T., and T. P. MAGILL: The incidence of neutralising antibodies for human influenza virus in the serum of human individuals of different ages. J. exp. Med. 63, 655—668 (1936).

$F_{106}$ FRANCIS, T., and T. P. MAGILL: Direct transmission of human influenza virus to mice. Proc. Soc. exp. Biol. (N.Y.) 36, 132—133 (1937).

$F_{107}$ FRANCIS, T., and T. P. MAGILL: Antigenic differences in strains of epidemic influenza virus. 2. Cross immunisation tests in mice. Brit. J. exp. Path. 19, 284—293 (1938).

$F_{108}$ FRANCIS, T., T. P. MAGILL, E. T. RICKARD, and D. M. BECK: Etiological and serological studies in epidemic influenza. Amer. J. publ. Hlth 27, 1141—1160 (1937).

$F_{109}$ FRANCIS, T., and E. MINUSE: Influence of saliva upon haemagglutination by influenza virus. Proc. Soc. exp. Biol. (N.Y.) 69, 291—294 (1948).

$F_{110}$ FRANCIS, T., and H. E. MOORE: A study of the neurotropic tendency in strains of the virus of epidemic influenza. J. exp. Med. 72, 717—728 (1940).

F₁₁₁ FRANCIS, T., H. E. PEARSON, J. E. SALK, and P. N. BROWN: Immunity in human subjects artificially infected with influenza virus type B. Amer. J. publ. Hlth **34**, 317—334 (1944).

F₁₁₂ FRANCIS, T., H. E. PEARSON, E. R. SULLIVAN, and P. M. BROWN: The effect of subcutaneous vaccination with influenza virus upon the virus inactivating capacity of nasal secretion. Amer. J. Hyg. **37**, 294—300 (1943).

F₁₁₃ FRANCIS, T., and K. PENTTINEN: The failure of merdicien to modify influenza virus infections. J. Immunol. **63**, 337—339 (1949).

F₁₁₄ FRANCIS, T., J. J. QUILLIGAN, and E. MINUSE: Resemblance of a strain of swine influenza virus to human A-prime strain. Proc. Soc. exp. Biol. (N.Y.) **71**, 216—220 (1949).

F₁₁₅ FRANCIS, T., J. J. QUILLIGAN, and E. MINUSE: Identification of another epidemic respiratory disease. Science **112**, 495—497 (1950).

F₁₁₆ FRANCIS, T., and J. E. SALK: A simplified procedure for the concentration and purification of influenza virus. Science **96**, 499—500 (1942).

F₁₁₇ FRANCIS, T., J. E. SALK, and W. M. BRACE: The protective effect of vaccination against epidemic influenza B. J. Amer. med. Ass. **131**, 275—278 (1946).

F₁₁₈ FRANCIS, T., J. E. SALK, H. E. PEARSON, and P. N. BROWN: Protective effect of vaccination against induced influenza A. Proc. Soc. exp. Biol. (N.Y.) **55**, 104—105 (1944).

F₁₁₉ FRANCIS, T., J. E. SALK, and J. J. QUILLIGAN: Experience with vaccination against influenza in the spring of 1947. Amer. J. publ. Hlth **37**, 1013—1016 (1947).

F₁₂₀ FRANCIS, T., and R. E. SHOPE: Neutralisation tests with sera of convalescent and immunised animals and the viruses of human and swine influenza. J. exp. Med. **63**, 645—653 (1936).

F₁₂₁ FRANCIS, T., and C. H. STUART-HARRIS: Studies on the nasal histology of epidemic influenza virus infection in the ferret. 1. The development and repair of the nasal lesion.

F₁₂₂ 2. Histological and serological observations on ferrets receiving repeated inoculations of epidemic influenza virus. J. exp. Med. **68**, 789—801, and 813—830 (1938).

F₁₂₃ FRANCIS, T., and M. VICENTE DE TORREGROSA: Combined infection of mice with H. influenzae and influenza virus by the intranasal route. J. infect. Dis. **76**, 70—77 (1945).

F₁₂₄ FRANKLIN, R. M.: The synthesis of fowl plague virus products in a proflavine-inhibited tissue culture system. Virology **6**, 525—539 (1958).

F₁₂₅ FRANKLIN, R. M., and P. M. BREITENFELD: The abortive infection of Earle's L cells by fowl plague virus. Virology **8**, 293—307 (1959).

F₁₂₆ FRANKLIN, R. M., and E. WECKER: Inactivation of some animal viruses by hydroxylamine and the structure of ribonucleic acid. Nature (Lond.) **184**, 343—345 (1959).

F₁₂₇ FRASER, K. B.: Antibody to purified virus inhibitor. Brit. J. exp. Path. **32**, 552—558 (1951).

F₁₂₈ FRASER, K. B.: Genetic interaction and interference between the MEL and NWS strains of influenza A virus. Brit. J. exp. Path. **34**, 319—328 (1953).

F₁₂₉ FRASER, K. B.: Features of the MEL-NWS recombination systems in influenza A virus. 1. The effect of age of the mouse on the intracerebral growth of the MEL strain of influenza A virus.

F₁₃₀ 2. The formation of NM recombinant strains of influenza A virus in mouse brain.

F₁₃₁ 3. The pathogenicity of recombinant Neuro-MEL viruses grown in mouse brain and chick embryo brain.

F₁₃₂ 4. Increments of virulence during successive cycles of double infection with two strains of influenza virus. Virology **9**, 168—177, 178—190, 191—201, and 202—214 (1959).

F₁₃₃ FRASER, K. B.: Immunofluorescence of abortive and complete infections of influenza A virus in hamster BHK 21 cells and mouse L cells. J. gen. Virol. **1**, 1—12 (1967).

F₁₃₄ FRASER, K. B., and F. M. BURNET: Studies on recombination with influenza viruses in the chick embryo. 2. Genetic interaction between influenza virus strains in the chick embryo. Aust. J. exp. Biol. med. Sci. **30**, 459—468 (1952).

F₁₃₅ FRASER, K. B., R. C. NAIRN, M. G. McENTEGART, and C. S. CHADWICK: Neurotropic and non-neurotropic influenza A infection of mouse brain studied with fluorescent antibody. J. Path. Bact. **78**, 423—433 (1959).

F₁₃₆ FRENCH, E. L.: The pyrogenic effect of the influenza-mumps group of viruses in the laboratory rabbit. Aust. J. exp. Biol. med. Sci. **30**, 479—488 (1952).

F₁₃₇ FRENCH, E. L., and J. K. DINCEN: Primary isolation of influenza A virus: The optimal conditions for amniotic infection. Aust. J. exp. Biol. med. Sci. **36**, 257—265 (1958).

F₁₃₈ FRENCH, H.: Ministry of Health Report No. 4. H.M.S.O. London, 1920.

F₁₃₉ FRIDMAN, E. A.: Some data on the problem of non-homogeneous structure of virus population (Russian). Virus infekt. **19**, 67—75 (1959).

F₁₄₀ FRIDMAN, E. A.: The heterogeneity of type A2 influenza virus population. Acta virol. **4**, 274—282 (1960).

F₁₄₁ FRIDMAN, E. A., I. M. ANSHELES, E. S. STENINA, T. N. DAVYOVA, B. I. MEITIN, and T. A. KLUSHINA: Analysis of epidemiological and immuno-virological data obtained during the 1957 and 1959 waves of influenza epidemic in Leningrad. "Problemy grippa i ostrykh respiratornykh zabolevanii." Tez. dokl. nauch. Sess. Inst. AMN SSSR, Moscow, 114—117 (1959).

F₁₄₂ FRIEDEWALD, W. F.: The immunological response to influenza virus infection as measured by the complement fixation test. Relation of the complement fixing antigen to the virus particle. J. exp. Med. **78**, 347—366 (1943).

F₁₄₃ FRIEDEWALD, W. F.: Enhancement of the immunising capacity of influenza virus vaccines with adjuvants. Science **99**, 453—454 (1944).

F₁₄₄ FRIEDEWALD, W. F.: Qualitative differences in the antigenic composition of influenza A virus strains. J. exp. Med. **79**, 633—647 (1944).

F₁₄₅ FRIEDEWALD, W. F.: Adjuvants in immunisation with influenza virus vaccines. J. exp. Med. **80**, 477—491 (1944).

F₁₄₆ FRIEDEWALD, W. F., and E. W. HOOK: Influenza virus infection in the hamster. A study of inapparent infection and virus adaptation. J. exp. Med. **88**, 343—353 (1948).

F₁₄₇ FRIEDEWALD, W. F., E. S. MILLER, and L. R. WHATLEY: The nature of non-specific inhibition of virus haemagglutination. J. exp. Med. **86**, 65—75 (1947).

F₁₄₈ FRIEDEWALD, W. F., and E. G. PICKELS: Size of infective particle and haemagglutinin of influenza virus as determined by centrifugal analysis. Proc. Soc. exp. Biol. (N.Y.) **52**, 261—262 (1943).

F₁₄₉ FRIEDEWALD, W. F., and E. G. PICKELS: Centrifugation and ultrafiltration studies on allantoic fluid preparations of influenza virus. J. exp. Med. **79**, 301—317 (1944).

F₁₅₀ FRIEDMAN, R. M., and J. A. SONNABEND: Inhibition of interferon action by puromycin. J. Immunol. **95**, 696—703 (1965).

F₁₅₁ FRISCH, A. W., and J. CARSON: Mode of inactivation of influenza virus by tannic acid. J. Bact. **66**, 576—580 (1953).

F₁₅₂ FRISCH-NIGGEMEYER, W.: Absolute amount of ribonucleic acid in viruses. Nature (Lond.) **178**, 307—308 (1956).

F₁₅₃ FRISCH-NIGGEMEYER, W.: Chemical composition and internal structure of influenza virus with particular regard to the so-called "soluble antigen" (German). Zbl. Bakt. I. Abt. Orig. **183**, 294—317 (1961).

F₁₅₄ FRISCH-NIGGEMEYER, W., and L. HOYLE: The nucleic acid and carbohydrate content of influenza virus A and of virus fractions produced by ether disintegration. J. Hyg. (Lond.) **54**, 201—212 (1956).

F₁₅₅ FROMMHAGEN, L. H., N. K. FREEMAN, and C. A. KNIGHT: The lipid constituents of influenza virus, chick allantoic membrane, and sedimentable allantoic protein. Virology 5, 173—174 (1958).

F₁₅₆ FROMMHAGEN, L. H., and C. A. KNIGHT: The polysaccharide and ribonucleic acid content of purified influenza virus. Virology 2, 430—431 (1956).

F₁₅₇ FROMMHAGEN, L. H., and C. A. KNIGHT: Column purification of influenza virus. Virology 8, 198—208 (1959).

F₁₅₈ FROMMHAGEN, L. H., C. A. KNIGHT, and N. K. FREEMAN: The ribonucleic acid, lipid, and polysaccharide constituents of influenza virus preparations. Virology 8, 176—197 (1959).

F₁₅₉ FROMMHAGEN, L. H., and E. H. LENNETTE: Spontaneous dissappearance of the slow-sedimenting component of influenza virus preparations. Virology 7, 247—248 (1959).

F₁₆₀ FUKUMI, H.: Studies on the P-Q phases of A/Asian/57 influenza viruses. Bull. Wld Hlth Org. 20, 421—434 (1959).

F₁₆₁ FULTON, F.: Growth cycle of influenza virus. Nature (Lond.) 164, 189 (1949).

F₁₆₂ FULTON, F.: The titration of influenza virus neutralising antibodies. J. Hyg. (Lond.) 50, 265—274 (1952).

F₁₆₃ FULTON, F., and P. ARMITAGE: Surviving tissue suspensions for influenza virus titrations. J. Hyg. (Lond.) 49, 247—262 (1951).

F₁₆₄ FULTON, F., and K. R. DUMBELL: The serological comparison of strains of influenza virus. J. gen. Microbiol. 3, 97—110 (1949).

F₁₆₅ FULTON, F., and A. ISAACS: Influenza virus multiplication in the chick chorioallantoic membrane. J. gen. Microbiol. 9, 119—131 (1953).

F₁₆₆ FURER, N. M., I. P. FOMINA, O. I. ARTANOMOVA, and T. I. BALEZINA: Antiviral action of antibiotics obtained from cultures of actinomyces violaceus. Antibiotiki, Moscow 4, 30—35 (1959),

F₁₆₇ FURESZ, J.: Studies on inhibitors of influenza virus haemagglutination in normal rabbit and ferret sera. Canad. J. Microbiol. 5, 505—511 (1959).

G₁ GAJDUSEK, D. C.: Suspended cell tissue cultures for study of virus growth kinetics. Proc. Soc. exp. Biol. (N.Y.) 83, 621—624 (1953).

G₂ GALEAZZI, F., and G. VISIO: Studies on the synergism between influenza virus and bacteria. 2. Organotropism of staphylococcus aureus in mice inoculated with influenza virus (Italian). Riv. Ist. sieroter. ital. 33, 47—58 (1958).

G₃ GALPER, G. S.: Experience in atabrine prophylaxis of influenza (Russian). Vo.-med. Zh. 11, 78 (1953).

G₄ GAMBURG, V. P.: New data concerning the thermostable serum inhibitor (Russian). "Problemy grippa i ostrykh respiratornykh zabolevanii." Tez. Dokl. nauch Sess. Inst. AMN SSSR, Moscow, 55—57 (1959).

G₅ GARD, S.: Haemagglutination as a diagnostic method. Theoretical aspects. Acta path. Suppl. 91, 107—117 (1951).

G₆ GARD, S.: Aspects of the formation of incomplete virus. 2nd Symp. Soc. gen. Microbiol. 211—244 (1953).

G₇ GARD, S., and P. VON MAGNUS: Studies on interference in experimental influenza. 2. Purification and centrifugation experiments. Ark. Kemi. Mineral. Geol. 24b, No. 8 (1947).

G₈ GARD, S., P. VON MAGNUS, A. SVEDMYR, and A. BIRCH-ANDERSEN: Studies on the sedimentation of influenza virus. Arch. ges. Virusforsch. 4, 591—611 (1952).

G₉ GAUSCH, C. R., and J. S. YOUNGER: A tissue culture color test for measuring influenza virus and antibody. Proc. Soc. exp. Biol. (N.Y.) 101, 853—856 (1951).

G₁₀ GAVRILOV, V. I.: Study of influenza A virus variability under experimental conditions (Russian). Probl. obshch. virol. Tez. Dokl. 6-y Sess. Inst. Virusol. AMN SSSR, Moscow, 37 (1953).

G₁₁ GAVRILOV, V. I.: Influenza A virus variation in passage through the bodies of immunised mice (Russian). Dissertation, Moscow (1954).

$G_{12}$ GAYDAMAKA, M. G.: Study of the effectiveness of specific substances in the prophylaxis of influenza during the 1957 epidemic in Kharkov (Russian). Aziatskii Gripp (sborn. nauch. rabot) Kiev 266—268 (1958).

$G_{13}$ GAYDAMAKA, M. G., R. M. MIKULINSKAYA, D. D. FYADINA, and A. S. DROMASHKO: Data on the epidemiological effectiveness of anti-influenza vaccination in Kharkov in 1957 (Russian). Vop. Virus. 5, 135—140 (1958).

$G_{14}$ GAYDAMAKA, M. G., G. P. VAGANOV, A. S. DROMASHKO, B. D. SHVETSKAYA, and D. P. FYALINA: Disease of the upper respiratory tract in horses following the human influenza epidemic in 1957. Bull Wld Hlth Org. 20, 505—508 (1959).

$G_{15}$ GENT, W. L. G., N. A. GREGSON, D. B. GAMMACK, and J. H. RAPER: The lipid-protein unit in myelin. Nature (Lond.) 204, 553—555 (1964).

$G_{16}$ GERBER, P., and E. ADAMS: Development of resistance of influenza B virus to polysaccharides. Science 128, 1571 (1958).

$G_{17}$ GERBER, P., D. HAMRE, and C. G. LOOSLI: Antigenic variants of influenza A virus (PR 8). 2. Serological and immunological characteristics of variants derived from variants. J. exp. Med. 103, 413—424 (1956).

$G_{18}$ GERBER, P., C. G. LOOSLI, and D. HAMRE: Antigenic variants of influenza A virus (PR 8). 1. Their development during serial passage in the lungs of partially immune mice. J. exp. Med. 101, 627—638 (1955).

$G_{19}$ GERMANOVA, K. I., and T. YA. GONCHARSKAYA: Antiviral properties of streptothricins A B C D and F. Abstr. 9th Int. Congr. Microbiol. 537 (1966).

$G_{20}$ GERNGROSS, O. G., T. V. PYSINA, and L. N. SEREBRYAKOVA: Etiology of influenza outbreaks in Vladivostock in 1957—59 and antigenic structure of the isolated strains (Russian). "Problemy grippa i ostrykh respiratornykh zabolevanii." Tez. dokl. nauch. Sess. Inst. AMN SSSR, Moscow, 24—26 (1959).

$G_{21}$ GERONE, P. J., T. G. WARD, and W. A. CHAPPEL: Combined infection in mice with influenza virus and diplococcus pneumoniae. Amer. J. Hyg. 66, 331—341 (1957).

$G_{22}$ GIERER, A., and G. SCHRAMM: Infectivity of ribonucleic acid from tobacco mosaic virus. Nature (Lond.) 177, 702—703 (1956).

$G_{23}$ GIERER, A., and G. SCHRAMM: Infectivity of ribonucleic acid from tobacco mosaic virus (German). Z. Naturforsch. 11b, 138—142 (1956).

$G_{24}$ GINSBERG, H. S.: Comparison of quantities of egg and mouse adapted influenza viruses required to infect each host. Proc. Soc. exp. Biol. (N.Y.) 84, 249—252 (1953).

$G_{25}$ GINSBERG, H. S.: Formation of non-infectious influenza virus in mouse lungs: Its dependence upon extensive pulmonary consolidation initiated by the viral inoculum. J. exp. Med. 100, 581—603 (1954).

$G_{26}$ GINSBERG, H. S.: Production of pulmonary lesions by influenza viruses in immunised mice. J. Immunol. 72, 24—29 (1954).

$G_{27}$ GINSBERG, H. S.: Suppression of influenza viral pneumonia in mice by the nonspecific action of Xerosin. J. Immunol. 75, 430—440 (1955).

$G_{28}$ GINSBERG, H. S., and J. R. BLACKBURN: Reactions of influenza viruses with guineapig polymorphonuclear leucocytes. 1. Virus-cell interaction. Virology 2, 618—636 (1956).

$G_{29}$ GINSBERG, H. S., and F. L. HORSFALL: A labile component of normal serum which combines with various viruses. Neutralisation of infectivity and inhibition of haemagglutination by the component. J. exp. Med. 90, 475—495 (1949).

$G_{30}$ GINSBERG, H. S., and F. L. HORSFALL: Quantitative aspects of the multiplication of influenza A virus in the mouse lung. Relation between the degree of viral multiplication and the extent of pneumonia. J. exp. Med. 95, 135—145 (1952).

$G_{31}$ GINSBERG, H. S., and A. T. WILSON: Inactivation of several viruses by liquid ethylene oxide. Proc. Soc. exp. Biol. (N.Y.) 73, 614—616 (1950).

G₃₂  GIORDANO, G., and G. PINO: Researches on anti-influenza vaccination. 2. Comparison of immune responses to one and two doses of vaccine given intradermally (Italian). Riv. Ist. sieroter. ital. 28, 501—509 (1953).

G₃₃  GIRARDI, A. J., H. McMICHAEL, and W. HENLE: The use of HeLa cells in suspension for the quantitative study of virus propagation. Virology 2, 532—544 (1956).

G₃₄  GITELMAN, A. K., O. BURDUCEA, and A. G. BUKRINSKAYA: Inhibitory effect of histones and actinomycin D on myxovirus reproduction with a special reference to input multiplicity of infection. Abstr. 9th int. Congr. Microbiol. 531 (1966).

G₃₅  GLOVER, R. E.: Spread of infection from the respiratory tract of the ferret. 2. Association of influenza A virus and streptococcus group C. Brit. J. exp. Path. 22, 98—107 (1941).

G₃₆  GOLBERT, Z. V., and O. P. PETERSON: Histopathologic changes in white mice after the intraperitoneal injection of large doses of influenza virus (Russian). Vop. Virus. 1, 190 (1948).

G₃₇  GOMPELS, A. E. H.: Antigenic relationships of swine influenza virus. J. gen. Microbiol. 9, 140—148 (1953).

G₃₈  GORBUNOVA, A. S.: Biological properties of the influenza viruses and problems of their classification (Russian). Zh. Mikrobiol. (Mosk.) 2, 39 (1953).

G₃₉  GORBUNOVA, A. S.: Antigenic relationships between influenza viruses of the types A, A₁ and A₂ (Russian). Vop. Virus. 4, 401—406 (1959).

G₄₀  GORBUNOVA, A. S., and V. I. GAVRILOV: Evaluation of the complement fixation reaction as a method of serological differentiation of influenza virus strains (Russian). Vop. Virus. 4, 55 (1954).

G₄₁  GORBUNOVA, A. S., and V. A. ISACHENKO: Peculiarities of identification of strains of A2 influenza viruses isolated in 1957 (Russian). Vop. Virus. 5, 48—53 (1958).

G₄₂  GORBUNOVA, A. S., A. N. LOZHKINA, V. M. STAKHANOVA, V. I. ISACHENKO, T. P. VORONINA, A. I. SIMAKOV, and A. A. KOPEIKIN: Antigenic pattern of A2 influenza viruses according to materials from the strain service of the Regional Influenza Centre of the USSR (Russian). Vop. Virus 5, 25—35 (1958).

G₄₃  GORBUNOVA, A. S., and S. E. MARENNIKOVA: The problem of the O and D variants of the influenza virus (Russian). Byull. eksp. Biol. Med. 10, 292 (1949).

G₄₄  GORBUNOVA, A. S., and S. E. MARENNIKOVA: The problem of the O and D variants of the influenza virus (Russian). Annot. rabot. Inst. virol. AMN SSSR Za 1949, Moscow (1950).

G₄₅  GORBUNOVA, A. S., E. V. MOLIBOG, F. K. GONSOVSKY, G. I. FJODOROVA, U. A. ABENOVA, and N. A. MAXIMOVITCH: Influenza virus strains different in their enzymic activity and character of their interaction with cells. Abstr. 9th int. Congr. Microbiol. 373 (1966).

G₄₆  GORDON, I.: Demonstration of antigenic differences between different strains of influenza B. J. Immunol. 44, 231—236 (1942).

G₄₇  GOREV, N. E.: The use of broth culture filtrates of pseudomonas fluorescens for the destruction of non-specific thermostable (At 56°C) influenza virus inhibitors in human and animal sera. Acta virol. 2, 171—178 (1958).

G₄₈  GOSTLING, J. V. T.: Growth of influenza A virus in cultured cells of embryo mouse lung. Brit. J. exp. Path. 39, 679—684 (1958).

G₄₉  GOTLIEB, T., and G. K. HIRST: The experimental production of combination forms of virus. 3. The formation of doubly antigenic particles from influenza A and B virus and a study of the ability of individual particles of X virus to yield two separate strains. J. exp. Med. 99, 307—320 (1954).

G₅₀  GOTLIEB, T., and G. K. HIRST: The experimental production of combination forms of virus. 6. Reactivation of influenza viruses after inactivation by ultraviolet light. Virology 2, 235—248 (1956).

G₅₁  GOTO, H.: Virological and immunological studies of Asian influenza in 1960 (Japanese). Sapporo Med. J. 24, 205—210 (1963).

$G_{52}$ GOTTSCHALK, A.: N-substituted isoglucosamine released from mucoproteins by the influenza virus enzyme. Nature (Lond.) 167, 845—847 (1951).

$G_{53}$ GOTTSCHALK, A.: The influenza virus enzyme and its mucoprotein substrate. Yale J. Biol. Med. 26, 352—364 (1954).

$G_{54}$ GOTTSCHALK, A.: The synthesis of 2-carboxypyrrole from D-glucosamine and pyruvic acid: Its bearing on the structure of neuraminic acid. Arch. Biochem. 69, 37—44 (1957).

$G_{55}$ GOTTSCHALK, A.: The influenza virus neuraminidase. Nature (Lond.) 181, 377—378 (1958).

$G_{56}$ GOTTSCHALK, A.: On the mechanism underlying initiation of influenza virus infection. Ergebn. Mikrobiol. 32, 1—22 (1959).

$G_{57}$ GOTTSCHALK, A.: The molecular structure of ovine submaxillary gland glycoprotein. Proc. Internat. Conf. "Salivary glands and their secretions", Washington, Pergamon Press 351—364 (1964).

$G_{58}$ GOTTSCHALK, A., and S. FAZEKAS DE ST GROTH: On the relationship between the indicator profile and prosthetic group of mucoproteins inhibitory for influenza virus haemagglutinin. J. gen. Microbiol. 22, 690—696 (1960).

$G_{59}$ GOTTSCHALK, A., and E. R. B. GRAHAM: A neuraminidase susceptible disaccharide from bovine salivary mucoprotein. Z. Naturforsch. 13b, 821—822 (1958).

$G_{60}$ GOTTSCHALK, A., and E. R. B. GRAHAM: 6-a-D-sialyl-N-acetylgalactosamine: The neuraminidase-susceptible prosthetic group of bovine salivary mucoprotein. Biochim. biophys. Acta (Amst.) 34, 380—391 (1959).

$G_{61}$ GOTTSCHALK, A., and P. E. LIND: Ovomucin, a substrate for the enzyme of influenza virus. 1. Ovomucin as an inhibitor of haemagglutination by heated Lee virus. Brit. J. exp. Path. 30, 85—92 (1949).

$G_{62}$ GOTTSCHALK, A., and P. E. LIND: Product of interaction between influenza virus enzyme and ovomucin. Nature (Lond.) 164, 232—233 (1949).

$G_{63}$ GOTTSCHALK, A., and B. T. PERRY: Ovomucin, a substrate for the enzyme of influenza virus. 3. Enzyme activity as an integral function of the influenza virus particle. Brit. J. exp. Path. 32, 408—413 (1951).

$G_{64}$ GRAHAM, A. F.: The chemical analysis of purified influenza virus A (PR 8 strain) containing radioactive phosphorus. Canad. J. Res. 28, 186—195 (1950).

$G_{65}$ GRAHAM, A. F.: The fractionation of phosphorus containing constituents in the allantoic membrane of the embryonated egg. Canad. J. Res. 28, 271—280 (1950).

$G_{66}$ GRAHAM, A. F., G. DEMPSTER, and B. BUCHNER: The toxicity of $P^{32}$ for normal and influenza infected embryos. J. Bact. 63, 426—427 (1952).

$G_{67}$ GRAHAM, A. F., and L. McCLELLAND: Uptake of radioactive phosphorus by influenza virus. Nature (Lond.) 163, 949 (1949).

$G_{68}$ GRAHAM, A. F., and L. McCLELLAND: The uptake of radioactive phosphorus by influenza virus A (PR 8 strain). Canad. J. Res. 28, 121—134 (1950).

$G_{69}$ GRANOFF, A.: Plaque formation with influenza strains. Virology 1, 252 (1955).

$G_{70}$ GRANOFF, A., and D. W. KINGSBURY: Effect of actinomycin D on the replication of Newcastle disease and influenza viruses. Ciba Found. Symp. "Cellular biology of myxovirus infections" 96—119 (1964).

$G_{71}$ GRANOFF, A., O. C. LIU, and W. HENLE: A small haemagglutinating component in preparations of Newcastle disease virus. Proc. Soc. exp. Biol. (N.Y.) 75, 684—691 (1950).

$G_{72}$ GREEN, I. J., M. LIEBERMAN, and W. J. MOGABGAB: The behavior of influenza viruses in various tissue culture systems. J. Immunol. 78, 233—239 (1957).

$G_{73}$ GREEN, R. H., A. F. RASMUSSEN, and J. E. SMADEL: Chemoprophylaxis of experimental influenza infection in eggs. Publ. Hlth Rep. (Wash.) 61, 1401—1403 (1946).

$G_{74}$ GREEN, M., and M. A. STAHMANN: Virus inhibition by some aminoacid analogues. Arch. Biochem. 55, 63—70 (1955).

G₇₅ GREEN, R. H., and D. W. WOOLLEY: Inhibition by certain polysaccharides of haemagglutination and of multiplication of influenza virus. J. exp. Med. 86, 55—64 (1947).

G₇₆ GRIST, N. R., J. B. LANDSMAN, and T. ANDERSON: Studies in the aetiology of pneumonia in Glasgow 1950—51. Lancet 640 (1962).

G₇₇ GRIEFF, D., H. BLUMENTHAL, M. CHIGA, and H. PINKERTON: The effects on biological materials of freezing and drying by vacuum sublimation. 2. Effect on influenza virus. J. exp. Med. 100, 89—101 (1954).

G₇₈ GROSS, W. O.: Properties of two strains of virus from the influenza epidemics of 1952 and 1953 (German). Z. Hyg. Infekt.-Kr. 138, 224—245 (1953).

G₇₉ GROSS, W. O.: Can an influenza virus after isolation be identified with the prevalent epidemic strain? (German). Z. Hyg. Infekt.-Kr. 138, 246—260 (1953).

G₈₀ GROSSBERG, S. E.: Cytocidal assay and propagation of human influenza viruses in cultured primary hamster kidney cells. Proc. Soc. exp. Biol. (N.Y.) 113, 546—553 (1963).

G₈₁ GROSSBERG, S. E.: Human influenza A viruses: Rapid plaque assay in hamster kidney cells. Science 144, 1246—1247 (1964).

G₈₂ GROSSBERG, S. E., E. W. HOOK, and R. R. WAGNER: Haemorrhagic encephalopathy in chicken embryos infected with influenza virus. 3. Viral interference at a distant site induced by prior allantoic infection. J. Immunol. 88, 1—8 (1962).

G₈₃ GROSSGEBAUER, K.: Dark field and phase contrast microscopic observations in connection with serological investigations on myxoviruses (German). Zbl. Bakt. I. Abt. Orig. 171, 430—439 (1958).

G₈₄ GROUPE, V., C. G. ENGLE, P. E. GAFFNEY, and R. A. MANAKER: Virucidal activity of representative anti-infective agents against influenza and vaccinia viruses. Appl. Microbiol. 3, 333—336 (1955).

G₈₅ GROUPE, V., and E. C. HERMANN: Modification of the neurotoxic effect of influenza virus in mice by Xerosin. J. Immunol. 74, 249—254 (1955).

G₈₆ GROUPE, V., and L. H. PUGH: Inactivation of influenza virus and of viral haemagglutinin by the ciliate tetrahymena geleii. Science 115, 307—308 (1952).

G₈₇ GROUPE, V., and L. H. PUGH: Interference between influenza virus and infectious bronchitis of chickens. J. Bact. 63, 295—296 (1952).

G₈₈ GRUBB, T. C., M. L. MIESSE, and B. PUETZER: The inactivation of influenza virus by certain vapors. J. Bact. 53, 61—66 (1947).

G₈₉ GRUNERT, R. R., J. W. McGAHEN, and W. L. DAVIES: The in vivo antiviral activity of 1-adamantanamine (Amantadine). 1. Prophylactic and therapeutic activity against influenza viruses. Virology 26, 262—269 (1965).

G₉₀ GULRAJANI, T. S.: Studies on respiratory diseases of pigs. 3. Persistence of influenza viruses in the respiratory tract. J. comp. Path. Therap. 61, 101—107 (1951).

G₉₁ GUSTYREV, V. F.: Immunologic and epidemiologic effectiveness of inoculations against influenza with live vaccine (Russian). "Problemy grippa i ostrykh respiratornykh zabolevanii." Tez. dokl. nauch. Sess. Inst. AMN SSSR, Moscow, 128—129 (1959).

G₉₂ GUTMAN, N. R., and L. MENTKEVICH: Significance of the A prime virus in the influenza epidemic process (Russian). Zh. Mikrobiol. (Mosk.) 9, 38 (1954).

G₉₃ GUZHIYENKO, G. N., and M. A. MOROZENKO: Treatment of influenza in children with dry influenza antiserum (Russian). "Virusnyye Infecktsii" Moscow, 207—212 (1953).

H₁ HAAS, R., and H. WULFF: The growth of influenza virus in tissue cultures of ferret kidney cells (German). G. Microbiol. 1, 65—71 (1955).

H₂ HAAS, R., and H. WULFF: Multiplication of influenza virus in cultures of calf kidney tissue (German). Z. Hyg. Infekt.-Kr. 143, 568—577 (1957).

H₃ HACHISUKA, Y., and N. KATO: The inhibitory effect of 4-aminopterin on influenza virus propagation in tissue culture. Nagoya J. med. Sci. 18, 29—31 (1955).

H₄ HAFF, R. F., P. W. SCHRIVER, C. G. ENGLE, and R. C. STEWART: Pathogenesis of influenza in ferrets. 1. Tissue and blood manifestations of disease. J. Immunol. 96, 659—667 (1966).

H₅ HAFF, R. F., P. W. SCHRIVER, and R. C. STEWART: Pathogenesis of influenza in ferrets. Nasal manifestations of disease. Brit. J. exp. Path. 47, 435—444 (1966).

H₆ HAFF, R. F., and R. C. STEWART: Role of sialic acid receptors in adsorption of influenza virus to chick embryo cells. J. Immunol. 94, 842—851 (1965).

H₇ HAHNEMANN, F. V., and V. REINECKE: *In ovo* production of interferon induced by influenza virus of varying degrees of incompleteness. Acta path. microbiol. scand. 63, 241—248 (1965).

H₈ HALE, W. M., and A. P. McKEE: The value of influenza vaccination when done at the beginning of an epidemic. Amer. J. Hyg. 42, 21—27 (1945).

H₉ HALE, W. M., and A. P. McKEE: The intracranial toxicity of influenza virus for mice. Proc. Soc. exp. Biol. (N.Y.) 59, 81—84 (1945).

H₁₀ HALLAUER, C.: Variants and mutants in viruses pathogenic for animals. Bull. schweiz. Akad. med. Wiss. 16, 173—184 (1960).

H₁₁ HALONEN, P.: Amantadine hydrochloride in the prophylaxis of respiratory virus infections. Abstr. 9th int. Congr. Microbiol. Moscow, 548 (1966).

H₁₂ HAMBURGER, V., and K. HABEL: Teratogenic and lethal effects of influenza A and mumps viruses on early chick embryos. Proc. Soc. exp. Biol. (N.Y.) 66, 608—617 (1947).

H₁₃ HAMRE, D., J. APPEL, and C. G. LOOSLI: Viraemia in mice with pulmonary influenza A virus infection. J. Lab. clin. Med. 47, 182—193 (1955).

H₁₄ HAMRE, D., C. G. LOOSLI, and R. C. EWERT: Antigenic variants of influenza A virus (PR 8 strain). 6. Results of repeated passages of the same variant in mice immunised with monovalent homologous vaccine or polyvalent vaccine. J. infect. Dis. 108, 151—162 (1961).

H₁₅ HAMRE, D., C. G. LOOSLI, and P. GERBER: Antigenic variants of influenza A virus (PR 8 strain). 3. Serological relationships of a line of variants derived in sequence in mice given homologous vaccine.

H₁₆ 4. Serological characters of a second line of variants developed in mice given polyvalent vaccine. J. exp. Med. 107, 829—844, and 845—855 (1958).

H₁₇ HANA, L., and L. HOYLE: The disintegration of the internal ribonucleoprotein of influenza virus A with the production of serologically distinct components. Acta virol. 10, 506—512 (1966).

H₁₈ HANA, L., O. KRIZANOVA, B. STYK, and F. SOKOL: Some data on the nature of the cofactor enhancing the activity of imperfect antibodies against A2 influenza virus strains. Acta virol. 5, 325 (1961).

H₁₉ HANA, L., and B. STYK: Some properties of influenza C virus inhibitor from rat serum. Acta virol. 3, 85—90 (1959).

H₂₀ HANA, L., and B. STYK: Influence of delipidisation on the haemagglutination inhibiting activity of rat serum against influenza type C viruses. Acta virol. 4, 392—393 (1960).

H₂₁ HANA, L., and B. STYK: Lipoids as inhibitors of haemagglutinating-inhibiting activity against "avid" strains of A2 influenza. Acta virol. 5, 190—192 (1961).

H₂₂ HANA, L., and B. STYK: Characterisation of antibodies, cofactor, and non-specific viral inhibitors on sephadex G-200. Acta virol. 6, 479 (1962).

H₂₃ HANA, L., and B. STYK: An attempt to differentiate Beta-inhibitor from cofactor occurring in bovine serum. Acta virol. 6, 77—83 (1962).

H₂₄ HANA, L., B. STYK, and D. KOCISKOVA: Estimation of the distribution and the character of inhibitors against "avid" A2 influenza virus strains in protein fractions of guineapig and horse sera obtained by paper electrophoresis. Acta virol. 4, 356—364 (1960).

H₂₅ HANA, L., B. STYK, and S. SCHRAMEK: Studies on the size of antibodies, cofactor, and non-specific inhibitors of influenza virus using separation on sephadex G-200. Science Tools 10, 4—7 (1963).

H₂₆ HANAN, R.: The effect of tocopheryl esters on influenza virus. Proc. Soc. exp. Biol. (N.Y.) **75**, 440—444 (1950).

H₂₇ HANIG, M., and H. BERNKOPF: The sedimentable components of influenza virus propagated in de-embryonated eggs. J. Immunol. **65**, 585—590 (1950).

H₂₈ HANIG, M., and J. M. STEIM: The difference in effective hydrodynamic diameters of "FAST" and "SLOW" influenza viruses propagated in de-embryonated eggs. Virology **12**, 499—510 (1960).

H₂₉ HANNOUN, C.: Action of purine and pyrimidine analogues on the multiplication of influenza virus in chick embryos (French). Atti Congr. Int. Microbiol. 6th Congr. Rome **1**, 431—432 (1953).

H₃₀ HANNOUN, C.: Effect of conjunctival instillation of influenza virus in rabbits and mice (French). Ann. Inst. Pasteur **96**, 170—177 (1959).

H₃₁ HANSON, R. J., J. E. KEMPF, and A. V. BOAND: Phagocytosis of influenza virus. 2. Its occurrence in normal and immune mice. J. Immunol. **79**, 422—427 (1957).

H₃₂ HARBOE, A.: The course in Norway of the influenza pandemic caused by the virus of type A/Asia/57. (Norwegian). Saertrykk Nordisk Med. **60**, 1701 (1958).

H₃₃ HARBOE, A.: HI antibodies in ferrets cross-infected with distantly related influenza viruses. Acta path. microbiol. scand. **57**, 211—221 (1963).

H₃₄ HARBOE, A.: The influenza virus haemagglutination inhibition by antibody to host material. Acta path. microbiol. scand. **57**, 317—330 (1963).

H₃₅ HARBOE, A.: The normal allantoic antigen which neutralises the influenza virus HI antibody to host material. Acta path. microbiol. scand. **57**, 488—492 (1963).

H₃₆ HARBOE, A., and R. REENAAS: The course of antibody development to influenza A2 strains. Acta path. microbiol. scand. **46**, 266—272 (1959).

H₃₇ HARBOE, A., R. REENAAS, and M. OPPEDAL: Studies on Francis inhibitor of influenza virus haemagglutination in rabbit serum fractions obtained by electrophoresis on starch grain. Acta path. microbiol. scand. **44**, 92—105 (1958).

H₃₈ HARBOE, R., F. ROST, K. JUUL, and T. MADLAND: Percentage of individuals presenting antibody to A/Asia/57 after an influenza outbreak in a military camp in October 1957. Arch. ges. Virusforsch. **8**, 446—452 (1958).

H₃₉ HARDY, P. H., and F. L. HORSFALL: Reactions between influenza virus and a component of normal allantoic fluid. J. exp. Med. **88**, 463—484 (1948).

H₄₀ HARE, R.: The effect of passive immunisation on experimental virus influenza in mice. J. Path. Bact. **49**, 411—428 (1939).

H₄₁ HARE, R., and M. CURL: Adsorption of influenza virus. Canad. J. Res. **25**, 43—52 (1947).

H₄₂ HARE, R., M. CURL, and L. McCLELLAND: The efficiency of the red cell adsorption and elution method for the preparation of influenza vaccine. Canad. J. publ. Hlth **37**, 284—291 (1941).

H₄₃ HARE, R., and D. M. MACKENZIE: The standardisation of influenza vaccine by red cell agglutination and antigenic tests in mice. Canad. J. publ. Hlth **38**, 141—148 (1947).

H₄₄ HARE, L., L. McCLELLAND, and J. MORGAN: A method for the concentration of influenza virus. Canad. J. publ. Hlth **33**, 325—331 (1942).

H₄₅ HARE, L., and W. C. RICKEN: Long-term variation in the titre of neutralising antibody for influenza virus in the sera of adults and children. J. Immunol. **40**, 253—266 (1941).

H₄₆ HARFORD, C. G., and A. HAMLIN: Effect of influenza virus on cilia and epithelial cells in the bronchi of mice. J. exp. Med. **95**, 173—190 (1952).

H₄₇ HARFORD, C. G., A. HAMLIN, and E. PARKER: Electron microscopy of early cytoplasmic changes due to influenza virus. J. exp. Med. **101**, 577—589 (1955).

H₄₈ HARFORD, C. G., and M. HARA: Pulmonary oedema in influenzal pneumonia of the mouse and the relation of fluid in the lung to the inception of pneumococcal pneumonia. J. exp. Med. **91**, 245—257 (1950).

$H_{49}$ HARFORD, C. G., V. LEIDLER, and M. HARA: Effect of the lesion due to influenza virus on the resistance of mice to inhaled pneumococci. J. exp. Med. 89, 53—68 (1949).

$H_{50}$ HARFORD, C. G., M. R. SMITH, C. McLEOD, and W. B. WOOD: Infection of rats with the virus of influenza. J. Immunol. 53, 163—169 (1946).

$H_{51}$ HARRIS, S., and W. HENLE: Lymphocytopenia in rabbits following intravenous injection of influenza virus. J. Immunol. 59, 9—20 (1948).

$H_{52}$ HATCH, L. A.: An analysis of 101 strains of influenza A virus isolated in England and Wales during the winter of 1955—56 by the laboratories of the Public Health Laboratory Service. Mth. Bull. Minist. Hlth Lab. Serv. 16, 138—141 (1957).

$H_{53}$ HAUKENES, G., A. HARBOE, and K. MORTENSSON-EGNUND: A uronic and sialic acid free chick allantoic mucopolysaccharide sulphate which combines with influenza virus HI-antibody to host material. 1. Purification of the substance.

$H_{54}$ 2. Chemical composition. Acta path. microbiol. scand. 64, 534—542 (1965) and 66. 510—518 (1966).

$H_{55}$ HAUSEN, P., H. HAUSEN, R. ROTT, C. SCHOLTISSEK, and W. SCHÄFER: Early events in the reproduction cycle of animal viruses. Viruses, Nucleic acids, and Cancer. 282—295 (1963), Univ. Texas.

$H_{56}$ HAUSLER, W. J., and E. C. DICK: The purification and concentration of influenza A virus (PR8) by a method of continuous extraction with glycine of a zinc-virus precipitate. J. infect. Dis. 107, 189—194 (1960).

$H_{57}$ HAUSSMAN, H. G., W. SIMROCK, and L. BETZ: Clinical and serological diagnosis of epidemic influenza (German). Z. klin. Med. 147, 51—72 (1950).

$H_{58}$ HEATH, R. B., and D. A. J. TYRRELL: The behaviour of some influenza viruses in tissue cultures of kidney cells of various species. Arch. ges. Virusforsch. 8, 577—591 (1959).

$H_{59}$ HEINMETZ, F.: Studies with the electron microscope on the interaction of influenza virus and red cells. J. Bact. 55, 823—831 (1948).

$H_{60}$ HELLER, L., A. ESPMARK, and P. VIRIDEN: Immunological relationship between infectious cough in horses and human influenza A. Arch. ges. Virusforsch. 7, 120—124 (1956).

$H_{61}$ HENDERSON, J. R., and J. E. KEMPF: Intracellular proteins associated with the synthesis of the influenza A virus. Virology 9, 72—83 (1959).

$H_{62}$ HENLE, G., and W. HENLE: Neurological signs in mice following intracerebral inoculation of influenza viruses. Science 100, 410—411 (1944).

$H_{63}$ HENLE, G., and W. HENLE: Studies on the toxicity of influenza viruses. 1. The effect of intracerebral injection of influenza viruses.

$H_{64}$ 2. The effect of intra-abdominal and intravenous injection of influenza viruses. J. exp. Med. 84, 623—637, and 639—660 (1946).

$H_{65}$ HENLE, G., A. GIRARDI, and W. HENLE: A non-transmissible cytopathogenic effect of influenza virus in tissue culture accompanied by formation of non-infectious haemagglutinins. J. exp. Med. 101, 25—41 (1955).

$H_{66}$ HENLE, W.: Studies on host-virus interaction in the chick embryo influenza virus system. 1. adsorption and recovery of seed virus.

$H_{67}$ 2. The propagation of virus in conjunction with the host cells.

$H_{68}$ 3. Development of infectivity, haemagglutination, and complement fixation activities during the first infectious cycle. J. exp. Med. 90, 1—11, 13—22, and 23—37 (1949).

$H_{69}$ HENLE, W.: Interference phenomena between animal viruses. A review. J. Immunol. 64, 203—236 (1950).

$H_{70}$ HENLE, W.: Status of vaccination against influenza. Mod. Med. 75—81 (1955).

$H_{71}$ HENLE, W.: The use of phosphorus 32 in the study of incomplete forms of influenza virus. Trans. N.Y. Acad. Sci. 18, 255—260 (1956).

$H_{72}$ HENLE, W.: Persistent infection of L cell cultures by myxoviruses. Ciba Found. Symp. "Cellular biology of myxovirus infections" 299—316 (1964).

H₇₃ HENLE, W., and L. A. CHAMBERS: The serological activity of extra embryonic fluids of chick infected with virus of influenza A. Proc. Soc. exp. Biol. (N.Y.) 46, 713—717 (1941).

H₇₄ HENLE, W., and G. HENLE: Interference of inactive virus with the propagation of the virus of influenza. Science 98, 87—89 (1943).

H₇₅ HENLE, W., and G. HENLE: Interference between inactive and active viruses of influenza. 1. The incidental occurrence and artificial induction of the phenomenon.

H₇₆ 2. Factors influencing the phenomenon. Amer. J. med. Sci. 207, 705—717 and 717—733 (1944).

H₇₇ 3. Cross-interference between various related and unrelated viruses.

H₇₈ 4. The nature of the interfering agent. Amer. J. med. Sci. 210, 362—369 and 369—374 (1945).

H₇₉ HENLE, W., and G. HENLE: The toxicity of influenza viruses. Science 102, 398—400 (1945).

H₈₀ HENLE, W., and G. HENLE: Effects of adjuvants on vaccination of human beings against influenza. Proc. Soc. exp. Biol. (N.Y.) 59, 179—181 (1945).

H₈₁ HENLE, W., and G. HENLE: The effect of ultraviolet irradiation on various properties of influenza viruses. J. exp. Med. 85, 347—364 (1947).

H₈₂ HENLE, W., and G. HENLE: Studies on the toxicity of influenza viruses. 3. Immunisation of mice against the toxic effects of influenza virus A. A new potency test for the assay of vaccines of influenza virus. J. Immunol. 59, 45—58 (1948).

H₈₃ HENLE, W., G. HENLE, F. DEINHARDT, and V. V. BERGS: Studies on persistent infections of tissue cultures. 4. Evidence for the production of an interferon in MCN cells by myxoviruses. J. exp. Med. 110, 525—541 (1959).

H₈₄ HENLE, W., and G. HENLE, V. GROUPE, and L. A. CHAMBERS: Studies on complement fixation with the viruses of influenza. J. Immunol. 48, 163—180 (1944).

H₈₅ HENLE, W., G. HENLE, B. HAMPILL, E. D. MARIS, and J. STOKES: Experiments on vaccination of human beings against epidemic influenza. J. Immunol. 53, 75—93 (1946).

H₈₆ HENLE, W., G. HENLE, and M. W. KIRBER: Interference between inactive and active viruses of influenza. 5. Effect of irradiated virus on the host cells. Amer. J. med. Sci. 214, 529—541 (1947).

H₈₇ HENLE, W., G. HENLE, and E. B. ROSENBERG: The demonstration of one-step growth curves of influenza virus through the blocking effect of irradiated virus on further infection. J. exp. Med. 86, 423—437 (1947).

H₈₈ HENLE, W., G. HENLE, J. STOKES, and E. P. MARIS: Experimental exposure of human subjects to the viruses of influenza. J. Immunol. 52, 145—165 (1946).

H₈₉ HENLE, W., and F. S. LIEF: Antigenic analysis of influenza viruses by complement fixation. 4. Strain-specific V antigens in the serodiagnosis of influenza. J. Immunol. 84, 491—500 (1960).

H₉₀ HENLE, W., F. S. LIEF, and A. FABIYI: Strain specific complement fixation test in antigenic analysis and serodiagnosis of influenza. Lancet 818—820 (1958).

H₉₁ HENLE, W., and O. C. LIU: Studies on host-virus interactions in the chick embryo influenza virus system. J. exp. Med. 94, 305—322 (1951).

H₉₂ HENLE, W., O. C. LIU, and N. B. FINTER: Studies on host-virus interactions in the chick embryo influenza virus system. 9. The period of liberation of virus from infected cells. J. exp. Med. 100, 53—70 (1954).

H₉₃ HENLE, W., O. C. LIU, K. PAUCKER, and F. S. LIEF: Studies on host-virus interactions in the chick embryo influenza virus system. 14. The relation between tissue bound and liberated virus materials under various conditions of infection. J. exp. Med. 103, 799—822 (1956).

$H_{94}$ HENLE, W., and K. PAUCKER: Interference between inactivated and active influenza viruses in the chick embryo. 1. A re-evaluation of factors of dosage and timing using infectivity titrations for assay. Virology 6, 181—197 (1958).

$H_{95}$ HENLE, W., and E. B. ROSENBERG: One-step growth curves of various strains of influenza A and B viruses and their inhibition by inactivated virus of the homologous type. J. exp. Med. 89, 279—285 (1949).

$H_{96}$ HENLE, W., and M. WIENER: Complement fixation antigens of influenza viruses types A and B. Proc. Soc. exp. Biol. (N.Y.) 57, 176—179 (1944).

$H_{97}$ HENNESSEN, W. A.: Antigenic analysis of influenza B strains isolated in 1952. Bull. Wld Hlth Org. 6, 481—485 (1952).

$H_{98}$ HENNESSY, A. V.: Prevention of influenza by vaccination in infants and children. Quart. Rev. Pediat. 17, 91—94 (1962).

$H_{99}$ HENNESSY, A. V., and F. M. DAVENPORT: Epidemiologic implications of the distribution by age of antibody response to experimental influenza virus vaccines. J. Immunol. 80, 114—121 (1958).

$H_{100}$ HENNESSY, A. V., and F. M. DAVENPORT: Relative merits of aqueous and adjuvant influenza vaccines when used in a two-dose schedule. Publ. Hlth Rep. (Wash.) 76, 411—419 (1961).

$H_{101}$ HENNESSY, A. V., and F. M. DAVENPORT: Relative antigenic potency in man of polyvalent influenza virus vaccines containing isolated haemagglutinins or intact viruses. J. Immunol. 97, 235—238 (1966).

$H_{102}$ HENNESSY, A. V., F. M. DAVENPORT, and T. FRANCIS: Studies of antibodies to strains of influenza virus in persons of different ages in sera collected in a postepidemic period. J. Immunol. 75, 401—409 (1955).

$H_{103}$ HENNESSY, A. V., F. M. DAVENPORT, R. J. M. HORTON, J. A. NAPIER, and T. FRANCIS: Asian influenza. Occurrence and recurrence, a community and family study. Milit. Medicine 129, 38—50 (1964).

$H_{104}$ HENNESSY, A. V., E. MINUSE, and F. M. DAVENPORT: A twenty-one-year experience with antigenic variation among influenza B viruses. J. Immunol. 94, 301—306 (1965).

$H_{105}$ HENNESSY, A. V., E. MINUSE, F. M. DAVENPORT, and T. FRANCIS: An experience with vaccination against influenza B in 1952 by use of monovalent vaccine. Amer. J. Hyg. 58, 165—173 (1953).

$H_{106}$ HENRY, C., and J. S. YOUNGER: Influence of normal animal sera on influenza viruses in cultures of trypsin-dispersed monkey kidney cells. J. Immunol. 78, 214—221 (1957).

$H_{107}$ HERS, J. F. P.: The histopathology of the respiratory tract in human influenza. II. Stenpert Kroese, Leiden (1955).

$H_{108}$ HERS, J. F. P., and J. MULDER: Rapid tentative post-mortem diagnosis of influenza with the aid of cytological smears of the tracheal epithelium. J. Path. Bact. 43, 329—332 (1951).

$H_{109}$ HERZBERG, K., G. MAY, and H. BECK: Immunity in influenza. 1. Behaviour of the PR8 and FM1 strains. Z. Immun.-Forsch. 112, 409—419 (1955).

$H_{110}$ HERZBERG, K., G. MAY, and H. BECK: The isolation of seven strains of influenza virus from the epidemic of 1955 (German). Z. Hyg. Infekt.-Kr. 142, 1—14 (1955).

$H_{111}$ HEYMANN, H., Z. R. GULICK, C. J. DE BOER, G. STEVENS, and R. L. MAYER: The inhibition of ribonuclease by acidic polymers and their use as possible antiviral agents. Arch. Biochem. 73, 366—383 (1958).

$H_{112}$ HILDEBRANDT, H. M., H. F. MAASSAB, and P. W. WILLIS: Influenza virus pericarditis. Amer. J. Dis. Child. 104, 579—582 (1962).

$H_{113}$ HILLEMAN, M. R.: System for measuring and designating antigenic components of influenza viruses with analyses of recently isolated strains. Proc. Soc. exp. Biol. (N.Y.) 78, 208—215 (1951).

$H_{114}$ HILLEMAN, M. R.: A pattern of antigenic variation. Fed. Proc. 11, 798—803 (1952).

$H_{115}$ HILLEMAN, M. R., E. L. BUESCHER, and J. E. SMADEL: Preparation of dried antigen and antiserum for the agglutination inhibition test for virus influenza. Publ. Hlth Rep. (Wash.) 66, 1195—1203 (1951).

$H_{116}$ HILLEMAN, M. R., F. J. FLATLEY, S. A. ANDERSON, H. L. LUECKING, and D. J. LEVINSON: Distribution and significance of Asian and other influenza antibodies in the human population. New Engl. J. Med. 258, 969—974 (1958).

$H_{117}$ HILLEMAN, M. R., and F. L. HORSFALL: Comparison of the antigenic patterns of influenza A virus strains determined by in ovo neutralisation and haemagglutination-inhibition. J. Immunol. 69, 343—356 (1952).

$H_{118}$ HILLEMAN, M. R., R. P. MASON, and E. L. BUESCHER: Antigenic pattern of strains of influenza A and B. Proc. Soc. exp. Biol. (N.Y.) 75, 829—835 (1950).

$H_{119}$ HILLEMAN, M. R., R. P. MASON, and N. G. ROGERS: Laboratory studies on the 1950 outbreak of influenza. Publ. Hlth Rep. (Wash.) 65, 771—777 (1950).

$H_{120}$ HILLEMAN, M. R., and J. H. WERNER: Influence of non-specific inhibitor on the diagnostic haemagglutination-inhibition test for influenza. J. Immunol. 71, 110—117 (1953).

$H_{121}$ HIMMELWEIT, F.: Influenza B virus isolated from a fatal case of pneumonia. Lancet 793—794 (1943).

$H_{122}$ HINZ, R. W., and J. T. SYVERTON: Mammalian cell cultures for study of influenza viruses. 1. Preparation of monolayer cultures with collagenase.

$H_{123}$ 2. Virus propagation. Proc. Soc. exp. Biol. (N.Y.) 101, 19—22, and 22—26 (1959).

$H_{124}$ HIRSCH, A.: Handbook of geographical and historical pathology. The New Sydenham Society. London Vol. 1 (1883).

$H_{125}$ HIRST, G. K.: Agglutination of red cells by allantoic fluid of chick embryos infected with influenza virus. Science 94, 22—23 (1941).

$H_{126}$ HIRST, G. K.: The quantitative determination of influenza virus and antibodies by means of red cell agglutination. J. exp. Med. 75, 47—64 (1942).

$H_{127}$ HIRST, G. K.: In ovo titrations of influenza virus and of neutralising antibodies in chick embryos. J. Immunol. 45, 285—292 (1942).

$H_{128}$ HIRST, G. K.: Direct isolation of influenza virus in chick embryos. J. Immunol. 45, 293—302 (1942).

$H_{129}$ HIRST, G. K.: Adsorption of influenza haemagglutinins and virus by red blood cells. J. exp. Med. 76, 195—209 (1942).

$H_{130}$ HIRST, G. K.: Adsorption of influenza virus on cells of the respiratory tract. J. exp. Med. 78, 99—109 (1943).

$H_{131}$ HIRST, G. K.: Studies of antigenic differences among strains of influenza A by means of red cell agglutination. J. exp. Med. 78, 407—423 (1943).

$H_{132}$ HIRST, G. K.: Direct isolation of influenza virus in chick embryos. Proc. Soc. exp. Biol. (N.Y.) 58, 155—157 (1945).

$H_{133}$ HIRST, G. K.: Studies on the mechanism of adaptation of influenza virus to mice. J. exp. Med. 86, 357—366 (1947).

$H_{134}$ HIRST, G. K.: Comparisons of influenza virus strains from three epidemics. J. exp. Med. 86, 367—381 (1947).

$H_{135}$ HIRST, G. K.: The nature of the virus receptors of red cells. 1. Evidence on the chemical nature of the virus receptors of red cells and of the existence of a closely analogous substance in normal serum.

$H_{136}$ 2. The effect of partial heat inactivation of influenza virus on the destruction of red cell receptors and the use of inactivated virus in the measurement of serum inhibition. J. exp. Med. 87, 301—314, and 315—328 (1948).

$H_{137}$ 3. Partial purification of the virus agglutination inhibitor in human plasma.

$H_{138}$ 4. Effect of sodium periodate on the elution of influenza virus from red cells. J. exp. Med. 89, 223—232, and 233—243 (1949).

$H_{139}$ HIRST, G. K.: Receptor destruction by viruses of the mumps-NDV-influenza group. J. exp. Med. 91, 161—175 (1950).

$H_{140}$ HIRST, G. K.: The relationship of the receptors of a new strain of virus to those of the mumps-NDV-influenza group. J. exp. Med. 91, 177—184 (1950).

$H_{141}$ HIRST, G. K.: Strain-specific elements in influenza antigens. J. exp. Med. 96, 589—603 (1952).

$H_{142}$ HIRST, G. K.: Studies of mixed infections with NDV, poliovirus and influenza. 9th Symp. Soc. gen. Microbiol. "Virus growth and variation" 82—101 (1959).

$H_{143}$ HIRST, G. K.: Genetic recombination with Newcastle disease virus, polioviruses and influenza. Cold Spr. Harb. Symp. quant. Biol. 27, 303—310 (1962).

$H_{144}$ HIRST, G. K., and T. GOTLIEB: The experimental production of combination forms of virus. 1. The occurrence of combination forms after simultaneous inoculation of the allantoic sac with two distinct strains of influenza virus.

$H_{145}$ 2. A study of serial passage in the allantoic sac of agents that contain the antigens of two distinct influenza A strains. J. exp. Med. 98, 41—52, and 53—70 (1953).

$H_{146}$ 5. Alterations in the virulence of neurotropic influenza virus as a result of mixed infection. Virology 1, 221—235 (1955).

$H_{147}$ HIRST, G. K., T. GOTLIEB, and A. GRANOFF: Studies on mixed infections with influenza virus. Ciba Found. Symp. "The nature of viruses" 191—201 (1957).

$H_{148}$ HIRST, G. K., and E. G. PICKELS: A method for titration of influenza haemagglutinins and influenza antibodies with the aid of a photoelectric densitometer. J. Immunol. 45, 273—283 (1942).

$H_{149}$ HIRST, G. K., N. PLUMMER, and W. F. FRIEDEWALD: Human immunity following vaccination with formalinised influenza virus. Amer. J. Hyg. 42, 45—56 (1945).

$H_{150}$ HIRST, G. K., E. R. RICKARD, and L. WHITMAN: A new method for concentrating influenza virus from allantoic fluid. Proc. Soc. exp. Biol. (N.Y.) 50, 129—133 (1942).

$H_{151}$ HIRST, G. K., E. R. RICKARD, L. WHITMAN, and F. L. HORSFALL: Antibody response of human beings following vaccination with influenza viruses. J. exp. Med. 75, 495—511 (1942).

$H_{152}$ HIRST, G. K., A. VILCHES, O. ROGERS, and C. L. ROBBINS: The effect of vaccination on the incidence of influenza B. Amer. J. Hyg. 45, 96—101 (1947).

$H_{153}$ HOBSON, D.: The strain-specific serological activity of a non-haemagglutinating fraction of influenza viruses. Brit. J. exp. Path. 47, 257—265 (1966).

$H_{154}$ HOFFMAN, C. E., E. M. NEUMAYER, R. F. HAFF, and R. A. GOLDSBY: Mode of action of the antiviral activity of amantadine in tissue culture. J. Bact. 90, 623—628 (1965).

$H_{155}$ HOLLAND, W. W., A. ISAACS, S. K. R. CLARKE, and R. B. HEATH: A serological trial of Asian influenza vaccine after the autumn epidemic. Lancet 820—822 (1958).

$H_{156}$ HOLLOS, I.: Inhibition of the multiplication of influenza virus by Francis inhibitor coupled to aniline derivatives. Acta microbiol. Acad. Sci. hung. 7, 231—242 (1960).

$H_{157}$ HOLTERMANN, O. A., W. D. HILLIS, and M. A. J. MOFFAT: The development of soluble (S) and viral (V) antigens of influenza A virus in tissue culture as studied by the fluorescent antibody technique. Acta path. microbiol. scand. 50, 398—429 (1960).

$H_{158}$ HOLTERMANN, O. A., and K. B. PETERSEN: On the strain specificity of the antibody response in influenza A, as measured by complement fixation using purified antigens, and its relation to in ovo neutralising antibodies. Acta path. microbiol. scand. 50, 430—442 (1960).

$H_{159}$ HOOK, E. W., and R. R. WAGNER: The resistance-promoting activity of endotoxins and other microbial products. 2. Protection against the neurotoxic action of influenza virus. J. Immunol. 83, 310—317 (1959).

$H_{160}$ HORIKAWA, T., and T. SUZUKI: Studies on the substrates having influenza virus adsorption capacities: 1. Combining factors of fowl erythrocytes. J. Mie Med. Coll. Otanicho 1, 1—4 (1950).

H₁₆₁  HORIKAWA, T., T. SUZUKI, and H. FUZITA: Isolation of influenza and New-castle disease viruses at interepidemic period. Mie med. J. 4, 95—102 (1954).

H₁₆₂  HORNE, R. W., and P. WILDY: Symmetry in virus architecture. Virology 15, 348—373 (1961).

H₁₆₃  HORSFALL, F. L.: Neutralisation of epidemic influenza virus. The linear relationship between the quantity of serum and the quantity of virus neutralised. J. exp. Med. 70, 209—222 (1939).

H₁₆₄  HORSFALL, F. L.: Viral multiplication. Fed. Proc. 8, 518—522 (1949).

H₁₆₅  HORSFALL, F. L.: On the reproduction of influenza virus. Quantitative studies with procedures which enumerate infective and haemagglutinating virus particles. J. exp. Med. 100, 135—161 (1954).

H₁₆₆  HORSFALL, F. L.: Reproduction of influenza viruses. Quantitative investigations with particle enumeration procedures on the dynamics of influenza A and B virus reproduction. J. exp. Med. 102, 441—473 (1955).

H₁₆₇  HORSFALL, F. L., and H. S. GINSBERG: An improved $CO_2$ cabinet for low temperature storage of infectious agents with gaseous $CO_2$ excluded from the specimen compartment. J. Bact. 61, 443—451 (1951).

H₁₆₈  HORSFALL, F. L., and E. H. LENNETTE: A complex vaccine effective against different strains of influenza virus. Science 91, 492—494 (1940).

H₁₆₉  HORSFALL, F. L., E. H. LENNETTE, and E. R. RICKARD: A complex vaccine against influenza A virus. Quantitative analysis of the antibody response produced in man. J. exp. Med. 73, 335—355 (1941).

H₁₇₀  HORSFALL, F. L., and I. TAMM: Variation demonstrable by methods other than serological. Bull. N.Y. Acad. Med. 28, 765—768 (1952).

H₁₇₁  HORVATH, S.: A new sensitive method of the rolling drum type for influenza virus titration. Acta microbiol. Acad. Sci. hung. 1, 481—494 (1954).

H₁₇₂  HORVATH, S.: Study on the cyclic multiplication of influenza and ND viruses in tissue cultures. Acta microbiol. Acad. Sci. hung. 5, 123—131 (1958).

H₁₇₃  HORVATH, S.: Tissue culture and influenza virus studies (German). Z. ges. inn. Med. 9, 316 (1958).

H₁₇₄  HORVATH, S., and V. BALAZS: Influenza in Szeged 1953—54. Acta microbiol. Acad. Sci. hung. 3, 241—246 (1956).

H₁₇₅  HORVATH, I., E. SZOLLOSY, and G. IVANOVICS: Distribution of the receptor substance of the influenza and related viruses in the tissue elements of different animal species. Acta physiol. hung. 2, 77—86 (1950).

H₁₇₆  HOTCHIN, J. E., S. COHEN, H. RUSKA, and C. RUSKA: The adsorption of influenza virus to erythrocytes in the electron microscope (German). Dsch. med. Wschr. 84, 849 (1959).

H₁₇₇  HOTCHIN, J. E., S. M. COHEN, H. RUSKA, and C. RUSKA: Electron microscopical aspects of haemadsorption in tissue cultures infected with influenza virus. Virology 6, 689—701 (1958).

H₁₇₈  HOTCHIN, J. E., R. DEIBEL, and L. M. BENSON: Location of noncytopathic myxovirus plaques by haemadsorption. Virology 10, 275—280 (1960).

H₁₇₉  HOTZ, G., and F. B. BANG: Electron microscope studies of ferret respiratory cells infected with influenza. Bull. Johns Hopk. Hosp. 101, 175—208 (1957).

H₁₈₀  HOTZ, G., and W. SCHAFER: Ultrahistological studies on the multiplication of the virus of fowl plague (German). Z. Naturforsch. 10b, 1—5 (1955).

H₁₈₁  HOWE, C.: The influenza virus receptor and blood group antigens of human erythrocyte stroma. J. Immunol. 66, 9—35 (1951).

H₁₈₂  HOWE, C., L. T. LEE, A. HARBOE, and G. HAUKENES: Immunochemical study of influenza virus and associated host tissue components. J. Immunol. 98, 543—557 (1967).

H₁₈₃  HOWE, C., H. M. ROSE, and L. T. LEE: Observations on the relationship between haemagglutinin and neuraminidase of influenza virus. Proc. Soc. exp. Biol. (N.Y.) 108, 420—425 (1961).

H₁₈₄  HOWE, C., H. M. ROSE, and L. SCHNEIDER: Enzymic action of influenza virus on human erythrocyte stroma components. Proc. Soc. exp. Biol. (N.Y.) **96**, 88—94 (1957).

H₁₈₅  HOYLE, L.: A report on the aetiology of the recent influenza epidemic. Mth. Bull. Minist. Hlth Lab. Serv. **3**, 58—61 (1944).

H₁₈₆  HOYLE, L.: An analysis of the complement fixation reaction in influenza. J. Hyg. (Lond.) **44**, 170—178 (1945).

H₁₈₇  HOYLE, L.: Technique of the complement fixation reaction in influenza. Mth. Bull. Minist. Hlth Lab. Serv. **7**, 114—116 (1948).

H₁₈₈  HOYLE, L.: The growth cycle of influenza A. A study of the relations between virus, soluble antigen and host cell in fertile eggs inoculated with influenza virus. Brit. J. exp. Path. **29**, 390—399 (1948).

H₁₈₉  HOYLE, L.: The effect of triphenylmethane dyes on the intracellular growth of influenza A. Brit. J. exp. Path. **30**, 123—129 (1949).

H₁₉₀  HOYLE, L.: Growth cycle of influenza A. Nature (Lond.) **164**, 1137 (1949).

H₁₉₁  HOYLE, L.: The multiplication of influenza viruses in the fertile egg. J. Hyg. (Lond.) **48**, 277—297 (1950).

H₁₉₂  HOYLE, L.: The multiplication of complement fixing antigen and red-cell agglutinin in the chorio-allantoic membrane of fertile eggs inoculated with influenza virus. J. Path. Bact. **64**, 419—422 (1952).

H₁₉₃  HOYLE, L.: Structure of the influenza virus. The relation between biological activity and chemical structure of virus fractions. J. Hyg. (Lond.) **50**, 229—245 (1952).

H₁₉₄  HOYLE, L.: The multiplication of the influenza virus considered in relation to the general problem of biological multiplication. 2nd Symp. Soc. gen. Microbiol. "The nature of virus multiplication." 225—248 (1952).

H₁₉₅  HOYLE, L.: The release of influenza virus from the infected cell. J. Hyg. (Lond.) **52**, 180—188 (1954).

H₁₉₆  HOYLE, L.: The use of radioactive influenza virus to determine the fate of the infecting particle on entry into the host cell. Ciba Foundation Symp. "The nature of viruses." 211—218 (1956).

H₁₉₇  HOYLE, L.: Nucleotide composition of influenza virus ribonucleic acid. Abst. VIIth int. Congr. Microbiol. 239 (1958).

H₁₉₈  HOYLE, L.: The nucleic acid of the influenza virus. Symp. VII, IVth int. Congr. Biochem., Vienna (1958).

H₁₉₉  HOYLE, L.: The entry of myxoviruses into the cell. Cold Spr. Harb. Symp. quant. Biol. **27**, 113—121 (1962).

H₂₀₀  HOYLE, L.: Effect of progressive iodination of influenza virus on its biological properties. Ciba Foundation Symposium, "Cellular biology of myxovirus infections". 152—162 (1964).

H₂₀₁  HOYLE, L., and S. P. DAVIES: Amino-acid composition of the protein components of influenza virus A. Virology **13**, 53—57 (1961).

H₂₀₂  HOYLE, L., and R. W. FAIRBROTHER: Isolation of the influenza virus and the relation of antibodies to infection and immunity. Brit. med. J. 655 (1937).

H₂₀₃  HOYLE, L., and R. W. FAIRBROTHER: Antigenic structure of influenza viruses; The preparation of elementary body suspensions and the nature of the complement fixing antigen. J. Hyg. (Lond.) **37**, 512—519 (1937).

H₂₀₄  HOYLE, L., and R. W. FAIRBROTHER: Further studies of complement fixation in influenza: Antigen production in egg-membrane culture and the occurrence of a zone phenomenon. Brit. J. exp. Path. 18, 425—429 (1937).

H₂₀₅  HOYLE, L., and R. W. FAIRBROTHER: Serological diagnosis of epidemic influenza by the complement fixation reaction. Brit. med. J. 991 (1947).

H₂₀₆  HOYLE, L., and N. B. FINTER: The use of influenza virus labelled with radiosulphur in studies of the early stages of the interaction of virus with the host cell. J. Hyg. (Lond.) **55**, 290—297 (1957).

H₂₀₇  HOYLE, L., and W. FRISCH-NIGGEMEYER: The disintegration of influenza virus particles on entry into the host cell. J. Hyg. (Lond.) **53**, 474—486 (1955).

H<sub>208</sub> HOYLE, L., and L. HANA: The chemical reactions of influenza virus proteins. J. Path. Bact. 92, 447—460 (1966).

H<sub>209</sub> HOYLE, L., R. W. HORNE, and A. P. WATERSON: The structure and composition of the myxoviruses. II. Components released from the influenza virus particle by ether. Virology 13, 448—459 (1961).

H<sub>210</sub> HOYLE, L., R. W. HORNE, and A. P. WATERSON: The structure and composition of the myxoviruses. III. The interaction of influenza virus particles with cytoplasmic particles derived from normal chorio-allantoic membrane cells. Virology 17, 533—542 (1962).

H<sub>211</sub> HOYLE, L., B. JOLLES, and R. B. MITCHELL: The incorporation of radioactive phosphorus in the influenza virus and its distribution in serologically active virus fractions. J. Hyg. (Lond.) 52, 119—127 (1954).

H<sub>212</sub> HOYLE, L., R. REED, and W. T. ASTBURY: Electron microscope studies of the structure of the influenza virus. Nature (Lond.) 171, 256 (1956).

H<sub>213</sub> HUANG, J. S., and F. B. BANG: The susceptibility of chick embryo organ cultures to influenza virus following excess vitamin A. J. exp. Med. 120, 129—147 (1964).

H<sub>214</sub> HUDSON, N. P., M. M. SIGEL, and F. S. MARKHAM: Antigenic relationship of British swine influenza strains to standard human and swine influenza viruses. The use of chicken and ferret antisera in red cell agglutination. J. exp. Med. 77, 467—471 (1943).

H<sub>215</sub> HULTON, J. V., and A. P. MCKEE: Fixation of "neutralised" influenza virus by susceptible cells. J. Bact. 63, 437—447 (1952).

H<sub>216</sub> HUMMELER, K., L. KRAVIS, and M. M. SIGEL: A rapid method of identification of newly isolated influenza viruses. J. Bact. 64, 253—257 (1952).

H<sub>217</sub> HURRELL, J. M. W.: Methods of storing viruses at low temperatures with particular reference to the myxovirus group. J. med. Lab. Technol. 24, 30—41 (1967).

H<sub>218</sub> HURST, E. W.: Approaches to the chemotherapy of virus diseases. J. Pharm. Pharmacol. 9, 273—292 (1957).

H<sub>219</sub> HURST, E. W., and R. HULL: The chemotherapy of virus diseases, with brief consideration of the influence of dietary, hormonal and other factors in virus infections. Pharmacol. Rev. 8, 199—263 (1956).

H<sub>220</sub> HYDE, R. A.: Behaviour of influenza A viruses in the rabbit, rat, and guinea-pig. Amer. J. Hyg. 36, 338—353 (1942).

H<sub>221</sub> HYDE, R. R., and A. CHAPMAN: Behaviour of certain filterable agents from cases of colds and influenza isolated directly from man to chick membranes. Amer. J. Hyg. 26, 116 (1937).

I<sub>1</sub> INGLOT, A., and F. M. DAVENPORT: Studies on the role of leucocytes in infection with influenza virus. J. Immunol. 88, 55—65 (1962).

I<sub>2</sub> ISAACS, A.: Reactivation of neutral mixtures of influenza virus and serum by virus inactivated by heat. Brit. J. exp. Path. 29, 529—530 (1948).

I<sub>3</sub> ISAACS, A.: The 1951 influenza viruses. Proc. roy. Soc. Med. 44, 801—803 (1951).

I<sub>4</sub> ISAACS, A.: P-Q variation in influenza viruses. Lancet 676—678 (1953).

I<sub>5</sub> ISAACS, A.: Recent evolution and spread of strains of influenza A virus. Lancet 960—961 (1956).

I<sub>6</sub> ISAACS, A.: Antibodies to influenza viruses in the sera of Nigerians. J. Hyg. (Lond.) 55, 485—488 (1957).

I<sub>7</sub> ISAACS, A.: Viral interference. 9th Symp. Soc. gen. Microbiol. "Virus growth and variation" 102—121 (1959).

I<sub>8</sub> ISAACS, A.: Production and action of interferon. Cold Spr. Harb. Symp. quant. Biol. "Basic mechanisms in animal virus biology" 343—349 (1962).

I<sub>9</sub> ISAACS, A., and C. H. ANDREWES: The spread of influenza. Evidence from 1950—1951. Brit. med. J. 921 (1951).

I<sub>10</sub> ISAACS, A., and I. ARCHETTI: Pathways of influenza spread. Lancet 457—459 (1954).

I₁₁ ISAACS, A., and A. BOZZO: The use of V. cholerae filtrates in the destruction of non-specific inhibitor in ferret sera. Brit. J. exp. Path. **32**, 325—335 (1951).

I₁₂ ISAACS, A., and D. C. BURKE: Mode of action of interferon. Nature (Lond.) **182**, 1073—1074 (1958).

I₁₃ ISAACS, A., D. C. BURKE, and L. FADEYEVA: Effect of interferon on the growth of viruses on the chick chorion. Brit. J. exp. Path. **39**, 447—458 (1958).

I₁₄ ISAACS, A., R. A. COX, and Z. ROTEM: Foreign nucleic acid as the stimulus to make interferon. Lancet 113—116 (1963).

I₁₅ ISAACS, A., R. DEPOUX, and P. FISET: The viruses of the 1952—53 influenza epidemic. Bull. Wld Hlth Org. **11**, 967—979 (1954).

I₁₆ ISAACS, A., and H. B. DONALD: Particle counts of haemagglutinating viruses. J. gen. Microbiol. **12**, 241—247 (1955).

I₁₇ ISAACS, A., and M. EDNEY: Variation in laboratory strains of influenza viruses. 1. Biological characters of the variants.

I₁₈ 2. Genetic aspects of the variations. Brit. J. exp. Path. **31**, 196—208, and 209—216 (1950).

I₁₉ ISAACS, A., and M. EDNEY: Interference between inactive and active influenza viruses in the chick embryo. 1. Quantitative aspects of interference.

I₂₀ 2. The site of interference.

I₂₁ 4. The early stages of virus multiplication and interference.

I₂₂ 5. The behaviour of different strains of challenge virus.

I₂₃ 6. The behaviour of different strains of interfering virus. Aust. J. exp. Biol. med. Sci. **28**, 219—230, 231—238, and 635—645 (1950). Also **29**, 169—178, and 179—186 (1951).

I₂₄ ISAACS, A., and F. FULTON: Growth of influenza virus in the chick chorion. Nature (Lond.) **171**, 90 (1953).

I₂₅ ISAACS, A., and F. FULTON: Interference in the chick chorion. J. gen. Microbiol. **9**, 132—139 (1953).

I₂₆ ISAACS, A., A. W. GLEDHILL, and C. H. ANDREWES: Influenza A viruses. Laboratory studies with special reference to the European outbreak of 1950—51. Bull. Wld Hlth Org. **6**, 287—315 (1952).

I₂₇ ISAACS, A., and G. HITCHCOCK: Role of interferon in recovery from virus infection. Lancet 69—71 (1960).

I₂₈ ISAACS, A., and J. LINDENMANN: Virus interference. 1. The interferon. Proc. roy. Soc. B **147**, 258—267 (1957).

I₂₉ ISAACS, A., J. LINDENMANN, and R. C. VALENTINE: Virus interference. 2. Some properties of interferon. Proc. roy. Soc. B **147**, 268—273 (1957).

I₃₀ ISAACS, A., G. NEGRONI, and D. A. J. TYRRELL: Infection of volunteers with Asian influenza Virus. Lancet 886—887 (1957).

I₃₁ ISAACS, A., and A. T. RODEN: Administration of live influenza virus to volunteers. Lancet 697—699 (1956).

I₃₂ ISAACS, A., and A. A. DE C. SAMPAIO: The effect of adaptation to mice on the antigenicity of influenza virus. J. Path. Bact. **45**, 613—615 (1953).

I₃₃ ISAACS, A., and J. D. STONE: Influenza virus-antibody union. Aust. J. exp. Biol. med. Sci. **27**, 423—428 (1949).

I₃₄ ISACHENKO, V. A.: The properties of avidity and sensitivity to inhibitors in influenza A2 viruses (Russian). "Problemy grippa i ostrykh respiratornykh zabolevanii." Tez. dokl. nauch. Sess. Inst. AMN SSSR, Moscow, 49—50 (1959).

I₃₅ ISACHENKO, V. A., G. A. SMIRNOVA, and A. S. GORBUNOVA: Isolation of an inhibitor-sensitive variant from an inhibitor-resistant culture of influenza A2 virus by gel-filtration in Sephadex (Russian). Vop. Virus. **1**, 97—99 (1965).

I₃₆ ISAYEVA, E. S., and KH. ZH. ZHUMATOV: Isolation of infectious RNA from type A and B influenza viruses and study of its properties. Abstr. Inst. All-Union Biochem. Congr. **11**, 168 (1964).

I₃₇ ISHIDA, N., and W. W. ACKERMANN: Growth characteristics of influenza virus. Properties of the initial cell-virus complex. J. exp. Med. **104**, 501—515 (1956).

I₃₈ ISHIDA, N., T. MORIZUKA, and Y. HANIEMA: Virus antigens in kidneys of mice infected with influenza A. Nature (Lond.) **201**, 421—422 (1964).

I₃₉ ISHIDA, N., and T. OSATO: Initial reactions during influenza A invasion into L cells. Growth characteristics of myxoviruses in tissue cultures. Tohoku J. exp. Med. **73**, 46—55 (1960).

I₄₀ ISHIDA, N., and H. TOZAWA: Specificity of a neuraminidase obtained from disrupted Sendai virus. Abstr. 9th Int. Congr. Microbiol. 498 (1966).

I₄₁ IVANOVICS, G., S. HORVATH, and E. SZOLLOSY: The influenza virus adsorbing capacity of the vascular endothelium of various mammals. Acta Microbiol. Acad. Sci. hung. **2**, 121—129 (1954).

I₄₂ IVANOVICS, G., S. HORVATH, and E. SZOLLOSY: On the adsorption capacity of the inner lining of the blood vessels for influenza and related viruses. Sond. Naturwissensch. **20**, 482 (1954) (German).


J₁ JAGGER, J., and E. C. POLLARD: Inactivation with fast charged particles of infectivity and haemagglutination in influenza A virus. Radiation Res. **4**, 1—19 (1956).

J₂ JAMES, S. M., and P. FISET: Serum inhibitors of Asian strains of influenza virus. Nature (Lond.) **184**, 1656—1657 (1959).

J₃ JAMESON, P., and A. S. LEVINE: Substrate specificity of neurotropic influenza virus neuraminidases. J. Bact. **90**, 563—564 (1965).

J₄ JANDASEK, L., and N. VOLAKOVA: Some properties of the haemagglutination inhibitor of type A2 influenza viruses contained in the serum of normal guineapigs. Acta virol. **4**, 7—16 (1960).

J₅ JENSEN, K. E.: Strain variations and cross-relationships in influenza virus. Amer. J. publ. Hlth **44**, 1167—1173 (1954).

J₆ JENSEN, K. E.: Immunologic significance of antigenic differences among strains of influenza A isolated during the same epidemic. J. Immunol. **78**, 373—379 (1957).

J₇ JENSEN, K. E.: New set of type A influenza viruses. J. Amer. med. Ass. **164**, 2025—2029 (1957).

J₈ JENSEN, K. E., F. M. DAVENPORT, A. V. HENNESSY, and T. FRANCIS: Characterisation of influenza antibodies by serum absorption. J. exp. Med. **104**, 199—209 (1956).

J₉ JENSEN, K. E., F. L. DUNN, and R. Q. ROBINSON: Influenza 1957. A variant and the pandemic. Progr. med. Virol. N.Y. **1**, 165—209 (1958).

J₁₀ JENSEN, E. M., E. E. FORCE, and J. B. UNGER: Inhibitory effect of ammonium ions on influenza virus in tissue cultures. Proc. Soc. exp. Biol. (N.Y.) **107**, 447—451 (1961).

J₁₁ JENSEN, K. E., and T. FRANCIS: Antigen-antibody precipitates in solid medium with influenza virus. J. Immunol. **70**, 321—325 (1953).

J₁₂ JENSEN, K. E., and T. FRANCIS: The antigenic composition of influenza virus measured by antibody-absorption. J. exp. Med. **98**, 619—639 (1953).

J₁₃ JENSEN, E. M., and O. C. LIU: Inhibitory effect of simple aliphatic amines on influenza virus in tissue culture. Proc. Soc. exp. Biol. (N.Y.) **112**, 456—459 (1963).

J₁₄ JENSEN, K. E., E. MINUSE, and T. FRANCIS: Serologic comparisons with lines of influenza virus isolated and serially transferred in different experimental hosts. J. Immunol. **78**, 356—364 (1957).

J₁₅ JENSEN, K. E., and W. D. PETERSON: Comparative measurements of antigenic differences among human and swine influenza viruses. J. Immunol. **78**, 365—372 (1957).

J₁₆ JERUSHALMY, Z., A. KOHN, and A. DE VRIES: Interaction of myxoviruses with human blood platelets in vitro. Proc. Soc. exp. Biol. (N.Y.) **106**, 462—466 (1961).

J₁₇ JOHNSON, R. B., and W. W. ACKERMANN: Intracellular pattern of nucleic acid in virus infection. Proc. Soc. exp. Biol. (N.Y.) **86**, 318—321 (1954).

$J_{18}$ JOHNSTON, E. J. D., R. G. KIDD, H. J. RYLANCE, and R. G. SOMMERVILLE: The inhibition of neuraminidase and antiviral action. Brit. J. Pharmacol. 27, 415—425 (1966).

$J_{19}$ JONES, M.: Adaptation of influenza virus to heat. Proc. Soc. exp. Biol. (N.Y.) 58, 315—319 (1945).

$J_{20}$ JONES, T. C., and F. D. MAURER: Neutralisation studies of the viruses of influenza A, influenza B, and swine influenza with equine influenza convalescent serums. Amer. J. vet. Res. 3, 179—182 (1942).

$J_{21}$ JORDAN, W. S., and S. G. GAYLIN: The antigenic variation of influenza B viruses. Demonstration of an antigenic spectrum by use of titre ratios. J. Immunol. 70, 393—399 (1953).

$K_1$ KAGANSKAYA, YE. S.: Experience in the use of the complement fixation reaction for the diagnosis of influenza (Russian). Zh. Mikrobiol. (Mosk.) 1, 41 (1952).

$K_2$ KAJI, M., R. OSEASOHN, W. S. JORDAN, and J. H. DINGLE: Isolation of Asian virus from extrapulmonary tissues in fatal human influenza. Proc. Soc. exp. Biol. (N.Y.) 100, 272 (1959).

$K_3$ KALININA, L. I., and Z. I. POSPELOVA: Experimental study of the prophylactic and therapeutic activity of influenza gamma globulin (Russian). "Problemy grippa i ostrykh respiratornykh zabolevanii." Tez. dokl. nauch. Sess. Inst. AMN SSSR, Moscow. 142—143 (1959).

$K_4$ KALTER, S. S.: The effect of age upon susceptibility to infection with influenza virus. J. Immunol. 63, 17—22 (1949).

$K_5$ KALTER, S. S.: Host growth and its relation to influenza virus infection. J. Immunol. 63, 29—35 (1949).

$K_6$ KALTER, S. S.: A rapid method for detection of influenza virus during epidemics. Proc. Soc. exp. Biol. (N.Y.) 74, 607—608 (1950).

$K_7$ KALTER, S. S.: Inapparent infections with influenza viruses. Proc. Soc. exp. Biol. (N.Y.) 76, 570—571 (1951).

$K_8$ KALTER, S. S., G. BERG, J. E. PRIER, K. A. JOCHEIM, and J. M. McKENNA: Virus proliferation in the mouse. 1. Growth curves of influenza (mouse-adapted PR8) and Theiler's mouse encephalomyelitis (GD VII) viruses. J. Immunol. 75, 410—416 (1955).

$K_9$ KALTER, S. S., H. L. CASEY, K. E. JENSEN, R. Q. ROBINSON, and R. H. GORRIE: Evaluation of laboratory diagnostic procedures with A/Asian influenza. Proc. Soc. exp. Biol. (N.Y.) 100, 367—370 (1959).

$K_{10}$ KALTER, S. S., O. D. CHAPMAN, D. A. FEELEY, and S. L. MacDOWELL: An epidemic of influenza A due to an atypical strain. The relationship of this strain to other influenza viruses. J. Immunol. 59, 147—157 (1948).

$K_{11}$ KALTER, S. S., H. J. SMOLIN, J. McELANKEY, and J. TEPPERMAN: Endocrines and their relation to influenza virus infection. J. exp. Med. 93, 529—538 (1951).

$K_{12}$ KALTER, S. S., and J. TEPPERMAN: Influenza virus proliferation in hypoxic mice. Science 115, 621—622 (1952).

$K_{13}$ KANTOROVICH: Methods of detecting influenzal antigen in the blood and nasopharyngeal discharges of sick people (Russian). Gripp. Trudy AMN SSSR 28, 104 (1953).

$K_{14}$ KAPLAN, M. M.: Discussion of paper on "Influenza of horses" by E. R. DOLL. Amer. Rev. resp. Dis. 83, No. 211, 48—53 (1961).

$K_{15}$ KAPLAN, M. M., and A. M. M. PAYNE: Serological survey in animals for type A influenza in relation to the 1957 pandemic. Bull. Wld Hlth Org. 20, 465—488 (1959).

$K_{16}$ KASEL, J. A., R. H. ALFORD, V. KNIGHT, G. H. WADDELL, and M. M. SIGEL: Experimental infection of human volunteers with equine influenza virus. Nature (Lond.) 206, 41—43 (1965).

$K_{17}$ KASS, E. H., F. A. NEVA, and M. FINLAND: Failure of ACTH to protect against acutely lethal toxins of influenza virus and rickettsiae. Proc. Soc. exp. Biol. (N.Y.) 76, 560 (1951).

$K_{18}$ KATES, M., A. C. ALLISON, D. A. J. TYRRELL, and A. T. JAMES: Lipids of influenza virus and their relation to those of the host cell. Biochim. biophys. Acta (Amst.) 52, 455—466 (1961).

$K_{19}$ KATES, M., A. C. ALLISON, D. A. J. TYRRELL, and A. T. JAMES: Origin of lipids in influenza virus. Cold Spr. Harb. Symp. quant. Biol. 27, 293—302 (1962).

$K_{20}$ KATHAN, R. H.: Kelp extracts as antiviral substances. Ann. N.Y. Acad. Sci. 130, 390—397 (1965).

$K_{21}$ KATHAN, R. H., L. J. M. RIFF, and M. REAL: Association between the erythrocyte haemagglutination inhibitor and the M-N blood group substances. Proc. Soc. exp. Biol. (N.Y.) 114, 90—92 (1963).

$K_{22}$ KATO, N., and H. HARA: The toxic effect on rabbits of influenza virus given intravenously. Brit. J. exp. Path. 42, 145—152 (1961).

$K_{23}$ KATO, N., and A. OKADA: The relation of the toxic agent to the subunits of influenza virus particles. Brit. J. exp. Path. 42, 253—265 (1961).

$K_{24}$ KATO, N., K. SHIMIZU, and H. HARA: Local skin reactivity to influenza virus. Brit. J. exp. Path. 42, 445—454 (1961).

$K_{25}$ KAVANAU, J. L.: Structure and function of biological membranes. Nature (Lond.) 198, 525—530 (1963).

$K_{26}$ KEIL, A. W., H. LIPPELT, F. MULLER, and G. BRAND: The O-D phase change in influenza A virus (German). Arch. ges. Virusforsch. 7, 36—46 (1956).

$K_{27}$ KELLY, R., and D. GREIFF: The level of lactic dehydrogenase activity as an indication of the growth of influenza virus in the embryonate egg. J. exp. Med. 113, 125—129 (1961).

$K_{28}$ KELLY, R. T., and D. GREIFF: Neuraminidase and neuraminidase-labile substrates in experimental influenza virus encephalitis. Biochim. biophys. Acta (Amst.) 110, 548—553 (1965).

$K_{29}$ KELLY, R., D. GREIFF, and D. ANDERSON: Histochemical and biochemical changes in the chorioallantoic membranes and allantoic fluids of embryonate eggs infected with influenza virus. J. Immunol. 89, 815—822 (1962).

$K_{30}$ KEMPF, J. E., and H. T. CHANG: Hypotensive action of influenza A virus on rats. Proc. Soc. exp. Biol. (N.Y.) 72, 272—275 (1948).

$K_{31}$ KEMPF, J. E., and E. T. HARKNESS: Observations on the toxin of influenza virus. Proc. Soc. exp. Biol. (N.Y.) 75, 80—83 (1950).

$K_{32}$ KHAYT, S. L., A. S. VINOGRADOVA, R. I. KOTINA, and B. YE. EDEL'MAN: Experiment in active immunisation of people against influenza (Russian). Zh. Mikrobiol. (Mosk.) 3—4, 93 (1942).

$K_{33}$ KHAYT, S. L., A. S. VINOGRADOVA, and N. V. NARTSISSOV: Survival of laboratory strains of influenza virus on the human nasopharyngeal mucosa (Russian). Zh. Mikrobiol. (Mosk.) 1, 4 (1949).

$K_{34}$ KHERSONSKAYA, R. YA.: Experience in the therapeutic use of ecmoline in influenza (Russian). Tr. Ob'yed. nauch. Sess. AMN SSSR, Moscow, 171—174 (1953).

$K_{35}$ KHERSONSKAYA, R. YA.: The treatment of influenza and its complications (Russian). "Problemy grippa i ostrykh respiratornykh zabolevanii." Tez. dokl. nauch. Sess. Inst. AMN SSSR, Moscow, 192—193 (1959).

$K_{36}$ KHOOBYARIAN, N., and D. L. WALKER: Induced resistance in mice to intravenous toxicity of influenza virus. Proc. Soc. exp. Biol. (N.Y.) 94, 74—78 (1957).

$K_{37}$ KHOOBYARIAN, N., and D. L. WALKER: Effect of cortisone on mouse resistance to intravenous toxicity of influenza virus. Proc. Soc. exp. Biol. (N.Y.) 94, 295—298 (1957).

$K_{38}$ KILBOURNE, E. D.: The influence of cortisone on experimental viral infection. 1. Prolongation of survival time and suppression of inflammation in chick embryos infected with influenza B virus. J. Immunol. 74, 57—62 (1955).

$K_{39}$ 2. Effects on antibody formation and acquired immunity. Proc. Soc. exp. Biol. (N.Y.) 90, 685—687 (1955).

$K_{40}$ 3. Effects on certain dynamics of influenza virus increase.

$K_{41}$    4. Negation of interference as the mechanism by which cortisone induces increased virus yields.

$K_{42}$    5. Inhibition of influenza virus synthesis. J. exp. Med. **106**, 851—862, 863—881, and 883—891 (1957).

$K_{43}$    KILBOURNE, E. D.: Inhibition of influenza virus multiplication with a glucose anti-metabolic (2 deoxy-D-glucose). Nature (Lond.) **183**, 271—272 (1959).

$K_{44}$    KILBOURNE, E. D.: The inaccessibility of genetically active but non-infective influenza virus to the intracellular action of ribonuclease. Virology **11**, 291—294 (1960).

$K_{45}$    KILBOURNE, E. D.: Influenza virus genetics. Progr. med. Virol. **5**, 79—126 (1963).

$K_{46}$    KILBOURNE, E. D., H. C. ANDERSON, and F. L. HORSFALL: Concurrent infection with influenza A and B viruses in a single epidemic of influenza. J. Immunol. **67**, 547—558 (1951).

$K_{47}$    KILBOURNE, E. D., and F. L. HORSFALL: Increased virus in eggs injected with cortisone. Proc. Soc. exp. Biol. (N.Y.) **76**, 116—118 (1951).

$K_{48}$    KILBOURNE, E. D., and F. L. HORSFALL: Mouse-egg neutralisation. Neutralisation in the mouse of influenza viruses not adapted to the mouse. J. Immunol. **67**, 431—436 (1951).

$K_{49}$    KILBOURNE, E. D., and J. S. MURPHY: Genetic studies of influenza viruses. 1. Viral morphology and growth capacity as exchangeable genetic traits. Rapid *in ovo* adaptation of early passage Asian strain isolates by combination with PR8. J. exp. Med. **111**, 387—406 (1960).

$K_{50}$    KILBOURNE, E. D., A. SUGIURA, and S. C. WONG: Serial multiplication of an influenza virus (NWS) in certain human diploid cell strains. Proc. Soc. exp. Biol. (N.Y.) **116**, 225—228 (1964).

$K_{51}$    KIRBER, M. W., and W. HENLE: A comparison of influenza complement fixation antigens derived from allantoic fluids and membranes. J. Immunol. **65**, 229—244 (1950).

$K_{52}$    KLEIN, M.: The mechanism of the virucidal action of ascorbic acid. Science **101**, 587—589 (1945).

$K_{53}$    KLEIN, M., H. BREWER, J. E. PEREZ, and B. DAY: The inactivation of influenza virus by mercurials and the reactivation by sodium thioglycolate and BAL. J. Immunol. **59**, 135—140 (1948).

$K_{54}$    KLEIN, M., S. CHAEVSKY, and M. MULLER: A rapid agglutination test for the detection of antibodies to viruses of influenza. J. Immunol. **97**, 131—137 (1966).

$K_{55}$    KLEIN, M., and J. E. PEREZ: The reversal *in vivo* by BAL of inactivation of influenza A virus and diplococcus pneumoniae by bichloride of mercury. J. Immunol. **60**, 349—358 (1948).

$K_{56}$    KLEIN, M., and D. A. STEVENS: *In vitro* and *in vivo* activity of synthetic detergents against influenza A virus. J. Immunol. **50**, 265—273 (1945).

$K_{57}$    KLEMPERER, H. G.: An effect of phloridzin on influenza virus elution and on neuraminidase activity. Virology **12**, 495—498 (1960).

$K_{58}$    KLEMPERER, H.: Glucose breakdown in chick embryo cells infected with influenza virus. Virology **13**, 68—77 (1961).

$K_{59}$    KLENK, E., H. FAILLARD, and H. LEMPFRID: The enzymatic activity of influenza virus (German). Hoppe-Seyler's Z. physiol. Chem. **301**, 235—246 (1955).

$K_{60}$    KLENK, E., and H. LEMPFRID: The nature of the cell receptors for influenza virus (German). Hoppe-Seyler's Z. physiol. Chem. **307**, 278—283 (1957).

$K_{61}$    KLOENE, W., F. B. BANG, M. R. COOPER, H. KULEMANN, M. OLA, K. V. SHAH, S. M. CHAKRABORTY, and H. G. PEREIRA: Isolation of a new antigenic type of influenza B virus. Virology **28**, 774—775 (1966).

$K_{62}$    KLUSKA, V., M. MACKU, and J. MENSIK: Demonstration of antibodies against swine influenza viruses in man. Čs. Pediat. **16**, 408—414 (1961).

$K_{63}$    KNIGHT, C. A.: A sedimentable component of allantoic fluid and its relationship to influenza viruses. J. exp. Med. **80**, 83—100 (1944).

K$_{64}$ KNIGHT, C. A.: The stability of the influenza virus in the presence of salts. J. exp. Med. 79, 285—290 (1944).

K$_{65}$ KNIGHT, C. A.: The preparation of highly purified PR8 influenza virus from infected mouse lungs. J. exp. Med. 83, 11—24 (1946).

K$_{66}$ KNIGHT, C. A.: Precipitin reactions of highly purified influenza viruses and related materials. J. exp. Med. 83, 281—294 (1946).

K$_{67}$ KNIGHT, C. A.: The nucleic acid and carbohydrate of influenza virus. J. exp. Med. 85, 99—116 (1947).

K$_{68}$ KNIGHT, C. A.: Aminoacid composition of highly purified particles of influenza A and B. J. exp. Med. 86, 125—129 (1947).

K$_{69}$ KNIGHT, C. A., and W. M. STANLEY: The effect of some chemicals on purified influenza virus. J. exp. Med. 79, 291—300 (1944).

K$_{70}$ KNOWLES, R. C., and H. C. KING: The 1963 equine influenza epizootic. J. Amer. vet.-med. Ass. 143, 1108—1110 (1963).

K$_{71}$ KNYASEVA, L. D.: Viroscopy as a method of early diagnosis of influenza (Russian). Probl. grippa Tez. dokl. ob'yed. nauch. Konf. Inst. AMN SSSR, Kiev, 57—58 (1955).

K$_{72}$ KOCH, A.: Effects of formaldehyde on influenza virus. 4. Effects on the titratable groups of the virus. Acta microbiol. Acad. Sci. hung. 9, 97—102 (1962).

K$_{73}$ KOCH, A., and E. CSONKA: Effects of formaldehyde on influenza virus. 2. Effects on the infectivity of the virus. Acta microbiol. Acad. Sci. hung. 5, 311—316 (1958).

K$_{74}$ KOCH, A., S. HORVATH, and G. IVANOVICS: An attempt to detect the virus infection of the chick embryo by registrating the refractive power of the allantoic fluid. Acta physiol. hung. 2, 525—531 (1951).

K$_{75}$ KOCISKOVA, D., B. STYK, and L. HANA: The relation of properdin and complement to the thermolabile co-factor of normal serum necessary for the reaction of specific antibody with influenza A2 virus. Acta virol. 5, 19—25 (1961).

K$_{76}$ KOEN, J. S.: A practical method for field diagnosis of swine diseases. Amer. J. Vet.-Med. 14, 468—470 (1919).

K$_{77}$ KOFMAN, S. YA.: Latest in the treatment of influenza (Russian). "Virusnyye Infektsii." Medgiz Leningrad, 213—218 (1953).

K$_{78}$ KOFMAN, S. YA., M. I. KAL'YANOVA, and S. M. LYUBINSKAYA: Experience in the prophylaxis of influenza at an industrial enterprise using influenza antiserum inhalations (Russian). "Virusnyye Infektsii." Tr. Inst. epidem. Mikrobiol. i gig. isneni Pastera i Inst. Eksper. med. AMN SSSR, Leningrad 13, 166 (1953).

K$_{79}$ KOHLER, H. R.: The pathogenesis of influenza infection of the chick embryo after inoculation of virus into the allantoic sac (German). Zbl. Bakt. I. Abt. Orig. 163, 92—117 (1955).

K$_{80}$ KOHN, A., and D. DANON: Rapid titration of viral haemagglutinins and haemagglutination inhibition antibodies with the aid of a fragiligraph. Experientia (Basel) 21, 296—297 (1965).

K$_{81}$ KOLCHURINA, A. A.: The study of genetic characteristics of influenza virus strains in connection with their pathogenicity. Abstr. 9th Int. Congr. Microbiol. Moscow 385 (1966).

K$_{82}$ KOLCHURINA, A. A., and V. M. BOLOTOVSKIY: Study of interaction of inhibitor sensitive and inhibitor resistant vaccine strains of virus of influenza A2 (Russian). Vop. Virus. 1, 61—66 (1965).

K$_{83}$ KOLDATITSKAYA, YE. A.: Cytological characteristics of the nasal secretion in influenza and seasonal catarrhs of the upper respiratory tract (Russian). Zh. Mikrobiol. (Mosk.) 6, 40 (1947), and 12, 39 (1948).

K$_{84}$ KOLDATITSKAYA, YE. A.: Cytological characteristics of the nasal secretion in influenza and in cold rhinitis (Russian). Vop. Virus. 1, 283 (1948).

K$_{85}$ KON, Y. S., S. G. FAIERSHTEIN, N. K. KULICHKOVA, A. S. SHAPIRO, and I. I. MOGILEVSKII: Preliminary epidemiological evaluation of live influenza vaccine of A2 type (Russian). Vop. Virus. 5, 122—125 (1958).

$K_{86}$  Kono, R., Y. Ashishara, M. Matumoto, T. Shimizu, and K. Okazaki: Studies on the experimental infection of swine with human influenza virus FM1 strain. Jap. J. exp. Med. 21, 419—424 (1951).

$K_{87}$  Koppel, Z., J. Vrtiak, M. Vasil, and S. Spiesz: Veterinarstri 6, 267 (1956).

$K_{88}$  Korbecki, M.: Changes in metabolic activity of KB cells infected with influenza virus A2. Bull. Acad. pol. Sci. 13, 391—393 (1965).

$K_{89}$  Kornyushenko, N. P.: Type B virus influenza (Russian). Vrach. delo. 5, 431 (1952).

$K_{90}$  Kornyushenko, N. P.: Material on the study of the immunobiological reactivity of the body in virus influenza with different types of course (Russian). Gripp. i OKVDP Tr. ob'yed. Sess. Inst. AMN SSSR 128 (1953).

$K_{91}$  Kornyushenko, N. P., and N. A. Maximovich: Intrauterine transmission of influenza infection in experimental animals. Acta virol. 5, 26—30 (1961).

$K_{92}$  Kornyushenko, N. P., M. K. Topchy, E. V. Sidorenko, and V. C. Baisheva: Study of various myxovirus components on cultivation in mouse fibroblast cells (line L) and on chick embryos. Abstr. 9th int. Congr. Microbiol. Moscow 504 (1966).

$K_{93}$  Korovin, A. A., and I. P. Dmitrovskaya: Experience in specific serum prophylaxis of influenza (Russian). "Gripp." Tr. AMN SSSR, Moscow 28, 226 (1953).

$K_{94}$  Korych, B., and V. Frankova: Possibility of employing porcine kidney tissue cultures for quantitative determination of myxovirus influenza C. (Czech.). Čs. Epidem. 15, 135—140 (1966).

$K_{95}$  Kosyakov, P. N., and Z. I. Rovnova: Qualitative differences in antigenic properties of type B influenza virus isolated in different periods. Probl. virol. (Lond.) 5, 789—795 (1960).

$K_{96}$  Kosyakov, P. N., and Z. I. Rovnova: Antigenic components of the host in influenza virus (Russian). Vop. Virus. 1, 17—23 (1965).

$K_{97}$  Kozinski, A. W.: A new method of absorption of influenza antibodies (Polish). Med. Sosw. Microbiol., Warsaw 3, 71—82 (1951).

$K_{98}$  Kozinski, A. W.: The influence of cellular fractions from Salm. Typhi upon the agglutination of erythrocytes by influenza virus (Polish). Med. Dosw. Microbiol. Warsaw 3, 83—94 (1951).

$K_{99}$  Kozinski, A. W., and E. Mikulaszek: The purification of an influenza virus inhibitor from egg white. General properties of the inhibitor (Polish). Med. Dosw. Microbiol. Warsaw 3, 242—254 (1951).

$K_{100}$  Kravchenko, A. G., and P. M. Sekretta: The problem of methods of experimental investigation of the influenza virus (Russian). Vop. Virus. 2, 178 (1949).

$K_{101}$  Kreuger, A. B., and others: Attempts to protect against influenza virus with various sulphonamides, acridines, and antibiotics. Science 98, 348—349 (1943).

$K_{102}$  Krizanova, O., and J. Lesko: Separation of gamma inhibitor on carboxymethyl-cellulose. Acta virol. 8, 179—182 (1964).

$K_{103}$  Krizanova, O., and F. Sokol: Purification of bovine serum beta-inhibitor by chromatographic methods and its reaction with A1 influenza virus. Acta virol. 10, 35—42 (1966).

$K_{104}$  Krizanova-Laucikova, O., J. Szanto, and P. Albrecht: Localisation of virus haemagglutination inhibitor in the chorioallantoic membrane. Virology 15, 501—503 (1961).

$K_{105}$  Krizanova-Laucikova, O., J. Szanto, and D. Kociskova: Purification and some properties of the thermostable inhibitor against avid A2 influenza viruses from horse serum. Acta virol. 5, 4—11 (1961).

$K_{106}$  Krizanova-Laucikova, O., J. Szanto, D. Kociskova, G. Ruttkay-Nedecky, and F. Sokol: Differences in the properties of two inhibitors against avid A2 influenza virus strains from horse serum. Acta virol. 5, 12—18 (1961).

K₁₀₇  KRIZANOVA-LAUCIKOVA, O., J. SZANTO, D. KOCISKOVA, G. RUTTKAY-NEDECKY, and F. SOKOL: Presence in horse serum of two inhibitors of avid A2 influenza virus strains. Nature (Lond.) **189**, 513 (1961).

K₁₀₈  KROEGER, A. V.: Evidence for the absence of anaphylactogenic host protein in highly purified PR8 influenza virus. J. Immunol. **89**, 136—144 (1962).

K₁₀₉  KROEGER, A. V., and J. E. KEMPF: Inactivation of the influenza virus by low voltage Roentgen rays. J. Bact. **77**, 237—246 (1959).

K₁₁₀  KRUEGER, A. P.: The effects of certain detergents on influenza virus (types A and B) U.S. Navy Med. Bull. **40**, 622—631 (1942).

K₁₁₁  KUCHLER, C., and W. KUCHLER: Studies on antiviral activity of virothricin. Acta virol. **10**, 195—199 (1966).

K₁₁₂  KUNDIN, W. D., M. L. ROBBINS, and P. K. SMITH: Inhibition of influenza virus multiplication in tissue cultures by structural analogues of aminoacids. Virology **7**, 1—27 (1959).

K₁₁₃  KUNDIN, W. D., M. L. ROBBINS, and P. K. SMITH: Antiviral activity of 3,5-diiodo-4-hydroxybenzene sulphonamide in chick embryo lung tissue culture. Proc. Soc. exp. Biol. (N.Y.) **116**, 425—427 (1964).

K₁₁₄  KUNZE, F.: An investigation of the action of normal human serum on haemagglutination by influenza virus (German). Z. Hyg. Infekt.-Kr. **129**, 233—252 (1949).

K₁₁₅  KUSHNAREVA, E. E.: Laboratory diagnosis of influenza by cytological examination of conjunctival secretions (Russian). Zh. Mikrobiol. (Mosk.) **30**, 25—26 (1959).


L₁  LACKOVIC, V., and L. BORECKY: The reticuloendothelial system and virus infection. 2. Production of interferon and antibody-like substances in mouse peritoneal cells infected with myxoviruses *in vivo*. Arch. ges. Virusforsch. **27**, 619—630 (1965).

L₂  LACORTE, J. G., E. MONTEIRO, and J. C. LOURES: The first strains of influenza virus isolated in Brazil (Portuguese). Hospital Rio de Janeiro **35**, 211—213 (1949).

L₃  LACORTE, J. G., E. MONTEIRO, and J. C. LOURES: Additional observations on the first strains of influenza virus isolated in Brazil (Portuguese). Hospital Rio de Janeiro **35**, 357—359 (1949).

L₄  LACORTE, J. G., E. MONTEIRO, and J. C. LOURES: Influence of X rays on the influenza virus (French). Abstr. 6th int. Congr. Microbiol. Rome **3**, 104—106 (1953).

L₅  LACORTE, J. G., E. MONTEIRO, and J. C. LOURES: Action of X rays on influenza virus (Portuguese). Hospital Rio de Janeiro **45**, 243—250, 487—492, and 779—783 (1954).

L₆  LACORTE, J. G., E. MONTEIRO, and J. C. LOURES: Resistance of influenza virus to intermittent action of X rays (Portuguese). Hospital Rio de Janeiro **48**, 437—443 (1955).

L₇  LACORTE, J. G., E. MONTEIRO, and J. C. LOURES: Neurotropism of influenza virus (Portuguese). Hospital Rio de Janeiro **48**, 203—207 (1955).

L₈  LACORTE, J. G., E. MONTEIRO, and J. C. LOURES: Action of X rays on the neurotropism of influenza virus DL strain of Rio de Janeiro (Portuguese). Hospital Rio de Janeiro **48**, 343—346 (1955).

L₉  LACORTE, J. G., E. MONTEIRO, and J. C. LOURES: Resistance of influenza virus to the silver oligodynamic action (Portuguese). Mem. Inst. Osw. Cruz **53**, 537—544 (1955).

L₁₀  LACORTE, J. G., E. MONTEIRO, and J. C. LOURES: The complement fixation test in studies of Asian influenza virus (Portuguese). Mem. Inst. Osw. Cruz **58**, 89—101 (1960).

L₁₁  LAFFERTY, K. J.: The interaction between virus and antibody. 1. Kinetic studies.

L₁₂      2. Mechanism of the reaction. Virology **21**, 61—75, and 76—90 (1963).

L₁₃ LAHELLE, O., and F. L. HORSFALL: Multiplication of influenza virus in dead chick embryos. Proc. Soc. exp. Biol. (N.Y.) **70**, 547—551 (1949).

L₁₄ LAIDLAW, P. P.: Epidemic influenza: A virus disease. Lancet 1118—1124 (1935).

L₁₅ LAIDLAW, P. P., W. SMITH, C. H. ANDREWES, and G. W. DUNKIN: Influenza: The preparation of immune sera in horses. Brit. J. exp. Path. **16**, 275—290 (1935).

L₁₆ LAMPSON, G. P., A. A. TYTELL, M. M. NIMES, and M. R. HILLEMAN: Purification and characterisation of chick embryo interferon. Proc. Soc. exp. Biol. (N.Y.) **112**, 469—478 (1963).

L₁₇ LANG, G., and C. G. WELLS: Wilmot virus. A new influenza virus infecting turkeys. Arch. ges. Virusforsch. **19**, 81—90 (1966).

L₁₈ LANGE, F. C.: Immunologic properties of haemagglutinating subunits and S antigen preparations obtained from influenza virus after treatment with ether. Arch. ges. Virusforsch. **19**, 70—74 (1966).

L₁₉ LANNI, F., and Y. T. LANNI: A quantitative theory of influenza virus haemagglutination-inhibition. J. Bact. **64**, 865—882 (1952).

L₂₀ LANNI, F., and Y. T. LANNI: Influenza virus as enzyme: Mode of action against inhibitory mucoprotein from egg white. Virology **1**, 40—57 (1955).

L₂₁ LANNI, F., Y. T. LANNI, and J. W. BEARD: Inhibitory effect of cow's milk on influenza virus haemagglutination. Proc. Soc. exp. Biol. (N.Y.) **72**, 227—232 (1949).

L₂₂ LANNI, Y. T., F. LANNI, and J. W. BEARD: Dialysable factor in allantoic fluid affecting virus haemagglutination-inhibition reactions. Proc. Soc. exp. Biol. (N.Y.) **78**, 229—235 (1950).

L₂₃ LANNI, F., Y. T. LANNI, and J. W. BEARD: The egg-white inhibitor of influenza virus haemagglutination. 3. Stability in relation to the preparation of a standard of activity.

L₂₄ 4. Production of modified (Weak) inhibitor by influenza viruses: An interpretation of virus-induced changes in the character of the inhibition curve.

L₂₅ 7. Inhibitor gradient for influenza viruses. J. Immunol. **66**, 269—279 (1950) and **66**, 169—179 and 213—224 (1951).

L₂₆ LANNI, F., D. G. SHARP, E. A. ECKERT, E. S. DILLON, D. BEARD, and J. W. BEARD: The egg-white inhibitor of influenza virus haemagglutination. 1. Preparation and properties of semi-purified inhibitor. J. biol. Chem. **179**, 1275—1287 (1949).

L₂₇ LARIN, N. M., M. P. COPPING, R. H. HERBST-LAIER, B. ROBERTS, and R. B. M. WENHAM: Antiviral activity of gliotoxin. Chemotherapia (Basel) **10**, 12—23 (1965).

L₂₈ LAUCIKOVA, O.: Purification and some properties of the inhibitor of virus haemagglutination from chorioallantoic membranes of the chick embryo. Acta virol. **1**, 2—11 (1957).

L₂₉ LAUCIKOVA, O.: The influence of the inhibitor of virus haemagglutination from chorioallantoic membrane on the multiplication of the influenza virus strain PR8. Acta virol. **2**, 7—11 (1958).

L₃₀ LAUCIKOVA, O.: Enzymatic split product of influenza virus haemagglutination inhibitor. Acta virol. **3**, 139—144 (1959).

L₃₁ LAUER, D. V., and G. M. LOZOVA: Pathological morphology of experimental influenza (Russian). Tr. Kiyevskogo sanitarno-bakteriolog. Inst. Kiev. 2 (1939).

L₃₂ LAUFFER, M. A., and G. L. MILLER: The mouse infectivity titration of influenza virus. J. exp. Med. **79**, 197—203 (1944).

L₃₃ LAUFFER, M. A., and W. M. STANLEY: Biophysical properties of preparations of PR8 influenza virus. J. exp. Med. **80**, 531—548 (1944).

L₃₄ LAUFFER, M. A., and M. WHEATLEY: Destruction and denaturation of influenza A virus. Arch. Biochem. **32**, 436—447 (1951).

L₃₅ LAVER, W. G.: A micromethod for the N-terminal aminoacid analysis proteins. Biochim. biophys. Acta (Amst). **53**, 469—475 (1961).

$L_{36}$ LAVER, W. G.: Purification, N-terminal aminoacid analysis, and disruption of an influenza virus. Virology 14, 499—502 (1961).

$L_{37}$ LAVER, W. G.: The structure of influenza viruses. 1. N-terminal aminoacid analyses.

$L_{38}$ 2. C-terminal aminoacid analyses.

$L_{39}$ 3. Disruption of the virus particle and separation of neuraminidase activity. Virology 18, 19—32 (1962); 20, 20—28, and 251—262 (1963).

$L_{40}$ LAVER, W. G.: Structural studies on the protein subunits from three strains of influenza virus. J. molec. Biol. 9, 109—124 (1964).

$L_{41}$ LAVER, W. G., and E. D. KILBOURNE: Identification in a recombinant influenza virus of structural proteins derived from both parents. Virology 30, 493—501 (1966).

$L_{42}$ LAVER, W. G., J. PYE, and G. L. ADA: The molecular size of neuraminidase from vibrio cholerae (strain 4Z). Biochim. biophys. Acta (Amst.) 81, 177—180 (1964).

$L_{43}$ LAVER, W. G., and R. G. WEBSTER: The structure of influenza viruses. 4. Chemical studies of the host antigen. Virology 30, 104—115 (1966).

$L_{44}$ LAVROV, S. V., and G. A. GALEGOV: Inhibitory effect of 5-fluorouracil on the strain WSN influenza A virus reproduction in tissue culture. Abstr. 9th int. Congr. Microbiol. Moscow 531—532 (1966).

$L_{45}$ LAZARUS, A. S.: Isolation of three strains of type B influenza virus incompletely related to LEE. Proc. Soc. exp. Biol. (N.Y.) 71, 317—319 (1949).

$L_{46}$ LECLERC, J.: Action of ribonuclease on the multiplication of influenza virus. Nature (Lond.) 177, 578—589 (1956).

$L_{47}$ LECLERC, J.: The action of ribonuclease on the influenza virus synthesis (French). Ann. Inst. Pasteur 93, 772—785 (1957).

$L_{48}$ LEDINKO, N.: Production of plaques with influenza viruses. Nature (Lond.) 175, 999—1000 (1955).

$L_{49}$ LEDINKO, N.: Studies with influenza virus B of recent human origin. 2. Genetic interaction with MIL strain of influenza B virus. J. Immunol. 74, 380—386 (1955).

$L_{50}$ LEDINKO, N.: An analysis of the process of adaptation of influenza virus B of recent human origin to the mouse lung. J. gen. Microbiol. 15, 47—60 (1956).

$L_{51}$ LEDINKO, N., V. V. BERGS, and W. HENLE: Evaluation of suspended chick embryo lung cells as a host system for influenza virus. Virology 4, 109—125 (1957).

$L_{52}$ LEDINKO, N., and B. PERRY: Studies with influenza virus B of recent human origin. 1. Adaptation to the mouse lung. J. Immunol. 74, 371—379 (1955).

$L_{53}$ LEHMANN-GRUBE, F.: A sensitive plaque assay for influenza viruses. Virology 21, 520—522 (1963).

$L_{54}$ LEHMANN-GRUBE, F.: Influenza viruses in cell cultures. 1. Preparation and use of fetal pig lungs for quantal assay.

$L_{55}$ 2. Use of calf kidney cells for quantal assay.

$L_{56}$ 3. Propagation in calf kidney cells. Arch. ges. Virusforsch. 14, 1—14, and 177—188 (1964); 17, 534—548 (1965).

$L_{57}$ LENNETTE, E. H., and F. L. HORSFALL: Studies on epidemic influenza virus. The nature and properties of the complement fixing antigen. J. exp. Med. 72, 233—246 (1940).

$L_{58}$ LENNETTE, E. H., and F. L. HORSFALL: Studies on influenza virus. The complement fixing antigen of influenza A and swine influenza viruses. J. exp. Med. 73, 581—599 (1941).

$L_{59}$ LEONIDOVA, S. L., M. I. MIKHAILOVA, V. V. RITOVA, and A. G. KALINKINA: Characteristics of virus strains contained in the anti-influenza vaccines in 1958 and 1959 (Russian). "Problemy grippa i ostrykh respiratornykh zabolevanii." Tez. dokl. nauch. Sess. Inst. AMN SSSR, Moscow 126—127 (1959).

$L_{60}$ LEPINE, P., V. SAUTTER, L. REINIC, and J. MAURIN: Influenza epidemic of 1948—49: Serological and experimental study of the virus (French). Ann. Inst. Pasteur 77, 108—126 (1949).

$L_{61}$ LEPINE, P., V. SAUTTER, L. REINIC, and J. MAURIN: The influenza epidemic in France. Studies of the antigenic type of the virus responsible for the epidemic. (French). Bull. Inst. nat. Hyg. (Paris) 4, 107—109 (1949).

$L_{62}$ LESHCHINSKAYA, YE. V.: Phagocytic activity of leucocytes in influenza and acute catarrhs of the upper respiratory tract (Russian). Zh. Mikrobiol. (Mosk.) 12, 26—28 (1955).

$L_{63}$ LESTER, W.: The influence of relative humidity on the infectivity of air-borne influenza virus (PR8). J. exp. Med. 88, 361—368 (1948).

$L_{64}$ LEUCHTENBERGER, C., R. LEUCHTENBERGER, T. BRUNNER, D. NORLIN, and S. WEISS: Transformation induced by PR8 virus in primary cultures of mouse kidney and bronchus, and production of malignant kidney tumours in mice by subcultures. Proc. nat. Acad. Sci. (Wash.) 53, 694—701 (1965).

$L_{65}$ LEVINE, A. S., P. H. BOND, and H. C. ROUSE: Modification of viral synthesis in tissue culture by substituting pyruvate for glucose in the medium. Proc. Soc. exp. Biol. (N.Y.) 93, 233—235 (1956).

$L_{66}$ LEVINE, A. S., P. H. BOND, A. R. SCALA, and M. D. EATON: Studies on the relationship of potassium to influenza virus multiplication. J. Immunol. 76, 386—392 (1956).

$L_{67}$ LEVINE, A. S., T. T. PUCK, and B. P. SAGIK: An absolute method for assay of virus haemagglutinins. J. exp. Med. 98, 521—531 (1953).

$L_{68}$ LEVY, A. H., and R. R. WAGNER: Increased avidity of antibody for Baltimore 1957 strains of Asian influenza virus (PQR variation). Proc. Soc. exp. Biol. (N.Y.) 98, 357—361 (1958).

$L_{69}$ LEWIS, P. A., and R. E. SHOPE: Swine influenza. 2. A haemophilic bacillus from the respiratory tract of infected swine. J. exp. Med. 54, 361—372 (1931).

$L_{70}$ LIBIKOVA, H., and D. BLASKOVIC: Properties of an influenza virus of type B isolated in Slovakia in the epidemic of 1949 (Czech.). Čas. Lék. čes. 89, 1224—1229 (1950).

$L_{71}$ LIEF, F. S.: Antigenic analysis of influenza viruses by complement fixation. 7. Further studies on production of pure anti-S serum and on specificity of type A S antigens. J. Immunol. 90, 172—177 (1963).

$L_{72}$ LIEF, F. S.: Toxicity, pathogenicity, and antigenicity of inhibitor-sensitive and nonsensitive substrains of influenza A2 isolates. J. Immunol. 92, 286—298 (1964).

$L_{73}$ LIEF, F. S., and D. COHEN: Equine influenza. Studies of the virus and of antibody patterns in convalescent, interepidemic and post-vaccination sera. Amer. J. Epidem. 82, 225—246 (1966).

$L_{74}$ LIEF, F. S., A. FABIYI, and W. HENLE: Antigenic analyses of influenza viruses by complement fixation. 1. The production of antibodies to the soluble antigen in guineapigs. J. Immunol. 80, 53—65 (1958).

$L_{75}$ LIEF, F. S., and W. HENLE: Studies on the soluble antigen of influenza virus. 1. The release of S antigen from elementary bodies by treatment with ether.

$L_{76}$ 2. A comparison of the effects of sonic vibration and ether treatment of elementary bodies.

$L_{77}$ 3. The decreased incorporation of S antigen into elementary bodies of increasing incompleteness. Virology 2, 753—771, 772—781, and 782—797 (1956).

$L_{78}$ LIEF, F. S., and W. HENLE: Methods and procedures for use of complement fixation technique in type and strain-specific diagnosis of influenza. Bull. Wld Hlth Org. 20, 411—420 (1959).

$L_{79}$ LIEF, F. S., and W. HENLE: Antigenic analysis of influenza viruses by complement fixation. 6. Implications of age distribution of antibodies to heterologous strains following infection and vaccination. J. Immunol. 85, 494—510 (1960).

L₈₀  LIEF, F. S., W. HENLE, and W. D. SCHRACK: Antigenic analysis of influenza viruses by complement fixation. 5. Homologous antibody responses in man following vaccination with A2 virus. J. Immunol. **85**, 483—493 (1960).

L₈₁  LIEF, F. S., M. OSTAPIAK, A. FABIYI, and W. HENLE: Antigenic analysis of influenza viruses by complement fixation. 3. Rapid identification of new isolates. J. Immunol. **81**, 478—483 (1958).

L₈₂  LIKAR, M.: Some factors influencing the development of soluble antigen for influenza A virus grown on chorioallantoic membranes of chick embryos. Acta med. iugosl. **9**, 19—27 (1955).

L₈₃  LIND, P. E., and F. M. BURNET: Back-recombination of influenza A strains obtained in recombination experiments. Aust. J. exp. Biol. med. Sci. **31**, 361—371 (1953).

L₈₄  LIND, P. E., and F. M. BURNET: Recombination between neurotropic and non-neurotropic influenza virus strains. Aust. J. exp. Biol. med. Sci. **32**, 437—452 (1954).

L₈₅  LIND, P. E., and F. M. BURNET: Recombination between virulent and non-virulent strains of influenza virus. 1. The significance of heterozygosis.

L₈₆  2. The behaviour of virulence markers on recombination. Aust. J. exp. Biol. med. Sci. **35**, 57—66, and 67—78 (1957).

L₈₇  LINDENMANN, J.: The staining of influenza virus. J. gen. Microbiol. **15**, 759—763 (1957).

L₈₈  LINDENMANN, J.: Resistance of mice to mouse-adapted influenza A virus. Virology **16**, 203—204 (1962).

L₈₉  LINDENMANN, J., D. C. BURKE, and A. ISAACS: Studies on the production mode of action and properties of interferon. Brit. J. exp. Path. **38**, 551—562 (1957).

L₉₀  LINDH, H. F., and M. FORBES: Protective effect of 2-diethylaminoethyl-4-methylpiperazine-1-carboxylate in mice infected with influenza virus. Proc. Soc. exp. Biol. (N.Y.) **121**, 65—68 (1966).

L₉₁  LINK, F., D. BLASKOVIC, and J. RAUS: Studies on the effect of antiviral substances on experimental virus infections. 1. The effect of urethane on influenza virus infection.

L₉₂  2. Dose response relationship between concentration and antiviral effect of urethane in mice. Acta virol. **5**, 89—90 and 245—249 (1961).

L₉₃  LINK, F., J. SZANTO, and O. KRIZANOVA: A quantitative assay of the *in vivo* protective effect of gamma inhibitor against inhibitor sensitive A2 influenza virus strains. Acta virol. **8**, 71—75 (1964).

L₉₄  LIPPELT, H., and G. BRAND: Enrichment of influenza virus on primary isolation (German). Arch. ges. Virusforsch. **5**, 277—287 (1954).

L₉₅  LIPPELT, H., and E. MANNWEILER: The behaviour of influenza virus in chick embryo kidney tissue cultures. 1. Virus adsorption (German). Arch. ges. Virusforsch. **10**, 636—646 (1961).

L₉₆  LIPPELT, H., and W. WIRTH: Unspecific inhibitors of influenza virus in normal guineapig serum (German). Arch. ges. Virusforsch. **9**, 497—509 (1959).

L₉₇  LISITSA, F. M.: Neurologic complications of influenza in 1957 (Russian). Klin. med. (Mosk.) **37**, 47—53 (1959).

L₉₈  LIU, C.: Studies on influenza infection in ferrets by means of fluorescein-labelled antibody. 1. The pathogenesis and diagnosis of the disease.

L₉₉  2. The role of soluble antigen in nuclear fluorescence and cross reactions. J. exp. Med. **101**, 665—675 and 677—685 (1955).

L₁₀₀  LIU, O. C., H. BLANK, J. SPIZIZEN, and W. HENLE: The incorporation of radioactive phosphorus into influenza virus. J. Immunol. **73**, 415—425 (1954).

L₁₀₁  LIU, O. C., J. E. CARTER, R. G. MALSBERGER, A. N. DESANCTIS, and B. HAMPILL Studies on the chemotherapy of viral infections. 2. The effect of caprochlorone on influenza virus infections in mice. J. Immunol. **78**, 222—227 (1957).

L₁₀₂  LIU, O. C., and W. HENLE: Studies on host-virus interactions in the chick embryo influenza virus system. 4. The role of inhibitors of haemagglutination in the evaluation of viral multiplication.

$L_{103}$  5. Simultaneous serial passage of the agents of influenza A and B in relation to variations in the growth cycle of influenza B virus.

$L_{104}$  7. Data concerning the significance of infectivity titration end points and the separation of clones at limiting dilution. J. exp. Med. 94, 269—289, and 291—304 (1951); 97, 889—902 (1953).

$L_{105}$  LIU, O. C., R. G. MALSBERGER, J. E. CARTER, A. N. DeSANCTIS, F. P. WIENER, and B. HAMPILL: Studies on the chemotherapy of viral infections. 1. The activity of caprochlorone on influenza viral infection in the de-embryonated egg. J. Immunol. 78, 214—221 (1957).

$L_{106}$  LIU, O. C., K. PAUCKER, and W. HENLE: Studies on host-virus interactions in the chick embryo influenza virus system. 13. Some aspects of non-infectious virus production. J. exp. Med. 103, 777—797 (1956).

$L_{107}$  LOBAN, K. M.: The problem of treating influenza with a combination of ecmoline and penicillin (Russian). Antibiotiki 181—188 (1954).

$L_{108}$  LOBODZINSKA, M., and B. KLUBINSKA: Segregation of influenza A, A2, and B virus strains into strains with varying sensitivity to horse serum inhibitor. Arch. Immunol. Terap. Dosw. Warsaw 8, 687—694 (1960).

$L_{109}$  LOBODZINSKA, M., and O. KRIZANOVA: Purification and properties of an influenza virus inhibitor isolated from mouse lungs. Acta virol. 10, 28—34 (1966).

$L_{110}$  LÖFFLER, H.: Haemagglutination by Asian influenza virus (German). Schweiz. Z. allg. Path. 21, 1002—1007 (1958).

$L_{111}$  LÖFFLER, H., G. HENLE, and W. HENLE: Attempts to influence the incomplete reproductive cycle of influenza virus in HeLa cells by antibodies. J. Immunol. 88, 763—776 (1962).

$L_{112}$  LOFSTROM, G.: Complement fixation and mouse protection tests in routine serologic studies on influenza. Acta path. microbiol. scand. 24, 315—325 (1947).

$L_{113}$  LOFSTROM, G.: Comparison between two Swedish influenza virus strains and virus A strain PR8. Acta med. scand. 133, 162—170 (1949).

$L_{114}$  LOOSLI, G. C., D. HAMRE, and P. GERBER: Antigenic variants of influenza A virus (PR8 strain). 5. Virulence, antigenic potency and cross-protection tests in mice of the original and second series. J. exp. Med. 107, 857—868 (1958).

$L_{115}$  LOOSLI, G. C., M. H. LEMON, O. H. ROBERTSON, and E. APPEL: Experimental air-borne influenza infection. 1. Influence of humidity on survival of virus in air. Proc. Soc. exp. Biol. (N.Y.) 53, 205—206 (1943).

$L_{116}$  LOOSLI, G. C., O. H. ROBERTSON, and T. T. PUCK: The production of experimental influenza in mice by inhalation of atmospheres containing influenza virus dispersed as fine droplets. J. infect. Dis. 72, 142—153 (1943).

$L_{117}$  LOW, I. E., M. D. EATON, and S. B. URETSKY: Formation of incomplete influenza virus in ascites tumor cells. Studies on complement-fixing S antigen. J. Immunol. 89, 414—421 (1962).

$L_{118}$  LOZHKINA, A. N.: Study of washings from volunteers vaccinated with type A prime influenza TchE virus (Russian). Vop. Virus. 4, 269 (1954).

$L_{119}$  LOZHKINA, A. N.: A study of A2 influenza virus strains isolated in Moscow in October-November 1957 (Russian). Vop. Virus. 5, 45—48 (1958).

$L_{120}$  LOZHKINA, A. N.: Study of the biological characteristics of A2 influenza virus strains in Moscow during January-March 1959 (Russian). "Problemy grippa i ostrykh respiratornykh zabolevanii." Tez. dokl. nauch. Sess. Inst. AMN SSSR, Moscow 17 (1959).

$L_{121}$  LOZHKINA, A. N., V. M. STAKHANOVA, and V. B. OLLI: Removal of non-specific inhibitors for the A2 strains from human and animal sera by V. cholerae filtrate (Russian). "Problemy grippa i ostrykh respiratornykh zabolevanii." Tez. dokl. nauch. Sess. Inst. AMN SSSR, Moscow 54 (1959).

$L_{122}$  LOZOVAYA, A. N.: The etiology of influenza outbreaks and experimental material on the adaptation of the influenza virus to new species of animals (Russian). Zh. Mikrobiol. (Mosk.) 7, 60 (1942).

$L_{123}$ LOZINSKAYA, T. M.: Biological, immunogenic and reactogenic properties of influenza virus thermovariants. Abstr. 9th int. Congr. Microbiol. Moscow 375 (1966).

$L_{124}$ LUKASHEVICH, S. I.: Methods of adaptation of type B influenza virus (Russian). Anno. rabot Inst. virusol. AMN SSSR za 1949 (1950).

$L_{125}$ LUKSCH, F.: Pathologic anatomy of influenza. Review based chiefly on German sources. Arch. Path. 5, 448—491 (1928).

$L_{126}$ LUSH, D., and F. M. BURNET: Influenza virus in the developing egg. 6. Complement fixation with egg membrane antigens. Aust. J. exp. Biol. med. Sci. 15, 375—383 (1937).

$L_{127}$ LUZYANINA, G. YA.: Material on the serological variability of the influenza virus (Russian). Bibl. kan'd. rapot. otdela. virusol. Inst. Eksper. med. AMN SSSR za 1947—1949, Leningrad 22 (1950).

$L_{128}$ LUZYANINA, G. YA.: Study of the antigenic and biological properties of influenza virus strains isolated in 1949 (Russian). Problemy grippa i OKVDP Tr. ol'yed. Sess. Inst. AMN SSSR, Moscow 10 (1952).

$L_{129}$ LUZYANINA, G. YA.: Change in the biological properties of the mouse strains influenza virus when passaged in white rats (Russian). Problemy grippa i OKVDP Tez. dokl. ob'yed. Sess. Inst. AMN SSSR, Moscow 13 (1952).

$L_{130}$ LUZYANINA, G. YA.: Material on the mechanism of serological variability of the influenza virus. (Russian). Trudy. AMN SSSR XXVIII, 20 (1953).

$L_{131}$ LUZYANINA, T. J.: The protective role of the thermolabile virus-neutralising inhibitors in resistance to influenza. Abstr. 9th int. Congr. Microbiol. Moscow 374 (1966).

$L_{132}$ LUZYANINA, T. I., A. L. SALMINEN, and A. A. SMORODINTSEV: Peculiarities of the interaction of the thermostable inhibitors from normal sera with type A2 influenza viruses of the 1957 epidemic. Acta virol. 4, 137—145 (1960).

$L_{133}$ LYARSKAYA, T. A.: The blood picture and rhinocytoscopic findings in influenza patients during the outbreak in the winter of 1959 (Russian). "Problemy grippa i ostrykh respiratornykh zabolevanii." Tez. dokl. nauch. Sess. Inst. AMN SSSR, Moscow 38—39 (1959).

$L_{134}$ LYCKE, E.: A factor in bovine amniotic fluid inhibiting influenza virus haemagglutination. Arch. ges. Virusforsch. 5, 425—431 (1954).

$L_{135}$ LYTLE, R. I., and R. L. WOOLRIDGE: Purification and stabilisation of influenza virus antigens for use in haemagglutination-inhibition or complement-fixation antibody titrations. J. Lab. clin. Med. 41, 290—295 (1953).

$M_1$ MAASSAB, H. F.: An infectious ribonucleic acid derived from influenza infected cells. Proc. nat. Acad. Sci. (Wash.) 45, 877—881 (1959).

$M_2$ MAASSAB, H. F.: The propagation of multiple viruses in chick kidney cultures. Proc. nat. Acad. Sci. (Wash.) 45, 1035—1039 (1959).

$M_3$ MAASSAB, H. F.: Further studies on the infectivity of ribonucleic acid prepared from influenza-infected cells. J. Immunol. 90, 265—270 (1963).

$M_4$ MAASSAB, H. F.: Adaptation of influenza virus to growth at $26°C$. Abstr. 9th int. Congr. Microbiol. Moscow 375 (1966).

$M_5$ MAASSAB, H. F., and W. W. ACKERMANN: Nucleic acid metabolism of virus-infected HeLa cells. Ann. N.Y. Acad. Sci. 81, 29—37 (1959).

$M_6$ MACPHERSON, I. A., J. F. WILKINSON, and R. H. A. SWAIN: The effect of Klebsiella aerogenes and Klebsiella cloacae polysaccharides on haemagglutination by and multiplication of the influenza group of viruses. Brit. J. exp. Path. 34, 603—615 (1953).

$M_7$ MADOFF, M. A., E. H. EYLAR, and L. WEINSTEIN: Serological studies of the neuraminidases of vibrio cholerae, diplococcus pneumoniae, and influenza virus. J. Immunol. 85, 603—613 (1960).

$M_8$ MAGILL, T. P.: The sorption of influenza virus by chicken erythrocytes. J. exp. Med. 94, 31—43 (1951).

$M_9$ MAGILL, T. P.: Propagation of influenza virus in "immune" environments. J. exp. Med. 102, 279—289 (1955).

$M_{10}$ MAGILL, T. P.: Indications of hereditary, spatial rearrangement of antigen complexes in the influenza virus. J. exp. Med. 114, 441—447 (1961).

$M_{11}$ MAGILL, T. P., and T. FRANCIS: Studies with human influenza virus cultured in artificial medium. J. exp. Med. 63, 803—811 (1936).

$M_{12}$ MAGILL, T. P., and T. FRANCIS: Antigenic differences in strains of epidemic influenza virus. 1. Cross-neutralisation tests in mice. Brit. J. exp. Path. 19, 273—284 (1938).

$M_{13}$ MAGILL, T. P., and A. C. JOTZ: A pattern of influenza virus variation. J. Bact. 64, 619—628 (1952).

$M_{14}$ MAGILL, T. P., N. PLUMMER, W. G. SMILLIE, and J. Y. SUGG: An evaluation of vaccination against influenza. Amer. J. Hyg. 42, 94—105 (1945).

$M_{15}$ MAGILL, T. P., and J. Y. SUGG: Antigenically different strains of virus from a localised influenza outbreak. Proc. Soc. exp. Biol. (N.Y.) 53, 104—106 (1943).

$M_{16}$ MAGILL, T. P., and J. Y. SUGG: The significance of antigenic differences among strains of the A group of influenza viruses. J. exp. Med. 80, 1—7 (1944).

$M_{17}$ MAGILL, T. P., and J. Y. SUGG: Physical-chemical factors in agglutination of sheep erythrocytes by influenza virus. Proc. Soc. exp. Biol. (N.Y.) 66, 89—92 (1947).

$M_{18}$ MAGILL, T. P., and J. Y. SUGG: The reversibility of the O-D type of influenza virus variation. J. exp. Med. 87, 535—546 (1948).

$M_{19}$ MAGNUS, P. VON: Studies on interference in experimental influenza. 1. Biological observations. Ark. Kemi Mineral. Geol. 24 B, No. 7 (1946).

$M_{20}$ MAGNUS, P. VON: Propagation of the PR8 strain of influenza virus in chick embryos. 1. The influence of various experimental conditions on virus multiplication.

$M_{21}$ 2. The formation of "incomplete" virus following the inoculation of large doses of seed virus.

$M_{22}$ 3. Properties of the incomplete virus produced in serial passages of undiluted virus.

$M_{23}$ 4. Studies on the factors involved in the formation of incomplete virus upon serial passage of undiluted virus. Acta path. microbiol. scand. 28, 250—277, and 278—293 (1951), also 29, 157—181 (1951), and 30, 311—335 (1952).

$M_{24}$ MAGNUS, P. VON: The in ovo production of incomplete virus by B/Lee and A/PR8 influenza viruses. Arch. ges. Virusforsch. 17, 414—423 (1965).

$M_{25}$ MAGRASSI, F.: Study of epidemic influenza in the autumn of 1948. Origin, clinical and etiological characteristics of the epidemic in Sardinia (Italian). Minerva Med. 1, 565—569 (1949).

$M_{26}$ MAGRASSI, F., P. ALTUCCI, G. A. BUONANNO, R. BUONICONTI, A. DI SIMONE, G. LORENZUTTI, V. PECORI, and U. SAPIO: The use in man of Xenalamine, a synthetic drug (a keto-aldehyde derivative of biphenyl) with protective activity in various virus infections. 2. Its chemotherapeutic activity in human influenzal infection (Italian). G. Mal. infett. 12, 131—142 (1960).

$M_{27}$ MAGRASSI, F., P. ALTUCCI, G. P. JORI, G. LORENZUTTI, U. SAPIO, and G. TARRO: Properties of an avirulent influenza A virus variant derived from drug-treated mice infected with PR8 virus. Arch. ges. Virusforsch. 18, 422—432 (1966).

$M_{28}$ MAGRASSI, F., G. CAVALLINI, and E. MASSARINI: Drugs active against influenza virus infection in mice. Arch. ges. Virusforsch. 10, 19—30 (1960).

$M_{29}$ MAKOWER, H., and Z. SKURSKA: Experimental investigations on haemagglutination by influenza viruses. 3. The influence of temperature on the Hirst-Salk test (Polish). Med. Dosw. i Microbiol. Warsaw 3, 277—288 (1951).

$M_{30}$ MAKSIANOVICH, M. I., and T. S. ILYINA: Isolation and study of a new type of influenza virus (Russian). Dokl. Akad. Nauk. Usbekskoy SSR 2, 47 (1951).

$M_{31}$ MAKSIANOVICH, M. I., N. A. LEONOVA, and T. KUZ'MINA: The etiology of the 1946 influenza outbreak in Tashkent (Russian). Zh. MEI 12, 76 (1948).

$M_{32}$ MALOMUZH, F. F.: Clinical aspects and histopathology of the mucous membrane of the upper respiratory tract in virus influenza and acute colds (Russian). Tr. Inst. ukha, gorla i nosa, Moscow 176 (1952).

$M_{33}$ MANIRE, G. P.: Studies on the toxicity for mice of incomplete influenza virus. Acta path. microbiol. scand. 40, 501—510 (1957).

$M_{34}$ MANIRE, G. P.: Studies on the modification of response of mice to toxic influenza virus preparations. Acta path. microbiol. scand. 40, 511—520 (1957).

$M_{35}$ MANNWEILER, E.: The influenza virus in tissue cultures (German). Zbl. Bakt. I. Abt. Orig. 176, 375—376 (1959).

$M_{36}$ MANNWEILER, E.: The behaviour of influenza viruses in chick embryo tissue cultures (German). 2. Virus multiplication.

$M_{37}$ 3. The passage of prototype strains of influenza virus, the cytopathic effect of A/Singapore/1/57 and comparative $ID_{50}$ titrations. Arch. ges. Virusforsch. 10, 647—660, and 661—671 (1961).

$M_{38}$ MARCUS, P. I., J. M. SALB, and V. G. SCHWARTZ: Nuclear surface N-acetyl-neuraminic acid terminating receptors for myxovirus attachment. Nature (Lond.) 208, 1122—1124 (1965).

$M_{39}$ MARCUSE, K., and G. M. BECKE: The behaviour of human red cells in the Hirst test with influenza virus (German). Arch. ges. Virusforsch. 4, 659—662 (1952).

$M_{40}$ MARENNIKOVA, S. S.: Variability of the influenza virus (Russian). Byull. éksp. Biol. Med. 8, 65 (1952).

$M_{41}$ MARENNIKOVA, S. S., N. I. KALININA, N. A. PONOMAREVA, and Z. I. POSPELOVA: Production and experimental study of anti-influenza gamma ad beta globulins (Russian). "Problemy grippa i ostrykh respiratornykh zabolevanii." Tez. dokl. nauch. Sess. Inst. AMN SSSR, Moscow 144 (1959).

$M_{42}$ MARINESCU, G., D. SARETEANU, E. NASTAC, N. DRAGANESCU, and B. MUSET: A study of the resistance to influenza infection in suckling mice born to immunised mice (Rumanian). Stud. Cercet. Inframicrobiol. 10, 315—326 (1959).

$M_{43}$ MARMION, B. P., C. C. CURTAIN, and J. PYE: The effect of human bronchial secretions (sputum) on the haemagglutinin and infectivity of influenza virus. Aust. J. exp. Biol. med. Sci. 31, 505—518 (1953).

$M_{44}$ MASUREL, N., and J. MULDER: Studies on the content of haemagglutinin-inhibiting antibody for swine influenza virus A. Verk. Inst. prev. Geneesk. 52, 1 (1962).

$M_{45}$ MATHEKA, H. D., and O. ARMBRUSTER: Gradient elution of influenza viruses from anion exchange resin and a demonstration of a biological difference between the three components obtained from the PR8 strain. Virology 6, 584—600 (1958).

$M_{46}$ MATHIEU, P., L. COLOBERT, O. CREACH, and O. FONTANGER: Kinetics of the hydrolysis of 3 lactose-lactaminic acid by influenza B (Lee strain). (French). Ann. Inst. Pasteur 101, 817—833 (1961).

$M_{47}$ MATSUMOTO, T., and I. NAGATA: Studies on the variation of influenza virus. 1. The haemagglutination-inhibiting action of normal mouse lung suspensions. Nagoya J. med. Sci. 14, 62—68 (1951).

$M_{48}$ MAWSON, J., and C. SWAN: Intranasal vaccination of humans with living attenuated influenza virus strains. Med. J. Aust. 1, 394—396 (1943).

$M_{49}$ MAXIMOVICH, N. A., and V. P. MITCHENKO: Study of cellular and viral nucleic acids in experimental influenza infection using fluorescence microscopy. Acta virol. 4, 227—232 (1960).

$M_{50}$ MAXIMOVICH, N. A., and O. G. PETROVSKAYA: Further studies of viral inclusions in experimental influenza using fluorescence microscopy. Acta virol. 6, 127—131 (1962).

$M_{51}$ MAYRON, L. W., B. ROBERT, R. J. WINZLER, and M. E. RAFELSON: Studies on the neuraminidase of influenza virus. 1. Separation and some properties of the enzyme from Asian and PR8 strains. Arch. Biochem. 92, 475—483 (1961).

$M_{52}$ MAZZACCA, G., and A. R. BIANCHI: Growth characteristics of A PR8 influenza virus in monkey kidney cell cultures (Italian). Rev. Ist. Sieroter. ital. 36, 312—321 (1961).

M53  Mc CARROLL, J. R., and E. D. KILBOURNE: Immunisation with Asian-strain influenza vaccine. Equivalence of the subcutaneous and intradermal routes. New Engl. J. Med. 259, 618—621 (1958).

M54  McCLELLAND, L., and R. HARE: The adsorption of influenza virus by red cells and a new in vitro method of measuring antibodies for influenza virus. Canad. J. publ. Hlth 32, 530—538 (1941).

M55  McCLELLAND, L., and C. E. VAN ROOYEN: Studies on the effect of some aromatic amidines and other chemical compounds on the growth of influenza virus in the embryonated egg. Canad. J. Res. E 27, 177—185 (1949).

M56  McCREA, J. F.: Non-specific serum inhibition of influenza haemagglutination. Aust. J. exp. Biol. med. Sci. 24, 283—291 (1946).

M57  McCREA, J. F.: Mucins and mucoids in relation to influenza virus infection. 2. Isolation and characterisation of the serum mucoid inhibitor of heated influenza virus. Aust. J. exp. Biol. med. Sci. 26, 355—370 (1948).

M58  McCREA, J. F.: Studies on influenza virus receptor substance and receptor substance analogues. 1. Preparation and properties of a homogeneous mucoid from the salivary gland of sheep.

M59  2. Isolation and purification of a mucoprotein receptor substance from human erythrocyte stroma treated with pentane. Biochem. J. 55, 132—140 (1953). and Yale J. Biol. Med. 26, 191—210 (1953).

M60  McCREA, J. F.: Turbidity-reducing activity of influenza virus and receptor-destroying enzyme on sheep submandibular mucoid. Nature (Lond.) 172, 912 (1953).

M61  McKEE, A. P.: Non-toxic influenza virus. J. Immunol. 66, 151—167 (1951).

M62  McKEE, A. P., and W. M. HALE: Reactivation of over-neutralised mixtures of influenza virus and antibody. J. Immunol. 54, 233—243 (1946).

M63  McKEE, A. P., and W. M. HALE: A search for human carriers of influenza virus. J. Immunol. 61, 369—372 (1949).

M64  McLEAN, I. W., D. BEARD, A. R. TAYLOR, D. G. SHARP, J. W. BEARD, A. E. FELLER, and J. H. DINGLE: Influence of temperature of incubation on the increase of influenza B virus (Lee strain) in the chorioallantoic fluid of chick embryos. J. Immunol. 48, 305—316 (1944).

M65  McQUEEN, J. L., and F. M. DAVENPORT: Experimental influenza in sheep. Proc. Soc. exp. Biol. (N.Y.) 112, 1004—1006 (1963).

M66  McQUEEN, J. L., F. M. DAVENPORT, R. J. KERRAN, and H. A. DAWSON: Studies of equine influenza in Michigan. 2. Epizootiology. Amer. J. Epidem. 83, 280—286 (1963).

M67  McQUEEN, J. L., F. M. DAVENPORT, and E. MINUSE: Studies of equine influenza in Michigan. 1. Etiology. Amer. J. Epidem. 83, 271—279 (1963).

M68  Medical Research Council Report: Antibody response and clinical reactions with saline and oil adjuvant influenza vaccines. Brit. med. J. 1229—1232 (1955).

M69  Medical Research Council: Clinical trials of influenza vaccine. Third progress report to the Medical Research Council by its Committee on clinical trials of influenza vaccine. Brit. med. J. ii, 1—7 (1957).

M70  MEDILL-BROWN, M., and B. A. BRIODY: Mutation and selection pressure during adaptation of influenza virus to mice. Virology 1, 301—312 (1955).

M71  MEENAN, P. N., M. R. BOYD, and R. MULVANEY: Human influenza virus in Domesticated animals. Brit. med. J. ii, 86—89 (1962).

M72  MEIKLEJOHN, G.: Observations on live influenza vaccine. J. Amer. med. Ass. 172, 1354—1356 (1960).

M73  MEIKLEJOHN, G., and A. J. MORRIS: Influenza vaccination. Ann. intern. Med. 49, 529—535 (1958).

M74  MELANDER, B.: N,N-anhydrobis-(B hydroxyethyl) biguanide hydrochloride (ABOB) in prophylaxis and suppression of experimental influenza. Antibiot. and Chemother. N.Y. 10, 34—45 (1960).

M75  MELLANBY, H., C. H. ANDREWES, J. A. DUDGEON, and D. G. MACKAY: Vaccination against influenza A. Lancet 978—982 (1948).

$M_{76}$  MELNIKOVA, N. A.: Isolation of influenza virus from nasopharyngeal washings of patients by means of passages through X-irradiated embryos (Russian). Vop. Virus. 5, 70—71 (1958).

$M_{77}$  MELNOTTE, P., C. JOLIBOIS, G. CATEIGNE, M. THIBON, J. CHEVE, and S. ADAMS: Anti-influenza vaccination (French). Sém. Hôp. Paris 35, 3353—3358 (1959).

$M_{78}$  MERIGAN, T. C., D. F. GREGORY, and J. K. PETRALLI: Physical properties of human interferon prepared in vitro and in vivo. Virology 29, 515—522 (1966).

$M_{79}$  MERKULOVA, Z. I.: Periods of preservation of the influenza viruses on surfaces and in air (Russian). Zh. Mikrobiol. (Mosk.) 10, 68 (1953).

$M_{80}$  MERKULOVA, S. T., and I. A. KAZANSKIY: Viruscopic method of diagnosing influenza in people (Russian). Annotatsii rabot Inst. virusol. AMN SSSR za 1949, Moscow 18 (1950).

$M_{81}$  MERKULOVA, S. T., I. A. KAZANSKIY, and V. KHOMAZYUK: Method of diagnosing influenza in people on the basis of viroscopy (Russian). Vop. Virus. 4, 183 (1954).

$M_{82}$  METCALF, T. G.: The serological reactivity of influenza virus precipitated and concentrated by zinc solutions. J. infect. Dis. 101, 40—47 (1957).

$M_{83}$  MEYER, H. M., M. R. HILLEMAN, M. L. MIESSE, I. P. CRAWFORD, and A. J. BANKHEAD: New antigenic variant in Far East influenza epidemic. Proc. Soc. exp. Biol. (N.Y.) 95, 609—616 (1957).

$M_{84}$  MILLER, F. A., W. A. RIGHTSEL, B. J. SLOAN, J. EHRLICH, J. C. FRENCH, and Q. R. BARTZ: Antiviral activity of Tenuazonic acid. Nature (Lond.) 200, 1338—1339 (1963).

$M_{85}$  MILLER, G. L.: A study of conditions for the optimum production of PR8 influenza virus in chick embryos. J. exp. Med. 79, 173—183 (1944).

$M_{86}$  MILLER, G. L.: Influence of pH and of certain other conditions on the stability of the infectivity and red cell agglutinating activities of influenza virus. J. exp. Med. 80, 507—520 (1944).

$M_{87}$  MILLER, G. L.: Improved measurement of influenza virus haemagglutinin titre. J. Immunol. 95, 336—344 (1965).

$M_{88}$  MILLER, G. L., M. H. LAUFFER, and W. M. STANLEY: Electrophoretic studies on PR8 influenza virus. J. exp. Med. 80, 549—559 (1944).

$M_{89}$  MILLER, G. L., and W. M. STANLEY: Quantitative aspects of the red blood cell agglutination test for influenza virus. J. exp. Med. 79, 185—195 (1944).

$M_{90}$  MILLER, H. K.: The nucleic acid content of influenza virus. Virology 2, 312—320, (1956).

$M_{91}$  MILLER, H. K., and R. W. SCHLESINGER: Differentiation and purification of influenza viruses by adsorption on aluminium phosphate. J. Immunol. 75, 155—160 (1955).

$M_{92}$  MILLS, R. F. N.: The effect of infecting the cells of the de-embryonated egg with influenza virus on their uptake of glucose and aminoacids. J. gen. Microbiol. 19, 473—481 (1958).

$M_{93}$  MIMS, C. A.: An analysis of the toxicity for mice of influenza virus. 1. Intracerebral toxicity. 2. Intravenous toxicity. Brit. J. exp. Path. 41, 586—598 (1960).

$M_{94}$  MIMS, C. A.: Aspects of the pathogenesis of virus diseases. Bact. Rev. 28, 30—71 (1964).

$M_{95}$  MINUSE, E., and F. M. DAVENPORT: Simultaneous recovery of type A1 and type C influenza viruses from a patient. J. Lab. clin. Med. 38, 747—750 (1951).

$M_{96}$  MINUSE, E., J. L. McQUEEN, F. M. DAVENPORT, and T. FRANCIS: Studies of antibodies to 1956 and 1963 equine influenza viruses in horses and man. J. Immunol. 94, 563—566 (1965).

$M_{97}$  MINUSE, E., J. J. QUILLIGAN, and T. FRANCIS: Type C influenza virus. 1. Studies of the virus and its distribution.

$M_{98}$  2. Intranasal inoculation of human individuals. J. Lab. clin. Med. 43, 31—42, and 43—47 (1954).

$M_{99}$ MINUSE, E., P. W. WILLIS, F. M. DAVENPORT, and T. FRANCIS: An attempt to demonstrate viraemia in cases of Asian influenza. J. Lab. clin. Med. **59**, 1016—1019 (1962).

$M_{100}$ MYAMOTO, H., K. AKAMA, and E. MORITA: The influence of nucleic acid on haemagglutination by influenza virus and its mechanism. Gunma J. med. Sci. **2**, 69—75 (1953).

$M_{101}$ MIZUTANI, H., T. BEALS, A. V. HENNESSY, and F. M. DAVENPORT: A simple procedure for purification of viral haemagglutinin. Virology **17**, 210—211 (1962).

$M_{102}$ MIZUTANI, H., and H. MIZUTANI: Action of phospholipase C on influenza virus. Nature (Lond.) **204**, 781—782 (1964).

$M_{103}$ MIZUTANI, H., and H. MIZUTANI: Small haemagglutinin of influenza virus and its responses to antibodies and inhibitor. Virology **26**, 761—763 (1965).

$M_{104}$ MOGABGAB, W. J.: Influenza C virus in monkey kidney tissue cultures. J. Bact. **83**, 209—210 (1962).

$M_{105}$ MOGABGAB, W. J., I. J. GREEN, and O. C. DIERKHISING: Primary isolation and propagation of influenza viruses in cultures of human embryonic renal tissue. Science **120**, 320—321 (1954).

$M_{106}$ MOGABGAB, W. J., I. J. GREEN, O. C. DIERKHISING, and I. A. PHILLIPS: Isolation and cytopathogenic effect of influenza B viruses in monkey kidney cultures. Proc. Soc. exp. Biol. (N.Y.) **89**, 654—659 (1955).

$M_{107}$ MOGABGAB, W. J., B. HOLMES, and W. PELON: Influenza A and B viruses in monkey kidney cultures. 1. Properties of the viruses. J. infect. Dis. **108**, 315—323 (1961).

$M_{108}$ MOGABGAB, W. J., and F. L. HORSFALL: Effect of sodium monofluoroacetate on the multiplication of influenza virus, mumps virus, and pneumonia virus of mice (PVM). J. exp. Med. **96**, 531—547 (1952).

$M_{109}$ MOGABGAB, W. J., W. PELON, G. E. BURCH, and B. HOLMES: Characteristics of the Asian strain of influenza A. Proc. Soc. exp. Biol. (N.Y.) **99**, 116—120 (1958).

$M_{110}$ MOGABGAB, W. J., G. I. SIMPSON, and I. J. GREEN: Growth characteristics and cytopathogenic effects of influenza A and B in cultures of human embryo tissues. J. Immunol. **76**, 314—327 (1956).

$M_{111}$ MONTO, A. S., and F. OLAZABAL: Asian influenza in the Panama canal zone: Isolation of a virus variant and protective effect of a vaccine containing A2/Japan/305/57. Amer. J. Epidem. **83**, 101—112 (1966).

$M_{112}$ MOORE, D. H., M. C. DAVIES, S. LEVINE, and M. E. ENGLERT: Correlation of structure with infectivity of influenza virus. Virology **17**, 470—479 (1962).

$M_{113}$ MORGAN, C., K. C. HSU, R. A. RIFKIND, A. W. KNOX, and H. M. ROSE: The application of ferritin-conjugated antibody to electron microscopic studies of influenza virus in infected cells. 1. The cellular surface.

$M_{114}$ 2. The interior of the cell. J. exp. Med. **114**, 825—832, and 833—836 (1961).

$M_{115}$ MORGAN, C., K. C. HSU, and H. M. ROSE: Structure and development of viruses as observed in the electron microscope. 7. Incomplete influenza virus. J. exp. Med. **116**, 553—564 (1962).

$M_{116}$ MORGAN, C., R. A. RIFKIND, and H. M. ROSE: The use of ferritin-conjugated antibodies in electron microscope studies of influenza and vaccinia viruses. Cold Spr. Harb. Symp. quant. Biol. "Basic mechanisms in animal virus biology" **28**, 57—66 (1962).

$M_{117}$ MORGAN, C., and H. M. ROSE: Electron-microscopic observations on adenoviruses and viruses of the influenza group. Symp. Soc. gen. Microbiol. "Virus growth and variation" 256—272 (1959).

$M_{118}$ MORGAN, C., H. M. ROSE, and D. H. MOORE: Structure and development of viruses observed in the electron microscope. 3. Influenza virus. J. exp. Med. **104**, 171—182 (1956).

$M_{119}$ MORGAN, C., H. M. ROSE, and D. H. MOORE: An evaluation of host cell changes accompanying viral multiplication as observed in the electron microscope. Ann. N.Y. Acad. Sci. **68**, 302—323 (1957).

$M_{120}$ MORGAN, H. R., M. W. BARNES, and M. FINLAND: Antigenic differences among influenza A viruses, including serologic response of patients. J. Lab. clin. Med. **33**, 1212—1219 (1948).

$M_{121}$ MORGAN, H. R., and M. FINLAND: Simplified technique for titrating influenza virus neutralising antibody in the chick embryo. Proc. Soc. exp. Biol. (N.Y.) **68**, 618—619 (1948).

$M_{122}$ MORI, R., J. H. SCHIEBLE, and W. W. ACKERMANN: Reaction of polyoma and influenza viruses with receptors of erythrocytes and host cells. Proc. Soc. exp. Biol. (N.Y.) **109**, 685—690 (1962).

$M_{123}$ MOROSOV, M. A.: Instructions on the use of the viroscopic method after M. A. MOROSOV (Russian). Ofitsial'noye izd. Min. zdravoskht. SSSR (1951).

$M_{124}$ MOROSOV, M. A.: A new method of viroscopy (Russian). Tez. dokl. 6-y Sess. Inst. virusol. AMN SSSR, Moscow 19 (1953).

$M_{125}$ MOROZENKO, M. A.: Serological characteristics of influenza in infancy. (Russian). Gripp. Tr. AMN SSSR, XXVIII, 164 and 209 (1953).

$M_{126}$ MOROZENKO, M. A.: Experimental data on the mechanism of occurrence of passive anti-influenzal immunity (Russian). Gripp. i OKVDP Tr. ob'yed. Sess. Inst. AMN SSSR, Moscow 139 (1953).

$M_{127}$ MOROZENKO, M. A., and L. S. VLASTELITSA: Virological and serological characteristics of type A influenza in adults and children (Russian). "Virusnyye Infektsi." Tr. Leningradskogo inst. epidem. i bakteriol. imeni. Pastera XIII, 64 (1950).

$M_{128}$ MOROZKIN, N. I., R. Ya. KHERSONSKAYA, and M. I. SLOBODYANYUK: Modern methods of treating influenza patients (Russian). Vrach. delo. 3, 229—323 (1954).

$M_{129}$ MORRISON, A. P., D. R. SHAW, A. S. KENNEY, and J. STOKES: Complement fixation studies on the sera of individuals vaccinated with active virus of human influenza. Amer. J. med. Sci. 197, 253—260 (1939).

$M_{130}$ MOSLEY, V. M., and R. W. G. WYCKOFF: Electron micrography of the virus of influenza. Nature (Lond.) **157**, 263 (1946).

$M_{131}$ MOTE, J. R.: Human and swine influenza. "Virus and Rickettsial diseases" Harvard Univ. Press (1943).

$M_{132}$ MOTE, J. R., and L. D. FOTHERGILL: The effect of human strains of haemophilus influenzae on influenza virus infection of swine. J. Bact. 40, 505—516 (1940).

$M_{133}$ MULDER, J.: Broad aspects of the problem of human virulence in influenza viruses. Ciba Found. Study group No. 4 "Virus virulence and Pathogenicity" 43—57 (1960).

$M_{134}$ MULDER, J., and L. M. BRANS: Studies on the antigenic behaviour of egg and egg-mouse-egg lines of strains of influenza virus with the aid of the haemagglutination inhibition test. Antonie v. Leeuwenhoek 18, 139—151 (1952).

$M_{135}$ MULDER, J., L. M. BRANS, and J. F. PH. HERS: Studies on the increase of antibodies in nasal secretions of human beings after intracutaneous vaccination with influenza virus, with the aid of the haemagglutination inhibition technique. Antonie v. Leeuwenhoek 18, 131—138 (1952).

$M_{136}$ MULDER, J., L. M. BRANS, and I. DeNOOYER: The antigenic composition of the influenza virus B strains isolated during the epidemic of influenza B in the winter 1951—52 in the Netherlands. Antonie v. Leeuwenhoek **19**, 189—196 (1953).

$M_{137}$ MULDER, J., L. M. BRANS, and N. MASUREL: Studies on the antigenic composition of the influenza virus A strains isolated in the Netherlands in the period 1947—1953. Onderz. Meded. u. Ned. Inst. v. praev. Geneesk. No. 15 II. E. Stenfert Kroese Leiden (1956).

$M_{138}$ MULDER, J., and W. R. O. GOSLINGS: Simplified micro-Hirst test. Acta brev. neerl. Physiol. **16**, 27—29 (1948).

$M_{139}$ MULDER, J., and N. MASUREL: Pre-epidemic antibody against the 1957 strain of Asiatic influenza in the serum of older persons living in the Netherlands. Lancet 810 (1958).

M$_{140}$ MULDER, J., I. DeNOOIYER, and L. M. BRANS: The influence of treating ferret influenza antisera with enzymes of crude filtrate of V Cholerae on the titre of the antibodies. Antonie v. Leeuwenhoek 18, 128—130 (1952).

M$_{141}$ MULDER, J., and C. H. STUART-HARRIS: Influenzal pneumonia: Causation and treatment. Bull. Wld Hlth Org. 8, 743—753 (1953).

M$_{142}$ MULDER, J., J. VAN DER VEEN, J. J. BRANS, and S. W. ENSERINK: Neutralisation of the non-specific inhibitor in influenza immune titrations after Hirst with an enzyme of vibrio cholerae. Acta brev. neerl. Physiol. 16, 57—61 (1948).

M$_{143}$ MULDER, J., J. VAN DER VEEN, J. J. BRANS, and S. W. ENSERINK: Rapid diagnosis of the sub-group of influenza A virus strains isolated during the epidemic of 1949 in the Netherlands. Antonie v. Leeuwenhoek 15, 125—128 (1949).

M$_{144}$ MULDER, J., J. VAN DER VEEN, L. M. BRANS, and S. W. ENSERINK: Antibody response against strains of influenza A virus in ferrets with basic immunity. Antonie v. Leeuwenhoek 15, 162—166 (1949).

M$_{145}$ MULDER, J., J. VAN DER VEEN, S. W. ENSERINK, and J. J. BRANS: Isolation of a strain of influenza A virus from the trachea in a case of influenzal pneumonia in the winter of 1947. Antonie v. Leeuwenhoek 14, 184—192 (1948).

M$_{146}$ MULDER, J., and G. J. VERDONK: Studies on the pathogenesis of a case of influenza A pneumonia of 3 days duration. J. Path. Bact. 61, 55—61 (1949).

M$_{147}$ MULLER, R. H., and H. M. ROSE: Concentration of influenza virus (PR8 strain) by a cation exchange resin. Proc. Soc. exp. Biol. (N.Y.) 80, 27—29 (1952).

M$_{148}$ MUNK, K., and W. SCHAFER: Properties of animal viruses studied with fowl plague as a model. 2. Serological studies on fowl plague virus and on its realtion with a normal cell protein (German). Z. Naturforsch. 6b, 372—379 (1951).

M$_{149}$ MURPHY, A. M.: The use of purified RDE in the destruction of non-specific inhibitor in serum. Aust. J. exp. Biol. med. Sci. 30, 363—368 (1952).

M$_{150}$ MURPHY, J. S., D. T. KARZON, and F. B. BANG: Studies of influenza A (PR8) infected tissue cultures by electron microscopy. Proc. Soc. exp. Biol. (N.Y.) 73, 596—599 (1950).

N$_1$ NAGLER, F. P., M. M. BURR, and A. L. GILLEN: The influenza virus epidemic in Canada during January—February 1951. Canad. J. publ. Hlth 42, 367—374 (1951).

N$_2$ NAGLER, F. P., C. E. VAN ROOYEN, and J. H. STURDY: An influenza virus epidemic at Victoria island N.W.T. Canada. Canad. J. publ. Hlth 40, 457—465 (1949).

N$_3$ NAKAGI, D., T. HAGA, K. IWASAKI, Y. NAKASI, and T. UCHIYAMA: Studies on the influenza virus. 4. Studies on the influenza virus vaccine. Immunisation experiment of man with vaccine inactivated with sodium ethylmercurithiosalycylate. Kitasato Arch. exp. Med. 27, 1—7 (1954).

N$_4$ NAKAGI, D., K. IWASAKI, S. HOMINI, Y. NAKASA, and T. UCHIYAMA: Studies on influenza virus. 3. Studies on the influenza virus vaccine. Immunisation experiment on mice with the influenza virus vaccine inactivated with sodium ethylmercurithiosalycylate. Kitasato Arch. exp. Med. 26, 209—215 (1953).

N$_5$ NARTSISSOV, N. V.: Complement fixation reaction in influenza (Russian). Zh. Mikrobiol. (Mosk.) 12, 47 (1948).

N$_6$ NAYAK, D. P., W. G. KELLEY, G. A. YOUNG, and N. R. UNDERDAHL: Progressive descending infection in mice determined by immunofluorescence. Proc. Soc. Biol. (N.Y.) 116, 200—206 (1964).

N$_7$ NAYAK, D. P., and A. F. RASMUSSEN: Influence of mitomycin C on the replication of influenza viruses. Virology 30, 673—683 (1966).

N$_8$ NECHAYEV, A. V.: Evaluation of the efficiency of specific serum therapy with the inhalation of influenza antiserum (Russian). Sovetsk. Med. 7, 25—29 (1940).

$N_9$ NECHAYEV, A. V., O. S. KORSHUNOVA, and M. I. BARU: Evaluation of the thera-peutic effect of inhalations of anti-influenzal immune serum in epidemic influenza (Russian). Arkh. biol. nauk. **52**, 155—161 (1938).

$N_{10}$ NEGRONI, G., and D. A. J. TYRRELL: Morphological observations on tissue cultures of epithelial cells infected with influenza A viruses. J. Path. Bact. **77**, 497—504 (1959).

$N_{11}$ NEKHLUDOVA, L. I.: Characteristic features of cases of influenza (Russian). Zh. Mikrobiol. (Mosk.) **6**, 39 (1951).

$N_{12}$ NEKHLUDOVA, L. I., G. F. KORNYEVA, N. V. PIKEL, and V. A. KUZNITSOVA: Peculiarities of influenza and influenza-like diseases in 1959 in Krasnoder (Russian). "Problemy grippa i ostrykh respiratornykh zabolevani." Tez. dokl. nauch. Sess. Inst. AMN SSSR, Moscow 101—103 (1959).

$N_{13}$ NELSON, M., and F. A. LEWIS: A relationship between swine and Asian strains of influenza A virus. Aust. J. exp. Biol. med. Sci. **36**, 505—510 (1958).

$N_{14}$ NELSON, A. A., and J. W. OLIPHANT: Histopathological changes in mice in-oculated with influenza virus. Publ. Hlth Rep. (Wash.) **54**, 2044—2054 (1939).

$N_{15}$ NEUMAYER, E. M., R. F. HAFF, and C. E. HOFFMANN: Antiviral activity of amantadine hydrochloride in tissue culture and in ovo. Proc. Soc. exp. Biol. (N.Y.) **119**, 393—396 (1965).

$N_{16}$ NEUMULLER, C.: Chemical studies on the (influenza) haemagglutination inhibitor of normal allantoic fluid. Arch. ges. Virusforsch. **5**, 242—249 (1953).

$N_{17}$ NICHOLAS, R. V., and W. HENLE: Vaccination as primary contact with influenza A and B viruses. Pediatrics 208—213 (1949).

$N_{18}$ NICOLLE, C., and C. LEBAILLY: Experimental researches in influenza (French). Ann. Inst. Pasteur **33**, 395 (1919).

$N_{19}$ NICOULESCO, I., L. CRETESCO, E. ZILISTEANU, and I. NAFTA: Separation of two types of influenza virus particles from the same antigenically homoge-neous viral population by means of ultraviolet rays (French). Abstr. 9th int. Congr. Microbiol. Moscow 390 (1966).

$N_{20}$ NIGG, C., J. H. CROWLEY, and D. E. WILSON: The use of chick embryo tissues and fluid as antigen in the complement fixation reaction in influenza. J. Immunol. **42**, 51—70 (1941).

$N_{21}$ NIGG, C., C. M. EKLUND, D. E. WILSON, and J. H. CROWLEY: Study of an epidemic of influenza B. Amer. J. Hyg. **35**, 265—284 (1942).

$N_{22}$ NIGG, C., D. E. WILSON, and J. H. CROWLEY: Studies on the cultivation of influenza virus. Amer. J. Hyg. **24**, 138—147 (1941).

$N_{23}$ NIHOUL, E.: The action of filtrates of cultures of V. cholerae on influenza virus receptors (French). C. R. Soc. Biol. (Paris) **145**, 1887—1891 (1951).

$N_{24}$ NIHOUL, E.: Mucopolysaccharide of egg white and inhibition of influenza virus haemagglutination (French). C. R. Soc. Biol. (Paris) **146**, 1275—1277 (1952).

$N_{25}$ NIHOUL, E.: Enzymatic action of influenza virus on ovomucin (French). C. R. Soc. Biol. (Paris) **146**, 1277—1280 (1952).

$N_{26}$ NOLL, H., T. AOYAGI, and J. ORLANDO: Intracellular synthesis of neuraminidase following infection of chorioallantoic membranes with influenza virus. Viro-logy **14**, 141—143 (1961).

$N_{27}$ NOLL, H., T. AOYAGI, and J. ORLANDO: The structural relationship of sialidase to the influenza virus surface. Virology **18**, 154—157 (1962).

$N_{28}$ NOLL, H., and J. S. YOUNGER: Virus-lipid interactions. 2. The mechanism of adsorption of lipophilic viruses to water-insoluble polar lipids. Virology **8**, 319—343 (1959).

$N_{29}$ NOVIN, J. S. F., J. A. ARMSTRONG, B. M. BALFOUR, and D. A. J. TYRRELL: Cellular changes accompanying the growth of influenza viruses in bovine cell cultures. J. Path. Bact. **84**, 1—18 (1962).

$O_1$ OBROSOVA-SEROVA, N. P.: Mutants of influenza A and A2 viruses induced by nitrous acid and preparation W4. Abstr. 9th int. Congr. Microbiol. Moscow 384 (1966).

$O_2$ O'Connell, C. J., A. L. Barron, F. Melgrom, and E. Witebsky: Haemadsorption to sections of virus-infected tissues. Proc. Soc. exp. Biol. (N.Y.) 117, 403—407 (1964).

$O_3$ O'Connell, C. J., and F. Milgrom: Demonstration of tissue receptors for viruses. Proc. Soc. exp. Biol. (N.Y.) 122, 1255—1258 (1966).

$O_4$ O'Connor, S., and R. R. Wagner: Age and susceptibility to neurotropic influenza virus. Proc. Soc. exp. Biol. (N.Y.) 86, 332—334 (1954).

$O_5$ Ogasawara, K., M. Aida, and I. Nagata: Lowered resistance to influenza infection of mice following immunisation with mercurial inactivated influenza virus. J. Immunol. 86, 599—605 (1961).

$O_6$ Okada, Y., and T. Suzuki: Interaction between influenza virus and Ehrlich's tumor cells. 2. Fate of the virus particles adsorbed onto Ehrlich's tumor cells. Med. J. Osaka Univ. 6, 987—994 (1956).

$O_7$ Oleinik, I. I., and V. N. Iagodinsky: Specific prophylaxis of influenza (Russian). "Problemy grippa i ostrykh respiratornykh zabolevanii." Tez. dokl. nauch. Sess. Inst. AMN SSSR, Moscow 131—132 (1959).

$O_8$ Olitsky, P. K., and F. L. Gates: Experimental studies of the nasopharyngeal secretions from influenza patients. J. exp. Med. 33, 713 (1921).

$O_9$ Onodera, Y., Y. Hinuma, and N. Ishida: Ineffectiveness of antibody on the intracellular development of influenza virus WS strain in Ehrlich ascites cells in mice. J. Immunol. 92, 648—656 (1964).

$O_{10}$ Opie, E. L.: Pathologic anatomy of influenza, based chiefly on American and British sources. Arch. Path. 5, 285—303 (1928).

$O_{11}$ Orlov, G. A., and M. M. Kanstler: Modern laboratory methods of diagnosing virus influenza (Russian). Sovetsk. Med. 11, 20 (1950).

$O_{12}$ Orlova, N. N.: Experimental evaluation of the efficiency of combined influenza vaccines (Russian). Vop. Virus. 1, 166 (1948).

$O_{13}$ Orlova, N. N.: Effect of stimulators on immunogenic properties of influenza formol vaccine (Russian). Tez. dokl. 3-y Sess. Inst. Virusol. AMN SSSR, Moscow 12 (1950).

$O_{14}$ Orlova, N. N.: Massive scale serum prophylaxis of influenza during the 1959 epidemic (Russian). "Problemy grippa i ostrykh respiratornykh zabolevani." Tez. dokl. nauch. Sess. Inst. AMN SSSR, Moscow 134—136 (1959).

$O_{15}$ Orlova, N. N., and L. I. Kalinina: Perfection of methods of producing influenza horse antiserum (Russian). Vop. Virus. 4, 356 (1954).

$O_{16}$ Orlova, N. N., and L. D. Knyazeva: Characteristics of an influenza outbreak in the summer of 1957 in Kemerovo. (Russian). Vop. Virus. 5, 19—21 (1958).

$O_{17}$ Orlova, T. G., and L. S. Diskina: Investigation on infectious RNA from A2 influenza virus (Russian). Proc. 3rd Conf. Inst. Viral Preparations, Moscow 40 (1962).

$O_{18}$ Orlova, T. G., J. N. Tatarinova, L. M. Mentkewicz, and V. M. Stakhanova: Studies on the infectious process in the chick embryo cells infected with influenza virus. Abstr. 9th int. Congr. Microbiol., Moscow 495 (1966).

$O_{19}$ Osato, T., and N. Ishida: The labelling of influenza virus with radiophosphorus in Maitland-type tissue culture. Growth characteristics of myxoviruses in tissue culture. Tohoku J. exp. Med. 72, 322—327 (1960).

$O_{20}$ Ostrovskaya, S. M., O. M. Chalkina, and S. B. Olekhnovich: The resistance of the influenza virus to various physical and chemical agents (Russian). Arkh. biol. nauk. 52, 19 (1938).

$O_{21}$ Overman, J. R., and W. F. Friedewald: Cold haemagglutination reaction with influenza virus. J. Immunol. 62, 415—424 (1949).

$P_1$ Paccaud, M. F., M. Couard, F. Burki, H. Gerber, and J. Lohrer: Outbreak of influenza A/Equi/2 in Switzerland. Nature (Lond.) 211, 101—102 (1964).

$P_2$ Pachaly, L., and R. Schuermann: Influenza and congenital malformations (German). Beitr. path. Anat. 121, 309—319 (1959).

$P_3$ Packaleu, T.: Quantitative studies with influenza A virus in tissue culture. Acta path. microbiol. scand. 23, 512—520 (1946).

P₄ PADGETT, B. L., and D. L. WALKER: Enzymatic variants of influenza virus. 1. Isolation and characterisation of slowly reacting enzymatic variants of influenza B virus.

P₅ 2. Effect of environmental factors on enzymic characteristics of a variant of influenza B virus. J. exp. Med. 106, 53—67 (1957), and 108, 651—664 (1958).

P₆ PANTHIER, R., G. CATEIGNE, and C. HANNOUN: Experimental evidence of inapparent influenzal infections in mice (French). C. R. Soc. Biol. (Paris) 142, 1470—1471 (1948).

P₇ PANTHIER, R., G. CATEIGNE, and C. HANNOUN: A new technique for the study of influenza virus. First practical application (French). Ann Inst. Pasteur 75, 338—350 (1948).

P₈ PANTHIER, R., G. CATEIGNE, and C. HANNOUN: Adaptation to the egg and mouse of a strain of virus recently isolated from a case of influenza (French). Bull. Inst. nat. Hyg. (Paris) 4, 109—112 (1949).

P₉ PANTHIER, R., G. CATEIGNE, and C. HANNOUN: Isolation of a strain of influenza virus. Reaction of young monkeys to the intranasal inoculation of this virus. Bull. Inst. nat. Hyg. (Paris) 4, 112—113 (1949).

P₁₀ PANTHIER, R., C. HANNOUN, and G. CATEIGNE: The interference phenomenon and the multiplication of influenza virus in the allantoic cavity of the chick embryo (French). C. R. Soc. Biol. (Paris) 142, 1215—1218 (1948).

P₁₁ PARIZH, B. M.: Freeze-drying of influenza virus. Abstr. 9th int. Congr. Microbiol., Moscow 388 (1966).

P₁₂ PARKER, E. R., W. B. DUNHAM, and W. J. MACNEAL: Resistance of the Melbourne strain of influenza virus to desiccation. J. Lab. clin. Med. 29, 37—42 (1944).

P₁₃ PARKER, E. R., and W. J. MACNEAL: Persistence of influenza virus on the human hand. J. Lab. clin. Med. 29, 121—126 (1944).

P₁₄ PARNES, V. A.: The contagiousness of experimental influenza in mice (Russian). Zh. Mikrobiol. (Mosk.) 9, 33 (1950).

P₁₅ PARODI, A., and J. FONTAINE: The advantages of Smorodintsev's influenza vaccine (French). Rev. Lyon Med. 8, 57—58 (1959).

P₁₆ PAUCKER, K.: Apparent antigenic change in a laboratory line of influenza A virus during continued propagation in chick embryos. J. Immunol. 85, 148—162 (1960).

P₁₇ PAUCKER, K., A. BIRCH-ANDERSEN, and P. VON MAGNUS: Studies on the structure of influenza virus. 1. Components of infectious and incomplete particles. Virology 8, 1—20 (1959).

P₁₈ PAUCKER, K., and W. HENLE: Studies on host-virus interaction in the chick embryo influenza virus system. 11. The effect of partial inactivation of standard seed virus at 37°C upon the progeny.

P₁₉ 12. Further analysis of yields derived from heat-inactivated standard seeds. J. exp. Med. 101, 479—492, and 493—506 (1955).

P₂₀ PAUCKER, K., and W. HENLE: Interference between inactivated and active influenza viruses in the chick embryo. 2. Interference by incomplete forms of influenza virus. Virology 6, 198—214 (1958).

P₂₁ PAUCKER, K., F. S. LIEF, and W. HENLE: Studies on the soluble antigen of influenza virus. 4. Fractionation of elementary bodies labelled with radioactive phosphorus. Virology 2, 798—810 (1956).

P₂₂ PAVLOVSKII, G. T.: Results of the rhinocytoscopic diagnosis of influenza in the 1959 epidemic (Russian). "Problemy grippa i ostrykh respiratornykh zabolevani." Tez. dokl. nauch. Sess. Inst. AMN SSSR, Moscow 37—38 (1959).

P₂₃ PAYNE, A. M. M.: Some aspects of the epidemiology of the 1957 influenza pandemic. Proc. roy. Soc. Med. 1009—1015 (1958).

P₂₄ PAYNE, A. M. M.: Global epidemiology today. Symp. "Perspectives in epidemiology" Yale J. Biol. 32, 4—15 (1959).

P$_{25}$ PAYZIN, S., and S. OKKAN: The 1949 influenza epidemic in Ankara, the nature of the epidemic and the virus. Turk Ijiyen ve Tecr. Biyoloji Derg. 9, 1—6 (1949).

P$_{26}$ PEARSON, H. E., and J. F. ENDERS: Cultivation of influenza A virus in roller tubes. Proc. Soc. exp. Biol. (N.Y.) 48, 140—143 (1941).

P$_{27}$ PECENKA, J., K. SKVRNOVA, I. HANA, A. IZBICKY, B. TUMOVA, O. MARKVART, L. LOUDA, Z. HARTL, J. HELCL, and V. KLEINBAUER: Contribution to the evaluation of vaccines against influenza (Czech). Čs. Epidem. 7, 365—373 (1958).

P$_{28}$ PELON, W., W. J. MOGABGAB, L. DIETLEIN, G. E. BURCH, and B. HOLMES: Antibody response to Asian influenza vaccination in man. Proc. Soc. exp. Biol. (N.Y.) 99, 120—124 (1958).

P$_{29}$ PENTTINEN, K.: The effect of low temperatures on vaccines of influenza virus and on purified influenza virus. J. Immunol. 64, 165—171 (1950).

P$_{30}$ PENTTINEN, K.: Effect of polyphloroglucinol phosphate on influenza virus and on chicken red cells. Ann. Med. exp. Fenn. 34, 88—94 (1956).

P$_{31}$ PENTTINEN, K., V. TOMMILA, and K. CANTELL: Variation in the results of haemagglutination and haemagglutination-inhibition tests with influenza and mumps viruses caused by red cells of different fowls. Ann. Med. exp. Fenn. 33, 159—168 (1955).

P$_{32}$ PEREIRA, M. S.: Typing of Q phase influenza A virus by serum neutralisation tests in monkey kidney cultures. Lancet 668—669 (1958).

P$_{33}$ PEREIRA, H. G., M. S. PEREIRA, and V. G. LAW: Antigenic variants of influenza A2 virus. Bull. Wld Hlth Org. 31, 129—132 (1964).

P$_{34}$ PEREIRA, H. G., and B. TUMOVA: Specific serum neutralisation of the reactivating activity of influenza A2 viruses. J. gen. Virol. 1, 131—133 (1967).

P$_{35}$ PEREIRA, H. G., B. TUMOVA, and V. G. LAW: Avian influenza A viruses. Bull. Wld Hlth Org. 32, 855—860 (1965).

P$_{36}$ PEREZ, J. E., and G. ARBONA: An outbreak of influenza due to type A-prime virus at Coamo Puerto Rico. Puerto Rico J. publ. Hlth 25, 377—381 (1950).

P$_{37}$ PEREZ, J. E., J. BARALT-PEREZ, and M. KLEIN: The reversal in vivo by BAL of HgCl$_2$-inactivated influenza A virus in the chick embryo. J. Immunol. 62, 405—413 (1949).

P$_{38}$ PERLINA, F. I.: Clinical syndromes of neurologic involvement during influenza (Russian). Sovetsk. Med. 24, 82—89 (1960).

P$_{39}$ PERRY, B., and F. M. BURNET: Recombination studies with two influenza virus B strains. Aust. J. exp. Biol. med. Sci. 31, 519—528 (1953).

P$_{40}$ PERRY, B. T., M. VAN DEN ENDE, and F. M. BURNET: Recombination with two influenza B strains in the de-embryonated egg. Aust. J. exp. Biol. med. Sci. 32, 469—479 (1954).

P$_{41}$ PERSHIN, G. N., N. S. BOGDENOVA, S. M. MAKIN, and V. M. LIKHOSLERSTOV: The antivirus activity of 2:6 dialkoxypyrans and 2-alkoxy △ 5-dihydropyrans (Cyclic acetals of derivatives of glutaric aldehyde) (Russian). Farmakol. i Toksikol. 28, 66 (1965).

P$_{42}$ Personnel of Naval Lab. Res. Unit 1.: Experimental human influenza. Amer. J. med. Sci. 207, 306—314 (1944).

P$_{43}$ Personnel of Naval Res. Unit 1.: The inhalatory route for prophylaxis and treatment of experimental influenza. 1. The distribution of inhaled material.

P$_{44}$ 2. Immune serum in prophylaxis and treatment. Amer. J. med. Sci. 207, 40—47, and 47—60 (1944).

P$_{45}$ Personnel of Naval Res. Unit 1.: Studies on the primary isolation of influenza viruses. U.S. Navy med. Bull. 46, 369—374 (1946).

P$_{46}$ PETERS, D.: Morphology of viruses pathogenic for man and animals (German). Zbl. Bakt. I. Abt. Orig. 176, 259—294 (1959).

P$_{47}$ PETERSEN, K. B., and P. VON MAGNUS: Isolation of a new virus from a child with an influenza-like disease. Danish med. Bull. 5, 157—159 (1958).

P$_{48}$ PETERSON, O. P.: Dynamics of influenza antigen in the respiratory tract of infected animals (Russian). Zh. Mikrobiol. (Mosk.) 10/11, 29 (1943).

P₄₉ PETERSON, O. P.: Methods of laboratory diagnosis of influenza and their significance in clinical practice (Russian). Dokl. 1-y Sess. Inst. Virusol. AMN SSSR po grippa, Moscow (1948).

P₅₀ PETERSON, O. P.: Concentration of the influenza virus in nasopharyngeal washings of patients (Russian). Vop. Virus. 1, 177 (1948).

P₅₁ PETERSON, O. P.: The effect of the temperature factor and ultraviolet irradiation of the influenza virus on the Hirst reaction (Russian). Vop. Virus. 2, 227 (1949).

P₅₂ PETERSON, O. P., and S. A. SEMASHKO: Quick diagnosis of influenza by the detection of the antigen in nasopharyngeal washings (Russian). Zh. Mikrobiol. (Mosk.) 3, 22 (1945).

P₅₃ PETERSON, O. P., and G. V. YEREMEYEV: Properties of soluble antigens of influenza virus (Russian). Vop. Virus. 2, 209 (1949).

P₅₄ PETERSON, W. D., F. M. DAVENPORT, and T. FRANCIS: A study in vitro of components in the transmission cycle of swine influenza virus. J. exp. Med. 114, 1023—1033 (1961).

P₅₅ PETRESCU, A., D. PRODAN, and V. GHEORGHE: Study on the influence of high frequency electric and magnetic fields on influenza virus type A. Morphologic alterations. Rev. roum. Inframicrobiol. 2, 339—342 (1965).

P₅₆ PETROV, YU. K.: Methods of production of highly active diagnostic, therapeutic, and prophylactic anti-influenza sera and sera against some influenza-like diseases (Russian). "Problemy grippa i ostrykh respiratornykh zabolevani." Tez. dokl. nauch. Sess. Inst. AMN SSSR, Moscow 218—219 (1959).

P₅₇ PETTIT, H., S. MUDD, and D. S. PEPPER: The Philadelphia and Alaska strains of influenza virus. Epidemic influenza in Alaska in 1935. J. Amer. med. Assoc. 106, 890—892 (1936).

P₅₈ PFEIFFER, R.: Provisional communication on the cause of influenza (German). Dtsch. med. Wschr. 18, 28 (1892).

P₅₉ PHILIP, R. N., J. A. BELL, D. J. DAVIS, M. O. BEEM, and P. M. BEIGELMANN: Epidemiological studies on influenza in familial and general population groups. 1. Preliminary report on studies with adjuvant vaccines. Amer. J. publ. Hlth 44, 34—42 (1954).

P₆₀ PICCIOTO, L.: Serologic response of subjects vaccinated with a single dose of Asian influenza vaccine (Italian). Riv. Ist. Sieroter. ital. 33, 80—83 (1958).

P₆₁ PIGAREVSKIY, V. YE.: Morphological study of the reactive processes in virus influenza (Russian). Probl. obshch. virusol. Tez. dokl. 6-y Sess. Inst. Virusol. AMN SSSR, Moscow 23 (1953).

P₆₂ PIGAREVSKIY, V. YE.: Cellular factors in the defence of the host in influenza. Acta virol. 1, 30—35 (1957).

P₆₃ PIGAREVSKIY, V. YE., and O. M. CHALKINA: New data on the methods of diagnosing influenza by means of rhinocytoscopy (Russian). Vop. Virus. 4, 210 (1957).

P₆₄ PILCHER, K. S., K. F. SOIKE, V. H. SMITH, F. TROSPER, and B. FOLSTON: Inhibition of multiplication of Lee influenza virus by canavanine. Proc. Soc. exp. Biol. (N.Y.) 88, 79—86 (1955).

P₆₅ PILLE, E. R.: Age characteristics of immunity in experimental influenza (Russian). Annot. rabot. Inst. virusol. AMN SSSR za 1949, Moscow 14 (1950).

P₆₆ PILLE, E. R.: Experimental data on age immunology in influenza (Russian). Tr. 3-y Sess. Inst. Virusol. AMN SSSR, Moscow 16 (1950).

P₆₇ PLACHINSKA, J.: Detection of influenza virus in cell culture by immuno-fluorescence. Med. Dosw. Mikrobiol. Warsaw 17, 245—249 (1965).

P₆₈ POLLEY, J. R.: Factors influencing inactivation of infectivity and haemagglutinin of influenza virus by gamma radiation. Canad. J. Microbiol. 7, 535—541 (1961).

P₆₉ POLLEY, J. R., M. BURR, and A. GILLEN: Preparation of a stable non-infective soluble influenza A antigen. Proc. Soc. exp. Biol. (N.Y.) 76, 330—332 (1951).

P₇₀　POLLEY, J. R., and M. M. GUERIN: The use of B-propiolactone for the prepara-
　　　tion of virus vaccines. 1. Selection of reaction conditions.

P₇₁　　　2. Antigenicity. Canad. J. Microbiol. 3, 863—870, and 871—877 (1957).

P₇₂　POLYAK, R. J., and A. A. SMORODINTSEV: Electrophoretic investigations on
　　　thermostable inhibitors of type A 2 influenza virus in normal animal sera.
　　　Acta virol. 5, 1—3 (1961).

P₇₃　PORTERFIELD, B. M.: The effect of calcium and electrolytes on the enzymic
　　　action of influenza viruses and V. cholerae extract. Brit. J. exp. Path. 33,
　　　196—201 (1952).

P₇₄　PORTOCALA, R., and M. ANDRIESCU: The reproduction of influenza virus by
　　　means of virus ribonucleic acid. 5. The influence of the purity of the phenol
　　　upon the activity of the ribonucleic acid. Stud. Cercet. Inframicrobiol. 12,
　　　77—81 (1961).

P₇₅　PORTOCALA, R., V. BOERU, and J. SAMUEL: Reproduction of influenza virus
　　　from purified virus ribonucleic acid (Rumanian). Stud. Cercet. Inframicro-
　　　biol. 10, 51—57 (1959).

P₇₆　PORTOCALA, R., V. BOERU, and J. SAMUEL: On the biosynthesis of influenza
　　　virus from a ribonucleic acid extract of the virus.

P₇₇　　　2. Properties of the recovered isolates and of the ribonucleic acid (French).
　　　C. R. Acad. Sci. (Paris) 249, 201—202, and 848—849 (1959).

P₇₈　PORTOCALA, R., V. BOERU, and J. SAMUEL: Biosynthesis of influenza virus
　　　starting from an ether-phenol virus extract. Acta virol. 3, 172—174 (1959).

P₇₉　PORTOCALA, R., V. BOERU, and J. SAMUEL: Role of ribonucleic acid in the
　　　infectivity of the influenza virus (Rumanian). Stud. Cercet. Inframicrobiol.
　　　11, 41—45 (1960).

P₈₀　PORTOCALA, R., V. BOERU, J. SAMUEL, and M. ANDRIESCU: The reproduction
　　　of influenza virus by means of virus ribonucleic acid. 4. Investigation of
　　　ribonuclease in fertile eggs of different ages. Stud. Cercet. Inframicrobiol.
　　　12, 71—75 (1961).

P₈₁　POTEL, J., S. HLAWATSCH, and L. DEGEN: The photometric haemagglutination
　　　test for the antigen determination of viruses and bacteria (German). Zbl.
　　　Bakt. I. Abt. Orig. 176, 331—334 (1959).

P₈₂　POWELL, W. F., and E. C. POLLARD: The effect of ionising radiation on the
　　　interfering property of influenza virus. Virology 2, 321—336 (1956).

P₈₃　POWELL, W. F., and R. B. SETLOW: The effect of monochromatic ultraviolet
　　　radiation on the interfering property of influenza virus. Virology 2, 337—343
　　　(1956).

P₈₄　PROSE, P. H., S. D. BALK, H. LIEBHABER, and S. KRUGMAN: Studies of a myxo-
　　　virus recovered from patients with infectious hepatitis. J. exp. Med. 122,
　　　1151—1160 (1965).

P₈₅　PRZESMYCKI, F., E. WALKOWSKA, H. DOBROWALSKA, Z. ZYCH, A. FELTNOWSKI,
　　　Z. PRZYBYLKIEWICZ, and J. ZANSKI: Anti-influenza vaccinations (Polish).
　　　Szczepieniz przeciwgrypowe Med. Dosw. Mikr. 4 (1954).

P₈₆　PUMPER, R. W., and H. M. YAMASHIRIYA: The isolation and virus cytopatho-
　　　genicity of a line of cells from human keratoacanthoma. Amer. J. Hyg. 72,
　　　284—288 (1960).

P₈₇　PYE, J.: Assay of inhibitors of influenza virus haemagglutination by electro-
　　　phoresis. Aust. J. exp. Biol. med. Sci. 33, 323—334 (1955).

P₈₈　PYE, J., H. F. HOLDEN, and H. B. DONALD: Sedimentation behaviour and
　　　electron microscopic examination of purified influenza virus. J. gen. Micro-
　　　biol. 14, 634—636 (1956).

Q₁　QUILLIGAN, J. J., R. D. BOCHE, E. J. CARRUTHERS, S. L. AXTELL, and J. C.
　　　TRIVEDI: Continuous cobalt-60 irradiation and immunity to influenza virus.
　　　J. Immunol. 90, 506—511 (1963).

Q₂　QUILLIGAN, J. J., and T. FRANCIS: Serological response to intranasal admini-
　　　stration of inactive influenza virus in children. J. clin. Invest. 26, 1079—
　　　1087 (1947).

$Q_3$  QUILLIGAN, J. J., T. FRANCIS, and E. MINUSE: Reactions to an influenza virus vaccine in infants and children. Amer. J. Dis. Child. **78**, 295—301 (1949).

$Q_4$  QUILLIGAN, J. J., T. FRANCIS, R. J. ROWE, D. G. TRAGGIS, J. D. ADCOCK, and H. KURTZ: The action of terramycin on the growth of strains of influenza, herpes simplex, and rabies viruses in chick embryos and mice. Ann. N.Y. Acad. Sci. **53**, 407—411 (1950).

$Q_5$  QUILLIGAN, J. J., E. MINUSE, and T. FRANCIS: Homologous and heterologous antibody response of infants and children to multiple injections of a single strain of influenza virus. J. clin. Invest. **27**, 572—579 (1948).

$Q_6$  QUILLIGAN, J. J., E. MINUSE, and T. FRANCIS: Type C influenza virus. 2. Intranasal inoculation of human beings. J. Lab. clin. Med. **43**, 43—49 (1954).

$R_1$  RAETTIG, H.: Investigations on immunity and serology in influenza (German).
$R_2$  Zbl. Bakt. I. Abt. Orig. **152**, 159—203, and 381—402 (1947).

$R_3$  RAETTIG, H.: Investigations on immunity to influenza virus. 2. The titre of haemagglutination-inhibition in normal human sera (German). Z. Immun.-Forsch. **106**, 146—153 (1949).

$R_4$  RAETTIG, H.: The possibilities of vaccination against influenza and the vaccine developed in the Robert Koch Institute (German). Zbl. Bakt. I. Abt. Orig. **167**, 85—86 (1958).

$R_5$  RAPPOPORT, R. S.: Characteristics of influenza B of children in an enclosed area (Russian). Bibl. i annot. rabot. otdela. virusol. Inst. Eksper. med. AMN SSSR za 1947—1949, Leningrad (1950).

$R_6$  RAPPOPORT, R. S., A. G. GULAMOV, and O. M. CHALKINA: Material on the virological and serological study of influenza type B (Russian). Gripp. Tr. AMN SSSR, Moscow XXVIII, 151 (1953).

$R_7$  RASKA, K.: The 1959 influenza epidemic (Czech). Čas. Lék. čes. **98**, 353—355 (1959).

$R_8$  RASMUSSEN, A. F., and J. C. STOKES: Chemical inhibition of the growth of the virus of influenza in the embryonated egg. J. Immunol. **66**, 237—247 (1951).

$R_9$  RATHOVA, V., L. BORECKY, and D. KOCISKOVA: An attempt to influence in vivo the serum levels of myxovirus inhibitors. 3. The effect of the administration of some vitamins on inhibitors and some other serum factors in guineapigs and mice (Czech). Bratislavske lekarske listy **43**, 31—40 (1963).

$R_{10}$  RATHOVA, V., D. KOCISKOVA, and O. KRIZANOVA: An attempt to prepare an immune serum against gamma-inhibitor of influenza virus. Acta virol. **8**, 551—554 (1964).

$R_{11}$  RATHOVA, V., D. KOCISKOVA, and O. KRIZANOVA: The use of indirect haemagglutination for evaluating antibody formation against purified gamma-inhibitor. Acta virol. **10**, 82 (1966).

$R_{12}$  RATNER, B., and S. UNTRACHT: Allergy to virus and rickettsial vaccines. 1. Allergy to influenza A and B vaccines in children. J. Amer. med. Ass. **132**, 899—905 (1946).

$R_{13}$  REDA, I. M., R. ROTT, and W. SCHAFER: Fluorescent antibody studies with NDV infected cell systems. Virology **22**, 422—425 (1964).

$R_{14}$  REED, L. J., and H. MUENCH: A simple method of estimating 50% end points. Amer. J. Hyg. **27**, 493—497 (1938).

$R_{15}$  REGINSTER, M.: Inactivation of influenza virus by caseinase C from streptomyces albus G culture filtrate. J. gen. Microbiol. **40**, 157—169 (1965).

$R_{16}$  REGINSTER, M.: Release of influenza virus neuraminidase by caseinase C of streptomyces albus G. J. gen. Microbiol. **42**, 323—331 (1966).

$R_{17}$  REGINSTER, M.: Effects of pronase on influenza virus. Acta virol. **10**, 111—116 (1966).

$R_{18}$  REIMER, C. B., R. S. BAKER, J. E. NEWLIN, and M. L. HAVENS: Influenza virus purification with the zonal ultracentrifuge. Science **152**, 1379—1381 (1966).

R₁₉ REINICKE, V., and F. V. HAHNEMANN: The influence of interferon and hydrocortisone on the production of incomplete influenza virus *in ovo*. Acta path. microbiol. scand. **63**, 51—58 (1965).

R₂₀ REZNIKOVA, O. YU., and A. P. MARISOVA: Material characterising the influenza morbidity rate in 1949 and the laboratory diagnosis of influenza (Russian). Zh. Mikrobiol. (Mosk.) **9**, 32 (1950).

R₂₁ RHEINS, M., and J. FINLAY: The effect of ultrasonic irradiation on certain properties of influenza virus. J. infect. Dis. **95**, 79—85 (1954).

R₂₂ RHIAN, M., A. S. EVANS, and J. L. MELNICK: The interaction of influenza virus and intact human erythrocytes observed by replica technique in the electron microscope. J. Immunol. **67**, 513—521 (1951).

R₂₃ RICE, C. E.: Studies of the complement fixation reaction in virus systems. 2. Activities of influenza virus antigens and antisera. J. Immunol. **56**, 343—356 (1947).

R₂₄ RICKARD, E. R., and T. FRANCIS: The demonstration of lesions and virus in the lungs of mice receiving large intraperitoneal inoculations of epidemic influenza virus. J. exp. Med. **67**, 953—972 (1938).

R₂₅ RICKARD, E. R., M. P. THIGPEN, and J. M. ADAMS: Antibody response to strains of influenza A and swine influenza viruses in the serums of infants experiencing their first infection with influenza virus A. J. infect. Dis. **76**, 203—207 (1945).

R₂₆ RICKARD, E. R., M. THIGPEN, and J. H. CROWLEY: The isolation of influenza A virus by the intra-allantoic inoculation of chick embryos with untreated throat washings. J. Immunol. **49**, 263—271 (1944).

R₂₇ RICKARD, E. R., M. THIGPEN, and J. H. CROWLEY: Vaccination against influenza at the university of Minnesota. Amer. J. Hyg. **42**, 12—20 (1945).

R₂₈ RIFKIND, R. A., K. C. HSU, C. MORGAN, B. C. SEEGAL, A. W. KNOX, and H. M. ROSE: Use of ferritin-conjugated antibody to localise antigen by electron microscopy. Nature (Lond.) **187**, 1094—1095 (1960).

R₂₉ RIGHTSEL, W. A., H. G. SCHNEIDER, B. J. SLOAN, P. R. GRAF, F. A. MILLER, Q. R. BARTZ, J. EHRLICH, and G. J. DIXON: Antiviral activity of gliotoxin and gliotoxin acetate. Nature (Lond.) **204**, 1333—1334 (1964).

R₃₀ RITOVA, V. V.: The etiology of influenza and of acute catarrh in young children (Russian). Problemy. grippa i OKVDP, Moscow II (1952).

R₃₁ RITOVA, V. V.: Influenza in infants (Russian). Tr. ob'yed. Sess. Inst. AMN SSSR, Moscow 133 (1953).

R₃₂ RITOVA, V. V.: Characterisation of influenza virus strains isolated in 1953 and improvement of the methods of isolating them (Russian). Vop. Virus. **4**, 45 (1954).

R₃₃ RITOVA, V. V.: Problems and perspectives in specific influenza prophylaxis (Russian). "Problemy grippa i ostrykh respiratornykh zabolevanii" Tez. dokl. nauch. Sess. Inst. AMN SSSR, Moscow 78—80 (1959).

R₃₄ RITOVA, V. V., K. F. ARTEMENKO, and A. K. SVETLOVA: The treatment of A2 influenza in children by a fluid anti-influenza A2 serum (Russian). Vop. Virus. **5**, 197—199 (1958).

R₃₅ RITOVA, V. V., and N. A. EVSTEYNEEVA: Live tissue culture influenza vaccine. Bull. Wld Hlth Org. **27**, 729—734 (1962).

R₃₆ RITOVA, V. V., and A. F. STEFANSKAYA: Doses of influenza vaccine for the vaccination of infants (Russian). Vop. Virus. **4**, 397 (1954).

R₃₇ RITOVA, V. V., and L. YA. ZAKSTEL'SKAYA: Improvement in the methods of preparing vaccine strains of the influenza A prime and B viruses (Russian). Zh. Mikrobiol. (Mosk.) **5**, 55 (1954).

R₃₈ ROBBINS, F. C., and J. F. ENDERS: Tissue culture techniques in the study of animal viruses. Amer. J. med. Sci. **220**, 316—338 (1950).

R₃₉ ROBERTS, D. H.: The isolation of an influenza A virus and a mycoplasma associated with duck sinusitis. Vet. Rec. **76**, 470—473 (1964).

R₄₀ ROBERTSON, O. H., C. G. LOOSLI, T. T. PUCK, E. BIGG, and B. F. MILLER: The protection of mice against infection with air-borne influenza virus by means of propylene glycol vapour. Science **93**, 612—613 (1941).

$R_{41}$ ROBINSON, R. Q., W. B. YARBROUGH, R. H. GORRIE, and W. R. DOWDLE: Antigenic relationship of 1961—62 type B influenza viruses to earlier type B strains. Proc. Soc. exp. Biol. (N.Y.) 112, 658—661 (1963).

$R_{42}$ RODRIGUES DA SILVA, G., and H. A. FELDMAN: Formalin-treated chicken erythrocytes as indicators of influenza A virus (Asian) and its antibody. Proc. Soc. exp. Biol. (N.Y.) 101, 241—245 (1959).

$R_{43}$ ROLLA, G.: Lack of chick allantoic antigen in influenza virus passed in chick kidney monolayers. Acta path. microbiol. scand. 62, 417—420 (1964).

$R_{44}$ ROLLA, G.: Investigations of some effects of human saliva on influenza virus. 3. Virus antibodies in saliva. Acta path. microbiol. scand. 65, 111—116 (1965).

$R_{45}$ ROLLY, H.: Substituted isonicotinic acid diphenylalkylamides with antiviral activity (German). Zbl. Bakt. I. Abt. Orig. 188, 335—344 (1963).

$R_{46}$ ROMANENKO, N. N.: Virological and serological characterisation of influenza in Leningrad (Russian). Tez. dokl. 3-y, Sess. Inst. Virusol. AMN SSSR, Moscow (1950).

$R_{47}$ ROMANENKO, N. N., and E. A. FRIDMAN: A new serological variety of the influenza B virus (Russian). Zh. Mikrobiol. (Mosk.) 7, 50 (1954).

$R_{48}$ ROMANENKO, N. N., and K. M. KHAYT: Experiments on the study of the mechanism of virus haemagglutination (Russian). Tr. Inst. Epidem. i bakteriol. imeni. Pastera, Vol. XIII, "Virusnyye Infektsii" Leningrad 75, 84 (1953).

$R_{49}$ ROMANOVSKA, E.: Studies on blood group antigens M and N. 5. The M and N blood group substances as inhibitors of influenza virus haemagglutination. Arch. Immun. Terap. Dosw. Warsaw 7, 749—757 (1959).

$R_{50}$ ROMVARY, J., G. TAKATSY, K. BARB, and E. FARKAS: Isolation of influenza virus strains from animals. Nature (Lond.) 193, 907—908 (1962).

$R_{51}$ ROMVARY, J., G. TAKATSY, and E. FARKAS: Serological evidence of the incidence of influenza equine A1 virus infections among horses in Hungary. Acta microbiol. Acad. Sci. hung. 12, 289—294 (1965).

$R_{52}$ ROSE, H. M., and A. GELLKORN: Inactivation of influenza virus with sulphur and nitrogen mustards. Proc. Soc. exp. Biol. (N.Y.) 65, 83—85 (1947).

$R_{53}$ ROSE, H. M., E. MOLLOY, and E. O'NEILL: Effect of penicillin on bacterial contamination of eggs and tissue cultures inoculated with unfiltered sputums. Proc. Soc. exp. Biol. (N.Y.) 60, 23—25 (1945).

$R_{54}$ ROSE, H. M., and C. MORGAN: Fine structure of virus-infected cells. Ann. Rev. Microbiol. 14, 217—240 (1960).

$R_{55}$ ROSE, M. A.: Influenza in horses. Vet. Rec. 77, 404 (1965).

$R_{56}$ ROSE, M. A.: Equine influenza viruses isolated at Cambridge in 1963 and 1965. Proc. roy. Soc. Med. 59, 5—6 (1966).

$R_{57}$ ROSENBAUM, M. J., and R. L. WOOLRIDGE: The use of soluble antigen for the serological diagnosis of influenza in vaccinated populations. J. infect. Dis. 99, 275—281 (1956).

$R_{58}$ ROSENBERG, M.: Morphology of some strains of influenza viruses of the A-Fe group as compared with influenza virus type A strain PR8 (Czech). Biol. Prace 4, 9—11 (1958).

$R_{59}$ ROSENBERG, A., C. HOWE, and E. CHARGAFF: Inhibition of influenza virus haemagglutination by a brain lipid fraction. Nature (Lond.) 177, 234—235 (1956).

$R_{60}$ ROTT, R., and R. DRZENIEK: Host-induced modification of lipid-containing RNA viruses. Abstr. 9th int. Congr. Microbiol. Moscow 467 (1966).

$R_{61}$ ROTT, R., R. DRZENIEK, S. M. SABER, and E. REICHERT: Blood group substances, Forssman and mononucleosis antigens in lipid-containing RNA viruses. Arch. ges. Virusforsch. 19, 273—288 (1966).

$R_{62}$ ROTT, R., S. SABER, and C. SCHOLTISSEK: Effect on myxovirus of mitomycin C, actinomycin D, and pre-treatment of the host cell with ultraviolet light. Nature (Lond.) 205, 1187—1190 (1965).

$R_{63}$ ROTT, R., and W. SCHAFER: Studies of the non-infectious haemagglutinating particle of influenza virus (German). Z. Naturforsch. 16b, 310—321 (1961).

$R_{64}$ Rott, R., and W. Schafer: Behaviour of the antigenic and other biological properties of human and animal viruses on treatment with nitrous acid (German). Z. Naturforsch. 17b, 160—164 (1962).

$R_{65}$ Rott, R., and W. Schafer: The various virus-specific units produced by myxo-virus-infected cells. Ciba Found. Symp. "Cellular biology of myxovirus infections" 27—50 (1964).

$R_{66}$ Rott, R., and C. Scholtissek: Investigations about the formation of incomplete forms of fowl plague virus. J. gen. Microbiol. 33, 303—312 (1963).

$R_{67}$ Rott, R., and C. Scholtissek: Influence of actinomycin on the growth of myxo-viruses (German). Z. Naturforsch. 19b, 316—323 (1964).

$R_{68}$ Rozee, K. R., G. L. Williams, and C. E. Van Rooyen: Detection and titration of Asian influenza A virus by HeLa cell and monkey kidney cell cultures. Science 128, 591—592 (1958).

$R_{69}$ Rubini, J. R., A. F. Rasmussen, and M. A. Stahnann: Inhibitory effect of synthetic lysine polypeptides on growth of influenza virus in embryonated eggs. Proc. Soc. exp. Biol. (N.Y.) 76, 662—665 (1951).

$R_{70}$ Ruttkay-Nedecky, G., and S. Ivanicova: Electrophoretic control of influenza A2 virus purification and demonstration of an additional virus-specific electrophoretic component in the purified virus preparations. Acta virol. 9, 508—518 (1965).

$R_{71}$ Rybakov, N.: Prevention of influenza with an atebrine aerosol (Russian). Med. rabotn. 50, 1274 (1954).

$R_{72}$ Rykowska, R., and L. Sawicki: The gamma inhibitor of normal horse serum. 2. The protective properties of the serum and its fractions against infection of mice and tissue cultures with the Singapore strain of influenza virus (Russian). Med. Dosw. i Mikrob. Warsaw 15, 47—54 (1963).

$R_{73}$ Rytel, M. W., R. E. Shope, and E. D. Kilbourne: An antiviral substance from penicillium funiculosum. 5. Induction of interferon by Helenine. J. exp. Med. 123, 577—584 (1966).

$S_1$ Sagik, B., T. Puck, and S. Levine: Quantitative aspects of the spontaneous elution of influenza virus from red cells. J. exp. Med. 99, 251—260 (1964).

$S_2$ Sakharov, P. P.: "Inheritance of acquired characteristics" (Russian). Moscow (1952).

$S_3$ Salk, J. E.: Partial purification of the virus of epidemic influenza by adsorption on calcium phosphate. Proc. Soc. exp. Biol. (N.Y.) 46, 709—712 (1941).

$S_4$ Salk, J. E.: A simplified procedure for titrating haemagglutinating capacity of influenza virus and the corresponding antibody. J. Immunol. 49, 87—98 (1944).

$S_5$ Salk, J. E.: The immunising effect of calcium phosphate adsorbed influenza virus. Science 101, 122—124 (1945).

$S_6$ Salk, J. E.: Variation in influenza viruses. A study of heat stability of the red cell agglutinating factor. Proc. Soc. exp. Biol. (N.Y.) 63, 134—139 (1946).

$S_7$ Salk, J. E.: Effect of formalin in increasing heat stability of influenza virus haemagglutinin. Proc. Soc. exp. Biol. (N.Y.) 63, 140—143 (1946).

$S_8$ Salk, J. E.: Studies on the antigenicity in man of calcium phosphate adsorbed influenza virus; with comments on the question of dose of virus needed in vaccines for human use. J. Immunol. 57, 301—321 (1947).

$S_9$ Salk, J. E., M. L. Bailey, and A. M. Laurent: The use of adjuvants in studies on influenza immunisation. 2. Increased antibody formation in human sub-jects inoculated with influenza virus vaccine in water-in-oil emulsion. Amer. J. Hyg. 55, 439—456 (1952).

$S_{10}$ Salk, J. E., M. Contakos, A. M. Laurent, M. Sorensen, and A. J. Rapalski: The use of adjuvants in studies on influenza immunisation. 3. Degree of persistence of antibody in human subjects 2 years after vaccination. J. Amer. med. Assoc. 151, 1169—1175 (1953).

$S_{11}$  SALK, J. E., and A. M. LAURENT: The use of adjuvants in studies on influenza immunisation. 1. Measurement in monkeys of the dimensions of antigenicity of virus-mineral oil emulsions. J. exp. Med. 95, 429—447 (1952).

$S_{12}$  SALK, J. E., A. M. LAURENT, and M. L. BAILEY: Direction of research on vaccination against influenza. New studies with immunologic adjuvants. Amer. J. publ. Hlth 41, 669—677 (1951).

$S_{13}$  SALK, J. E., W. J. MENKE, and T. FRANCIS: A clinical, epidemiological, and immunological evaluation of vaccination against epidemic influenza. Amer. J. Hyg. 42, 57—93 (1945).

$S_{14}$  SALK, J. E., H. E. PEARSON, P. N. BROWN, and T. FRANCIS: Protective effect of vaccination against induced influenza B. Preliminary report. Proc. Soc. exp. Biol. (N.Y.) 55, 106—107 (1944).

$S_{15}$  SALTSAM, J. H., F. LANNI, and J. W. BEARD: Inhibitory effect of human saliva on haemagglutination by influenza virus A (PR8) and the swine influenza virus. J. Immunol. 63, 261—267 (1949).

$S_{16}$  SAMPAIO, A. A. C.: Inhibitors of influenza virus haemagglutination in normal animal sera. Bull. Wld Hlth Org. 6, 467—472 (1952).

$S_{17}$  SAMPAIO, A. A. C.: Antigenic analysis of influenza viruses by the haemagglutination-inhibition technique. Results of the use of different animal sera. Bull. Wld Hlth Org. 6, 473—480 (1952).

$S_{18}$  SAMPAIO, A. A. C., and A. ISAACS: The action of trypsin on normal serum inhibitors of influenza virus haemagglutination. Brit. J. exp. Path. 34, 152—158 (1953).

$S_{19}$  SAMVELOVA, S. A.: Experience in mass prophylactic vaccination against influenza (Russian). Zh. Mikrobiol. (Mosk.) 9, 9 (1954).

$S_{20}$  SAMVELOVA, S. A., and G. YA. KUZMINSKAYA: Peculiarities of the epidemiology of influenza in Moscow in 1957 (Russian). Vop. Virus. 5, 8—14 (1958).

$S_{21}$  SANDERS, M., I. KIEM, and D. LAGUNOFF: Cultivation of viruses. A critical review. Arch. Path. 56, 148—225 (1953).

$S_{22}$  SASLAW, S., and H. CARLISLE: Aerosol exposure of monkeys to influenza virus Proc. Soc. exp. Biol. (N.Y.) 119, 838—843 (1965).

$S_{23}$  SASLAW, S., H. E. WILSON, C. A. DOAN, O. C. WOOLPERT, and J. L. SCHWAB: Reactions of monkeys to experimentally induced influenza virus A infection. An analysis of the relative roles of humoral and cellular immunity under conditions of optimal and deficient nutrition. J. exp. Med. 84, 113—125 (1946).

$S_{24}$  SAWICKI, L., S. BARON, and A. ISAACS: Influence of increased oxygenation on influenza virus infection in mice. Lancet 680—682 (1961).

$S_{25}$  SAWICKI, L., and H. GRZELAKOWA: The growth of influenza virus in monkey kidney tissue culture investigated with the haemadsorption technique (Polish). Med. Dosw. i. Mikrob. Warsaw 12, 241—249 (1960).

$S_{26}$  SCHAEFFER, M.: Preparation of purified influenza virus. Proc. Soc. exp. Biol. (N.Y.) 51, 32—34 (1942).

$S_{27}$  SCHAEFFER, M., F. F. SILVER, and C. C. PI: Studies on the chemotherapy of virus infections. 1. General anaesthetics and other drugs ineffective against experimental virus infection in mice. J. Immunol. 63, 109—115 (1949).

$S_{28}$  SCHÄFER, W.: Properties of animal viruses. 1. The complement fixing antigen of classical fowl plague (German). Z. Naturforsch. 6b, 207—212 (1951).

$S_{29}$  SCHÄFER, W.: Comparative sero-immunologic studies on the viruses of influenza and classical fowl plague (German). Z. Naturforsch. 10b, 81—91 (1955).

$S_{30}$  SCHÄFER, W.: Sero-immunologic studies on incomplete forms of the virus of classical fowl plague (German). Arch. exp. Vet.-Med. 9, 218—230 (1955).

$S_{31}$  SCHÄFER, W.: The significance of electron microscopy for virus studies (German). Dtsch. tierärztl. Wschr. 63, 15—16 (1956).

$S_{32}$  SCHÄFER, W.: Units isolated after splitting fowl plague virus. Ciba Found. Symp. "The nature of viruses." 91—103 (1957).

S₃₃ SCHÄFER, W.: Some observations concerning the reproduction of RNA-containing animal viruses. 9th Symp. Soc. gen. Microbiol. "Virus growth and variation." 61—81 (1959).

S₃₄ SCHÄFER, W.: Structure of some animal viruses and significance of their components. Bact. Rev. 27, 1—17 (1963).

S₃₅ SCHÄFER, W., and K. MUNK: Purification and properties of soluble antigens of classical fowl plague (German). Z. Naturforsch. 7b, 573—574 (1952).

S₃₆ SCHÄFER, W., and K. MUNK: Properties of animal viruses. 4. Studies of the latent phase in the growth of fowl plague virus (German). Z. Naturforsch. 7b, 608—619 (1952).

S₃₇ SCHÄFER, W., K. MUNK, and O. ARMBRUSTER: Properties of animal viruses. 3. Studies on the physicochemical and morphological properties of fowl plague virus (German). Z. Naturforsch. 7b, 29—33 (1952).

S₃₈ SCHÄFER, W., and R. ROTT: The preparation of virus vaccines with hydroxylamine. Course of inactivation and the effect of hydroxylamine upon various properties of certain viruses (German). Z. Hyg. Infekt.-Kr. 148, 256—268 (1962).

S₃₉ SCHÄFER, W., and G. SCHRAMM: On the isolation and characterisation of the virus of classical fowl plague (German). Z. Naturforsch. 5b, 91—102 (1950).

S₄₀ SCHÄFER, W., G. SCHRAMM, and E. TRAUB: Studies of the virus of atypical fowl plague (German). Z. Naturforsch. 4b, 157—167 (1949).

S₄₁ SCHÄFER, W., and W. ZILLIG: On the structure of the elementary bodies of fowl plague virus. 1. Preparation, physicochemical and biological properties of split products (German). Z. Naturforsch. 9b, 779—788 (1954).

S₄₂ SCHÄFER, W., W. ZILLIG, and K. MUNK: Isolation and characterisation of the non-infectious haemagglutinin of fowl plague (German). Z. Naturforsch. 9b, 329—340 (1954).

S₄₃ SCHAFFER, F. L.: Physical and chemical properties and infectivity of RNA from animal viruses. Cold Spr. Harb. Symp. quant. Biol. "Basic mechanisms in animal biology." 27, 89—99 (1962).

S₄₄ SCHILD, G. C., and C. H. STUART-HARRIS: Serological epidemiological studies with influenza A viruses. J. Hyg. (Lond.) 63, 479—490 (1965).

S₄₅ SCHILD, G. C., and R. N. P. SUTTON: Inhibition of influenza viruses in vitro and in vivo by 1-adamantanamine hydrochloride. Brit. J. exp. Path. 46, 263—273 (1965).

S₄₆ SCHLESINGER, R. W.: Production of incomplete influenza virus in mouse brain. Proc. Soc. Amer. Bacteriologists 71 (1950).

S₄₇ SCHLESINGER, R. W.: Incomplete growth cycle of influenza virus in mouse brain. Proc. Soc. exp. Biol. (N.Y.) 74, 541—548 (1950).

S₄₈ SCHLESINGER, R. W.: Immunological barriers to viral growth. Ann. N.Y. Acad. Sci. 54, 953—959 (1952).

S₄₉ SCHLESINGER, R. W.: The relation of functionally deficient forms of influenza virus to viral development. Cold Spr. Harb. Symp. quant. Biol. 18, 55—59 (1953).

S₅₀ SCHLESINGER, R. W.: Interference between animal viruses. "The viruses." Acad. Press N.Y. 3, 157—194 (1959).

S₅₁ SCHLESINGER, R. W., and H. V. KARR: Influenza virus and its mucoprotein substrate in the chorioallantoic membrane of the chick embryo. 1. Characterisation and quantitative assay of soluble substrate and studies on its relation to allantoic cells.

S₅₂ 2. Stepwise inactivation of substrate and its relation to the mode of viral multiplication. J. exp. Med. 103, 309—332 and 333—349 (1956).

S₅₃ SCHLESINGER, R. W., and G. H. WERNER: Quantitative and morphological aspects of the incomplete growth cycle of influenza virus in mouse brain. Fed. Proc. 11, 480 (1952).

S₅₄ SCHMIDT, G.: Experiences with Drescher's photometric method for the determination of viruses in the case of different myxoviruses (German). Zbl. Bakt. I. Abt. Orig. 176, 334—340 (1959).

$S_{55}$  SCHMIDT, B., and K. GROSSGEBAUER: Observations on mixtures of virus and red cells in sedimentation tubes (German). Z. Hyg. Infekt.-Kr. 144, 549—554 (1958).

$S_{56}$  SCHMIDT, B., and K. GROSSGEBAUER: Myxovirus haemagglutination experiments with erythrocytes from different species (German). Z. Hyg. Infekt.-Kr. 146, 26—47 (1959).

$S_{57}$  SCHMIDT, B., D. HARTMANN, and K. GROSSGEBAUER: The isolation of different haemagglutinin fractions from influenza virus with DEAE cellulose ion exchanger (German). Arch. ges. Virusforsch. 10, 361—367 (1960).

$S_{58}$  SCHMIDT, J. R. A., and A. F. RASMUSSEN: Inhibitory effect of cobaltous ions on multiplication of influenza virus. Proc. Soc. exp. Biol. (N.Y.) 81, 244—246 (1952).

$S_{59}$  SCHOLTENS, R. G., and J. H. STEELE: U.S. epizootic of equine influenza 1963. Epizootiology. Publ. Hlth Rep. (Wash.) 79, 393—398 (1964).

$S_{60}$  SCHOLTISSEK, C.: Biochemical studies on the cytopathic effect of influenza viruses. Abstr. 9th int. Congr. Microbiol. Moscow 373 (1966).

$S_{61}$  SCHOLTISSEK, C., R. DRZENIEK, and R. ROTT: Stepwise inactivation of an influenza virus by serial undiluted passages. Virology 30, 313—318 (1966).

$S_{62}$  SCHOLTISSEK, C., and R. ROTT: Influence of p-fluorophenylalanine on the production of viral ribonucleic acid and on the utilisability of viral protein during multiplication of fowl plague virus. Nature (Lond.) 191, 1023—1024 (1961).

$S_{63}$  SCHOLTISSEK, C., and R. ROTT: Studies on the growth of fowl plague virus (German). Z. Naturforsch. 16b, 109—115 (1961).

$S_{64}$  SCHOLTISSEK, C., and R. ROTT: Connection between the synthesis of ribonucleic acid and protein in the growth of a virus of the influenza group. (Fowl plague virus.) (German). Z. Naturforsch. 16b, 663—673 (1961).

$S_{65}$  SCHOLTISSEK, C., and R. ROTT: Synthesis of viral ribonucleic acid by a chemically inactivated influenza virus. Nature (Lond.) 199, 200—201 (1963).

$S_{66}$  SCHOLTISSEK, C., and R. ROTT: Binding of proflavine to deoxyribonucleic acid and its biological significance. Nature (Lond.) 204, 39 (1964).

$S_{67}$  SCHOLTISSEK, C., and R. ROTT: Behaviour of virus-specific activities in tissue cultures infected with myxoviruses after chemical changes of the viral ribonucleic acid. Virology 22, 169—176 (1964).

$S_{68}$  SCHOLTISSEK, C., R. ROTT, P. HAUSEN, H. HAUSEN, and W. SCHÄFER: Comparative studies of RNA and protein synthesis with a myxovirus and a small polyhedral virus. Cold Spr. Harb. Symp. quant. Biol. 27, 245—257 (1962).

$S_{69}$  SCHOYEN, R., A. HARBOE, and L. WANG: Influenza virus HI antibody combining with normal chick material and produced in hens immunised with chick-grown influenza virus. Acta path. microbiol. scand. 68, 103—107 (1966).

$S_{70}$  SCHULMAN, J. L., and E. D. KILBOURNE: Experimental transmission of influenza virus infection in mice. 1. The period of transmissibility.

$S_{71}$  2. Some factors affecting the incidence of transmitted infection. J. exp. Med. 118, 257—266 and 267—275 (1963).

$S_{72}$  SCHULMAN, J. L., and E. D. KILBOURNE: Induction of viral interference in mice by aerosols of inactivated influenza virus. Proc. Soc. exp. Biol. (N.Y.) 113, 431—435 (1963).

$S_{73}$  SCHULMAN, J. L., and E. D. KILBOURNE: Induction of partial specific heterotypic immunity in mice by a single infection with influenza A virus. J. Bact. 89, 170—174 (1965).

$S_{74}$  SEAL, J. R., and B. F. GUNDELFINGER: Efficacy of immunisation against Asian influenza in the United States Navy. U.S. armed Forces med. J. 9, 1720—1735 (1958).

$S_{75}$  SEGAL, L. S.: The efficacy of immunisation with live allantoic influenza vaccine in the epidemic experiment (Russian). Dissertation, Moscow (1952).

$S_{76}$  SELIMOV, M. A.: A comparative study of the haemagglutinating properties of mumps and influenza (Russian). Byull. éksp. Biol. med. 38, 60 (1954).

$S_{77}$ SELIVANOV, YA. M.: Peculiarities of the spread of influenza in Byelorussia in 1957 (Russian). Vop. Virus. 5, 51—53 (1958).

$S_{78}$ SELLERS, T. F., J. SCHULMAN, C. BOUVIER, R. McCUNE, and E. D. KILBOURNE: The influence of influenza virus infection on exogenous staphylococcal and endogenous murine bacterial infection of the bronchopulmonary tissues of mice. J. exp. Med. 114, 237—255 (1961).

$S_{79}$ SEMASHKO, S. A.: Treatment of virus influenza with atabrine (Russian). Annot. rabot. Inst. Virusol. AMN SSSR Leningrad 40 (1952).

$S_{80}$ SEN, H. G., G. W. KELLEY, N. R. UNDERDAHL, and G. A. YOUNG: Transmission of swine influenza virus by lungworm migration. J. exp. Med. 113, 517—520 (1961).

$S_{81}$ SEREDA, V. N.: Influence of influenza virus on the electrokinetic potential of erythrocytes (Russian). Problemy grippa i OKVDP. Tez. dokl. ob'yed. Sess. Inst. AMN SSSR, Leningrad 40 (1952).

$S_{82}$ SERY, V., J. RADOVSKY, V. TUMOVA, F. LOBKOWICZ, and S. CHOBOT: The use of horse anti-influenza serum in the prophylaxis of influenza (Czech). Čs. Epidem. 6, 309—317 (1957).

$S_{83}$ SETH, R., and S. L. KALRA: Viruses in respiratory infections in Delhi. Indian. J. med. Res. 53, 1109—1111 (1965).

$S_{84}$ SETO, J. T.: Sialidase (Neuraminidase) activity of standard and incomplete virus. Biochim. biophys. Acta (Amst.) 90, 420—422 (1964).

$S_{85}$ SETO, J. T.: Effect of sialidase antiserum on toxicity of influenza virus. Proc. Soc. exp. Biol. (N.Y.) 118, 1043—1046 (1965).

$S_{86}$ SETO, J. T., R. DRZENIEK, and R. ROTT: Isolation of a low molecular weight sialidase (Neuraminidase) from influenza virus. Biochim. biophys. Acta (Amst.) 113, 402—404 (1966).

$S_{87}$ SETO, J. T., B. J. HICKEY, and A. F. RASMUSSEN: Sialidase activity and related properties of influenza A2 viruses. Virology 9, 598—611 (1959).

$S_{88}$ SETO, J. T., and Y. HOKAMA: Disc electrophoresis analysis of sialidase from influenza virus. Ann. N.Y. Acad. Sci. 121, 640—650 (1964).

$S_{89}$ SETO, J. T., Y. NISHI, B. J. HICKEY, and A. F. RASMUSSEN: Relation of sialidase of influenza A viruses to virus particles as determined by electron microscopy. Virology 13, 13—18 (1961).

$S_{90}$ SETO, J. T., and R. ROTT: Functional significance of sialidase during influenza virus multiplication. Virology 30, 731—737 (1966).

$S_{91}$ SHADRIN, A. S.: Use of blood donors serum and of gamma globulin for influenza prophylaxis (Russian). "Problemy grippa i ostrykh respiratornykh zabolevanii" Tez. dokl. nauch. Sess. Inst. AMN SSSR, Moscow 138—139 (1959).

$S_{92}$ SHAKHMALIYEVA, Z. M., and K. M. ODINA: The course of infection in white mice infected with A and A2 influenza virus (Russian). "Problemy grippa i ostrykh respiratornykh zabolevanii" Tez. dokl. nauch. Sess. Inst. AMN SSSR, Moscow 178—179 (1959).

$S_{93}$ SHAKHNAZAROVA, N. G., and P. P. SAKHAROV: The neurotoxic influenza factor (Russian). Byull. éksp. Biol. Med. 22, 48 (1946).

$S_{94}$ SHARP, D. G., F. LANNI, and J. W. BEARD: The egg-white inhibitor of influenza virus haemagglutination. 2. Electron microscopy of the inhibitor. J. biol. Chem. 185, 681—688 (1950).

$S_{95}$ SHARP, D. G., A. R. TAYLOR, I. W. McLEAN, D. BEARD, J. W. BEARD, A. E. FELLER, and J. H. DINGLE: Isolation and characterisation of influenza virus B (Lee strain). J. Immunol. 48, 129—153 (1944).

$S_{96}$ SHAW, D. R., A. S. KENNEY, and J. STOKES: Serologic and immunologic studies relative to the viruses of human and swine influenza. Amer. J. med. Sci. 197, 247—253 (1939).

$S_{97}$ SHEAR, H. H., H. D. HEATH, D. T. IMAGAWA, M. H. JONES, and J. M. ADAMS: Neutralisation of teratogenic and lethal effects of influenza A virus in chick embryos. Proc. Soc. exp. Biol. (N.Y.) 89, 523—528 (1955).

S₉₈   SHEFFIELD, F. W., W. SMITH, and G. BELYAVIN: Purification of influenza virus by red-cell adsorption-elution. Brit. J. exp. Path. **35**, 214—222 (1954).

S₉₉   SHELOKOV, A., J. E. VOGEL, and L. CHI: Haemadsorption (adsorption-haemagglutination) test for viral agents in tissue culture with special reference to influenza. Proc. Soc. exp. Biol. (N.Y.) **97**, 802—809 (1958).

S₁₀₀  SHIMOJO, H., A. SUGIURA, T. AKAO, and C. ENOMOTO: Studies of a non-specific haemagglutination inhibitor of influenza A/Asian/57 virus. Bull. Inst. publ. Hlth Tokyo 7, 219—224 (1958).

S₁₀₁  SHIMSHELEVICH, S. B., and D. N. FADEYEVA: Treatment of influenza with ecmoline (Russian). Novisti med. **25**, 21—24 (1952).

S₁₀₂  SHIMSHELEVICH, S. B., and F. V. SMOLYANSKAYA: Outbreaks of influenza B in a children's group (Russian). Zh. Mikrobiol. (Mosk.) 1, 29 (1949).

S₁₀₃  SHIOTA, O.: Studies on the incomplete form of influenza virus. 1. Relations between growth curve of the virus and the yield of incomplete particles.

S₁₀₄  2. The morphological characteristics of so-called incomplete particles of the virus.

S₁₀₅  3. Nucleic acid content of complete and incomplete particles (Japanese). Virus **6**, 141—145, 193—206, and 207—213 (1956).

S₁₀₆  SHISKINA, O. I., and I. A. YURIKAS: Comparative evaluation of the methods of laboratory diagnosis in type B influenza (Russian). Tr. AMN SSSR, Moscow XXVIII, 122 (1953).

S₁₀₇  SHOPE, R. E.: Swine influenza. 1. Experimental transmission and pathology.

S₁₀₈  3. Filtration experiments and etiology. J. exp. Med. **54**, 349—360, and 373—380 (1931).

S₁₀₉  SHOPE, R. E.: The infection of mice with swine influenza. J. exp. Med. **62**, 561—572 (1935).

S₁₁₀  SHOPE, R. E.: The incidence of neutralising antibodies for swine influenza virus in the sera of human beings of different ages. J. exp. Med. **63**, 669—684 (1936).

S₁₁₁  SHOPE, R. E.: Immunological relationship between the swine and human influenza viruses in swine. J. exp. Med. **66**, 151—168 (1937).

S₁₁₂  SHOPE, R. E.: Serological evidence for the occurrence of infection with human influenza viruses in swine. J. exp. Med. **67**, 739—748 (1938).

S₁₁₃  SHOPE, R. E.: Serological studies of swine influenza viruses. J. exp. Med. **69**, 847—856 (1939).

S₁₁₄  SHOPE, R. E.: An intermediate host for the swine influenza virus. Science **89**, 441—442 (1939).

S₁₁₅  SHOPE, R. E.: The swine lungworm as a reservoir and intermediate host for swine influenza virus. 1. The presence of swine influenza virus in healthy and susceptible pigs.

S₁₁₆  2. The transmission of swine influenza virus by the swine lungworm.

S₁₁₇  3. Facts influencing transmission of the virus and the provocation of influenza.

S₁₁₈  4. The demonstration of masked swine influenza virus in lungworm larvae and swine under natural conditions. J. exp. Med. **74**, 41, and 49 (1941); and **77**, 111, and 127 (1943).

S₁₁₉  SHOPE, R. E.: The provocation of masked swine influenza virus by infection with human influenza virus. T. Diergeneesk. **76**, 414—420 (1951).

S₁₂₀  SHOPE, R. E.: An antiviral substance from penicillium funiculosum. 1. Effect upon infection in mice with swine influenza virus and Columbia SK encephalomyelitis virus.

S₁₂₁  2. Effect of Helenine upon infection of mice with Semliki forest virus.

S₁₂₂  3. General properties and characteristics of Helenine.

S₁₂₃  4. Inquiry into the mechanism by which Helenine exerts its antiviral effect. J. exp. Med. **97**, 601—625, 627—638, and 636—650 (1953), and J. exp. Med. **123**, 213—227 (1966).

S₁₂₄  SHOPE, R. E.: The swine lungworm as a reservoir and intermediate host for swine influenza virus. 5. Provocation of swine influenza by exposure of prepared swine to adverse weather. J. exp. Med. **102**, 567—572 (1955).

S₁₂₅  SHOPE, R. E., and T. FRANCIS: The susceptibility of swine to the virus of human influenza. J. exp. Med. 64, 791 (1936).

S₁₂₆  SHROIT, I. G., and V. S. DAKOVA: Data on the epidemiology, laboratory diagnosis and prophylaxis of influenza in the Moldavian SSSR (Russian). "Problemy grippa i ostrykh respiratornykh zabolevanii" Tez. dokl. nauch. Sess. Inst. Inst. AMN SSSR, Moscow 103—104 (1959).

S₁₂₇  SHUBLADZE, A. K.: The susceptibility of animals to influenza (Russian). Dokl. Akad. Nauk. SSSR 32, 674 (1941).

S₁₂₈  SHUBLADZE, A. K., and V. P. SOLOV'YEV: Study of the phenomenon of erythrocyte agglutination by the influenza virus (Russian). Zh. Mikrobiol. (Mosk.) 10—11, (17) 21 (1943).

S₁₂₉  SIDORENKO, E. V.: The importance of the complement fixation reaction in the diagnosis of fatal influenza cases (Russian). Gripp. Sborn. nauch. rabot. Kiev 3, 58—64 (1959).

S₁₃₀  SIEGERT, R., and P. BRAUNE: The pyrogens of myxoviruses. 1. Induction of hypothermia and its tolerance.

S₁₃₁  2. Resistance of influenza A pyrogens to heat, ultraviolet, and chemical treatment. Virology 24, 209—217, and 218—224 (1964).

S₁₃₂  SIGEL, M. M.: Differences in the haemagglutinating and antigenic properties of strains of influenza virus isolated in one outbreak. J. Immunol. 62, 81—95 (1949).

S₁₃₃  SIGEL, M. M., E. G. ALLEN, D. J. WILLIAMS, and A. J. GIRARDI: Immunologic response of hamsters to influenza virus strains. Proc. Soc. exp. Biol. (N.Y.) 72, 507—510 (1949).

S₁₃₄  SIGEL, M. M., A. W. KITTS, A. B. LIGHT, and W. HENLE: The recurrence of influenza A-prime in a boarding school after 2 years. J. Immunol. 64, 33—38 (1950).

S₁₃₅  SIGEL, M. M., F. W. SHAFFER, M. W. KIRBER, A. B. LIGHT, and W. HENLE: Influenza A in a vaccinated population. J. Amer. med. Assoc. 136, 437—441 (1948).

S₁₃₆  SIGURDSSON, B., and O. BJARNASON: Immunological studies on an epidemic of influenza in Iceland. Amer. J. publ. Hlth 36, 130—134 (1946).

S₁₃₇  SILVER, R. K., G. BRAUN, F. ZILLIKEN, G. H. WERNER, and P. GYORGI: Factors in human milk interfering with influenza virus activities. Science 123, 932—933 (1956).

S₁₃₈  SIMPSON, R. W.: Genetic studies with influenza A virus. Ciba Found. Symp. "Cellular biology of myxovirus infections" 187—217 (1964).

S₁₃₉  SIMPSON, R. W., and R. E. HAUSER: Structures associated with influenza virus suspensions treated with phospholipase C. Virology 27, 642—646 (1965).

S₁₄₀  SIMPSON, R. W., and R. E. HAUSER: Influence of lipids on the viral phenotype. 1. Interaction of myxoviruses and their lipid constituents with phospholipases. Virology 30, 684—697 (1966).

S₁₄₁  SIMPSON, R. W., and G. K. HIRST: Genetic recombination among influenza viruses. 1. Cross reactivation of plaque-forming capacity as a method for selecting recombinants from the progeny of crosses between influenza A strains. Virology 15, 436—451 (1961).

S₁₄₂  SIMPSON, G. I., and W. J. MOGABGAB: Reactivity of influenza A and B viruses propagated in cultures of human embryo tissues and in the chick embryo. J. Immunol. 78, 456—464 (1957).

S₁₄₃  SINIAK, K. M., S. D. KLYUSKO, M. B. MAKSIMOVICH, N. G. EZHOVA, A. A. OIKHBERG, and I. A. VINOGRADOV: Efficacy of blood donors anti-influenza serum in the prophylaxis of influenza in children's homes and schools (Russian). "Problemy grippa i ostrykh respiratornykh zabolevanii" Tez. dokl. nauch. Sess. Inst. AMN SSSR, Moscow 141—142 (1959).

S₁₄₄  SINCOVICS, J.: Virus neutralisation experiments with lymphoid cell and lymph node extracts. Acta microbiol. Acad. Sci. hung. 2, 385—400 (1955).

S₁₄₅  SINZ, W.: Experimental observations on the interference of influenza viruses (German). Zbl. Bakt. I. Abt. Orig. 197, 165—195 (1965).

S_146    SIVERSKAYA, P. I., and A. V. CHUDNOVA: Antibodies to influenza virus in the blood of donors and in sera produced industrially (Russian). "Problemy grippa i ostrykh respiratornykh zabolevanii" Tez. dokl. nauch. Sess. Inst. AMN SSSR, Moscow 157—159 (1959).

S_147    SJOBERG, B.: Flumidin, a new chemotherapeutic agent for the prophylaxis and suppression of influenza. Svenska Läk.-Tidu. 42, (1959).

S_148    SJOBERG, B.: Experiments on prophylaxis and suppression of epidemic influenza with N', N'-anhydrobis-(B-hydroxyethyl) biguanide-HCl (ABOB). A double-blind study. Antibiot. et Chemother. (Basel) 10, (1960).

S_149    SKLYANSKAYA, E., and O. PETERSON: Study of the nature of early increase in RNA synthesis in the mouse lung infected *in vivo* with influenza virus. Abstr. 9th int. Congr. Microbiol., Moscow 506 (1966).

S_150    SKURSKA, Z., H. MAKOWER, A. SYPULOWA, M. LOBODZINSKA, and T. KIDAN-KIEWICZ: Early and late strains of influenza virus in tissue cultures of the chicken embryo (Polish). Arch. Immun. ter. Dosw. 8, 101—110 (1960).

S_151    SLEPUSHKIN, A. N.: Comparative studies of the biological properties of vaccine strains of the influenza virus (Russian). Dissertation, Moscow (1954).

S_152    SLEPUSHKIN, A. N.: The efficacy of prophylactic measures during influenza outbreaks in 1957 (Russian). Vop. Virus. 5, 103—109 (1958).

S_153    SLEPUSHKIN, A. N.: Some epidemiological and etiological peculiarities of the 1957 influenza epidemic (Russian). Vop. Virus. 5, 33—36 (1958).

S_154    SLEPUSHKIN, A. N.: The effect of a previous attack of A1 influenza on susceptibility to A2 virus during the 1957 outbreak. Bull. Wld Hlth Org. 20, 297—301 (1959).

S_155    SLEPUSHKIN, A. N.: Data on the efficiency of live influenza vaccine. Abstr. 9th int. Congr. Microbiol., Moscow 379—380 (1966).

S_156    SLOBODIANIUK, M. I.: Rhinocytoscopic investigations in A/Asia/57 influenza (Russian). Aziatiksii Gripp (Sborn nauch. rabot.) Kiev 201—205 (1958).

S_157    SMADEL, J. E.: Influenza vaccine. Publ. Hlth Rep. (Wash.) 73, 2 (1958).

S_158    SMART, K. M., and E. D. KILBOURNE: The influence of cortisone on experimental viral infection. 6. Inhibition by hydrocortisone of interferon synthesis in the chick embryo.

S_159    7. Kinetics of interferon formation and its inhibition with hydrocortisone in relation to viral strain and virulence. J. exp. Med. 123, 299—307, and 309—325 (1966).

S_160    SMITH, D. H., and H. R. MORGAN: A cytopathogenic effect of influenza virus (PR8 strain) for mouse fibroblastic (L) cells *in vitro*. J. Immunol. 85, 180—189 (1960).

S_161    SMITH, W.: Cultivation of the virus of influenza. Brit. J. exp. Path. 16, 508—512 (1935).

S_162    SMITH, W.: The complement fixation reaction in influenza. Lancet 1256—1259 (1936).

S_163    SMITH, W.: Host and tissue specificity in infective disease. Proc. roy. Soc. Med. 42, 11—18 (1949).

S_164    SMITH, W.: The structural and functional plasticity of influenza virus. Lancet 885 (1952).

S_165    SMITH, W.: Progress in viral immunology. Brit. med. Bull. 9, 176—179 (1953).

S_166    SMITH, W., and C. H. ANDREWES: Serological races of influenza virus. Brit. J. exp. Path. 19, 293—314 (1938).

S_167    SMITH, W., C. H. ANDREWES, and P. P. LAIDLAW: A virus obtained from influenza patients. Lancet 66—68 (1933).

S_168    SMITH, W., C. H. ANDREWES, and P. P. LAIDLAW: Influenza. Experiments on the immunisation of ferrets and mice. Brit. J. exp. Path. 16, 291—302 (1935).

S_169    SMITH, W., G. BELYAVIN, and F. W. SHEFFIELD: A host-protein component of influenza viruses. Nature (Lond.) 172, 669 (1953).

S_170    SMITH, W., G. BELYAVIN, and F. W. SHEFFIELD: The host-tissue component of influenza viruses. Proc. roy. Soc. B. 143, 504—522 (1955).

S₁₇₁  SMITH, W., and A. COHEN: The enzymic activity of influenza viruses. Brit. J. exp. Path. **37**, 612—624 (1956).

S₁₇₂  SMITH, W., H. COHEN, G. BELYAVIN, and J. C. N. WESTWOOD: The effects of ethyl ether on some biological properties of influenza virus. Brit. J. exp. Path. **34**, 512—524 (1953).

S₁₇₃  SMITH, W., and C. H. STUART-HARRIS: Influenza infection of man from the ferret. Lancet 121—123 (1936).

S₁₇₄  SMITH, W., and M. A. WESTWOOD: Factors involved in influenza haemagglutination reactions. Brit. J. exp. Path. **30**, 48—61 (1949).

S₁₇₅  SMITH, W., and J. C. N. WESTWOOD: Influenza virus haemagglutination. The mechanism of the Francis phenomenon. Brit. J. exp. Path. **31**, 725—738 (1950).

S₁₇₆  SMITH, W., J. C. N. WESTWOOD, and G. BELYAVIN: Influenza. A study of four virus strains isolated in 1951. Lancet 1189 (1951).

S₁₇₇  SMITH, W., M. A. WESTWOOD, J. C. N. WESTWOOD, and G. BELYAVIN: Spontaneous mutation of influenza virus A during routine egg passage. Brit. J. exp. Path. **32**, 422—432 (1951).

S₁₇₈  SMORODINTSEV, A. A.: Problem of the etiology and specific prophylaxis of influenza in the light of the latest data (Russian). Arkh. biol. nauk. **52**, 3—18 (1938).

S₁₇₉  SMORODINTSEV, A. A.: Etiology and prophylaxis of influenza and seasonal catarrhs (Russian). Zh. Mikrobiol. (Mosk.) 10/11, 4 (1943).

S₁₈₀  SMORODINTSEV, A. A.: The latest in the epidemiology and specific prophylaxis of influenza (Russian). Vop. Virus. **1**, 102 (1948).

S₁₈₁  SMORODINTSEV, A. A.: Methods of laboratory diagnosis of influenza (Russian). Zh. Mikrobiol. (Mosk.) **12**, 31 (1948).

S₁₈₂  SMORODINTSEV, A. A.: Problems of current importance in the laboratory diagnosis, epidemiology, and specific prophylaxis of influenza (Russian). Vestn. AMN SSSR 3, 16 (1951).

S₁₈₃  SMORODINTSEV, A. A.: Mechanism of virus destruction in the bodies of resistant animals (Russian). "Sovremennyye voprosy meditinskoy nauchi." Moscow (1951).

S₁₈₄  SMORODINTSEV, A. A.: New system of performing the complement fixation reaction for the rapid diagnosis of influenza and other virus infections (Russian). Probl. grippa i OKVDP Tez. dokl. ob'yed. Sess. AMN SSSR, Moscow 27 (1952).

S₁₈₅  SMORODINTSEV, A. A.: Results of specific influenza prophylaxis with live vaccine (Russian). Gripp. Tr. AMN SSSR, Moscow XXVIII, 217 (1953).

S₁₈₆  SMORODINTSEV, A. A.: Problems of controlling influenza in the light of the latest data (Russian). Zh. Mikrobiol. (Mosk.) **9**, 22 (1954).

S₁₈₇  SMORODINTSEV, A. A.: Morphological studies on reactive processes in viral influenza in the respiratory tract of white mice exposed to the effect of roentgen rays (Russian). Vop. Virus. **2**, 290—296 (1957).

S₁₈₈  SMORODINTSEV, A. A.: Specific prophylaxis and treatment of influenza. Proc. 3rd int. Meet. Biol. Standard. Opatija 463—480 (1957).

S₁₈₉  SMORODINTSEV, A. A.: The effect of X ray irradiation on the course of experimental influenza infection in white mice and rats. Acta virol. **1**, 145—155 (1957).

S₁₉₀  SMORODINTSEV, A. A.: Results and problems of the specific prevention and treatment of influenza (Russian). Vest. Akad. med. nauk. SSSR **13**, 20—30 (1958).

S₁₉₁  SMORODINTSEV, A. A.: Results of the study of etiology and laboratory diagnosis of type A2 influenza during the 1957 and 1959 epidemics (Russian). "Problemy grippa i ostrykh respiratornykh zabolevanii" Tez. dokl. nauch. Sess. Inst. AMN SSSR, Moscow 8—12 (1959).

S₁₉₂  SMORODINTSEV, A. A.: Specific prophylaxis of influenza with a live vaccine (Russian). "Problemy grippa i ostrykh respiratornykh zabolevanii" Tez. dokl. nauch. Sess. Inst. AMN SSSR, Moscow 82—84 (1959).

$S_{193}$ SMORODINTSEV, A. A.: Cellular and humoral factors of defence against virus infection. Ciba Found. Symp. "Cellular biology of myxovirus infections" 317—343 (1964).

$S_{194}$ SMORODINTSEV, A. A., G. L. ALEXANDROVA, T. J. LOUSANINA, M. A. MOROZENKO, and A. A. SELIVANOV: Virological and serological characteristics of the 1957 influenza pandemic. Ann. Inst. Pasteur 97, 385—397 (1959).

$S_{195}$ SMORODINTSEV, A. A., G. I. ALEXANDROVA, B. A. MIKUTSKAYA, R. A. PLESHANOVA, E. A. SIROTENKO, T. E. BRAUN, and E. W. KALLAS: Results and perspectives of live influenza vaccine for the active immunisation of children. Abstr. 9th int. Congr. Microbiol., Moscow 380—381 (1966).

$S_{196}$ SMORODINTSEV, A. A., and O. M. CHALKINA: Active immunisation against influenza with vaccine from live attenuated virus (Russian). Sov. vrach. sbornik. 15, 16 (1949).

$S_{197}$ SMORODINTSEV, A. A., and O. M. CHALKINA: Material on the specific prophylaxis of influenza (Russian). Probl. grippa i ostrykh katar. verkhn. dykh. put. Tr. ob'yed. Sess. Inst. AMN SSSR, Moscow 59 (1952).

$S_{198}$ SMORODINTSEV, A. A., and O. M. CHALKINA: Results of specific influenza prophylaxis with live vaccine (Russian). "Virusnyye Infektsii" Tr. Leningrad Inst. Epidem. Mikrobiol. i gig. imeni Pastera i Inst. ekspor. med. AMN SSSR, Leningrad 142—160 (1953).

$S_{199}$ SMORODINTSEV, A. A., O. M. CHALKINA, and M. ANSHELES: Experience in the use of live vaccine against influenza (Russian). Zh. Mikrobiol. (Mosk.) 10, 52 (1953).

$S_{200}$ SMORODINTSEV, A. A., A. I. DROBYSHEVSKAYA, S. M. OSTROVSKAYA, and O. I. SHISHKINA: The etiology of influenzal diseases (Russian). Sov. vrach. zhurn. 16, 1455 (1936), also (English) Lancet 2, 1381 (1936).

$S_{201}$ SMORODINTSEV, A. A., A. I. DROBYSHEVSKAYA, S. M. OSTROVSKAYA, and O. I. SHISHKINA: Experimental material on the etiology of epidemic influenza (Russian). Sov. vrach. zhurn. 6, 403—419 (1937).

$S_{202}$ SMORODINTSEV, A. A., A. I. DROBYSHEVSKAYA, and O. I. SHISHKINA: On the aetiology of the 1936 influenza epidemic in Leningrad. Lancet 1383—1385 (1936).

$S_{203}$ SMORODINTSEV, A. A., A. G. GULAMOV, and O. M. CHALKINA: Experience in the specific prophylaxis of experimental influenza with inhalation of influenza antiserum (Russian). Sov. vrach. zhurn. 4, 255 (1940).

$S_{204}$ SMORODINTSEV, A. A., and O. I. SHISHKINA: Experimental analysis of the prophylactic and therapeutic effect of influenza immune serum on virus infection of mice with different methods of application of it (Russian). Arkh. biol. nauk. 52, 132—154 (1938).

$S_{205}$ SMORODINTSEV, A. A., and O. I. SHISHKINA: Natural and acquired immunity to influenza. Amer. Rev. soviet Med. 3, 400—413 (1946).

$S_{206}$ SMORODINTSEV, A. A., and O. I. SHISHKINA: Mechanism of acquired and natural immunity to influenza infection (Russian). Arkh. biol. nauk. 59, 20 (1940), also Zh. Mikrobiol. (Mosk.) 43, 49 (1945), Zh. Mikrobiol. (Mosk.) 26 (1946), Zh. Mikrobiol. (Mosk.) 2 (1947).

$S_{207}$ SMORODINTSEV, A. A., and O. I. SHISHKINA: The effect of normal laboratory animal sera on the influenza virus. Zh. Microbiol. (Mosk.) 6, 16 (1951).

$S_{208}$ SMORODINTSEV, A. A., and O. I. SHISHKINA: The effect of tissue enzymes on the influenza virus (Russian). Gripp. Tr. AMN SSSR XXXVIII, 5 (1953).

$S_{209}$ SMORODINTSEV, A. A., and A. A. YABROV: The mechanism of enhanced activity of anti-influenza virus neutralising antisera on their interaction with native serum from normal animals. Acta virol. 7, 193—208 (1963).

$S_{210}$ SMORODINTSEV, A. A., and V. M. ZHDANOV: Results and immediate tasks of the study of live influenza vaccine (Russian). Vop. Virus. 2, 67—72 (1957).

$S_{211}$ SMORODINTSEVA, O. A.: The effect of ultraviolet rays on the influenza virus (Russian). Zh. Mikrobiol. (Mosk.) 10—11, 17 (1943).

$S_{212}$ SMORODINTSEVA, O. A.: The effect of ultraviolet rays and ozone on the influenza virus (Russian). Gripp. Tr. AMN SSSR, Moscow XXVIII, 49 (1953).

$S_{213}$ SOHIER, R.: Some conceptions on the epidemiology and prophylaxis of virus infections (French). Rev. Hyg. Méd. soc. 2, 699—748 (1954).

S$_{214}$ Sokol, F., and S. Schramek: An improved method for the isolation of ribonucleic acid from myxoviruses. Acta virol. 8, 193—199 (1964).

S$_{215}$ Sokol, F., and J. Szurman: Isolation of ribonucleic acid from purified influenza virus by phenol extraction. Acta virol. 3, 175—180 (1959).

S$_{216}$ Sokolov, M. I.: Concentration and purification of influenza virus by adsorption onto erythrocytes (Russian). Zh. Mikrobiol. (Mosk.) 1, 41 (1949).

S$_{217}$ Sokolov, M. I.: The efficacy of active immunisation against influenza (Russian). Vrach. delo. 10, 929 (1950).

S$_{218}$ Sokolov, M. I.: Live dry influenza vaccine (Russian). Zh. MEI 10, 57 (1953).

S$_{219}$ Sokolov, M. I.: Active influenza immunisation (Russian). Medgiz., Moscow (1954).

S$_{220}$ Sokolov, M. I.: Protective vaccination against influenza (Russian). Vrach. delo. 10, 893 (1954).

S$_{221}$ Sokolov, M. I., and K. S. Kulikova: Immunogenic properties of a concentrated vaccine against influenza (Russian). Medgiz., Moscow (1954).

S$_{222}$ Sokolov, M. I., K. S. Kulikova, S. Ya. Kholeva, and N. B. Azadova: Factors facilitating preservation of vitality of influenza viruses during drying and prolonged preservation (Russian). Vop. Virus. 3, 239—241 (1958).

S$_{223}$ Sokolov, M. I., K. S. Kulikova, and A. H. Samvelova: Intranasal immunisation with live influenza vaccine (Russian). Zh. MEI 7, 28 (1951).

S$_{224}$ Sokolov, M. I., T. M. Lozinskaya, N. P. Obrosova-Serova, and R. Y. Podchernyeva: Experimental data on inherited variation of influenza viruses. Abstr. 9th int. Congr. Microbiol., Moscow 374 (1966).

S$_{225}$ Sokolov, M. I., S. B. Shimshelevich, and K. S. Kulikova: Specific prophylaxis of influenza (Russian). Zh. Mikrobiol. (Mosk.) 6, 45 (1951).

S$_{226}$ Sokolov, M. I., and P. Tu: Experimental studies on the method of obtaining live influenzal vaccine. I. Biological and immunological variability of influenza virus A 57 during its adaptation to chick embryo and mice (Russian). Zh. Mikrol. Moskva. 30, 11—16 (1959).

S$_{227}$ Sokolova, N. N., and I. N. Gaylonskaya: Comparative study of survival of liquid and powder for vaccines in human tissues (Russian). Vop. Virus. 4, 367 (1953).

S$_{228}$ Sokovikh, L. I., and A. S. Gorbunova: Sensibilizin; a peculiar influenza virus antigen in passive haemagglutination test. Abstr. 9th int. Congr. Microbiol., Moscow 389 (1966).

S$_{229}$ Solov'yev, V. D.: The problem of local antibody production in lung tissue (Russian). Annot. rabot. Inst. virusol. AMN SSSR Za, 1949, Moscow 13 (1950).

S$_{230}$ Solov'yev, V. D.: Current problems of the epidemiology and inoculation prophylaxis of influenza (Russian). Vo.-med. Zh. 3, 18 (1954).

S$_{231}$ Solov'yev, V. D.: The current state of the problems of the etiology and epidemiology of influenza (Russian). Zh. Mikrobiol. (Mosk.) 8, 3 (1956).

S$_{232}$ Solov'yev, V. D., and A. K. Alekseyeva: Comparison of the sensitivity of SOT's and monkey kidney epithelial cells to type A2 influenza virus (Russian). Vop. Virus. 4, 136—142 (1959).

S$_{233}$ Solov'yev, V. D., and A. K. Alekseyeva: The method of detection of influenza virus in tissue cultures (Russian). Zh. Mikrol. Moskva. 30, 16—20 (1959).

S$_{234}$ Solov'yev, V. D., and N. R. Gutman: Variability of influenza A prime virus in the course of adaptation to chick embryos (Russian). Zh. Mikrobiol. (Mosk.) 10, 65 (1953).

S$_{235}$ Solov'yev, V. D., and S. S. Marennikova: Variability of the influenza virus (Russian). Byull. éksp. Biol. Med. 12, 434 (1949), 9, 59 (1952), 12, 58—64 (1953).

S$_{236}$ Solov'yev, V. D., and S. S. Marennikova: Adaptation of the influenza virus to the body of a new host (Russian). Annot. rabot. Inst. Virusol. AMN SSSR za, 1949, Moscow (1950).

$S_{237}$ SOLOV'YEV, V. D., and S. S. MARENNIKOVA: Variation of the influenza virus under natural conditions and in experiment (Russian). Problemy grippa i OKVDP. Tez. dokl. ob'yed. Sess. Inst. AMN SSSR, Moscow (1952).

$S_{238}$ SOLOV'YEV, V. D., S. S. MARENNIKOVA, and N. R. GUTMAN: Variability of the influenza virus under natural conditions (Type A prime influenza virus) (Russian). Zh. Mikrobiol. (Mosk.) 8, 52 (1952).

$S_{239}$ SOLOV'YEV, V. D., and L. M. MENTKEVICH: The effect of colchicine on viral interference and interferon formation. Acta virol. 9, 308—312 (1965).

$S_{240}$ SOLOV'YEV, V. D., V. V. RITOVA, and D. I. DUDKINA: Significance of passive immunity to influenza in infancy (Russian). Pediatr. 5, 28 (1949).

$S_{241}$ SOLOV'YEV, V. D., A. K. SHUBLADZE, and N. MEL'TSER: Study of the phenomenon of erythrocyte agglutination by the influenza virus (Russian). Zh. Mikrobiol. (Mosk.) 3, 22 (1943).

$S_{242}$ SOLOV'YEV, V. D., and M. I. SOKOLOV: Immunological effectiveness of influenza vaccine (Russian). Zh. Mikrobiol. (Mosk.) 12, 77 (1948).

$S_{243}$ SOLOV'YEV, V. D., and L. YA. ZAKSTELSKAYA: Serological characterisation of virus influenza B (Russian). Zh. Mikrobiol. (Mosk.) 9, 20 (1950).

$S_{244}$ SOLOV'YEV, V. D., L. YA. ZAKSTELSKAYA, and B. N. TARASOV: Significance of the toxic factor in the pathogenesis of influenzal infection (Russian). Annot. rabot. Inst. Virusol. AMN SSSR za, 1949 g, Moscow 13 (1950).

$S_{245}$ SOMMA, R. E., H. C. TOSI, A. VALTONA, R. SALSAMUNDI, and E. GIANBRUNO: Rev. urug. Pat. clin. 1, 43 (1963).

$S_{246}$ SOVINOVA, O., and J. LUDVIK: Electron microscopic study of the influenza virus A/Equi/Prague/56. Acta virol. 3, 59—60 (1959).

$S_{247}$ SOVINOVA, O., B. TUMOVA, F. POUTSKA, and J. NEMEC: Isolation of a virus causing respiratory disease in horses. Acta virol. 2, 52—61 (1958).

$S_{248}$ SPENCE, L.: Kaolin treatment of sera for removal of non-specific inhibitors to Asian strains of influenza virus. Proc. Soc. exp. Biol. (N.Y.) 103, 425—427 (1960).

$S_{249}$ SPRECHER-GOLDBERGER, S.: The existence of two different metabolic patterns for the synthesis of RNA viruses. Arch. ges. Virusforsch. 18, 198—209 (1966).

$S_{250}$ SPROSSIG, M., and H. MEICKE: Inactivation of influenza viruses by α-dicarbonyl compounds. Acta virol. 7, 472—474 (1963).

$S_{251}$ SPROSSIG, M., and H. URBACH: A serologic study of influenza virus strains of the Asian type (German). Acta biol. med. germ. 1, 641—651 (1958).

$S_{252}$ STAIGER, H. R.: Plaque-type recombination in fowl plague virus. Virology 22, 419—422 (1964).

$S_{253}$ STAKHANOVA, V. M.: Different sensitivity of human and chicken erythrocytes to the action of non-specific inhibitors of haemagglutination (Russian). Vop. Virus. 5, 77—80 (1958).

$S_{254}$ STAKHANOVA, V. M.: The antigenic structure of influenza-like respiratory viruses of horses, swine and ducks (Russian). "Problemy grippa i ostrykh. respiratornykh zabolovanii" Tez. dokl. nauch. Sess. Inst. AMN SSSR, Moscow 48—49 (1959).

$S_{255}$ STAKHANOVA, V. M., A. N. LOZHKINA, and A. S. GORBUNOVA: The effect of potassium periodate on thermostable serum inhibitors to A2 virus (Russian). "Problemy grippa i ostrykh. respiratornykh zabolevanii" Tez. dokl. nauch. Sess. Inst. AMN SSSR, Moscow 55 (1959).

$S_{256}$ STAKHANOVA, V. M., E. M. ZHANTIEVA, and V. V. VSAHKOVA: Mechanism of myxovirus genetic interaction. Abstr. 9th int. Congr. Microbiol., Moscow 477—478 (1966).

$S_{257}$ STANLEY, E. D., R. E. MULDOON, L. W. AKERS, and G. G. JACKSON: Evaluation of antiviral drugs: The effect of amantadine on influenza in volunteers. Ann. N.Y. Acad. Sci. 130, 44—51 (1965).

$S_{258}$ STANLEY, W. M.: An evaluation of methods for the concentration and purification of influenza virus. J. exp. Med. 79, 255—266 (1944).

$S_{259}$ STANLEY, W. M.: The size of the influenza virus. J. exp. Med. 79, 267—283 (1944).

$S_{260}$ STANLEY, W. M.: The preparation and properties of influenza virus vaccines concentrated and purified by differential centrifugation. J. exp. Med. **81,** 193—218 (1945).

$S_{261}$ STANLEY, W. M.: The precipitation of purified concentrated influenza virus and vaccine by calcium phosphate. Science **101,** 332—335 (1945).

$S_{262}$ STENKULA, F.: Results of treatment of influenza type B with N, N-anhydrobis-(B-oxyethyl) biguanide-HCl (ABOB) (German). Fortschr. Med. **11,** 318 (1959).

$S_{263}$ STERN, H., and K. C. TIPPETT: Primary isolation of influenza viruses at 33°C. Lancet 1301—1302 (1963).

$S_{264}$ STEWART, R. B., and P. H. FRICKEY: Studies on cytotropism in animal viruses. 3. Growth of influenza virus in epithelial-like and fibroblastic cells derived from chick embryo lung. J. Bact. **92,** 972—977 (1966).

$S_{265}$ STEWART, R. B., and H. R. MORGAN: Studies on cytotropism in animal viruses. 1. Growth of influenza virus in lung cells derived from hatched chicks and chick embryos. J. Immunol. **82,** 264—273 (1959).

$S_{266}$ 2. A comparison of the growth of four influenza virus strains in vivo and in vitro in hatched chick and chick embryo lung tissue. J. Immunol. **85,** 465—468 (1960).

$S_{267}$ STEWART, F. S., and J. J. QUILLIGAN: A method for estimating the capacity of influenza virus to effect antigenic modification of red blood cells. J. Immunol. **66,** 411—419 (1951).

$S_{268}$ STOCK, C. C., and T. FRANCIS: Additional studies of the inactivation of the virus of epidemic influenza by soaps. J. Immunol. **47,** 303—308 (1943).

$S_{269}$ STOKES, J., and D. R. SHAW: Production of passive immunity against influenza virus by introducing immune serum into the respiratory tract. Amer. J. Dis. Child. **58,** 653 (1939).

$S_{270}$ STONE, J. D.: Enzymic modification of the reaction between influenza virus and susceptible tissue cells. Nature (Lond.) **159,** 780—781 (1947).

$S_{271}$ STONE, J. D.: Prevention of virus infection with enzyme of V. cholerae. 1. Studies with viruses of the mumps-influenza group in chick embryos.

$S_{272}$ 2. Studies with influenza virus in mice. Aust. J. exp. Biol. med. Sci. **26,** 49—64, and 287—298 (1948).

$S_{273}$ STONE, J. D.: Tryptic inactivation of the receptor destroying enzyme of V. cholerae and of the enzymic activity of influenza virus. Aust. J. exp. Biol. med. Sci. **27,** 229—244 (1949).

$S_{274}$ STONE, J. D.: Inhibition of influenza virus haemagglutination by mucoids. 1. Conversion of virus to indicator for inhibitor.

$S_{275}$ 2. Differential behaviour of mucoid inhibitors with indicator viruses. Aust. J. exp. Biol. med. Sci. **27,** 337—352, and 557—567 (1949).

$S_{276}$ STONE, J. D.: Adsorptive and enzymic behaviour of influenza virus during the O-D change. Brit. J. exp. Path. **32,** 367—376 (1951).

$S_{277}$ STONE, J. D., and G. L. ADA: Additive effect of viruses of the influenza group and the receptor destroying enzyme of V Cholerae. Brit. J. exp. Path. **31,** 275—284 (1950).

$S_{278}$ STONE, J. D., and G. L. ADA: Electrophoretic studies of virus-red cell interaction; relationship between agglutinability and electrophoretic mobility. Brit. J. exp. Path. **33,** 428—439 (1952).

$S_{279}$ STONE, J. D., and F. M. BURNET: The action of halogens on influenza virus with special reference to the action of iodine vapour on virus mists. Aust. J. exp. Biol. med. Sci. **23,** 205—212 (1945).

$S_{280}$ STRANDLI, O. K., K. MORTENSSEN-EGNUND, and A. HARBOE: Purification of the normal allantoic fluid antigen which reacts with influenza virus HI antibody to host material. Acta path. microbiol. scand. **60,** 265—270 (1964).

$S_{281}$ STRAUB, M.: The microscopical changes in the lungs of mice infected with influenza virus. J. Path. Bact. **45,** 75—98 (1937).

S₂₈₂ STRAUB, M.: The histology of catarrhal influenzal bronchitis and collapse of the lung in mice infected with influenza virus. J. Path. Bact. 50, 31—36 (1940).

S₂₈₃ STRAUB, M., and J. MULDER: Epithelial lesions in the respiratory tract in human influenzal pneumonia. J. Path. Bact. 40, 429—434 (1948).

S₂₈₄ STUART-HARRIS, C. H.: Transmission of influenza virus to hedgehogs. Brit. J. exp. Path. 17, 324—328 (1936).

S₂₈₅ STUART-HARRIS, C. H.: Influenza virus infection of rats and guinea pigs. Brit. J. exp. Path. 18, 485—492 (1937).

S₂₈₆ STUART-HARRIS, C. H.: A neurotropic strain of human influenza virus. Lancet 497—499 (1939).

S₂₈₇ STUART-HARRIS, C. H.: Observations on the agglutination of fowl red cells by influenza viruses. Brit. J. exp. Path. 24, 33—40 (1943).

S₂₈₈ STUART-HARRIS, C. H.: Influenza epidemics and the influenza viruses. Lectures 1 and 2. Brit. med. J. 209—216, and 251—257 (1945).

S₂₈₉ STUART-HARRIS, C. H.: Twenty years of influenza epidemics. Amer. Rev. resp. Dis. 83, 54—57 (1961).

S₂₉₀ STUART-HARRIS, C. H.: "Influenza and other virus infections of the respiratory tract." Edward Arnold London 2nd Edit. (1965).

S₂₉₁ STUART-HARRIS, C. H., C. H. ANDREWES, and W. SMITH: Study of epidemic influenza with special reference to the 1936—37 epidemic. Med. Res. Counc. Rep. No. 228 (1938).

S₂₉₂ STUART-HARRIS, C. H., and T. FRANCIS: Studies on the nasal histology of epidemic influenza virus infection in the ferret. 2. The resistance of regenerating respiratory epithelium to reinfection and to physicochemical injury. J. exp. Med. 68, 803—812 (1938).

S₂₉₃ STUART-HARRIS, C. H., and M. H. MILLER: Vagaries of the agglutination-inhibition reaction with newly isolated strains of influenza virus. Brit. J. exp. Path. 28, 394—403 (1947).

S₂₉₄ STUART-HARRIS, C. H., W. SMITH, and C. H. ANDREWES: The influenza epidemic of January—March 1939. Lancet 205—211 (1939).

S₂₉₅ STULBERG, C. S., and R. SCHAPIRA: Virus growth in tissue culture fibroblasts. 1. Influenza A and herpes simplex viruses. J. Immunol. 70, 51—59 (1953).

S₂₉₆ STYK, B.: Epidemic outbreak of influenza due to type C influenza virus (Czech). Čs. Epidem. 3, 137—145 (1953).

S₂₉₇ STYK, B.: Non-specific inhibitors in normal rat sera for the influenza C type virus. The relation between the influenza C type virus, Newcastle disease, and epidemic parotitis. Folia biol. (Praha) 1, 207—213 (1955).

S₂₉₈ STYK, B.: Selection and preparation of strains of influenza virus for a live vaccine of Smorodintsev type (Czech). Čs. Epidem. 6, 303—308 (1957).

S₂₉₉ STYK, B.: Co-factor and specific antibodies against influenza viruses. 1. Method of co-factor titration, co-factor content of various animal sera.

S₃₀₀ 2. The effect of substances destroying non-specific inhibitors on the co-factor and on specific A 2 antibodies. Acta virol. 5, 334—341, and 342—450 (1961).

S₃₀₁ 3. The potentiating effect of co-factor on specific antibodies of early immune and hyperimmune sera and the differences in the character of these antibodies. Acta virol. 6, 327—337 (1962).

S₃₀₂ STYK, B.: Sensibilisation by means of potassium periodate of inhibitor resistant A 2 influenza virus strains to non-specific serum inhibitors. Nature (Lond.) 193, 954—955 (1962).

S₃₀₃ STYK, B.: Effect of some inhibitor-destroying substances on the non-specific inhibitor of C influenza virus present in normal rat serum. Acta virol. 7, 88—89 (1963).

S₃₀₄ STYK, B.: Viral antibodies and the serum co-factor (Czech). Biol. Prace 11, 1—142 (1965).

S₃₀₅ STYK, B., and L. HANA: Investigations into specific antibodies against A 2 influenza viruses and the thermolabile co-factor by means of paper electrophoresis. Acta virol. 4, 365—370 (1960).

$S_{306}$ STYK, B., and L. HANA: The effect of inhibitor-destroying substances on the co-factor and specific antihaemagglutinins against A 2 influenza viruses. Acta virol. **5**, 194 (1961).

$S_{307}$ STYK, B., and L. HANA: Formation of 19 S and 7 S type viral antibodies. Symp. "Molecular and cellular basis of antibody formation" Prague 261—268 (1964).

$S_{308}$ STYK, B., and L. HANA: Immunodiffusion studies on the reaction of influenza virus with specific antibody and non-specific serum beta inhibitor. Acta virol. **10**, 281—290 (1966).

$S_{309}$
$S_{310}$ STYK, B., L. HANA, F. FRANEK, F. SOKOL, and J. MENSIK: Investigations on co-factor and influenzal antibodies by density gradient zonal centrifugation. Acta virol. **6**, 478 (1962), and **7**, 25—36 (1963).

$S_{311}$ STYK, B., L. HANA, and B. RADA: Antiviral and co-factor-like activities of some dextran sulphates. Acta virol. **7**, 480 (1963).

$S_{312}$ STYK, B., L. HANA, V. RATHOVA, and D. KOCISKOVA: Investigations of the complement components and the antibody co-factor by gel filtration on Sephadex G-200. Science Tools **12**, 37 (1965).

$S_{313}$ STYK, B., and D. KOCISKOVA: *in vivo* and *in vitro* effects of dextran sulphates, heparin, and pelantan on guinea pig complement, antibody co-factor, and non-specific viral inhibitors. J. Hyg. Epidem. **8**, 485—496 (1964).

$S_{314}$ STYK, B., D. KOCISKOVA, and L. HANA: Co-factor and specific antibodies against influenza viruses. 8. On the different nature of antibody co-factor and complement. Acta virol. **8**, 496—506 (1964).

$S_{315}$ STYK, B., V. RATHOVA, and D. BLASCOVIC: Thermostability of specific haemagglutination-inhibiting antibodies against the FE influenza virus and their reactivation by addition of fresh serum. Acta virol. **2**, 179—187 (1958).

$S_{316}$ SUBRAHMANYAN, T., and C. MIMS: Fate of intravenously administered interferon and the distribution of interferon during virus infections in mice. Information Exchange Group No. 6, Memo 116 (1965).

$S_{317}$ SUGG, J. Y.: An influenza virus pneumonia of mice that is non-transferable by serial passage. J. Bact. **57**, 399—403 (1949).

$S_{318}$ SUGG, J. Y.: The variation of antigenic pattern and mouse virulence in an influenza virus culture. J. Bact. **58**, 399—406 (1949).

$S_{319}$ SUGG, J. Y.: The relation of the concentration of unadapted and adapted influenza virus in the mouse lung to the death or survival of the infected host. J. Bact. **60**, 489—497 (1950).

$S_{320}$ SUGG, J. Y.: Further studies on serial passage of a mixture of influenza viruses in embryonated eggs. Proc. Soc. exp. Biol. (N.Y.) **76**, 199—202 (1951).

$S_{321}$ SUGG, J. Y.: Effect of intranasally administered immune serum on influenza virus present in mouse lung. Proc. Soc. exp. Biol. (N.Y.) **77**, 728—731 (1951).

$S_{322}$ SUGG, J. Y., and R. CLEELAND: Differences in trypsin susceptibility among influenza viruses and relationship of the susceptibility to the antigenic type or subtype of the virus. J. Immunol. **88**, 369—376 (1962).

$S_{323}$ SUGG, J. Y., and T. P. MAGILL: Susceptibility of convalescent ferrets to re-infection with influenza virus in absence of specific antibodies. Proc. Soc. exp. Biol. (N.Y.) **65**, 233—235 (1947).

$S_{324}$ SUGG, J. Y., and T. P. MAGILL: The serial passage of mixtures of different strains of influenza virus in embryonated eggs and in mice. J. Bact. **56**, 201—206 (1948).

$S_{325}$ SUGG, J. Y., and D. NAGAKI: Tissue tropism of a strain of influenza A virus in ferrets. J. Immunol. **74**, 46—50 (1955).

$S_{326}$ SUGIURA, A., and E. D. KILBOURNE: Genetic studies of influenza viruses. 2. Plaque formation by influenza viruses in a clone of a variant human heteroploid cell line.

$S_{327}$ 3. Production of plaque-type recombinants with A and A 1 strains. Virology **26**, 478—488 (1965), and **29**, 84—91 (1966).

$S_{328}$ SUGIURA, A., H. SHIMOJO, and C. ENOMOTO: Studies of the non-specific haemagglutination inhibitor against influenza A 2 viruses (gamma inhibitor). 2. Physicochemical properties of the gamma inhibitor.

$S_{329}$ 3. Destruction of gamma inhibitor by virus enzyme. Jap. J. exp. Med. 31, 159—167, and 169—177 (1961).

$S_{330}$ SULKIN, S. E.: The effect of environmental temperature on experimental influenza in mice. J. Immunol. 51, 291—300 (1945).

$S_{331}$ SULKIN, S. E., J. E. SMITH, and D. D. DOUGLASS: Experimental study of an institutional outbreak of epidemic influenza. J. infect. Dis. 69, 278—284 (1941).

$S_{332}$ SVEC, F. A., and G. F. FORSTER: Inhibition of the Hirst haemagglutination reaction by pneumococcal extracts, normal serums, and blood cell esterases. Proc. Soc. exp. Biol. (N.Y.) 66, 20—23 (1947).

$S_{333}$ SVEDMYR, A.: Studies on a factor in normal allantoic fluid inhibiting influenza virus haemagglutination. Ark. Kemi Mineral. Geol. 24b, 1—8 (1947).

$S_{334}$ SVEDMYR, A.: Studies on a factor in normal allantoic fluid inhibiting influenza virus haemagglutination. 1. Occurrence, physicochemical properties and mode of action.

$S_{335}$ 2. Virus-inhibitor interaction.

$S_{336}$ 3. Precipitation-dissolution reaction in mixtures of active virus and inhibitor.

$S_{337}$ 4. The effect of active virus, proteolytic enzymes and periodate on the inhibitor.

$S_{338}$ 5. Precipitation reaction in mixtures of inactive virus and inhibitor. Brit. J. exp. Path. 29, 295—308, and 309—321 (1948), also 30, 237—247, 248—254, and 255—266 (1949).

$S_{339}$ SVEDMYR, A.: Studies on a factor in normal allantoic fluid inhibiting influenza virus haemagglutination. Upsala Läk.-Fören. Förh. 54, 365—380 (1949).

$S_{340}$ SVEDMYR, A., P. VON MAGNUS, and E. A. FREUNDT: Influenza in Scandinavia during 1950—51. Acta path. microbiol. scand. 29, 96—108 (1951).

$S_{341}$ SVET-MOLDAVSKY, G. I., and I. Z. GHENDON: Demonstration of influenza virus using gel diffusion method of precipitation directly in tissue culture. Acta virol. 2, 120—122 (1958).

$S_{342}$ SVET-MOLDAVSKY, G. YA., and L. S. LOZOVSKAYA: Change in the properties of erythrocytes of animals after injection of influenza virus into them (Russian). Problemy. grippa i OKVDP. Tez. dokl. ob'yed. Sess. Inst. AMN SSSR, Moscow 47 (1952).

$S_{343}$ SZANTO, J.: The cultivation of some A-FE influenza virus strains in tissue cultures. 2. Comparison of the multiplication of various strains of A-FE influenza in tissue cultures (Czech). Biol. Prace 4, 63—71 (1958).

$S_{344}$ SZANTO, J., O. KRIZANOVA, and F. LINK: Interaction of gamma-inhibitor with A 2 influenza virus. Acta virol. 6, 524—530 (1962).

$S_{345}$ SZANTO, J., O. LAUCIKOVA, D. KOCISKOVA, and V. RATHOVA: Preparation of anti-inhibitor serum and its effect on the multiplication of influenza virus. Folia microbiol. (Praha) 5, 105—111 (1960).

$S_{346}$ SZANTO, J., O. LAUCIKOVA, D. KOCISKOVA, and V. RATHOVA: The effect of anti-inhibitor serum on the multiplication of influenza virus. Virology 10, 149 (1960).

$S_{347}$ SZANTO, J., and N. VALENTOVA: The cultivation of some A-FE influenza virus strains in tissue cultures (Czech). Biol. Prace 4, 59—62 (1958).

$S_{348}$ SZOLLOSY, E., G. IVANOVICS, and S. HORVATH: Distribution of the receptor substance of influenza and related viruses in tissue elements of different animal species. 2. The virus-adsorbing capacity of the respiratory tract of various vertebrates. Acta physiol. hung. 3, 431—440 (1952).

$S_{349}$ SZURMAN, J.: An attempt to isolate ribonucleic acid from chorioallantoic membranes of chick embryos infected with influenza virus (Polish). Med. Dosw. Mikrob. 12, 175—181 (1960).

$T_1$ TAKAGI, A., K. TANAKA, and E. TAKAO: On the influences of ox bile and one of its components, mucin, on the haemagglutinability of influenza viruses. Yonago Acta med. 1, 87—93 (1954).

$T_2$ TAKANO, K., J. WARREN, K. E. JENSEN, and A. L. NEAL: Nucleic acid induced resistance to viral infection. J. Bact. 90, 1542—1547 (1965).

$T_3$ TAKÁTSY, G.: Purified precipitated virus obtained by a new simple method. Acta med. Acad. Sci. hung. 3, 185—191 (1952).

$T_4$ TAKÁTSY, G., and K. BARB: On the normal serum inhibitors for the avid Asian strains of influenza virus. Acta virol. 3, 71—77 (1959).

$T_5$ TAKÁTSY, G., K. BARB, and E. FARKAS: The antigenic structure of the influenza virus strains isolated in Hungary in 1957. Acta microbiol. Acad. Sci. hung. 5, 287—296 (1958).

$T_6$ TAKÁTSY, G., K. BARB, and E. FARKAS: A new erythrocyte receptor with exclusive affinity to avid Asian strains of influenza virus. Acta virol. 3, 79—84 (1959).

$T_7$ TAKÁTSY, G., and J. FURÉSZ: On the antigenic variation of the influenza A virus. Arch. ges. Virusforsch. 7, 344—354 (1957).

$T_8$ TAKÁTSY, G., and M. HAMAR: Recent studies of the antigenic structure of influenza virus by the antibody absorption test. Acta microbiol. Acad. Sci. hung. 3, 203—212 (1955).

$T_9$ TAKEMOTO, K.: Use of a cation exchange resin for isolation of influenza virus. Proc. Soc. exp. Biol. (N.Y.) 85, 670—672 (1954).

$T_{10}$ TAKEMOTO, K. K., and A. M. LERNER: Human amnion cell cultures; susceptibility to viruses and use in primary virus isolations. Proc. Soc. exp. Biol. (N.Y.) 94, 179—182 (1957).

$T_{11}$ TAKEMOTO, K. K., R. K. LYNT, W. P. ROWE, R. J. HUEBNER, J. A. BELL, G. W. MELLIN, and D. J. DAVIS: Primary isolation of influenza A, B, and C viruses in monkey kidney tissue cultures. Proc. Soc. exp. Biol (N.Y.) 89, 308—311 (1955).

$T_{12}$ TAKEMOTO, K. K., M. L. ROBBINS, and P. K. SMITH: The experimental chemotherapy of influenza viruses. J. Immunol. 72, 139—145 (1954).

$T_{13}$ TAMM, I.: Inhibition of influenza and mumps virus multiplication by 456 or 567 trichloro-1-b-D-ribofuranosylbenzimidazole. Science 120, 847—848 (1954).

$T_{14}$ TAMM, I.: Influenza virus-erythrocyte interaction. 1. Reversible reaction between Lee virus and cat erythrocytes.

$T_{15}$ 2. Effect of egg passage on the reaction between a recently recovered influenza A virus and sheep erythrocytes. J. Immunol. 73, 180—189, and 190—198 (1954).

$T_{16}$ TAMM, I., and D. J. FLUKE: The effect of monochromatic ultraviolet irradiation on infectivity and haemagglutinating ability of the influenza virus type A strain PR8. J. Bact. 59, 449—461 (1950).

$T_{17}$ TAMM, I., K. FOLKERS, and F. L. HORSFALL: Inhibition of influenza virus multiplication by 2, 5-dimethylbenzimidazole. Yale J. Biol. Med. 24, 559—567 (1952).

$T_{18}$ TAMM, I., K. FOLKERS, and F. L. HORSFALL: Inhibition of influenza virus multiplication by alkyl derivatives of benzimidazole. 1. Kinetic aspects of inhibition by 2, 5-dimethylbenzimidazole as measured by infectivity titrations.

$T_{19}$ 2. Measurement of inhibitory activity by haemagglutination titration. J. exp. Med. 98, 219—227, and 229—243 (1953).

$T_{20}$ TAMM, I., K. FOLKERS, C. H. SHUNK, D. HEYL, and F. L. HORSFALL: Inhibition of influenza virus multiplication by alkyl derivatives of benzimidazole. 3. Relationship between inhibitory activity and chemical structure. J. exp. Med. 98, 245—259 (1953).

$T_{21}$ TAMM, I., K. FOLKERS, C. H. SHUNK, and F. L. HORSFALL: Inhibition of influenza virus multiplication by N-glycosides of benzimidazoles. J. exp. Med. 99, 227—249 (1954).

T$_{22}$ TAMM, I., H. S. GINSBERG, and F. L. HORSFALL: Antigenic similarity between the first and later egg passage strains of freshly recovered influenza A viruses. Proc. Soc. exp. Biol. (N.Y.) 81, 94—98 (1952).

T$_{23}$ TAMM, I., and F. L. HORSFALL: Characterisation and separation of an inhibitor of viral haemagglutination present in urine. Proc. Soc. exp. Biol. (N.Y.) 74, 108—114 (1950).

T$_{24}$ TAMM, I., E. D. KILBOURNE, and F. L. HORSFALL: Comparison of influenza B virus strains from the 1950 epidemic with strains from earlier epidemics. Proc. Soc. exp. Biol. (N.Y.) 75, 89—92 (1950).

T$_{25}$ TAMM, I., and D. A. J. TYRRELL: Influenza virus multiplication in the chorio-allantoic membrane in vitro. Kinetic aspects of inhibition by 5,6 dichloro 1-B-D-ribofuranosylbenzimidazole. J. exp. Med. 100, 541—562 (1954).

T$_{26}$ TANG, F. F.: The transmission of influenza virus WS to the Chinese mink and to David's squirrel. Brit. J. exp. Path. 19, 179—183 (1938).

T$_{27}$ TANI, T., and T. HAYASHI: Proliferation and preservation of egg adapted influenza viruses. Jap. J. med. Sci. Biol. 7, 435—443 (1954).

T$_{28}$ TATENO, I., and O. KITAMOTO: Rapid diagnosis of influenza by a fluorescent antibody technique. 2. Relation between immunocytological serological and clinical findings.

T$_{29}$ 3. Some new aspects of influenza infection revealed by the fluorescent antibody technique (Japanese). Jap. J. exp. Med. 35, 401—410, and 411—417 (1965).

T$_{30}$ TATENO, I., O. KITAMOTO, and J. A. KAWAMURA: Diverse immunocytologic findings of nasal smears in influenza. New Engl. J. Med. 274, 237—242 (1966).

T$_{31}$ TATENO, I., S. SUZUKI, A. KAWAMURA, and S. NAKAMURA: Rapid diagnosis of influenza by a fluorescent antibody technique. 1. Some basic information (Japanese). Jap. J. exp. Med. 35, 383—400 (1965).

T$_{32}$ TAVERNE, J., J. H. MARSHALL, and F. FULTON: The purification and concentration of viruses and soluble antigens on calcium phosphate. J. gen. Microbiol. 19, 451—461 (1958).

T$_{33}$ TAYLOR, A. R.: Chemical analysis of the influenza viruses A (PR8 strain) and B (Lee strain) and the swine influenza virus. J. biol. Chem. 153, 675—686 (1944).

T$_{34}$ TAYLOR, A. R., D. G. SHARP, D. BEARD, J. W. BEARD, J. H. DINGLE, and A. E. FELLER: Isolation and characterisation of influenza A virus (PR8 strain). J. Immunol. 47, 261—282 (1943).

T$_{35}$ TAYLOR, A. R., D. G. SHARP, I. W. MCLEAN, D. BEARD, and J. W. BEARD: Concentration and purification of influenza virus for the preparation of vaccines. J. Immunol. 50, 291—316 (1945).

T$_{36}$ TAYLOR, C. E.: Interference between influenza and equine encephalitis viruses in tissue culture. J. Immunol. 71, 125—133 (1953).

T$_{37}$ TAYLOR, R. M.: Experimental influenza with influenza A virus in mice. The increase in intrapulmonary virus after inoculation and the influence of various factors thereon. J. exp. Med. 73, 43—55 (1941).

T$_{38}$ TAYLOR, R. M.: Passive immunisation against experimental infection of mice with influenza A virus. Comparative effect of immune serum administered intranasally and intraabdominally. J. Immunol. 41, 453—462 (1941).

T$_{39}$ TAYLOR, R. M.: Studies on survival of influenza virus between epidemics and antigenic variations of the virus. Amer. J. publ. Hlth 39, 171—178 (1949).

T$_{40}$ TAYLOR, R. M.: A further note on 1233 (influenza C) virus. Arch. ges. Virusforsch. 4, 485—500 (1951).

T$_{41}$ TAYLOR, R. M., M. DREGUSS, and F. DE RITIS: Antigenic behaviour of certain Hungarian strains of epidemic influenza virus. Amer. J. Hyg. 31, 36—45 (1940).

T$_{42}$ TAYLOR, R. M., and A. S. PARODI: Use of hamster (Cricetus auratus) for detection of influenza virus in throat washings. Proc. Soc. exp. Biol. (N.Y.) 49, 105—108 (1942).

$T_{43}$ TAYLOR, R. M., A. S. PARODI, R. B. FERNANDEZ, and R. J. CHALVO: An aetiological study of influenza A in Argentina during 1941; comparison of the epidemiology of types A and B (Spanish). Rev. Inst. Bact. Buenos Aires 11, 44—57 (1942).

$T_{44}$ TERZIN, A. L.: Glycerine as a stabiliser of some complement fixing antigens of viral and rickettsial origin. Proc. Soc. exp. Biol. (N.Y.) 84, 215—218 (1953).

$T_{45}$ TERZIN, A. L., M. N. BORDJOSK, M. V. MILOVANOVIC, L. V. STOJKOVIC, and M. M. DIMIC: Some viral, rickettsial and leptospiral infections diagnosed in Serbia. A serological study. J. Hyg. (Lond.) 52, 129—150 (1954).

$T_{46}$ THIBON, M.: Action of benzododecinium on influenza virus (French). Ann. Inst. Pasteur 99, 414—420 (1960).

$T_{47}$ THOMSON, D., and R. THOMSON: Influenza. Influenza with special reference to the complications and sequelae, bacteriology of influenzal pneumonia, pathology, epidemiological data, prevention and treatment. Monograph 16, parts 1 and 2. Ann. Pickett-Thomson Res. Lab. 1933 and 1934.

$T_{48}$ TITOVA, S. M.: Experimental influenzal infection in dogs (Russian). Vop. Med. Virus 4, 114 (1954).

$T_{49}$ TODD, C., and J. P. RICE: Fowl plague. Med. Res. Counc. "System of Bacteriology" 7, 219—231 (1930).

$T_{50}$ TOMMILA, V.: Studies of the demonstration of influenza viruses by their capacity to inactivate inhibitors of virus haemagglutination. Ann. Med. exp. Fenn. 34, Suppl. 2 (1956).

$T_{51}$ TOPCIU, V., P. DIOSI, G. BRAN, A. ELIAS, O. NEVINGLOVSCI, and N. CSAKY: Investigations on the nature of influenza haemagglutination inhibitors found in animal sera. Abstr. 9th int. Congr. Microbiol. Moscow 391 (1966).

$T_{52}$ TOVARNITSKIY, V. I.: Methods of purification and concentration of viruses (Russian). Vop. Virus. 1, 24 (1948).

$T_{53}$ TOVARNITSKIY, V. I.: Chemistry and biochemistry of the influenza virus. (Russian). Uspekhi khim. 20, 231 (1951).

$T_{54}$ TOVARNITSKIY, V. I., and O. M. CHALKINA: Purification and concentration of the influenza virus (Russian). Zh. Mikrobiol. (Mosk.) 10—11, 13 (1943).

$T_{55}$ TOVARNITSKIY, V. I., and M. I. KARLINA: Virus-protein complexes (Russian). Biokhimiya 2, 10 (1945); Vop. Virus. 1, 59 (1948); Vop. Virus. 3, 20 (1950).

$T_{56}$ TOVARNITSKIY, V. I., and O. I. SHISHKINA: Electrophoretic investigations on the influenza virus (Russian). Zh. Mikrobiol. (Mosk.) 10—11, 16 (1943).

$T_{57}$ TOYOSHIMA, S., S. KANO, and T. UEDA: Search for effective synthetic agents against influenza virus. Keio J. Med. (Tokyo) 12, 149—159 (1963).

$T_{58}$ TRAUB, E.: Active immunity of chickens from virus infected eggs. (German). Zbl. Bakt. I. Abt. Orig. 164, 412—423 (1955).

$T_{59}$ TRAVASSOS, J., and M. BRUNO-LOBO: Studies on Asian influenza in Rio de Janiero. 5. Antibodies in the pre-epidemic period (Portuguese). An. Microbiol. (Rio de J.) 6, 107—150 (1958).

$T_{60}$ TRAVER, M. L., R. L. NORTHROP, and D. L. WALKER: Site of intracellular antigen production by myxoviruses. Proc. Soc. exp. Biol. (N.Y.) 104, 268—273 (1960).

$T_{61}$ TSAUZMER, G. M., L. Y. EBERT, S. B. NOZIK, and N. G. TSAUZMER: Serological diagnosis of influenza in Cheliabinsk, in 1957 (Russian). Vop. Virus. 5, 81—83 (1958).

$T_{62}$ TSUNODA, A., H. F. MAASSAB, K. W. COCHRAN, and W. C. EVELAND: Antiviral activity of a-methyl-1-adamantane-methylamine hydrochloride. "Antimicrobial agents and chemotherapy." Amer. Soc. Microbiol. 553—560 (1965).

$T_{63}$ TSURUMI, M., K. OGASAWARA, and K. FUJIL: On the isolation of the influenza virus. Kitasato Arch. exp. Med. 18, 39—51 (1941).

$T_{64}$ TSURUMI, M., K. OGASAWARA, and H. TAKAGI: Studies on influenza virus. Nagoya, J. med. Sci. 13, 61—67 (1939).

$T_{65}$  TSVETKOVA, I. V., and M. A. LIPKIND: Neuraminidase activity of myxoviruses in their interaction with susceptible and nonsusceptible cells. Abstr. 9th int. Congr. Microbiol. Moscow 387 (1966).

$T_{66}$  TUMOVA, B. K.: The biological properties of influenza virus and their variability (Czech). Čs. Epidem. 3, 218—230 (1954).

$T_{67}$  TUMOVA, B., D. FEDOVA, D. BOSCOVA, J. VOLENIKOVA, V. PROCHAZKOVA, and J. LUDVIK: The incidence and spread of a new variant of type B influenza virus in the population of Czechoslovakia in 1959—61. 2. Properties of the strains isolated. Acta virol. 7, 156—175 (1963).

$T_{68}$  TUMOVA, B., and O. FISERO-SOVINOVA: Properties of influenza viruses A/Asia/57 and A/Equi/Praha/56. 1. Agglutination of red blood cells. Bull. Wld Hlth Org. 20, 445—454 (1959).

$T_{69}$  TUMOVA, B., and O. FISERO-SOVINOVA: The comparison of some propertis of influenza A-57/FE and A/Equi/Praha/56 viruses. Abstr. 7th int. Congr. Microbiol. Uppsala 280 (1958).

$T_{70}$  TUMOVA, B., and H. G. PEREIRA: Genetic interaction between influenza A viruses of human and animal origin. Virology 27, 253—261 (1965).

$T_{71}$  TUSHINSKIY, M. D., V. D. STAVSKAYA, A. I. IAROSHEVSKIY, E. F. DAVIDENKOVA, F. S. SKARLATO, E. L. KAN, and E. A. SKRIABINA: Clinical aspects of pandemic of influenza in 1957 (Russian). Klin. Med. (Mosk.) 36, 43—48 (1958).

$T_{72}$  TWYBLE, E., and H. C. MASON: Haemagglutination by products of influenza virus using infected mouse lung and chick embryo as the sources of virus. J. Immunol. 49, 73—85 (1944).

$T_{73}$  TYRRELL, D. A. J.: Separation of inhibitors of haemagglutination and specific antibodies for influenza viruses by starch zone electrophoresis. J. Immunol. 72, 494—502 (1954).

$T_{74}$  TYRRELL, D. A. J.: New tissue culture systems for influenza, Newcastle disease and vaccinia viruses. J. Immunol. 74, 293—305 (1955).

$T_{75}$  TYRRELL, D. A. J., M. L. BYNOE, and B. HOORN: Studies on the antiviral activity of 1-adamantanamine. Brit. J. exp. Path. 46, 370—375 (1965).

$T_{76}$  TYRRELL, D. A. J., and A. H. CAMERON: The pathogenicity of influenza viruses for suckling mice. J. Path. Bact. 74, 37—48 (1957).

$T_{77}$  TYRRELL, D. A. J., and B. HOORN: The growth of some myxoviruses in organ cultures. Brit. J. exp. Path. 46, 514—518 (1965).

$T_{78}$  TYRRELL, D. A. J., and F. L. HORSFALL: Neutralisation of viruses by homologous immune serum. 1. Quantitative studies on factors which affect the neutralisation reaction with Newcastle disease, influenza A and bacterial virus T 3. J. exp. Med. 97, 845—861 (1953).

$T_{79}$  TYRRELL, D. A. J., and F. L. HORSFALL: Disruption of influenza virus. Properties of degradation products of the virus particle. J. exp. Med. 99, 321—342 (1954).

$T_{80}$  TYRRELL, D. A. J., I. TAMM, O. C. FORSSMAN, and F. L. HORSFALL: A new count of allantoic cells of the 10-day chick embryo. Proc. Soc. exp. Biol. (N.Y.) 86, 594—598 (1954).

$T_{81}$  TYRRELL, D. A. J., and R. C. VALENTINE: The assay of influenza virus particles by haemagglutination and electron microscopy. J. gen. Microbiol. 16, 668—675 (1957).

$U_1$  UHLER, M., and S. GARD: Lipid content of standard and incomplete influenza A virus. Nature (Lond.) 173, 1041 (1954).

$U_2$  UNANOV, S. S., L. V. VASIL'EV, and A. A. LEBEDOV: The epidemiological efficacy of the monovalent A2 influenza vaccine (Russian). Zh. Mikrobiol. (Mosk.) 30, 31—37 (1959).

$U_3$  UNDERDAHL, N. R., and G. A. YOUNG: Effect of dietary intake of fat-soluble vitamins on the intensity of experimental swine influenza virus infection of mice. Virology 2, 415—429 (1956).

U₄ UNDRITS, V. F., and A. M. KOLDOBSKIY: The problem of the differential diagnosis of influenza and the condition of the ENT organs in this disease (Russian). "Virusnyye Infektsii." Tr. Leningrad Inst. Epid. Mikrobiol. i g. Imuni Pastera Leningrad 13, 232 (1953).

U₅ UTZ, J. P.: Studies on the inactivation of influenza and Newcastle disease viruses by a specific lipid fraction of normal animal sera. J. Immunol. 63, 273—279 (1949).

U₆ UYS, C. J., and W. B. BECKER: The pathology of tern virus infection of chickens. S. Afr. J. Lab. clin. Med. 9, 93—94 (1963).

V₁ VALENTINE, R. C.: Structure and particle counts of the influenza virus and the adenovirus. 4th int. Congr. Electron Microscopy (Berlin) 577—586 (1958).

V₂ VALENTINE, R. C., and A. ISAACS: The structure of influenza virus filaments and spheres. J. gen. Microbiol. 16, 195—204 (1957).

V₃ VALENTINE, R. C., and A. ISAACS: The structure of viruses of the Newcastle disease-mumps-influenza (myxovirus) group. J. gen. Microbiol. 16, 680—685 (1957).

V₄ VAN DEN ENDE, M.: The utilisation of aminoacid solutions by Virus-infected eggs, studied by paper chromatography. J. gen. Microbiol. 4, 277—279 (1950).

V₅ VAN DEN ENDE, M., H. S. MYERS, D. McKENZIE, A. J. WALT, R. HOFFENBERG, and A. SWANEPOEL: The 1950 influenza epidemic in Capetown. S. Afr. med. J. 25, 445—449 (1951).

V₆ VAN DEN ENDE, M., G. S. TURNER, G. SELZER, and W. T. NAUDE: Antigenic comparison of influenza virus strains isolated in Capetown during 1952. S. Afr. med. J. 27, 975—980 (1953).

V₇ VAN ROOYEN, C. E., L. McCLELLAND, and E. K. CAMPBELL: Influenza in Canada during 1949. Studies on Eskimos. Canad. J. publ. Hlth 40, 447—456 (1949).

V₈ VAN ROOYEN, C. E., and A. J. RHODES: "Virus diseases of man." Oxford Univ. Press. Humphrey Milford, London (1940).

V₉ VAN DER VEEN, J., and J. MULDER: Studies on the antigenic composition of human influenza virus A strains. Onderz. Meded. Inst. Praev. Geneesk. No. 6, Stenfert Kroese, Leiden (1950).

V₁₀ VASHKOV, V. I., and A. K. ASTAFYEVA: Viricidal properties of lactic acid and glycerin vapours (Russian). Tr. Tsentral'nogo nauchno-issledovatel'skogo dezinfektsionogo. Inst. Moscow 6, 104 (1950).

V₁₁  VASHKOV, V. I., and A. K. ASTAFYEVA: Viricidal and bactericidal effect of certain
V₁₂ chemicals in air (Russian). Tr. Tsentral'nogo nauchno-issledovatel'skogo dezinfektsionogo. Inst. Moscow 7, 20 (1951) and 8, 27 (1954).

V₁₃ VASILYEVA, L. V.: Dynamics of influenzal antibodies against Type A and B in the adult population (Russian). Zh. Mikrobiol. (Mosk.) 3, 26 (1954).

V₁₄ VAYNBERG, B. G., and L. M. KHARMATS: Influenza virus reproduction in the body of Suslik (Russian). Zh. Mikrobiol. (Mosk.) 12, 28 (1949).

V₁₅ VDOVINA, L. A.: An outbreak of influenza in the Taldy-Kurgan district. (Russian). Zdravookhr. Kazakhstana 18, 30—31 (1958).

V₁₆ VEERARAGHAVAN, N., M. W. KIRTIKAR, and T. SREEVALSAN: Studies on the cultivation of influenza virus in vitro. Bull. Wld Hlth Org. 24, 711—722 (1961).

V₁₇ VEERARAGHAVAN, N., and T. SREEVALSAN: Evaluation of some methods of concentration and purification of influenza virus. Bull. Wld Hlth Org. 24, 695—702 (1961).

V₁₈ VERLINDE, J. D., and O. MAKSTANIEKS: An experimental investigation into the prophylactic value of influenza vaccine as used during the winter of 1952—53 (Dutch). Ned. T. Geneesk. 98, 2589—2591 (1954).

V₁₉ VERLINDE, J. D., O. MAKSTANIEKS, and C. A. G. NASS: Results of a protective vaccine against influenza in the winter 1952—53 (Dutch). Ned. T. Geneesk. 98, 559—564 (1954).

$V_{20}$ VERSTEEG, J., R. P. MOUTON, and J. D. VERLINDE: Asian influenza in horses (Dutch). T. Diergeneesk. **83**, 608—613 (1958).

$V_{21}$ VIEUCHANGE, J.: Protection of white mice against experimental influenza (French). C. R. Soc. Biol. (Paris) **134**, 391 (1939).

$V_{22}$ VIEUCHANGE, J.: On the susceptibility of certain species of monkeys to the influenza virus (French). Bull. Acad. Méd. (Paris) **121**, 100 (1939).

$V_{23}$ VILCHES, A., and G. K. HIRST: Interference between neurotropic and other unrelated viruses. J. Immunol. **57**, 125—140 (1947).

$V_{24}$ VINSON, J. W., and F. T. WALSH: The effect of terramycin on the PR8 strain of influenza A virus in chick embryos and mice. J. Immunol. **65**, 129—134 (1950).

$V_{25}$ VOGEL, L., and A. SHELOKOV: Adsorption-haemagglutination test for influenza virus in monkey kidney tissue cultures. Science **126**, 358—359 (1957).

$V_{26}$ VOLAKOVA, N., and L. JANDASEK: A thermostable inhibitor of newly isolated influenza virus strains in guineapig serum. Acta virol. **3**, 109—112 (1959).

$V_{27}$ VOLUKSKAYA, E. N.: Carbohydrate metabolism in experimental influenzal infection (Russian). Probl. obsch. virusol. Tez. Dokl. 6-y Sess. Inst. Virusol. AMN SSSR Moscow **60**, also Vop. Virus. **4**, 138 (1954).

$V_{28}$ VOLUKSKAYA, E. N., V. I. TOVARNITSKY, and A. M. BYELIKOVA: The effect of the properdin system on the influenza virus. J. Hyg. epidem. (Praha) **2**, 404—407 (1958).

$V_{29}$ VONKA, V.: Plaque technique with influenza A2 virus in rubber stoppered bottle cultures from monkey kidney cells. Arch. ges. Virusforsch. **15**, 514—524 (1965).

$V_{30}$ VOORSPUIJ, A. J. Z.: A new and simple method for the purification and concentration of influenza virus. Experientia (Basel) **12**, 474 (1949).

$V_{31}$ VOROB'YEVA, M. M.: The role of polyphosphates in influenza virus multiplication (Russian). Probl. obshch. virusol. Tez. dokl. 6-y Sess. Inst. virusol. AMN SSSR Moscow **1** (1953).

$V_{32}$ VOROB'YEVA, M. M.: The significance of adenosine triphosphoric acid in influenza virus multiplication (Russian). Dissertation, Moscow (1954).

$V_{33}$ VOSS, H., and E. WENGEL: Forms of influenza virus (German). Zbl. Bakt. I. Abt. Orig. **162**, 225—235 (1955).

$V_{34}$ VOTH, D. W., and H. A. FELDMAN: Aqueous, repository, and placebo influenza vaccines in an elderly population. Amer. J. publ. Hlth **53**, 1512 (1963).

$W_1$ WADDELL, G. H., M. B. TEIGLAND, and M. M. SIGEL: A new influenza virus associated with equine respiratory disease. J. Amer. vet. med. Ass. **143**, 587—590 (1963).

$W_2$ WAGNER, R. R.: Acquired resistance in mice to the neurotoxic action of influenza viruses. Brit. J. exp. Path. **33**, 157—167 (1952).

$W_3$ WAGNER, R. R.: Influenza virus multiplication in homologously transplanted mouse fetal tissues. J. Immunol. **71**, 253—260 (1953).

$W_4$ WAGNER, R. R.: A pantropic strain of influenza virus. Generalised infection and viraemia in the infant mouse. Virology **1**, 497—515 (1955).

$W_5$ WAGNER, R. R.: Influenza virus infection of transplanted tumours. 2. The relation of host immunity to virus action.

$W_6$ 3. Effect of passive immunisation on infections of the Ehrlich ascites tumour. J. Immunol. **71**, 439—445 and 446—454 (1955).

$W_7$ WAGNER, R. R.: Studies on the pathogenesis of influenzal pneumonitis. Intranasal vs. intravenous infection of mice. Yale. J. Biol. Med. **28**, 598—614 (1956).

$W_8$ WAGNER, R. R.: Viral interference. Some considerations of basic mechanisms and their potential relationship to host resistance. Bact. Rev. **24**, 151—166 (1960).

$W_9$ WAGNER, R. R., I. L. BENNETT, and V. S. LeQUIRE: The production of fever by influenza viruses. 1. Factors influencing the febrile response to single injections of virus.

$W_{10}$   2. Tolerance in rabbits to the pyrogenic effect of influenzal viruses. J. exp. Med. 90, 321—334 and 335—347 (1949).

$W_{11}$   WALKER, D. L., and F. L. HORSFALL: Lack of identity in neutralising and haemagglutination-inhibiting antibodies against influenza viruses. J. exp. Med. 91, 65—86 (1950).

$W_{12}$   WALKER, D. L., and N. KHOOBYARIAN: Reduced resistance in mice to the intravenous toxicity of influenza virus following cortisone administration. Rep. Naval Med. Res. Inst. Bethesda, Maryland, 15, 173—180 (1957).

$W_{13}$   WALLACE. D. G., and R. E. KISSLING: Studies in swine of Asian influenza virus. Bull. Wld Hlth Org. 20, 455—463 (1959).

$W_{14}$   WALLIS, C., N. SAKURADA, and J. L. MELNICK: Influenza vaccine prepared by photodynamic inactivation of virus. J. Immunol. 91, 677—682 (1963).

$W_{15}$   WALTER, G., and O. JARNIN: Investigations on the increase in agglutinability of red blood corpuscles produced by enzymes from viruses and bacteria (German). Z. Hyg. Infekt.-Kr. 131, 407—422 (1950).

$W_{16}$   WANG, C. I.: The relation of infectious and haemagglutinin titres to the adaptation of influenza virus to mice. J. exp. Med. 88, 515—519 (1948).

$W_{17}$   WANG, S.-P., and T. Y. LIN: Laboratory studies on some biological properties of the Far East influenza virus. J. infect. Dis. 103, 178—182 (1958).

$W_{18}$   WARD, T. G., and B. E. EDDY: An antigenically distinct subtype of influenza virus A which is virulent for mice in primary passage of allantoic fluid. Science 112, 501—503 (1950).

$W_{19}$   WARD, T. G., and M. SALCH: Laboratory studies on the relation between influenza antibody and immunity in the mouse. Amer. J. Hyg. 61, 82—88 (1955).

$W_{20}$   WARREN, J., A. NEAL, and D. RENNELS: Adsorption of myxoviruses on magnetic iron oxides. Proc. Soc. exp. Biol. (N.Y.) 121, 1250—1253 (1966).

$W_{21}$   WATERSON, A. P.: Two kinds of myxovirus. Nature (Lond.) 193, 1163—1164 (1962).

$W_{22}$   WATERSON, A. P., J. M. W. HURRELL, and K. E. JENSEN: The fine structure of influenza A, B, and C viruses. Arch. ges. Virusforsch. 12, 487—495 (1960).

$W_{23}$   WATSON, B. K.: Immunocytological effects of varied inocula of influenza virus. J. exp. Med. 114, 13—27 (1961).

$W_{24}$   WATSON, B. K., and A. H. COONS: Studies of influenza virus infection in the chick embryo using fluorescent antibody. J. exp. Med. 99, 419—427 (1954).

$W_{25}$   WEBSTER, R. G.: The immune response to influenza virus. 1. Effect of the route and schedule of vaccination on the time course of the immune response as measured by three serological methods. Immunology 9, 501—519 (1965).

$W_{26}$   WEBSTER, R. G.: Original antigenic sin in ferrets. The response to sequential infections with influenza viruses. J. Immunol. 97, 177—183 (1966).

$W_{27}$   WEBSTER, R. G., and W. G. LAVER: Influenza virus subunit vaccines. Immunogenicity and lack of toxicity for rabbits of ether and detergent disrupted virus. J. Immunol. 96, 596—605 (1966).

$W_{28}$   WEBSTER, R. G., and W. G. LAVER: Influenza virus subunit vaccines. Abstr. 9th int. Congr. Microbiol. Moscow 380 (1966).

$W_{29}$   WECKER, E.: The distribution of P32 in fowl plague virus by different ways of production (German). Z. Naturforsch. 12b, 208—210 (1957).

$W_{30}$   WECKER, E., and W. SCHAFER: Uptake of radioactive phosphorus in fowl plague virus (German). Z. Naturforsch. 11b, 181—187 (1956).

$W_{31}$   WEINBERGER, H. L., E. L. BUESCHER, and V. M. EDWARDS: Extent of antigenic variation of A2 influenza viruses from prototype strains. J. Immunol. 94, 47—54 (1965).

$W_{32}$   WELLER, T. H., and J. F. ENDERS: Production of haemagglutinin by mumps and influenza A viruses in suspended cell tissue cultures. Proc. Soc. exp. Biol. (N.Y.) 69, 124—128 (1948).

$W_{33}$   WELLS, R. J. H.: An outbreak of fowl plague in turkeys. Vet. Rec. 75, 783—786 (1963).

$W_{34}$ WELLS, W. F., and W. HENLE: Experimental air-borne disease. Quantitative inoculation by inhalation of influenza virus. Proc. Soc. exp. Biol. (N.Y.) 48, 298—301 (1941).

$W_{35}$ WEN, H. S., and S.-P. WANG: Mouse adaptation of the Asian influenza virus. J. infect. Dis. 105, 9—17 (1959).

$W_{36}$ WERNER, G. H.: Quantitative studies on influenza virus infection of the chick embryo by the amniotic route. J. Bact. 71, 505—515 (1956).

$W_{37}$ WERNER, G. H., and R. W. SCHLESINGER: Morphological and quantitative comparison between infectious and non-infectious forms of the influenza virus. J. exp. Med. 100, 203—215 (1954).

$W_{38}$ WESSLEN, N., S. HERMODSSON, and L. PHILIPSON: Interference between influenza and poliomyelitis viruses in tissue culture. Arch. ges. Virusforsch. 9, 31—44 (1959).

$W_{39}$ WESTWOOD, J. C. N.: A study of chick embryo lesions produced by influenza viruses. Brit. J. exp. Path. 33, 610—619 (1952).

$W_{40}$ WHEELOCK, E. F., and I. TAMM: Mitosis and division in HeLa cells infected with influenza or Newcastle disease viruses. Virology 8, 532—536 (1959).

$W_{41}$ WHIPPLE, H. E., and P. E. VAN REYEN: Antiviral substances. Ann. N.Y. Acad. Sci. 130, 1—482 (1965).

$W_{42}$ WHITE, D. O.: The mechanism of natural resistance of the allantois to influenza infection. Virology 9, 680—690 (1959).

$W_{43}$ WHITE, D. O.: Some aspects of influenza virus multiplication in the surviving allantois-on-shell system. Virology 10, 21—28 (1960).

$W_{44}$ WHITE, D. O., and I. M. CHEYNE: Early events in the eclipse phase of influenza and parainfluenza virus infection. Virology 29, 49—59 (1966).

$W_{45}$ WHITE, D. O., H. M. DAY, E. J. BATCHELDER, I. M. CHEYNE, and A. J. WANSBROUGH: Delay in the multiplication of influenza virus. Virology 25, 289—302 (1965).

$W_{46}$ WHITE, D. O., and S. FAZEKAS DE ST GROTH: Variation of host resistance to influenza viruses in the allantois. J. Hyg. (Lond.) 57, 123—133 (1959).

$W_{47}$ WHITE, J.: The position of Taylor's 1233 strain of virus in the Human red cell receptor gradient. Brit. J. exp. Path. 34, 668—673 (1953).

$W_{48}$ WHITTEN, W. K.: Inactivation of gonadotrophins. 1. Inactivation of serum gonadotrophin by influenza virus and receptor-destroying enzyme of V. cholerae.

$W_{49}$ 2. Inactivation of pituitary and chorionic gonadotrophins by influenza virus and receptor-destroying enzyme. Aust. J. Sci. Res. Ser. B 1, 271—277 and 388—390 (1948).

$W_{50}$ WIELGOSZ, G. S.: Pyruvate metabolism of chorioallantoic membrane infected with influenza A. Virology 3, 475—484 (1957).

$W_{51}$ WIENER, M., W. HENLE, and G. HENLE: Studies on the complement fixation antigens of influenza viruses types A and B. J. exp. Med. 83, 259—279 (1941).

$W_{52}$ WILCOX, W. C., E. M. WOOD, J. O. OH, N. B. EVERETT, and C. A. EVANS: Morphological and functional changes in corneal epithelium caused by the toxic effects of influenza and Newcastle disease viruses. Brit. J. exp. Path. 39, 601—609 (1958).

$W_{53}$ WILLIAMS, R. C., and R. W. G. WYCKOFF: Electron shadow micrography of virus particles. Proc. Soc. exp. Biol. (N.Y.) 58, 265—270 (1945).

$W_{54}$ WILLIAMSON, A. P., L. SIMONSEN, and R. J. BLATTNER: Specific organ defects in early chick embryos following inoculation of influenza virus. Proc. Soc. exp. Biol. (N.Y.) 92, 334—337 (1956).

$W_{55}$ WINTERNITZ, M. C., I. M. WASON, and F. P. McNAMARA: "The pathology of influenza." Yale Univ. Press. New Haven, Connecticut (1920).

$W_{56}$ WITZLER, J. D., and A. V. HENNESSY: The epidemiological significance of serologically determined attack rates of Asian influenza among personnel of naval vessels. University of Michigan Med. Bull. 24, 133—137 (1958).

W₅₇ WOLFF, H. L.: The use of the haemagglutination-inhibition test in epidemiological influenza virus studies. J. Hyg. (Lond.) 47, 396—397 (1949).

W₅₈ WONG, S. C., and E. D. KILBOURNE: Changing viral susceptibility of a human cell line in continuous cultivation. 1. Production of infective virus in a variant of the Chang conjunctival cell following infection with swine or NWS influenza viruses. J. exp. Med. 113, 95—110 (1961).

W₅₉ WOOD, T. R.: Methods useful in evaluating 1-adamantanamine hydrochloride, a new orally active synthetic antiviral agent. Ann. N.Y. Acad. Sci. 130, 419—431 (1965).

W₆₀ WOODHOUR, A. F., and K. E. JENSEN: Adsorption characteristics of influenza viruses to cholesterol and fatty acids. Proc. Soc. exp. Biol. (N.Y.) 100, 561—564 (1959).

W₆₁ WOODHOUR, A. F., K. E. JENSEN, and J. WARREN: Antibody response to influenza vaccines combined with hexadecylamine. Proc. Soc. exp. Biol. (N.Y.) 103, 200—204 (1960).

W₆₂ WOODHOUR, A. F., K. E. JENSEN, and J. WARREN: Development and application of new parenteral adjuvants. 5. Comparative potencies of influenza vaccines emulsified in various oils. J. Immunol. 86, 681—689 (1961).

W₆₃ WOODHOUR, A. F., D. P. METZGAR, G. P. LAMPSON, R. A. MACHLOWITZ, A. A. TYTELL, and M. R. HILLEMAN: New metabolisable immunologic adjuvant for human use. 4. Development of highly purified influenza virus vaccine in adjuvant 65. Proc. Soc. exp. Biol. (N.Y.) 123, 778—782 (1966).

W₆₄ WOODHOUR, A. F., D. P. METZGAR, T. B. STEIN, A. A. TYTELL, and M. R. HILLEMAN: New metabolisable immunologic adjuvant for human use. 1. Development and animal immune response. Proc. Soc. exp. Biol. (N.Y.) 116, 516—523 (1964).

W₆₅ WOOLLEY, D. W.: Purification of an influenza virus substrate, and demonstration of its competitive antagonism to apple pectin. J. exp. Med. 89, 11—22 (1949).

W₆₆ WOOLLEY, J. G., H. W. BOND, and T. D. PERRINE: The effect of chemical substances in mice infected with St. Lous encephalitis, PR8 strain of influenza or the Lansing strain of poliomyelitis virus. J. Immunol. 68, 621—626 (1952).

W₆₇ WOOLRIDGE, R. L., and Y. E. CRAWFORD: Serological diagnosis of influenza. Antibody levels in relation to the concentration of virus used in the haemagglutination-inhibition test. J. infect. Dis. 91, 159—164 (1952).

W₆₈ WOOLRIDGE, R. L., J. L. DEMEIO, J. E. WHITESIDE, and J. R. SEAL: Epidemic influenza B and C in Navy recruits 1953—54. A. antigenic studies on influenza virus type B. Proc. Soc. exp. Biol. (N.Y.) 88, 430—435 (1955).

W₆₉ WYCKOFF, R. W. G.: Electron microscopy of chick embryo membrane infected with PR8 influenza. Nature (Lond.) 168, 651—652 (1951).

W₇₀ WYCKOFF, R. W. G.: Formation of the particles of influenza virus. J. Immunol. 70, 187—196 (1953).

Y₁ YAKHNO, M. A.: Changes in A2/57 influenza virus during adaptation to mice (Russian). "Problemy grippa i ostrykh respiratornykh zabolevanii." Tez. dokl. nauch. Sess. Inst. AMN SSSR, Moscow 181—182 (1959).

Y₂ YAKHNO, M. A., and L. YA. ZAKSTELSKAYA: Agglutinating activity of type A2 influenza viruses for animal and human erythrocytes (Russian). Vop. Virus. 5, 68—70 (1958).

Y₃ YAKOVLEV, A. I., and S. G. ZVYAGIN: The effect of phytoncides of onion and garlic on the influenza virus (Russian). Annot. rabot. Inst. Virusol. AMN, SSSR za 1949g, Moscow 6 (1950).

Y₄ YEGOROV, L. I., and O. D. NOVOSHILOV: Experience in treatment of patients with virus influenza (Russian). Vo.-med. Zh. 8, 74—75 (1953).

Y₅ YEREMEYEV, G. V.: Purification and concentration of the influenza A and B viruses (Russian). Tez. dokl. ob'yed Sess. Inst. AMN, SSSR, Moscow 33 (1952).

$Y_6$ YEREMEYEV, G. V.: Mechanism of elution of influenza virus from erythrocytes (Russian). Problemy grippa i OKVDP. Tez. dokl. ob'yed Sess. Inst. AMN SSSR, Moscow 37 (1952).

$Y_7$ YEREMEYEV, G. V.: Rules and regulations of influenza virus adsorption by erythrocytes (Russian). Problemy grippa i OKVDP. Tez. dokl. ob'yed. Sess. Inst. AMN SSSR, Moscow 35 (1952).

$Y_8$ YEREMEYEV, G. V., and O. M. CHALKINA: New method of purification and concentration of type A influenza virus (Russian). Gripp. i OKVDP, Moscow 43 (1953).

$Y_9$ YEREMEYEV, G. V., and N. N. ORLOVA: Immunogenic properties of the separate fractions of the influenza virus (Russian). Vop. Virus. 2, 209 (1949).

$Y_{10}$ YERMOL'YEVA, Z. V., A. K. SHUBLADZE, S. YA. GAYDAMOVICH, and L. K. VALEDINSKAYA: The study of ecmolino in experimental influenza (Russian). Zh. Mikrobiol. (Mosk.) I, 64—67 (1952).

$Y_{11}$ YOUNG, G. A., and W. R. UNDERDAHL: Neutralisation and haemagglutinin inhibition of swine influenza virus by serum from suckling swine and milk from their dams. J. Immunol. 65, 369—373 (1950).

$Y_{12}$ YURIKAS, I. A.: Complement fixation reaction in the diagnosis of influenza (Russian). Bibl. i annot. rabot. otdela virusol. Inst. Eksper. Med. AMN SSSR za 1947—49 Leningrad (1950).

$Y_{13}$ YURIKAS, I. A.: Early diagnosis of influenza by the method of detecting the specific antigen in the washing from the pharynx by the complement fixation reaction (Russian). Gripp. Tr. AMN SSSR 28, 138 (1953).

$Z_1$ ZAKSTELSKAYA, L. YA.: The toxic properties of influenza virus (Russian). Tez. Dokl. 3-y Sess. Inst. AMN SSSR, Moscow 7 (1950).

$Z_2$ ZAKSTELSKAYA, L. YA.: Recovery of the virus from the urine in patients with epidemic influenza (Russian). Gripp. i OKVDP. Tr. ob'yed. Sess. Inst. AMN SSSR, Moscow, 72—74 (1953).

$Z_3$ ZAKSTELSKAYA, L. YA.: A simplified neutralisation test to show the presence of antibodies against a new strain of influenza virus (A/Asia/57) (Russian). Vop. Virus. 2, 373—375 (1957).

$Z_4$ ZAKSTELSKAYA, L. YA.: Experimental study of the efficacy of intranasal and subcutaneous methods of immunisation in influenza (Russian). Gripp. i OKVDP. Tr. ob'yed Sess. Inst. AMN SSSR, Moscow 135 (1953).

$Z_5$ ZAKSTELSKAYA, L. YA.: Phagocytic activity of leucocytes in influenza (Russian). Vop. Virus. 4, 133 (1954).

$Z_6$ ZAKSTELSKAYA, L. YA., V. A. EFIMOVA, and S. L. KHAYT: Humoral indices of the immunity to A2 influenza viruses and special aspects of their study (Russian). "Problemy grippa i ostrykh respiratornykh zabolevanii." Tez. dokl. nauch. Sess. Inst. AMN SSSR, Moscow 152—154 (1959).

$Z_7$ ZAKSTELSKAYA, L. YA., and V. V. RITOVA: Change in the antigenic structure of the influenza A and A prime viruses over the course of 5 years (Russian). Zh. Mikrobiol. (Mosk.) 7, 43 (1954).

$Z_8$ ZAKSTELSKAYA, L. YA., and V. V. RITOVA: Type specificity of immunological reactions in influenza (Russian). Zh. MEI 9, 49 (1954).

$Z_9$ ZAKSTELSKAYA, L. YA., and M. A. YAKHNO: On labile antibodies against A2 influenza viruses. Acta virol. 5, 329—333 (1961).

$Z_{10}$ ZALMANSON, E. S., T. I. BALESINA, N. I. KORABELNIKOVA, E. P. PILATSKAYA, Z. W. ERMOLIEVA, and L. L. FADEEVA: On the possibility of stimulation of interferon production in the human body and its practical use. Abstr. 9th int. Congr. Microbiol. Moscow 543 (1966).

$Z_{11}$ ZELLAT, J., and W. HENLE: Further studies in passive protection against the virus of influenza by the intranasal route. J. Immunol. 42, 239—249 (1941).

$Z_{12}$ ZEYTLENOK, N. A., and O. S. KORSHUNOVA: Isolation of influenza virus from patients by means of direct infection of mice and chick embryos (Russian). Arkh. biol. nauk. 59, 38 (1940).

$Z_{13}$ ZEYTLENOK, N. A., G. F. NESTEROVA, N. R. KRIGSGABER, M. V. RUBTSOV, L. N. YAKHONTOV, and V. M. BERNFELD: The study on antiviral and anticellular activity of styrylquinoline and related compounds in connection with their chemical structure. Abstr. 9th int. Congr. Microbiol. Moscow 532—533 (1966).

$Z_{14}$ ZEYTLENOK, N. A., and V. A. VAGZDANOVA: Experience in the study of the physiology of multiplication of Type A virus in the developing chick embryo (Russian). Probl. obsheh. Virusol. Tez. dokl. 5—6 (1953).

$Z_{15}$ ZHANTIEVA, E. M., and V. M. STAKHANOVA: Studies on RNA synthesis in cells infected with fowl plague virus (Russian). Byull. éksp. Biol. Med. 59, 66—68 (1965).

$Z_{16}$ ZHDANOV, V. M.: The perspectives of vaccination against influenza (Russian). Probl. grippa i ostrykh Katar. verkhn. dykhatel. put. Tez. dokl. ob'yed. Sess. Inst. AMN SSSR, Moscow 61 (1952).

$Z_{17}$ ZHDANOV, V. M.: Certain rules and regulations in virus variability (Russian). Tez. dokl. 6-y Sess. Inst. Virusol. AMN SSSR, Moscow 31 (1953).

$Z_{18}$ ZHDANOV, V. M.: Lines of variation of influenza viruses (Russian). Zh. Mikrobiol. (Mosk.) 2, 34 (1953).

$Z_{19}$ ZHDANOV, V. M.: Results of three years work of the influenza department. (Russian). Vop. Virus. 4, 5 (1954).

$Z_{20}$ ZHDANOV, V. M.: Certain problems of influenza epidemiology (Russian). Zh. Mikrobiol. (Mosk.) 9, 56 (1954).

$Z_{21}$ ZHDANOV, V. M.: Certain problems of the variation and evolution of influenza viruses (Russian). Nasledstvennost' i Izmenchivost 1, 317—323 (1959).

$Z_{22}$ ZHDANOV, V. M.: A suggested genetic map of influenza virus. Lancet 738—740 (1965).

$Z_{23}$ ZHDANOV, V. M.: The third type of nucleic acid-directed protein synthesis. Lancet 1017—1018 (1966).

$Z_{24}$ ZHDANOV, V. M., S. A. DEMIDOVA, and G. YA. SVET-MOLDAVSKIY: The inhibition of virus haemagglutination by S Typhosa Vi antigen (Russian). "Problemy grippa i ostrykh respiratornykh zabolevanii." Tez. dokl. nauch. Sess. Inst. AMN SSSR, Moscow, 52—54 (1959).

$Z_{25}$ ZHDANOV, V. M., V. P. HAMBURG, and G. YA. SVET-MOLDAVSKIY: Antigenicity of the inhibitor of influenza virus strain A/Asia/57. J. Immunol. 82, 9—11 (1959).

$Z_{26}$ ZHDANOV, V. M., L. B. MEKLER, S. M. KLEMINKO, and V. C. NAUMOVA: Chemical nature of substances forming the surface of influenza virus. Nature (Lond.) 198, 1326—1327 (1963).

$Z_{27}$ ZHDANOV, V. M., and I. I. NIKOLAYEV: The result of study of influenza vaccines (Russian). Zh. Mikrobiol. (Mosk.) 10, 3 (1953).

$Z_{28}$ ZHDANOV, V. M., and V. V. RITOVA: The pathogenesis of influenza (Russian). Klin. Med. (Moskva) 37, 45—48 (1959).

$Z_{29}$ ZHDANOV, V. M., V. V. RITOVA, A. V. ORLOVA, N. N. SOKOLOVA, and L. A. GOLYGINA: The characteristics of influenza virus strains isolated in 1957. Lancet 735—736 (1957).

$Z_{30}$ ZHDANOV, V. M., and G. A. SMIRNOVA: On the nature of the inactivating effect of animal tissue extracts against myxovirus haemagglutinin. Acta virol. 9, 137—143 (1965).

$Z_{31}$ ZHDANOV, V. M., and N. N. SOKOLOVA: Immunological characteristic of the associated influenza-diphtheria-pertussis vaccine; experimental studies (Russian). "Problemy grippa i ostrykh respiratornykh zabolevanii." Tez. dokl. nauch. Sess. Inst. AMN SSSR, Moscow 132—133 (1959).

$Z_{32}$ ZHDANOV, V. M., and V. D. SOLOVYEV: The influenza problem (Russian). Zh. Mikrobiol. (Mosk.) 3, 6 (1951).

$Z_{33}$ ZHDANOV, V. M., and V. D. SOLOVYEV: Principles of vaccination against influenza (Russian). Zh. Mikrobiol. (Mosk.) 1, 36 (1952).

Z₃₄    ZHDANOV, V. M., V. D. SOLOVYEV, and F. G. EPSHTEIN: "The study of influenza" (Translated from Ucheniye o grippe, Moscow Medgiz 1958). U.S. Dept. of Health, Education and Welfare (1960).

Z₃₅    ZHILKINA, A. S.: The study of the etiology and pathogenesis of "contagious coryza" and the so-called "sporadic" influenza in inter-epidemic periods (Russian). Tez. dokl. Sess. Tsentral. Inst. ukha, gorla i nosa, Moscow 5 (1952).

Z₃₆    ZHUMATOV, KH., and E. S. ISAEVA: Isolation of infectious RNA from type A and B viruses and study of its properties (Russian). Abstr. Inst. All-Union Biochem. Congr. 11, 168 (1964).

Z₃₇    ZHUMATOV, KH. ZH., and E. K. RESHETNIKOVA: Influenza morbidity in the collectives of the city of Alma Ata and the effectiveness of certain prophylactic measures (Russian). Zdravookhr. Karzakhstana 18, 16—23 (1958).

Z₃₈    ZIEGLER, J. E., and F. L. HORSFALL: Interference between influenza viruses. 1. The effect of active virus upon the multiplication of influenza viruses in the chick embryo.

Z₃₉    2. The effect of virus rendered non-infective by ultraviolet radiation upon the multiplication of influenza viruses in the chick embryo. J. exp. Med. 79, 361—377 and 379—400 (1944).

Z₄₀    ZILBER, L. A.: Certain lines of study of the influenza problem (Russian). Zh. Mikrobiol. (Mosk.) 12, 19 (1949).

Z₄₁    ZILBER, L. A., Z. L. BAYDAKOVA, and N. A. BLYUMBERG: The mechanism of antivirus immunity (Russian). Zh. Mikrobiol. (Mosk.) 1—2, 81 (1946).

Z₄₂    ZILBER, L. A., L. I. FALKOVICH, and YE. A. ARKHINA: Study of the virus of epidemic influenza (Russian). Zh. Mikrobiol. (Mosk.) 18, 554 (1937).

Z₄₃    ZILLIG, W., W. SCHAFER, and S. ULLMAN: On the structure of the elementary bodies of fowl plague virus (German). Z. Naturforsch. 10b, 199—206 (1955).

Z₄₄    ZILLIKEN, F., G. H. WERNER, R. K. SILVER, and P. GYORGY: Studies on the enzymatic properties of influenza viruses. The action of influenza B virus and RDE on the haemagglutinin-inhibitor of human Meconium. Virology 3, 464—474 (1957).

Z₄₅    ZIMMERMANN, T., and W. SCHÄFER: Effect of p-fluorophenylalanine on fowl plague virus multiplication. Virology 11, 676—698 (1966).

## ADDENDA

XB₁    BEARE, A. S., D. A. J. TYRRELL, J. C. MCDONALD, T. M. POLLOCK, C. E. D. TAYLOR, C. H. L. HOWELLS, and L. E. TYLER: Reactions and antibody response to live influenza vaccine prepared from the Iksha A 2 strain. J. Hyg. (Lond.) 65, 245—254 (1967).

XB₂    BRANDON, F. B., F. COX, G. O. LEASE, E. A. TIMM, E. QUINN, and I. W. MCLEAN: Respiratory virus vaccines. 3. Some biological properties of Sephadex purified ether-extracted influenza virus antigens. J. Immunol. 98, 800—805 (1967).

XB₃    BRYANS, J. T., W. W. ZENT, R. R. GRUNERT, and D. C. BOUGHTON: 1-adamantanamine hydrochloride prophylaxis for experimentally induced A/Equi/2 influenza virus infection. Nature (Lond.) 212, 1542—1544 (1966).

XC₁    CAMERON, T. P., R. H. ALFORD, J. A. KASEL, E. W. HARVEY, R. J. BYRNE, and V. KNIGHT: Experimental equine influenza in Chincoteague ponies. Proc. Soc. exp. Biol. (N.Y.) 124, 510—515 (1967).

XC₂    CHELI, R., and M. U. DIANZANI: Adsorption of PR 8 influenza virus on the red cells of guineapigs treated with haemolytic serum. Acta Haemat. 14, 15—22 (1955).

XD₁    DARDANONI, L., and C. SPANO: Antigenic analysis of A 2 influenza virus isolated from 1957—1963 in the district of Palermo (Italian). Riv. Ist. sieroter. ital. 39, 243 (1964).

XD₂    DORSET, M., C. N. MCBRYDE, and W. B. NILES: Remarks on "Hog flu". J. Amer. vet. med. Ass. 62, 162—171 (1922—23).

XF₁    FAZEKAS DE ST GROTH, S., A. ISAACS, and M. EDNEY: Multiplication of influenza virus under conditions of interference. Nature (Lond.) 170, 573 (1952).

XF₂ FINLAND, M., M. W. BARNES, and E. B. WELLS: Influenza virus infections in Boston. September 1947—August 1949. J. Lab. clin. Med. **37**, 88—103 (1951).

XF₃ FINTER, N. B.: Interferons. Frontiers of Biology North Holland Publishing Co., Amsterdam (1966).

XF₄ FLETCHER, R. D., J. E. HIRSCHFIELD, and M. FORBES: A common mode of action of ammonium ions and various amines. Nature (Lond.) **207**, 664—665 (1965).

XF₅ FRANCIS, T.: Recent advances in the study of influenza. J. Amer. med. Assoc. **105**, 251—254 (1935).

XF₆ FRANCIS, T.: Epidemiological studies in influenza. Amer. J. publ. Hlth **27**, 211—225 (1937).

XF₇ FRANCIS, T.: Immunology of experimental influenza. Amer. J. Hyg. **28**, 63—79 (1938).

XF₈ FRANCIS, T., and T. P. MAGILL: Vaccination of human beings with the virus of influenza. Proc. Soc. exp. Biol. (N.Y.) **33**, 604—606 (1935—36).

XF₉ FRANCIS, T., and T. P. MAGILL: Direct isolation of human influenza virus in tissue cultures and in egg membranes. Proc. Soc. exp. Biol. (N.Y.) **36**, 134—135 (1937).

XF₁₀ FRANCIS, T., and T. P. MAGILL: The antibody response of human subjects vaccinated with the virus of human influenza. J. exp. Med. **65**, 251—259 (1937).

XG₁ GREEN, R. G.: On nature of filterable viruses. Science **82**, 443—445 (1935).

XH₁ HALLAUER, C.: Immunity studies in fowl plague. 4. On inherited immunity (German). Z. Hyg. Infekt.-Kr. **118**, 605 (1936).

XH₂ HOBSON, D., E. A. GOULD, and H. I. FLOCKTON: Laboratory studies on a strain of Asian influenza virus used as a living vaccine. J. Hyg. (Lond.) **65**, 255—262 (1967).

XH₃ HORNE, R. W., A. P. WATERSON, P. WILDY, and A. E. FARNHAM: The structure and composition of myxoviruses. 1. Electron microscope studies of the structure of myxovirus particles by negative staining techniques. Virology **11**, 79—98 (1960).

XH₄ HOYLE, L.: An improved technique for inoculation of fertile eggs. J. Path. Bact. **59**, 480—481 (1947).

XK₁ KNIGHT, C. A.: Titration of influenza virus in chick embryos. J. exp. Med. **79**, 487—495 (1944).

XL₁ LAIDLAW, P. P.: The Rede Lecture. Cambridge University Press (1938).

XL₂ LOBODINSKA, M., A. MORAWIECKI, and Z. SKURSKA: Studies on the inhibition of haemagglutination and infection of the influenza virus by orozomucoid and its polymers. Abstr. 9th int. Congr. Microbiol. Moscow **533** (1966).

XL₃ LURIA, S. E.: Reactivation of irradiated bacteriophage by transfer of self-reduplicating units. Proc. nat. Acad. Sci. (Wash.) **33**, 253—264 (1947).

XM₁ MCDONALD, J. C., A. J. ZUCKERMANN, A. S. BEARE, and D. A. J. TYRRELL: Trials of live influenza vaccine in the Royal Air Force. Brit. med. J. i, 1036 (1962).

XN₁ NIKOLAYEV, I. I., and S. A. SAMVELOVA: The problem of the effectiveness of intranasal vaccination against virus influenza with type A and B vaccines (Russian). Gripp i OKVDP. Tr. ob'yed. nauch. Sess. AMN SSSR Moscow **83** (1953).

XO₁ OKUNO, Y., and K. NAKAMURA: Prophylactic effectiveness of live influenza vaccine in 1965. Bikens J. **9**, 89 (1966).

XP₁ PONS, M. W.: Studies on influenza virus ribonucleic acid. Virology **31**, 523—531 (1967).

XR₁ RITOVA, V. V., A. M. ZHUKOVSKY, and N. A. YEVSTIGNEYEVA: Comparative study of the immunological properties of live influenza vaccine in volunteers. J. Hyg. Epidem. (Praha) **7**, 272 (1963).

XS₁ SCADDING, J. G.: Lung changes in influenza. Quart. J. Med. **30**, 425—465 (1937).

XS₂ SCHMIDT, S., J. OERSKOV, and E. STEENBERG: Studies on experimental fowl plague (German). VI. Mitt. Z. Hyg. Infekt.-Kr. **118**, 455 (1936).

XS₃ SCHULMAN, J. L.: Experimental transmission of influenza virus infection in mice. 3. Differing effects of immunity induced by infection and by inactivated influenza virus vaccine on transmission of infection.

XS₄ 4. Relationship of transmissibility of different strains of virus and recovery of airborne virus in the environment of infector mice. J. exp. Med. 125, 467—478 and 479—488 (1967).

XS₅ SOLOVIEV, V. D., T. G. ORLOVA, L. A. PORUBEL, and I. N. VASILIEVA: Study of the genetic characters of vaccine strains of type A 2 influenza virus. Probl. Virol. 6, 743 (1961).

XS₆ STARKE, G., I. SIEBELIST, and E. HELLER: Studies on the question of the use of live influenza vaccine. 1. Results of tests of living influenza A 2 vaccine in a controlled experiment on volunteers (German). J. Hyg. Epidem. (Praha) 8, 37 (1964).

XT₁ TAKATSY, G., and K. BARB: New inhibitor and erythrocyte receptor substances for certain Asian strains of influenza virus. Nature (Lond.) 183, 52—53 (1959).

XZ₁ ZHDANOV, V. M., and V. D. SOLOVIEV: Some results of the study of Asian influenza. Amer. Rev. resp. Dis. 83, 211 178—187 (1961).